(*continued on back*)

Stochastic Modelling
and Analysis:
A Computational Approach

Stochastic Modelling and Analysis:
A Computational Approach

HENK C. TIJMS
Vrije Universiteit, Amsterdam

JOHN WILEY & SONS
Chichester · New York · Brisbane · Toronto · Singapore

Library of Congress Cataloging in Publication Data:

Tijms, H. C.
   Stochastic modelling and analysis.   *1507838*
   Includes bibliographies and indexes.
   1. Stochastic systems.   2. Stochastic analysis.
   I. Title
   QA402.T488   1986   519.2   85-22696

   ISBN 0 471 90911 4

British Library Cataloguing in Publication Data:

Tijms, Henk C.
   Stochastic modelling and analysis: a computational
   approach.—(Wiley series in probability and
   mathematical statistics)
   1. Mathematical models   2. Stochastic processes
   I. Title
   519.2   QA402

   ISBN 0 471 90911 4

Printed and bound in Great Britain.

*To my wife Els
and my sons
Steven and Jurgen*

Contents

Preface

Many real-world phenomena require the analysis of a system in a probabilistic rather than a deterministic setting. Stochastic models are becoming increasingly important for understanding or making a performance evaluation of complex systems in a broad spectrum of fields such as operations research, computer science, telecommunication and engineering.

The main concern of this text is the application of stochastic models to practical situations involving uncertainty and dynamism. Much emphasis is put on the basic concepts and use of models. We believe that major strengths of the book are a careful discussion of useful practical computational methods for solving stochastic models and a wide variety of realistic examples illustrating the basic models and the associated solution methods. Here we occasionally set foot in the rapidly developing and fascinating area of what may be called 'experimental' calculus for stochastic models—an approach that makes judicious use of high-speed computers in order to obtain insights into the theory or heuristic hints on how to tackle unsolved problems.

This textbook has grown out of many years of active research on stochastic models and teaching operations research and computer science students. In teaching these students I became aware that they benefit best from examples and problems in order to understand and appreciate the use of stochastic models. In the presentation of the material I have deliberately chosen to omit proofs that are not necessary for an understanding of the theory, but I have tried to give as many explanations as possible to provide useful insights into the working of the theory.

Although the basic concepts of useful stochastic models are at the core simple and intuitive, in fact many students find it difficult to translate a specific applied probability problem into an appropriate stochastic model. The student can only acquire the skills of modelling new situations by a considerable amount of practice in solving problems on his own. For this reason, the text contains many interesting and thought-provoking problems referring to a wide range of practical application areas such as inventory/production control, maintenance, reliability and queueing. A significant number of the exercises ask the student to write a computer code for the calculation of a solution to the problem considered. It is our conviction that, besides learning the art of modelling, another essential part of the student's education is the testing of the solution method developed for the problem under consideration. The student must discover by experience that the actual implementation of a proposed algorithm is often not quite as simple as it seems. The criterion we adopted was to never give an algorithm unless it was thoroughly tested. In fact problems in applied mathematics have not been solved satisfactorily until numerical answers can be calculated with sufficient ease and accuracy.

The prerequisite of the book is only calculus and a first course in probability, but the book requires some maturity of the reader. It is appropriate for senior undergraduate and graduate level courses for students in operations research, engineering and computer science. At the same time, the book contains much material not to be found in other texts and should also be appealing to practitioners and researchers. It is organized in such a way that several one-semester or one-quarter courses on stochastic models can be given on the basis of its contents. The various chapters are independent of each other to a large extent, although a working knowledge of chapter 2 on Markov chains is highly recommended before studying chapters 3 and 4 on Markovian decision processes and applied queueing models. Also, the reader should have a good working knowledge of the Poisson process which is discussed in some detail in chapter 1 on renewal processes. A few sections, being rather specialized, have been marked with asterisks; these sections can be omitted without loss of continuity. The ends of examples and proofs have been marked with the symbol □.

In undertaking this book, there were a number of individuals who have had an impact on the development of the resulting manuscript. My interest in stochastic operations research was aroused by Gijs De Leve and I learned much from his intuitive, probabilistic thinking. For many years I have had the privilege of doing joint research with such distinguished people as Ton De Kok, Awi Federgruen, Harrie Groenevelt, Arie Hordijk, Paul Schweitzer, Luuk Seelen, Frank Van Der Duyn Schouten and Michiel Van Hoorn—some of them former students of mine and now respected colleagues. Further, I would like to acknowledge useful discussions with John Bather, Wim Cohen, Eric Denardo, Paul Kühn, Isaac Meilijson, Sheldon Ross, Ed Silver, Nico Van Dijk, Uri Yechiali, Doug White and Ward Whitt. Also, Ton De Kok, Sheldon Ross, Paul Schweitzer and Nico Van Dijk have read portions of the manuscript and provided valuable suggestions. A special note of appreciation is in order to my students Arnout Eikeboom and Matthieu Van Der Heijden for their effective development and testing of computer codes and for providing helpful comments on earlier drafts of the manuscript. Also, Marc Salomon helped with plotting the figures. The invaluable help of graduate students was part of the excellent research environment at Vrije Universiteit . A marvellous typing job was done by Gloria Wirz-Wagenaar who accurately and quickly typed the various versions of the manuscript.

CHAPTER 1

Renewal processes with applications to inventory/production and reliability

1.0 INTRODUCTION

Renewal theory provides elegant and powerful tools for analysing stochastic processes that regenerate themselves from time to time. The long-run behaviour of a regenerative stochastic process can be studied in terms of its behaviour during a single regeneration cycle. In many applied probability problems, such as in inventory, queueing and reliability, regenerative stochastic processes are prevalent. Renewal theory and regenerative processes are the subjects of sections 1.1 and 1.2, while section 1.3 deals with renewal reward processes being regenerative processes on which a reward structure is superimposed. The emphasis is on basic results and how to apply them to specific problems rather than on proofs, although some proofs that provide insights are sketched. In section 1.4 special attention is paid to the Poisson process which is the renewal process mostly used in practical applications. The Poisson process arises not only in numerous real-world phenomena, but allows as well for tractable analysis because of its memoryless property. The important role of this characteristic property of the Poisson process is demonstrated by giving several illuminating examples. Simple control rules for production/inventory, queueing and maintenance problems can often be studied elegantly by the use of renewal theory. This is shown in section 1.5 in which we analyse several control rules for utilizing idle times in stochastic service systems. As a prelude to section 1.7, we consider in section 1.6 the practically important compound Poisson process. In section 1.7 we discuss a continuous review (s, S) inventory system with compound Poisson demand and positive lead times for the replenishment orders. Assuming a commonly used service level constraint on the long-run fraction of demand to be met directly from stock on hand, we derive practically useful results for the control parameters s and S by employing only simple renewal-theoretic methods. A fundamental result that is frequently used in inventory and queueing applications is the property that Poisson arrivals see time averages. This result is discussed in some detail in section 1.8. In production/inventory systems and queueing systems it is often important to have asymptotic estimates for the probability of stock-out occurrence and the waiting time probability. Such estimates are obtained in section 1.9 by using renewal-theoretic methods. In the final section 1.10 we analyse a production/inventory model with compound Poisson demand and a variable production rate, where it is required that a certain grade of customer service is provided. Using asymptotic results from renewal theory, we derive tractable and accurate approximations to the switching levels for restarting and stopping production. Section 1.10 as well as section 1.7 illustrate the simplicity of analysis to be achieved by a general renewal-theoretic approach to hard individual problems.

1.1 RENEWAL THEORY

Renewal theory concerns the study of stochastic processes counting the number of events that take place as a function of time. Here the interoccurrence times between successive events are independent and identically distributed random variables. For instance the events could be the arrivals of customers to a waiting line or the successive replacements of light bulbs. Although renewal theory originated from the analysis of replacement problems for components such as light bulbs, the theory has many applications to quite a wide range of practical probability problems. In inventory, queueing and reliability problems the analysis is often based on an appropriate identification of embedded renewal processes for the specific problem considered. For example in a queueing process the embedded events could be the arrivals of customers who find the system empty or in an inventory process the embedded events could be the replenishments of stock when the inventory position drops at or below the reorder point.

Formally, let $X_1, X_2, \ldots$ be a sequence of non-negative, independent random variables having a common probability distribution function

$$F(x) = P\{X_k \leqslant x\}, \qquad x \geqslant 0, k = 1, 2 \ldots.$$

It is assumed that

$$0 < E(X_1) < \infty.$$

The random variable X_n denotes the interoccurrence time between the $(n-1)$th and nth event in some specific probability problem. Letting

$$S_0 = 0, \qquad S_n = \sum_{i=1}^{n} X_i, \qquad n = 1, 2, \ldots,$$

we have that S_n is the epoch at which the nth event occurs. Defining for each $t \geqslant 0$,

$$N(t) = \text{the largest integer } n \geqslant 0 \text{ for which } S_n \leqslant t,$$

the random variable $N(t)$ represents the number of events up to time t.

Definition 1.1

The counting process $\{N(t), t \geq 0\}$ is called the renewal process.

It is said that a renewal occurs at time t if $S_n = t$ for some n. For each $t \geq 0$, the number of renewals up to time t is finite with probability 1. This is an immediate consequence of the strong law of large numbers implying that $S_n/n \rightarrow E(X_1)$ with probability 1 and thus $S_n \leq t$ only for finitely many n. We give some examples of a renewal process.

Example 1.1 A replacement problem

Suppose we have an infinite supply of electric bulbs, where the burning times of the bulbs are independent and identically distributed random variables. If the bulb in use fails, it is immediately replaced by a new bulb. Letting

$$X_i = \text{the burning time of the } i\text{th bulb}$$

then

$$N(t) = \text{the total number of bulbs to be used up to time } t. \qquad \square$$

Example 1.2 An inventory problem

Consider a periodic review inventory system for which the demands for a single product in the successive weeks $n = 1, 2, \ldots$ are independent and identically distributed random variables having a continuous distribution. Letting

$$X_i = \text{the demand in the } i\text{th week}$$

then

$$1 + N(x) = \text{the number of weeks until depletion of the current inventory } x. \qquad \square$$

The renewal function

In the sequel it is assumed that the common probability distribution function F of the interoccurrence times between consecutive renewals has a probability density f. This assumption will only be needed in the derivation of limiting results based on the key renewal theorem to be stated later. An important role in renewal theory is played by the *renewal function* defined by

$$M(t) = E(N(t)), \qquad t \geq 0.$$

Defining for $n = 1, 2, \ldots$ the probability distribution function

$$F_n(t) = P\{S_n \leq t\}, \qquad t \geq 0,$$

where $F_1(t) = F(t)$, the renewal function $M(t)$ can be represented as

$$M(t) = \sum_{n=1}^{\infty} F_n(t), \qquad t \geq 0. \tag{1.1}$$

Noting that for each non-negative, integer-valued random variable U holds $E(U) = \sum_{k \geq 0} P\{U > k\}$, the equation (1.1) follows directly from the basic relationship

$$N(t) \geq n \qquad \text{if and only if } S_n \leq t. \tag{1.2}$$

A computationally more useful result for $M(t)$ is the integral equation

$$M(t) = F(t) + \int_0^t M(t-x) f(x) \, dx, \qquad t \geq 0. \tag{1.3}$$

It is instructive to derive this equation. Fix $t > 0$. To compute $E(N(t))$, we condition on the time X_1 of the first renewal. Noting that after each renewal the process probabilistically starts over, we have under the condition that $X_1 = x$,

$$N(t) = \begin{cases} 0 & \text{if } x > t, \\ 1 + N(t-x) & \text{if } x \leq t. \end{cases}$$

Next, by the law of total expectation (see appendix A),

$$E(N(t)) = \int_0^{\infty} E(N(t) | X_1 = x) f(x) \, dx = \int_0^t \{1 + E(N(t-x))\} f(x) \, dx,$$

which gives (1.3). This integral equation for the renewal function $M(t)$ can explicitly be solved only for special cases (see exercise 1.4). In particular, for the exponential density $f(t) = \lambda e^{-\lambda t}$ we have $M(t) = \lambda t$ for all $t \geq 0$. In practice the integral equation (1.3) can be solved numerically by some discretization method such as the repeated Simpson rule with an appropriate starting procedure (see, for example, chapter 11 in Delves and Walsh, 1974) provided the interval $[0, t]$ is not too large. Such a method is easy to program and should be used in conjunction with the asymptotic expansion to be discussed next.

Asymptotic expansion of the renewal function

For larger values of t a useful asymptotic expansion of the renewal function $M(t)$ can be given. Before stating this result, we give some asymptotic results for the random variable $N(t)$. Intuitively, it will be clear that

$$\lim_{t \to \infty} \frac{N(t)}{t} = \frac{1}{E(X_1)} \qquad \text{with probability 1,} \tag{1.4}$$

that is the long-run average number of renewals per unit time equals the inverse of the mean interoccurrence time between two consecutive renewals. To verify this, note that by the definition of $N(t)$,

$$S_{N(t)} \leqslant t < S_{N(t)+1} \qquad \text{for all } t \geqslant 0.$$

Also, note that $N(t) \to \infty$ as $t \to \infty$ with probability 1, by $P\{S_n < \infty\} = 1$ for all n. Dividing each term of the above inequality by $N(t)$, letting $t \to \infty$ and noting that by the strong law of large numbers $S_n/n \to E(X_1)$ with probability 1, we obtain the result (1.4). In addition to this result, it can be shown with the relationship (1.2) and the central limit theorem that (see for example, Ross, 1983)

$$\lim_{t \to \infty} P\left\{ \frac{N(t) - t/\mu_1}{\sigma_1 \sqrt{t/\mu_1^3}} \leqslant x \right\} = \frac{1}{\sqrt{2\pi}} \int_{-\infty}^{x} e^{-u^2/2} \, du, \qquad x \geqslant 0,$$

assuming that $\mu_1 = E(X_1)$ and $\sigma_1^2 = \text{var}(X_1)$ are finite. That is, $N(t)$ is asymptotically normal distributed with mean t/μ_1 and variance $\sigma_1^2 t/\mu_1^3$.

We return to the renewal function $M(t)$. In view of (1.4), it will not be surprising that

$$\lim_{t \to \infty} \frac{M(t)}{t} = \frac{1}{E(X_1)}.$$

A proof of this result can be found, for example, in Ross (1983). This asymptotic result can be sharpened further. Assuming that $E(X_1^2) < \infty$, we have the asymptotic expansion

$$\lim_{t \to \infty} \left\{ M(t) - \frac{t}{E(X_1)} \right\} = \frac{E(X_1^2)}{2E^2(X_1)} - 1, \tag{1.5}$$

where $E^2(X_1)$ is the usual notation for the squared expectation of X_1. Below we give an elegant proof of (1.5) based on the famous key renewal theorem. The key renewal theorem and the asymptotic expansion (1.5) require that the probability distribution function F of the interoccurrence times has a density on some interval. We now state without proof the key renewal theorem (see Feller, 1971, for a proof).

Key Renewal Theorem

Suppose $a(t), t \geqslant 0$, is a finite sum of monotone, integrable functions. Let the function $Z(t)$, $t \geqslant 0$, be defined by the integral equation

$$Z(t) = a(t) + \int_0^t Z(t - x) f(x) \, dx, \qquad t \geqslant 0, \tag{1.6}$$

where $f(x)$ is the density of the interoccurrence time X_1. Then

$$\lim_{t \to \infty} Z(t) = \frac{1}{E(X_1)} \int_0^{\infty} a(x) \, dx. \tag{1.7}$$

The integral equation (1.6) is called a *renewal equation*. This equation has a *unique* solution that is bounded on finite intervals. Denoting by the renewal density $m(t)$ the derivative of the renewal function $M(t)$, the renewal equation (1.6) has the unique solution (see Feller, 1971)

$$Z(t) = a(t) + \int_0^t a(t-x)m(x)\,dx, \qquad t \geqslant 0. \tag{1.8}$$

We now verify the asymptotic expansion (1.5) for the renewal function. For that purpose, define the function

$$Z(t) = M(t) - \frac{t}{E(X_1)}, \qquad t \geqslant 0.$$

From (1.3), we find that $Z(t)$ satisfies the renewal equation (1.6) with

$$a(t) = F(t) - 1 + \frac{1}{E(X_1)} \int_t^\infty (x-t)f(x)\,dx, \qquad t \geqslant 0.$$

This function $a(t)$ is the sum of two monotone functions. For any non-negative random variable U with probability distribution function $U(x) = P\{U \leqslant x\}$, the following useful formula applies:

$$E(U^k) = k \int_0^\infty x^{k-1}\{1 - U(x)\}\,dx, \qquad k = 1, 2, \ldots, \tag{1.9}$$

as is readily verified by partial integration. Using (1.9), we find

$$\int_0^\infty a(t)\,dt = \frac{E(X_1^2)}{2E(X_1)} - E(X_1).$$

Next, by applying the key renewal theorem, we get the desired result (1.5).

It turns out that already for relatively small values of t the asymptotic expansion in (1.5) is useful for practical purposes provided the coefficient of variation of the interoccurrence time X_1 is not too large. To illustrate this, we have computed for several probability distribution functions F the critical number t_α defined by

$$\left| \frac{M(t) - M_{\text{asy}}(t)}{M(t)} \right| \leqslant \alpha \qquad \text{for} \quad t \geqslant t_\alpha,$$

where $M_{\text{asy}}(t)$ denotes the asymptotic expansion in (1.5). Here α is varied as $0.05, 0.01$ and 0.005. In all cases we have taken $E(X_1) = 1$. The squared coefficient of variation $c_X^2 = \text{var}(X_1)/E^2(X_1)$ is varied as $\frac{1}{3}, \frac{1}{2}, \frac{7}{10}, \frac{9}{10}, 1\frac{1}{4}, 2\frac{1}{2}, 5$ and 10. The value $c_X^2 = \frac{1}{3}$ corresponds to the Erlang-3 distribution, the values $c_X^2 = \frac{1}{2}$, $\frac{7}{10}$ and $\frac{9}{10}$ correspond to mixtures of Erlang-1 and Erlang-2 distributions with the same scale parameters, while the other values of c_X^2 correspond to hyper-exponential distributions of order two with balanced means (see appendix B). For

Table 1.1 The critical numbers t_α for the asymptotic expansion of $M(t)$

$\alpha\backslash C^2$	$\frac{1}{3}$	$\frac{1}{2}$	$\frac{7}{10}$	$\frac{9}{10}$	$1\frac{1}{4}$	$2\frac{1}{2}$	5	10
0.05	0.60	0.64	0.66	0.48	0.98	2.68	5.27	10.23
0.01	0.72	0.91	1.06	1.00	2.01	4.63	8.79	16.83
0.005	0.75	1.04	1.25	1.27	2.53	5.56	10.43	19.88

each of these distributions the associated renewal function can explicitly be given (see exercise 1.4). In Table 1.1 we give the critical numbers t_α. The following conclusion may be drawn from our numerical investigations. As a rule of thumb, the asymptotic expansion (1.5) for $M(t)$ may be used for $t \geqslant \theta E(X_1)$ where

$$\theta = \begin{cases} 1 & \text{when} \quad 0 < c_X^2 \leqslant 1 \\ 1\frac{1}{2}c_X^2 & \text{when} \quad c_X^2 > 1. \end{cases}$$

The limiting distribution of the residual life

In many practical probability problems an important quantity is the residual (excess) life at time t defined by

$$\gamma_t = S_{N(t)+1} - t$$

(see Figure 1.1 in which a renewal epoch is denoted by an $\times$). For the replacement problem of example 1.1, the random variable γ_t denotes the residual lifetime of the light bulb in use at time t. The limiting distribution of γ_t is of great importance in many applications. This limiting distribution can easily be obtained from the key renewal theorem. For that purpose we first derive a renewal equation for $P\{\gamma_t > u\}$ for fixed u. By conditioning on the time of the first renewal and noting that after each renewal the renewal process probabilistically starts over, it follows that

$$P\{\gamma_t > u \mid X_1 = x\} = \begin{cases} P\{\gamma_{t-x} > u\} & \text{if } x \leqslant t, \\ 0 & \text{if } t < x \leqslant t + u, \\ 1 & \text{if } x > t + u. \end{cases}$$

Thus, by the law of total probability (see appendix A),

$$P\{\gamma_t > u\} = \int_0^t P\{\gamma_{t-x} > u\} f(x)\,dx + \int_{t+u}^\infty f(x)\,dx, \qquad t \geqslant 0. \qquad (1.10)$$

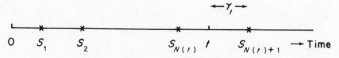

Figure 1.1 The residual life variable

For fixed u the equation (1.10) is a renewal equation for the function $Z(t) = P\{\gamma_t > u\}$, $t \geq 0$. Hence, by (1.8), for each $u \geq 0$,

$$P\{\gamma_t > u\} = 1 - F(t + u) + \int_0^t \{1 - F(t + u - x)\} m(x) \, dx.$$

This result is only tractable for special cases in which the renewal density $m(t)$ can explicitly be given. In particular, when

$$F(t) = 1 - e^{-\lambda t}, \qquad t \geq 0,$$

we have $m(t) = \lambda$ for all $t \geq 0$ (see exercise 1.4) and so we find

$$P\{\gamma_t > u\} = e^{-\lambda u}, \qquad u \geq 0, \tag{1.11}$$

regardless of t. This remarkable result states that for exponentially distributed interoccurrence times the residual life at any time t has the same exponential distribution as the original life regardless of how long ago the last renewal occurred. The exponential distribution is the only continuous distribution having this property of lack of memory.

In general, the distribution of the residual life γ_t is only tractable for t sufficiently large. By applying the key renewal theorem to the renewal equation (1.10) for $Z(t) = P\{\gamma_t > u\}$, $t \geq 0$, we obtain

$$\lim_{t \to \infty} P\{\gamma_t > u\} = \frac{1}{E(X_1)} \int_u^\infty \{1 - F(x)\} \, dx, \qquad u \geq 0,$$

or

$$\lim_{t \to \infty} P\{\gamma_t \leq u\} = \frac{1}{E(X_1)} \int_0^u \{1 - F(x)\} \, dx, \qquad u \geq 0. \tag{1.12}$$

In many practical applications this asymptotic expansion gives a useful approximation to the distribution of γ_t already for moderate values of t (say, $t \geq \theta E(X_1)$ with $\theta = 1$ for $0 < c_X^2 \leq 1$ and $\theta = 1\frac{1}{2} c_X^2$ otherwise). The limiting distribution of the residual life is called the *equilibrium excess distribution* and has applications in a wide variety of contexts. The equilibrium excess distribution can be given the following interpretation. Suppose that an outside observer considers the process after it has been in operation for a very long time, where the observer has no information about the past history of the process. Then the observer can predict the residual life of the item in use according to the equilibrium excess distribution. Denoting by γ_∞ the random variable having the equilibrium excess distribution as probability distribution, we have by using (1.9)

$$E(\gamma_\infty) = \frac{E(X_1^2)}{2E(X_1)}, \qquad E(\gamma_\infty^2) = \frac{E(X_1^3)}{3E(X_1)}. \tag{1.13}$$

It is noteworthy that the average residual life $E(\gamma_\infty)$ is smaller than the average original life $E(X_1)$ only if the coefficient of variation of X_1 is less than 1 (verify!).

Example 1.3   Approximations to an (s, S) inventory system

Suppose a periodic review inventory system for which the demands $X_1, X_2, \ldots$ for a single product in the successive weeks $1, 2, \ldots$ are independent random variables having a common probability density $f(x)$ with finite first two moments. Any demand exceeding the current inventory is backlogged until inventory becomes available by the arrival of a replenishment order. The inventory position is reviewed at the beginning of each week and is controlled by an (s, S) rule with $0 \leqslant s < S$. Under this control rule a replenishment order of size $S - x$ is placed when at a review the inventory level x is at or below the reorder point s; otherwise, no ordering is done. We assume instantaneous delivery of every replenishment order.

We are interested in the average frequency of ordering and the average order size. Since the inventory process starts from scratch on each time the inventory position is ordered up to the level S, it suffices to consider the behaviour of the inventory level only between two consecutive orderings. Thus the operating characteristics can be calculated by using a renewal model in which the weekly demand sizes represent the interoccurrence times of renewals. Noting that the weekly demand has a continuous distribution, it follows that the number of weeks between two consecutive orderings equals the number of weeks needed for a cumulative demand exceeding $S - s$. Also, the order size is the sum of $S - s$ and the undershoot of the reorder point s at the epoch of ordering (cf. Figure 1.2 in which a renewal occurrence is denoted by an $\times$).

The number of weeks needed for a cumulative demand exceeding $S - s$ is equal to $1 + N(S - s)$, while the undershoot of the reorder point s is just the excess variable γ_{S-s}. Hence the average number of weeks between two consecutive replenishment orders equals $1 + M(S - s)$ and the average order size is given by $S - s + E(\gamma_{S-s})$. Assuming that $S - s$ is sufficiently large relative to the average weekly demand, an application of the asymptotic results (1.5) and (1.13) gives

$$\text{The average frequency of ordering} \approx \left(\frac{S-s}{E(X_1)} + \frac{E(X_1^2)}{2E^2(X_1)} \right)^{-1}$$

and

$$\text{The average order size} \approx S - s + \frac{E(X_1^2)}{2E(X_1)}. \qquad \square$$

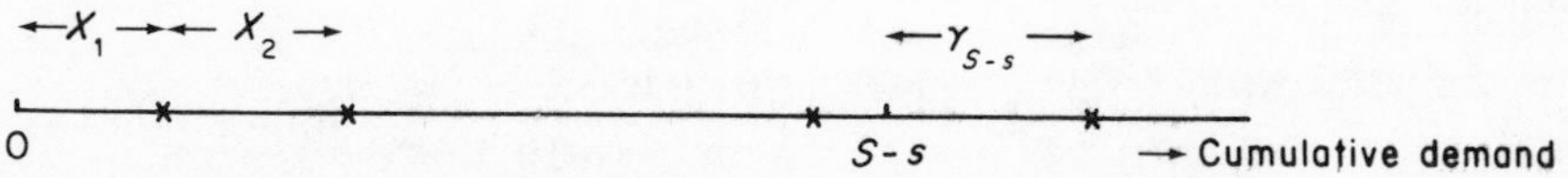

Figure 1.2 The inventory process modelled as a renewal process

1.2. RENEWAL REWARD PROCESSES

Many practical probability problems can be analysed by using renewal processes on which a proper reward structure is imposed. Let $\{N(t), t \geq 0\}$ be a renewal process for which a reward R_n is earned during the interoccurrence time X_n between the $(n-1)$th and nth renewal. It is assumed that $R_1, R_2,\ldots$ are independent and identically distributed random variables. In applications the reward R_n will often be composed of a fixed reward and a reward rate. We allow that R_n may depend on the length X_n of the nth renewal interval. For ease we suppose that the rewards are non-negative. Also, it is assumed that $E(R_1) < \infty$ and $0 < E(X_1) < \infty$.

Defining $R(t), t \geq 0$, as the cumulative reward earned up to time t, we have the following extremely useful result:

$$\lim_{t \to \infty} \frac{R(t)}{t} = \frac{E(R_1)}{E(X_1)} \qquad \text{with probability 1.} \tag{1.14}$$

Also, we have

$$\lim_{t \to \infty} \frac{E(R(t))}{t} = \frac{E(R_1)}{E(X_1)}. \tag{1.15}$$

In words, by defining a cycle as the time elapsed between two consecutive renewals, the relation (1.14) states that with probability 1

$$
\begin{aligned}
&\text{The long-run average reward per unit time} \\
&= \frac{E(\text{rewards earned during one cycle})}{E(\text{length of one cycle})}
\end{aligned}
\tag{1.16}
$$

It is remarkable that the outcome of the long-run average actual reward per unit time can be predicted with probability 1. That is if we are now going to run the system for an infinitely long time period, then we can say beforehand that with probability 1 in the long run the average reward per unit time will be equal to the constant given in (1.16). Such a statement offers holdfast, as opposed to a statement like $E(R) = 1$ which actually says very little about the possible values of the variate R.

The proof of relation (1.14) is very simple. To prove (1.14), observe that by the non-negativeness of the rewards,

$$\sum_{i=1}^{N(t)} R_i \leq R(t) \leq \sum_{i=1}^{N(t)+1} R_i,$$

and so

$$\frac{N(t)}{t} \frac{1}{N(t)} \sum_{i=1}^{N(t)} R_i \leq \frac{R(t)}{t} \leq \frac{N(t)+1}{t} \frac{1}{N(t)+1} \sum_{i=1}^{N(t)+1} R_i.$$

By the strong law of large numbers, $(1/n)\sum_{i=1}^{n} R_i \to E(R_1)$ with probability 1.

Also, $N(t) \to \infty$ as $t \to \infty$ and so the above inequalities together with (1.4) imply (1.14). A proof of (1.15) can be found in Ross (1983).

We now give some simple applications of the above results. More elaborate applications can be found in sections 1.4 and 1.5.

Example 1.4   An alternating production process

Suppose a production facility is alternately on and off so that production times and idle times succeed each other. Let P_1, P_2,... be the successive production times and let I_1, I_2,... be the successive idle times. It is assumed that $\{(P_n, I_n), n \geqslant 1\}$ constitutes a sequence of independent random variables. We are interested in P_{on}, the long-run fraction of time the facility is in production. To determine P_{on}, let us say that a renewal occurs each time the production is restarted. Also, imagine that a reward at rate 1 is incurred when the facility is producing. Then P_{on} can be interpreted as the long-run average reward per unit time and so, by (1.16), we have that, with probability 1,

$$P_{on} = \frac{E(P_1)}{E(P_1) + E(I_1)}. \qquad \square$$

Example 1.5   An age-replacement rule

Consider a replacement problem for light bulbs (machines) for which an age-replacement rule is used. The age-replacement rule prescribes to replace the bulb upon failure or upon reaching the age T, whichever occurs first. The critical level T is a control parameter. The lifetimes $X_1, X_2, \ldots$ of the bulbs are independent random variables having a common probability distribution function F with density f. We incur a fixed cost of $c_1 > 0$ for each planned replacement and a fixed cost of $c_2 > c_1$ for each failure replacement. We want to determine the long-run average cost per unit time.

Let us say that a renewal occurs each time a new bulb is installed. Then the cycle length is distributed as $\min(T, X_1)$ and, by conditioning on X_1, we find

$$\text{The expected cycle length} = T \int_T^\infty f(x)\,dx + \int_0^T x f(x)\,dx$$

$$= \int_0^T \{1 - F(x)\}\,dx.$$

In the same way, by conditioning on X_1, we find

$$\text{The expected cost per cycle} = c_1 \int_T^\infty f(x)\,dx + c_2 \int_0^T f(x)\,dx.$$

Together these two relations and (1.16) imply that, with probability 1,

$$\text{The long-run average cost per unit time} = \frac{c_1\{1 - F(T)\} + c_2 F(T)}{\int_0^T \{1 - F(x)\}\, dx}$$

Assume now that the failure rate function $r(x)$ defined by

$$r(x) = \frac{f(x)}{1 - F(x)}, \qquad x \geq 0,$$

is continuous and strictly increasing to infinity. Note that $r(x)\Delta x$ for Δx small gives approximately the probability that an item of age x will fail in the next time Δx (cf. also appendix B). Then, by putting the derivative of the average cost function equal to zero, it is readily verified that the minimizing value of T is the unique solution to the equation

$$r(T) \int_0^T \{1 - F(x)\}\, dx - F(T) = \frac{c_1}{c_2 - c_1}.$$

In general a numerical procedure should be used to solve this equation.   □

Example 1.6 *A repair problem for a deteriorating item*

Consider an item that deteriorates by incurring an amount of damage each period. The damages incurred in the successive periods are independent random variables having a common exponential density $f(x) = \alpha e^{-\alpha x}$, $x \geq 0$. The successive damages accumulate. At the end of each period the item is inspected. The item has to be repaired at a high cost of c_0 when the inspection reveals an accumulated damage larger than a given value z_0. However, a preventive repair at a lower cost of $c_1 > 0$ is possible when the accumulated damage at inspection is at or below z_0. Each repair takes a negligible time and after a repair the item can be regarded as without damage. The following control-limit rule is used. The item is repaired at each inspection that reveals an accumulated damage larger than the repair limit z with $0 < z \leq z_0$. We wish to determine the repair limit z for which the long-run average cost per period is minimal.

Under a given control rule with repair limit z, we can say that a cycle starts each time a repair is completed. The length of a cycle is given by the number of periods needed for a cumulative damage exceeding z. Thus, denoting by M the renewal function associated with the damage density f, it follows that

$$\text{The expected length of a cycle} = 1 + M(z).$$

The repair cost incurred at the end of a cycle is equal to c_0 when the excess in damage of the repair level z is larger than $z_0 - z$ and is equal to c_1 otherwise. Denoting this excess by γ_z, it follows that

$$\text{The expected cost incurred per cycle} = c_0 P\{\gamma_z > z_0 - z\} + c_1 P\{\gamma_z \leq z_0 - z\}.$$

Since the damage density f is exponential with mean $1/\alpha$, we have by exercise 1.4 and relation (1.11) that $M(z) = \alpha z$ and $P\{\gamma_z > x\} = e^{-\alpha x}$ for all z. Hence, under the control rule with repair limit z, we have that, with probability 1,

$$\text{The long-run average cost per period} = \frac{c_0 P\{\gamma_z > z_0 - z\} + c_1 P\{\gamma_z \leqslant z_0 - z\}}{1 + M(z)}$$

$$= \frac{(c_0 - c_1)e^{-\alpha(z_0 - z)} + c_1}{1 + \alpha z}.$$

It is a matter of simple analysis to verify that the average cost per period is minimal for the unique solution z with $0 < z < z_0$ of the equation

$$\alpha z e^{-\alpha(z_0 - z)} = \frac{c_1}{c_0 - c_1},$$

provided $\alpha z_0 > c_1/(c_0 - c_1)$; otherwise the optimal value of z equals z_0. This equation has to be solved numerically, e.g. by the standard Newton–Raphson iteration method. $\qquad\square$

Example 1.7 *Reliability of a one-unit operating system with a single spare part*

A computer system consists of one operating unit and has a single spare part as protection against failures. The time to failure of the operating unit has a general probability distribution function G. If the operating unit fails, it is replaced immediately by the spare if available. The failed unit is sent to a repair facility. This facility can repair only one unit at a time. The repair time has a general distribution function H. The operating times and the repair times are mutually independent. The computer system is down when both units are broken down. The performance measure of interest is the long-run fraction of time the system is down. To find this measure, note that a new cycle begins each time a new operating unit is installed and the other unit enters repair. Denote by the generic variables L and R the life length of an operating unit and the repair time of a failed unit. Some reflections show that the length of a cycle is distributed as $\max(L, R)$, while the downtime during a cycle is distributed as $\max(L, R) - L$. Since $P\{\max(L, R) \leqslant x\} = G(x)H(x)$, we obtain by using (1.9)

$$E(\text{length of a cycle}) = \int_0^\infty \{1 - G(x)H(x)\} \, dx$$

and

$$E(\text{downtime during a cycle}) = \int_0^\infty \{1 - G(x)H(x)\} \, dx - \int_0^\infty \{1 - G(x)\} \, dx.$$

Hence,

The long-run fraction of time the system is down $= \dfrac{\int_0^\infty G(x)\{1 - H(x)\}\,\mathrm{d}x}{\int_0^\infty \{1 - G(x)H(x)\}\,\mathrm{d}x}$.

$\square$

1.3   REGENERATIVE PROCESSES

Numerous stochastic processes arising, for example, in queueing and inventory systems have the property of regenerating themselves at certain points in time, so that the behaviour of the process after the regeneration epoch is a probabilistic replica of the behaviour starting at time zero and is independent of the behaviour before the regeneration epoch. Such stochastic processes are called *regenerative processes*. As an illustration, consider the stochastic process describing the behaviour of the inventory level in the (s, S) inventory system given in example 1.3. This stochastic process is regenerative with the epochs at which the inventory level is ordered up to S as regeneration epochs. In a natural way, the behaviour of a regenerative process over time can be related to the behaviour over a single cycle where a cycle is defined as the time elapsed between two consecutive regeneration epochs (renewal occurrences). In particular, limiting distributions of regenerative processes can be studied in terms associated with a single cycle. This renewal theoretic approach is very useful in specific problems.

Consider now a regenerative process $\{X(t), t \geq 0\}$ with a continuous time parameter but with discrete state space $\{0, 1, \ldots\}$. As an example, $X(t)$ is the number of customers in a queueing system at time t. In applications, one is often interested in the long-run fraction of time the system will spend in some state j. This fraction will be characterized by using renewal reward processes. Define the random variables:

T = the length of one cycle,
T_j = the amount of time the system spends in state j during one cycle.

It is assumed that $0 < E(T) < \infty$. Fix now state j and imagine the following reward structure. A reward at rate 1 is earned during the time the system is in state j and otherwise no rewards are earned. Clearly, for this reward structure, the long-run average reward per unit time represents the long-run fraction of time the system will spend in state j. Also, the expected reward per cycle equals $E(T_j)$. Now, by invoking relation (1.16) for renewal reward processes, we find that, with probability 1,

The long-run fraction of time the system will spend in state $j = \dfrac{E(T_j)}{E(T)}$.   (1.17)

We state without proof that in case the cycle length T has a probability density on some interval the following additional result holds (cf. Ross, 1983):

$$\lim_{t \to \infty} P\{X(t) = j\} \text{ exists and equals } \frac{E(T_j)}{E(T)}. \tag{1.18}$$

Next we state a basic formula for the long-run average value of the process $\{X(t)\}$. With probability 1,

$$\lim_{t \to \infty} \frac{1}{t} \int_0^t X(u)\, du = \frac{E(\int_0^T X(u)\, du)}{E(T)}. \tag{1.19}$$

To see this, imagine that for each state i the system earns a reward at rate i whenever the system is in that state. Then the left side of (1.19) represents the long-run average reward per unit time, while the numerator in the right side of (1.19) represents the expected reward earned during one cycle. By invoking relation (1.16) for renewal reward processes, we get (1.19).

So far, we have considered regenerative processes with a continuous time parameter. Similar results apply to regenerative processes with a discrete time parameter. As an illustration, we consider the following example.

Example 1.3   (continued) Approximations to an (s, S) inventory system

Suppose we wish to determine the long-run average on-hand inventory at the beginning of each week just after any ordering when the delivery of any replenishment order is instantaneous. It is assumed that the parameters s and S of the (s, S) rule used are both positive. Define

$X_n = $ the on-hand inventory at the beginning of the nth week
      just after any ordering.

The stochastic process $\{X_n, n = 1, 2, \ldots\}$ with discrete time parameter regenerates itself each time a replenishment order is placed. Denote by the cycle length N the number of weeks between two successive orderings. Assuming that a cycle starts at the beginning of the first week, we have that, in the long run,

The average on-hand inventory at the beginning of a week just

$$\text{after any ordering} = \frac{E(\sum_{k=1}^{N} X_k)}{E(N)}$$

with probability 1. It was already shown in section 1.1 that

$$E(N) = 1 + M(S - s),$$

where $M(x)$ is the renewal function associated with the probability distribution function of the weekly demand. The numerator of the above ratio may be given the following cost interpretation. Assuming that a holding cost of x is incurred during a week when the inventory equals x at the beginning of that week just after delivery of an order (if any), we have

$$E(\sum_{k=1}^{N} X_k) = \text{the expected holding costs incurred during one cycle.}$$

To evaluate these costs, we define the cost function

$h(x) =$ the expected holding costs incurred until the next replenishment order is placed when at the beginning of the present week the on-hand inventory equals x, $x > s$.

In particular,

$h(S) =$ the expected holding costs incurred during one cycle.

The parametrization with respect to the starting inventory enables us to calculate the desired costs $h(S)$ from the solution to a renewal equation. Supposing the current inventory position equals x and by conditioning on the demand to occur in the next week, it follows that

$$h(x) = x + \int_0^{x-s} h(x-y)f(y)\,dy, \qquad x > s,$$

where $f(y)$ denotes the probability density of the weekly demand. This equation is a renewal equation and may be written into the standard form (1.6) by the transformation $t = x - s$ and $Z(t) = h(t + s)$. It follows from (1.8) that the solution to the renewal equation for $h(x)$ is given by

$$h(x) = x + \int_0^{x-s} (x-y)m(y)\,dy$$

$$= x + sM(x-s) + \int_0^{x-s} M(y)\,dy, \qquad x > s,$$

where the latter equality is obtained by partial integration. In particular,

$$h(S) = S + sM(S-s) + \int_0^{S-s} M(y)\,dy$$

implying that

The average on-hand inventory at the beginning of a week just after any ordering

$$= \frac{S + sM(S-s) + \int_0^{S-s} M(y)\,dy}{1 + M(S-s)}.$$

This expression is in general not tractable enough for practical purposes. However, it will be shown below that

$$\lim_{x \to \infty} \left[\int_0^x M(y)\,dy - \left\{ \frac{x^2}{2\mu_1} + \left(\frac{\mu_2}{2\mu_1^2} - 1 \right)x + \frac{\mu_2^2}{4\mu_1^3} - \frac{\mu_3}{6\mu_1^2} \right\} \right] = 0, \qquad (1.20)$$

where μ_k denotes the kth moment of the weekly demand X_1. This asymptotic expansion in conjunction with (1.5) yields the simple approximation

The average on-hand inventory at the beginning of a week just after any ordering

$$\approx \left(\frac{S-s}{\mu_1} + \frac{\mu_2}{2\mu_1^2}\right)^{-1} \left\{ S + \left(\frac{S-s}{\mu_1} + \frac{\mu_2}{2\mu_1^2} - 1\right)s + \frac{(S-s)^2}{2\mu_1} \right.$$
$$\left. + \left(\frac{\mu_2}{2\mu_1^2} - 1\right)(S-s) + \frac{\mu_2^2}{4\mu_1^3} - \frac{\mu_3}{6\mu_1^2} \right\},$$

provided $S - s$ is sufficiently large relative to the average weekly demand μ_1.

To verify the asymptotic expansion of $\int_0^x M(y)\,dy$, note that the relation (1.5) suggests that, for some constant c,

$$\int_0^x M(y)\,dy \approx \frac{x^2}{2\mu_1} + x\left(\frac{\mu_2}{2\mu_1^2} - 1\right) + c \qquad \text{for large } x.$$

To determine the constant c, define the function

$$\bar{M}(x) = \int_0^x M(y)\,dy - \frac{x^2}{2\mu_1} - x\left(\frac{\mu_2}{2\mu_1^2} - 1\right), \qquad x \geqslant 0.$$

By integrating both sides of the equation (1.3) over x, we get, after interchanging the order of integration, a renewal equation for the function $\int_0^x M(y)\,dy$. Next it follows that $\bar{M}(x)$ satisfies the renewal equation

$$\bar{M}(x) = a(x) + \int_0^x \bar{M}(x-y)f(y)\,dy, \qquad x \geqslant 0,$$

where after some manipulations involving (1.9) the function $a(x)$ can be written as

$$a(x) = \left(\frac{\mu_2}{2\mu_1^2} + \frac{x}{\mu_1}\right) \int_x^\infty \{1 - F(y)\}\,dy - \frac{1}{\mu_1} \int_x^\infty y\{1 - F(y)\}\,dy.$$

Using (1.9), we find

$$\int_0^\infty a(x)\,dx = \frac{\mu_2^2}{4\mu_1^2} - \frac{\mu_3}{6\mu_1}.$$

Next, by applying the key renewal theorem, we obtain the desired result (1.20). $\qquad \square$

1.4 POISSON PROCESS

The Poisson process is an extremely useful renewal process for modelling purposes in numerous practical applications, e.g. to model arrival processes for queueing systems or demand processes for inventory systems. It is empirically found that in a wide variety of circumstances the arising stochastic processes can quite well be represented by a Poisson process. An intuitive explanation of this fact can be based on the phenomenon that in the situation of many individual events, each having a small probability of occurrence, the actual number of events occurring approximately follows a Poisson process.

There are several equivalent definitions of the Poisson process. We give the following definition.

Definition   1.2

A renewal process $\{N(t), t \geq 0\}$ with interoccurrence times $X_1, X_2, \ldots$ between successive events is called a Poisson process if the interoccurrence times have a common exponential probability density

$$f(t) = \lambda e^{-\lambda t}, \qquad t \geq 0.$$

The parameter λ is called the intensity (arrival rate) of the Poisson process.

In the remainder of this section we use for the Poisson process $\{N(t)\}$ the terminology 'arrivals' instead of 'renewals'.

We shall first characterize the distribution of the counting variable $N(t), t > 0$. For that purpose we define, for $k = 1, 2, \ldots$,

$$S_k = \sum_{i=1}^{k} X_i = \text{the epoch at which the } k\text{th arrival occurs}$$

It is well known from probability theory that the sum of k independent random variables having a common exponential distribution has an Erlang-k distribution. Thus the random variable S_k has the probability density

$$f_k(t) = \lambda^k \frac{t^{k-1}}{(k-1)!} e^{-\lambda t}, \qquad t \geq 0, \tag{1.21}$$

and the probability distribution function

$$F_k(t) = 1 - \sum_{j=0}^{k-1} e^{-\lambda t} \frac{(\lambda t)^j}{j!}, \qquad t \geq 0. \tag{1.22}$$

Incidentally, observe from (1.21) and (1.22) the useful identity

$$\int_0^t \lambda^k \frac{x^{k-1}}{(k-1)!} e^{-\lambda x} \, dx = 1 - \sum_{j=0}^{k-1} e^{-\lambda t} \frac{(\lambda t)^j}{j!}, \qquad t \geq 0, k = 1, 2, \ldots. \tag{1.23}$$

From (1.22) and the relation

$$P\{N(t) = j\} = P\{S_j \leq t < S_{j+1}\} = P\{S_{j+1} > t\} - P\{S_j > t\},$$

it follows that, for each $t \geqslant 0$,

$$P\{N(t)=j\} = e^{-\lambda t}\frac{(\lambda t)^j}{j!}, \qquad j=0,1,\ldots. \qquad (1.24)$$

That is for each $t \geqslant 0$ the random variable $N(t)$ representing the total number of arrivals up to time t has a Poisson distribution with mean λt.

In many practical applications the calculation of (cumulative) Poisson probabilities is required. The probabilities $p_j(t) = e^{-\lambda t}(\lambda t)^j/j!$ can be calculated recursively from

$$p_0(t) = e^{-\lambda t} \qquad \text{and} \qquad p_j(t) = \frac{\lambda t}{j}p_{j-1}(t), \qquad j=1,2,\ldots.$$

Here it is assumed that λt is not such a large number that exponent underflow occurs when evaluating $p_0(t) = e^{-\lambda t}$; otherwise one may apply the recursion scheme $q_j(t) = \ln(\lambda t/j) + q_{j-1}(t)$ for $j \geqslant 1$, starting with $q_0(t) = -\lambda t$, and next calculate $p_j(t)$ from $p_j(t) = e^{q_j(t)}$ for those values of j for which no exponent underflow occurs on the used computer.

The memoryless property of the Poisson process

Next we derive a memoryless property that is characteristic for the Poisson process. For each $t > 0$, define the residual life variable

$\gamma_t =$ the amount of time from epoch t until the first arrival after epoch t.

In view of (1.12), we have, independently of t, that

$$P\{\gamma_t \leqslant u\} = 1 - e^{-\lambda u}, \qquad u \geqslant 0. \qquad (1.25)$$

That is at each point of time the waiting time until the next arrival has the same exponential distribution as the original interarrival time, regardless of how long ago the last arrival occurred. The Poisson process is the only renewal process with this memoryless property. The lack of memory of the Poisson process explains the mathematical tractability of this process, since in specific problems the analysis does not require a state variable indicating the time elapsed since the last arrival.

The lack of memory of the Poisson process is also reflected in the exponential distribution. Suppose that the random variable Y (say, the lifetime of a light bulb) is exponentially distributed with mean $1/\mu$. Then, independently of t,

$$P\{Y > t + s \mid Y > t\} = P\{Y > s\} = e^{-\mu s}, \qquad s \geqslant 0, \qquad (1.26)$$

i.e. the residual life of the bulb has the same exponential distribution as the original lifetime regardless of how long the bulb has already been in use.

The relation (1.26) follows directly from

$$P\{Y > t + s \mid Y > t\} = \frac{P\{Y > t + s, Y > t\}}{P\{Y > t\}} = \frac{e^{-\mu(t+s)}}{e^{-\mu t}}.$$

We can restate (1.26) by saying that the exponential distribution has a constant failure rate. That is, for Δt small,

$$P\{Y \in (t, t + \Delta t) \mid Y > t\} = \mu \Delta t + o(\Delta t) \qquad (1.27)$$

regardless of t. Here $o(h)$ is the standard notation for some unspecified function $g(h)$ having the property that $\lim_{h \to 0} g(h)/h = 0$, that is $o(h)$ is negligibly small compared with h as $h \to 0$. We obtain the characterization (1.27) of the exponential distribution from (1.26) and the series expansion

$$1 - e^{-h} = h - \frac{h^2}{2!} + \frac{h^3}{3!} - \cdots = h + o(h) \qquad \text{for } h \text{ small.} \qquad (1.28)$$

In addition to the memoryless property (1.26) we mention the following useful facts for the exponential distribution. If Y_1 and Y_2 are two independent exponentials with respective means $1/\mu_1$ and $1/\mu_2$, then

$$P\{\min(Y_1, Y_2) > t\} = e^{-(\mu_1 + \mu_2)t} \qquad \text{and} \qquad P\{Y_1 < Y_2\} = \frac{\mu_1}{\mu_1 + \mu_2}. \qquad (1.29)$$

That is the minimum of two independent exponentials Y_1 and Y_2 with failure rates μ_1 and μ_2 is exponentially distributed with failure rate $\mu_1 + \mu_2$, and with probability $1/(\mu_1 + \mu_2)$ the lifetime Y_1 expires before Y_2 does.

Example 1.8 A production line with two stations in series

Consider a production line with two stations in series. The items produced at station 1 are passed to station 2 where they are assembled. There is a buffer space at station 2. Each station can handle only one item at a time. The production time of an item at station 1 is exponentially distributed with mean $1/\mu_1$, while the assembly time of an item at station 2 is exponentially distributed with mean $1/\mu_2$. The production times and the assembly times are assumed to be mutually independent. Suppose that presently an item is in production at station 1, while the buffer at station 2 holds $K \geqslant 1$ items including the one in assembly. We wish to determine the probability that the buffer becomes empty before the current production at station 1 is completed. To compute this probability, let us say that an event of type 1 occurs when a production at station 1 is completed and an event of type 2 occurs when an assembly at station 2 is finished. First, consider the special case of $K = 1$ item in the buffer. By the memoryless property of the exponential distribution the time until the first event of type 1 (2) is exponentially distributed with mean $1/\mu_1 (1/\mu_2)$. Hence, by (1.29), $\mu_2/(\mu_1 + \mu_2)$ gives the probability that the first event of type 2 occurs before the first event of type 1. Next we consider the case of $K = 2$ items in the buffer. In order to have two events of type 2 occur before the first event of type 1, it is necessary that the first event is of type 2, and this occurs with probability $\mu_2/(\mu_1 + \mu_2)$. If the first event is of type 2, then at the epoch of this event the remaining production time at station 1 is again exponentially distributed with mean $1/\mu_1$ by the lack of memory of the exponential distribution. Thus, given that the first event is of type 2, the conditional probability that the second event is also of type 2 equals also $\mu_2/(\mu_1 + \mu_2)$. Hence the desired probability that two events of type 2 occur before the first event of type 1 is $\{\mu_2/(\mu_1 + \mu_2)\}^2$. In general, with K units in the buffer at station 2 we find by continuing the above reasoning that $\{\mu_2/(\mu_1 + \mu_2)\}^K$ is the probability that the buffer becomes empty before the current production at station 1 is completed. $\qquad\square$

Example 1.9 *A queueing system with Poisson arrivals and exponential services*

Consider a single-server station at which customers arrive according to a Poisson process with rate λ. The service times of the customers are independent random variables having a common exponential distribution with mean $1/\mu$, and are also independent of the arrival process. The number of customers in the system changes at the arrival and service completion epochs. We are interested in the transition mechanism of this process. Let us say that a transition occurs each time an arrival or service completion occurs. Suppose that at the present moment a customer is in service. Define now Z as the time until the next transition and let A be the event that the next transition is an arrival. We first derive $P\{Z > t\}$ and $P(A)$. For that purpose, let X be the time until the next arrival and let Y be the time until the next service completion. Then $P\{Z > t\} = P\{\min(X, Y) > t\}$ and $P(A) = P(X < Y)$. Now, by the lack of memory of the Poisson process and the exponential distribution, the waiting time X and the remaining service time Y have exponential distributions with respective means $1/\lambda$ and $1/\mu$. Also, X and Y are independent. Hence, by (1.29),

$$P\{Z > t\} = e^{-(\lambda + \mu)t} \quad \text{and} \quad P(A) = \frac{\lambda}{\lambda + \mu}.$$

At first sight one might guess that the random variable Z and the event A are stochastically dependent. However, this is not the case. Using the memoryless property, we verify that

$$P\{Z > t, A\} = P\{Z > t\}\, P(A),$$

showing that the random variable Z and the event A are stochastically independent. The proof is simple. Using a standard conditioning argument,

$$P\{Z > t, A\} = P\{Y > X > t\} = \int_t^\infty P\{Y > X > t \mid X = x\} \lambda e^{-\lambda x} dx$$

$$= \int_t^\infty e^{-\mu x} \lambda e^{-\lambda x} dx = \frac{\lambda}{\lambda + \mu} e^{-(\lambda + \mu)t} = P(A)P\{Z > t\}.$$

Thus, independently of t,

$$P(A \mid Z = t) = P(A),$$

i.e. the next transition will be generated by an arrival (service completion) with a probability $\lambda/(\lambda + \mu)\{\mu/(\lambda + \mu)\}$ independently of the time until the next transition. This remarkable result, saying that the next state of the process is independent of the time until the next transition, is generally true for situations in which the times between transitions are generated by competing Poisson processes which are independent of each other. This phenomenon will underly the construction of continuous-time Markov chains in chapter 2. □

Example 1.10   A stochastic clearing system

In a communication system messages requiring transmission arrive according to a Poisson process with rate λ. Every T time units the system is inspected and at each inspection time all messages present (if any) are cleared from the system. A fixed cost of $K > 0$ is incurred for each inspection. Also, for each message there is a holding cost of $h > 0$ per unit time the message has to wait before it is cleared from the system. We wish to determine the optimal value of T for which the long-run average cost per unit time is minimal.

We first derive with the theory of renewal reward processes the average cost per unit time when the inspection period has length T. To do this, we observe that the stochastic process describing the number of messages in the system regenerates itself each time the system is cleared from all messages present. This fact uses the lack of memory of the arrival process of messages so that at any clearing epoch it is not relevant how long ago the last message arrived. Hence we can say that a cycle is completed each time the system is cleared from all messages present. We have

The expected length of a cycle $= T$.

To find the expected holding cost incurred during a cycle, denote by S_k the arrival epoch of the kth message and let $\delta(x) = 1$ for $x > 0$ and $\delta(x) = 0$ otherwise. The probability density of S_k is given by (1.21). Assuming that at epoch 0 the system is empty of messages and using $\sum_{k=1}^\infty f_k(t) = \lambda$, we find

The expected holding cost incurred during $[0, T)$

$$= hE\left\{\sum_{k=1}^{\infty} (T - S_k)\delta(T - S_k)\right\}$$

$$= h\sum_{k=1}^{\infty} \int_0^T (T - t)f_k(t)\,dt$$

$$= h\int_0^T (T - t)\lambda\,dt$$

$$= h\frac{\lambda}{2}T^2. \tag{1.30}$$

Hence the expected cost per cycle equals $K + \frac{1}{2}h\lambda T^2$ and so, by (1.16), we have that, with probability 1,

$$\text{The long-run average cost per unit time} = \frac{1}{T}(K + \frac{1}{2}h\lambda T^2).$$

The average cost per unit time is minimal for

$$T^* = \sqrt{\frac{2K}{h\lambda}}. \qquad \square$$

Further properties of the Poisson process

We return to the Poisson process and discuss some further properties. The memoryless property has several important consequences. First, the number of arrivals in any time interval $(t, t + s]$ is independent of the number of arrivals up to time t and only depends on the length s of the interval. That is,

$$P\{\text{there occur } k \text{ arrivals in } (t, t + s]\} = e^{-\lambda s}\frac{(\lambda s)^k}{k!}, \qquad k = 0, 1, \ldots, \quad (1.31)$$

independently of what happened up to time t. Second, we can characterize the Poisson process by the fact that its 'failure rate' is constant and is equal to λ. By (1.25) and (1.28) we have for Δt small that

$$P\{\text{an arrival occurs in } (t, t + \Delta t]\} = \lambda\Delta t + o(\Delta t), \tag{1.32}$$

independently of what happened up to time t. Here the standard symbol $o(h)$ denotes some unspecified function $g(h)$ having the property that $\lim_{h \to 0} g(h)/h = 0$. We note that the Poisson process could be defined equivalently by the following conditions:

(a) The probability of a renewal in $(t, t + \Delta t]$ is $\lambda\Delta t + o(\Delta t)$.

(b) The probability of more than one renewal in $(t, t + \Delta t]$ is $o(\Delta t)$.
(c) The occurrence of renewals in $(t, t + \Delta t]$ is independent of the history of the process up to time t.

This definition of a Poisson process will be generalized in chapter 2 when defining a continuous-time Markov chain.

Next we mention the following two facts for a Poisson process (cf. Ross, 1985):

(a) Suppose that $\{N_1(t), t \geqslant 0\}$ and $\{N_2(t), t \geqslant 0\}$ are independent Poisson processes with respective rates λ_1 and λ_2; then the sum $\{N_1(t) + N_2(t), t \geqslant 0\}$ of the two processes is a Poisson process with rate $\lambda_1 + \lambda_2$.
(b) Suppose that $\{N(t), t \geqslant 0\}$ is a Poisson process with rate λ and that each arrival is marked with probability p independently of all other arrivals. Let $N_1(t)$ and $N_2(t)$ denote respectively the number of marked and unmarked arrivals occurring up to time t. Then $\{N_1(t), t \geqslant 0\}$ and $\{N_2(t), t \geqslant 0\}$ are both Poisson processes with respective rates λp and $\lambda(1 - p)$. Also, these two processes are independent.

The next example, being a prelude to example 1.14, introduces a useful technique that is based on the 'failure rate' representation of the Poisson process.

Example 1.11   The average workload in a continuous production process

Consider a processing plant at which batches of fluid material arrive according to a Poisson process with rate λ. The batch amounts $Y_1, Y_2, \ldots$ are independent random variables having a common probability distribution function F with density f and finite first two moments, and are also independent of the arrival process. The material is processed at a constant rate of $\sigma > 0$ whenever the inventory of unprocessed material is positive. It is assumed that $\sigma > \lambda E(Y_1)$, that is the processing rate is larger than the average amount of work offered to the plant per unit time. We are interested in the long-run average inventory of unprocessed material.

The process describing the inventory of unprocessed material regenerates itself each time the inventory becomes zero, since the Poisson process describing the arrival of batches is memoryless. Define a cycle as the time elapsed between two consecutive epochs at which the system becomes empty. Imagine that the system incurs a holding cost at rate hx whenever the inventory of unprocessed material equals x. Then the average cost per unit time equals h times the average inventory of unprocessed material. Each cycle consists of an idle period I and a busy period B for the plant. Since the arrival process of batches is a Poisson process, we have

$$E(I) = \frac{1}{\lambda}.$$

To analyse the busy period, define for each $x > 0$ the basic functions

$\tau(x) = $ the expected amount of time until the system becomes empty, given that at epoch 0 the inventory of unprocessed material is x,

$c(x) = $ the expected costs incurred until the system becomes empty, given that at epoch 0 the inventory of unprocessed material is x.

Then, by conditioning on the size of the batch whose arrival initiates the busy period, it follows that

$$E(\text{length of one busy period}) = \int_0^\infty \tau(x)f(x)\,dx$$

and

$$E(\text{costs incurred during one busy period}) = \int_0^\infty c(x)f(x)\,dx.$$

Hence, with probability 1,

$$\text{The long-run average cost per unit time} = \frac{\displaystyle\int_0^\infty c(x)f(x)\,dx}{1/\lambda + \displaystyle\int_0^\infty \tau(x)f(x)\,dx} \tag{1.33}$$

This formula with $h = 1$ in $c(x)$ gives the average inventory of unprocessed material. The functions $\tau(x)$ and $c(x)$ can be found by the following powerful argument. Fix $x > 0$ and suppose that the current inventory of unprocessed material is $x + \Delta x$ with Δx small. We condition on what happens in the first $\Delta x/\sigma$ time units. An amount of Δx will be processed in a time $\Delta x/\sigma$. Since the arrival process of batches is a Poisson process, a batch will arrive in the first $\Delta x/\sigma$ time units with probability $\lambda \Delta x/\sigma + o(\Delta x)$. In case of an arrival in this time interval, the inventory of unprocessed material will jump to $x + y$ on condition that the size of the arriving batch is y; otherwise the inventory decreases to x. Thus

$$\tau(x + \Delta x) = \frac{\Delta x}{\sigma} + \lambda\frac{\Delta x}{\sigma}\int_0^\infty \tau(x + y)f(y)\,dy + \left(1 - \lambda\frac{\Delta x}{\sigma}\right)\tau(x) + o(\Delta x),$$

$$c(x + \Delta x) = hx\frac{\Delta x}{\sigma} + \lambda\frac{\Delta x}{\sigma}\int_0^\infty c(x + y)f(y)\,dy + \left(1 - \lambda\frac{\Delta x}{\sigma}\right)c(x) + o(\Delta x).$$

Subtracting $\tau(x)$, respectively $c(x)$, from both sides of the corresponding equation, dividing the resulting equations by Δx and letting $\Delta x \to 0$, we find

$$\tau'(x) = \frac{1}{\sigma} + \frac{\lambda}{\sigma} \int_0^\infty \tau(x+y)f(y)\,dy - \frac{\lambda}{\sigma}\tau(x), \qquad x > 0,$$

$$c'(x) = \frac{hx}{\sigma} + \frac{\lambda}{\sigma} \int_0^\infty c(x+y)f(y)\,dy - \frac{\lambda}{\sigma}c(x), \qquad x > 0.$$

We can argue that $\tau(x)$ is linear in x and $c(x)$ is quadratic in x. Then, substituting these functional forms in the above integro-differential equations and equating coefficients, we obtain after some algebra that, for all $x \geq 0$,

$$\tau(x) = \frac{x}{\sigma - \lambda E(Y_1)}, \qquad c(x) = \frac{h}{2\{\sigma - \lambda E(Y_1)\}}\left\{x^2 + \frac{\lambda x E(Y_1^2)}{\sigma - \lambda E(Y_1)}\right\}. \quad (1.34)$$

To verify that $\tau(x)$ is linear in x and $c(x)$ is quadratic in x, we first observe that the time to empty the system and the holding cost incurred during this time are not changed by the following way of processing the material. Let the current inventory of unprocessed material be equal to $x + y$ and suppose that this inventory is divided into two amounts x and y. First, we process the amount x and any newly arriving batch until only the amount y is left. This takes a time having an expectation $\tau(x)$, and the expected holding cost incurred during this time is $c(x) + h y \tau(x)$. Next we process the amount y and any newly arriving batch until the system is empty. This takes a time having an expectation $\tau(y)$, and the expected holding cost incurred during this time is $c(y)$. Thus we find $\tau(x+y) = \tau(x) + \tau(y)$ and $c(x+y) = c(x) + h y \tau(x) + c(y)$ for all $x, y > 0$. From these relations we can derive the desired functional forms. We omit the details.

It follows from (1.33) and (1.34) with $h = 1$ that, with probability 1,

The long-run average inventory of unprocessed material

$$= \frac{\lambda E(Y_1^2)}{2\{\sigma - \lambda E(Y_1)\}}. \qquad (1.35)$$

Similarly, by assuming a cost at rate 1 whenever the inventory of unprocessed material is positive, it is readily verified that, with probability 1,

The long-run fraction of time the processing plant is idle

$$= \frac{1/\lambda}{1/\lambda + \int_0^\infty \tau(x)f(x)\,dx} = 1 - \frac{\lambda E(Y_1)}{\sigma}. \qquad (1.36)$$

The formulae (1.35) and (1.36) with $\sigma = 1$ apply in fact to the following standard queueing system. Customers arrive at a single-server station according to a Poisson process with rate λ. The service times of the customers are independent and identically distributed random variables $Y_1, Y_2, \ldots$, and are also independent of the arrival process. The server can handle only one customer at a time and works at a unit rate whenever the system is not empty. Identify now the server with the processor and the service time of a customer with a batch of work for the server. Then formula (1.35) with $\sigma = 1$ is the famous *Pollaczek–Khintchine formula* for the average amount of uncompleted work in the system and the formula (1.36) with $\sigma = 1$ gives the fraction of time the system is empty. □

A theoretical explanation of the occurrence in practice of Poisson processes

We conclude this section with a theoretical explanation of the occurrence in practice of processes closely approximating to Poisson processes. For that purpose, consider n independent renewal processes $\{N_k(t), t \geq 0\}$, $k = 1, \ldots, n$. Define the superposition process $\{N(t), t \geq 0\}$ by $N(t) = \sum_{k=1}^{n} N_k(t)$. For example $N_k(t)$ is the number of ships from some harbour k that arrive during $(0, t]$ in the harbour of Rotterdam, and $N(t)$ is the sum of these arrivals up to time t. For the renewal process $\{N_k(t)\}$, denote by F_k the probability distribution function of the interoccurrence time between two consecutive arrivals and denote by α_k the first moment of F_k. Assuming that at the present moment the processes are already in progress for a very long time and that no information is known about the histories of the processes up to now, we define W_k as the waiting time until the next renewal in the process $\{N_k(t)\}$. Also, let $W = \min(W_1, \ldots, W_n)$ be the waiting time until the next event in the superposition process $\{N(t)\}$. Suppose now that the number n of renewal processes gets very large and that in each individual renewal process the renewals occur very rarely so that the pooled effect is due to many small causes. More precisely, it is supposed that the arrival rates $1/\alpha_k$ get very small such that

$$\frac{1}{\alpha_1} + \cdots + \frac{1}{\alpha_n} \to \lambda \quad \text{as} \quad n \to \infty.$$

Then it can heuristically be argued that

$$P\{W > u\} \approx e^{-\lambda u} \quad \text{for all} \quad u \geq 0, \tag{1.37}$$

supporting the claim that the superposition process $\{N(t)\}$ is approximately a Poisson process. To argue (1.37), note first that

$$P\{W > u\} = P\{W_1 > u\} \cdots P\{W_n > u\},$$

since $W_1, \ldots, W_n$ are independent. Since the processes are assumed to be in progress for a very long time, we have by (1.12) that

$$P\{W_k \leq u\} \approx \frac{1}{\alpha_k} \int_0^u \{1 - F_k(x)\} \, dx.$$

Now fix u. For n sufficiently large, the average interarrival times α_k will be very large and so, for any k, $F_k(x) \approx 0$ for $0 \leqslant x \leqslant u$. Thus, for n large enough, $P\{W_k \leqslant u\} \approx u/\alpha_k$ and so

$$P\{W > u\} \approx \left(1 - \frac{u}{\alpha_1}\right) \cdots \left(1 - \frac{u}{\alpha_n}\right) \approx \left\{1 - \left(\frac{u}{\alpha_1} + \cdots + \frac{u}{\alpha_n}\right)\right\}$$
$$\approx e^{-u(1/\alpha_1 + \cdots + 1/\alpha_n)},$$

which yields (1.37). This limiting result provides a theoretical explanation of the occurrence of Poisson processes in a wide variety of circumstances. For example the number of calls received at a large telephone exchange is the superposition of the individual calls of many subscribers each calling infrequently. Thus the overall number of calls can be expected to closely be a Poisson process. Similarly, a Poisson demand process for a given product can be expected if the demands are the superposition of the individual requests of many customers each asking infrequently for that product.

*1.5 SOME CONTROL MODELS ANALYSED BY RENEWAL REWARD PROCESSES

This section illustrates the usefulness of renewal reward processes by giving several applications to control problems arising in production and queueing systems.

Example 1.12   The N-policy for starting up and shutting down a production process

Consider a production facility at which production orders arrive according to a Poisson process with rate λ. The production times $Y_1, Y_2, \ldots$ of the orders are independent random variables having a common probability distribution function F with finite first two moments. Also, the production process is independent of the arrival process. The facility can only work on one order at a time. It is assumed that $E(Y_1) < 1/\lambda$, that is the average production time per order is less than the mean interarrival time between two consecutive orders. The facility operates only intermittently and is shut down when no orders are present any more. A fixed setup cost of $K > 0$ is incurred each time the facility is reopened. Also, a holding cost of $h > 0$ per unit time is incurred for each order present. In view of this cost structure, the facility is only turned on when enough orders have accumulated. The so-called N-policy reactivates the facility as soon as N orders are present. For ease we assume that it takes a zero setup time to restart production. We want to determine the value of the control parameter N which minimizes the long-run average cost per unit time.

To analyse this problem, we first observe that for a given N-policy the process describing jointly the number of orders present and the status of the facility (on or off) regenerates itself each time and facility is turned on. Define a cycle as the time elapsed between two consecutive reactivations of the facility. Clearly, each cycle consists of a busy period B with production and an idle period I with no production. We deal separately with the idle and the busy period. Using the memoryless property of the Poisson process, the length of the idle period is the sum of N exponential random variables each having mean $1/\lambda$. Hence

$$E(I) = N/\lambda$$

and

$$E(\text{holding cost incurred during } I) = h\left(\frac{N-1}{\lambda} + \cdots + \frac{1}{\lambda}\right) = \frac{h}{2\lambda} N(N-1).$$

To deal with the busy period, we define for $n = 1, 2, \ldots$ the quantities

$t_n = $ the expected time until the facility becomes empty, given
      that at epoch 0 a production starts with n orders present,

and

$h_n = $ the expected holding costs incurred until the facility
       becomes empty, given that at epoch 0 a production starts
       with n orders present.

These quantities are in fact independent of the control rule considered. Now, by (1.16), we have under an N-policy that, with probability 1,

$$\text{The long-run average cost per unit time} = \frac{(h/2\lambda)N(N-1) + K + h_N}{N/\lambda + t_N}. \tag{1.38}$$

To find the functions t_n and h_n, define, for $j = 0, 1, \ldots,$

$a_j = $ the probability that j orders arrive during the production time
     of a single order.

Note that this definition uses the lack of memory of the Poisson process (why?). Assume for ease that the production time Y_1 has a probability density f. By conditioning on the production time and using (1.31), it follows that

$$a_j = \int_0^\infty e^{-\lambda y} \frac{(\lambda y)^j}{j!} f(y)\,dy, \qquad j = 0, 1, \ldots .$$

It is readily verified that

$$\sum_{j=1}^\infty j a_j = \lambda E(Y_1) \qquad \text{and} \qquad \sum_{j=1}^\infty j^2 a_j = \lambda^2 E(Y_1^2) + \lambda E(Y_1). \tag{1.39}$$

We now derive recursion relations for the quantities t_n and h_n. Suppose that at epoch 0 a production starts with n orders present and denote by v the number of new orders arriving during the production time of the first order. Then the time to empty the system equals the first production time plus the time to empty the system when $n - 1 + v$ orders are present. Thus

$$t_n = E(Y_1) + \sum_{j=0}^{\infty} t_{n-1+j} a_j, \qquad n = 1, 2, \ldots,$$

where $t_0 = 0$. Similarly, using that by (1.30) the expected holding cost incurred during the first production equals $hnE(Y_1) + \frac{1}{2}h\lambda E(Y_1^2)$, it follows that

$$h_n = hnE(Y_1) + \frac{1}{2}h\lambda E(Y_1^2) + \sum_{j=0}^{\infty} h_{n-1+j} a_j, \qquad n = 1, 2, \ldots,$$

where $h_0 = 0$. We can argue that t_n is linear in n and h_n is quadratic in n. Then, substituting these functional forms in the above recursion relations and using (1.39), we find after some algebra that for $n = 1, 2, \ldots,$

$$t_n = \frac{nE(Y_1)}{1 - \lambda E(Y_1)},$$

$$h_n = \frac{h}{1 - \lambda E(Y_1)} \left[\frac{1}{2}n(n-1)E(Y_1) + nE(Y_1) + \frac{\lambda n E(Y_1^2)}{2\{1 - \lambda E(Y_1)\}} \right]. \qquad (1.40)$$

To verify that t_n is linear in n and h_n is quadratic in n, we first observe that t_n and h_n do not depend on the specific order in which the production orders are coped with during the production process. Imagine the following production discipline. The n initial orders $O_1, \ldots, O_n$ (say) are separated. Order O_1 is produced first, after which all orders (if any) are produced that have arrived during the production time of O_1, and this way of production is continued until the facility is free of all orders but $O_2, \ldots, O_n$. Next this procedure is repeated with order O_2, etc. Thus we find that $t_n = nt_1$, proving that t_n is linear in n. Using the same arguments and noting that $h_1 + h(n-k)t_1$ gives the expected holding cost incurred during the time to free the system of order O_k and its direct descendants until only the orders $O_{k+1}, \ldots, O_n$ are left, it follows that

$$h_n = \sum_{k=1}^{n} \{h_1 + h(n-k)t_1\} = nh_1 + \frac{1}{2}hn(n-1)t_1.$$

Combining the above results and letting $\rho = \lambda E(Y_1)$, we find for an N-policy that, with probability 1,

The long-run average cost per unit time

$$= \frac{\lambda(1-\rho)K}{N} + h\left\{ \rho + \frac{\lambda^2 E(Y_1^2)}{2(1-\rho)} + \frac{N-1}{2} \right\}. \qquad (1.41)$$

It is noteworthy that this expression needs from the production time only the first two moments. Also note that, by putting $K = 0$ and $h = 1$ in (1.41), we find that, with probability 1,

The long-run average number of orders in the system
$$= \rho + \frac{\lambda^2 E(Y_1^2)}{2(1 - \rho)} + \frac{N - 1}{2}. \tag{1.42}$$

For the special case of $N = 1$ this formula reduces to the famous Pollaczek–Khintchine formula for the average number of customers in the system for the standard single-server queue with Poisson arrivals and general service times. In addition to (1.42), it is interesting to note that (cf. also 1.36),

The long-run fraction of time the facility is in production
$$= \frac{E(B)}{E(I) + E(B)} = \rho \tag{1.43}$$

independently of the control parameter N. Heuristically, this result could be seen by noting that $\rho = \lambda E(Y_1)$ is the average amount of work per unit time offered to the production facility.

The optimal value of N can be obtained by differentiating the right side of (1.41) in which we take N as a continuous variable. Since the average cost is convex in N, it follows that the average cost is minimal for one of the two integers nearest to

$$N = \sqrt{\frac{2\lambda(1 - \rho)K}{h}}. \tag{1.44}$$

$\square$

Example 1.13   The T-policy for utilizing idle time in a production process

We consider again the production system from example 1.12 except that the system is now controlled in a different way when it becomes idle. Each time the production facility becomes empty of orders, the facility is used during a period of fixed length T for some other work in order to utilize the idle time. After this vacation period the facility is reactivated for servicing the orders only when at least one order is present; otherwise the facility is used again for some other work during a vacation period of length T. This utilization of idle time is continued until at least one order is present after the end of a vacation period. This control policy is called the T-policy. The cost structure is the same as in example 1.12. We wish to determine the value of T for which the long-run average cost per unit time is minimal.

We first analyse a given T-policy. Under a T-policy the process describing the number of jobs in the system is regenerative. In view of the lack of memory of the arrival process of orders, the epochs at which a vacation period starts can be taken as regeneration epochs. Thus it is convenient to define a cycle as the time elapsed between two consecutive epochs at which a vacation period starts.

Suppose that at epoch 0 a cycle starts and denote by v the number of orders arriving during the vacation period $(0, T]$. Under the condition that $v = n$ the expected length of the first cycle equals $T + t_n$, where $t_0 = 0$ and t_n is given in (1.40) for $n \geq 1$. Noting that $e^{-\lambda T}(\lambda T)^n/n!$ is the probability that n jobs arrive during a time T, it follows that

$$\text{The expected length of one cycle} = T + \sum_{n=1}^{\infty} t_n e^{-\lambda T}\frac{(\lambda T)^n}{n!}$$

$$= T + \frac{\lambda T E(Y_1)}{1 - \lambda E(Y_1)} = \frac{T}{1 - \lambda E(Y_1)}.$$

By (1.30), the expected holding cost incurred during the first T time units of the first cycle equals $\frac{1}{2}h\lambda T^2$. Under the condition that $v = n$ the expected holding cost incurred in the remaining part of the cycle equals h_n, where $h_0 = 0$ and h_n is given in (1.40) for $n \geq 1$. It now follows after some algebra that

The expected cost incurred during one cycle

$$= K + \tfrac{1}{2}h\lambda T^2 + \sum_{h=1}^{\infty} h_n e^{-\lambda T}\frac{(\lambda T)^n}{n!}$$

$$= K + \frac{h}{1 - \lambda E(Y_1)}\left[\tfrac{1}{2}\lambda T^2 + \lambda T E(Y_1)\right.$$

$$\left. + \frac{\lambda^2 T E(Y_1^2)}{2\{(1 - \lambda E(Y_1)\}}\right].$$

Hence under a T-policy we have that, with probability 1,

$$\text{The long-run average cost per unit time} = \frac{K\{1 - \lambda E(Y_1)\}}{T} + \tfrac{1}{2}h\lambda T + h\lambda E(Y_1)$$

$$+ \frac{h\lambda^2 E(Y_1^2)}{2\{1 - \lambda E(Y_1)\}}.$$

The long-run average cost per unit time is minimal for

$$T^* = \sqrt{\frac{2K\{1 - \lambda E(Y_1)\}}{h\lambda}}. \qquad \square$$

Example 1.14 *The D-policy for controlling a continuous production process*

We consider the same model as in example 1.11 except that the processing plant is now closed down each time the inventory of unprocessed material becomes zero. The processing plant is controlled by a so-called *D*-policy which prescribes to reopen the plant when the inventory of unprocessed material exceeds the critical level *D*. The setup time to restart the processing is negligible. Also, the following cost structure is assumed. A fixed setup cost of $K > 0$ is incurred each time the plant is reopened and a non-negative holding cost at rate hx whenever the inventory of unprocessed material is x. We want to determine the D-policy for which the long-run average cost per unit time is minimal.

An exact analysis of this problem leads to rather intractable results. However, by invoking asymptotic expansions from renewal theory, we find a simple approximate solution for the D-policy. It will turn out that the approximate solution is very close to the exact solution provided the relevant values of D are sufficiently large compared with the average batch size.

Since the arrival process of batches is a Poisson process, we have for a given D-policy that the process describing jointly the inventory of unprocessed material and the status of the plant (on or off) regenerates itself each time the plant is shut down. Define a cycle as the time elapsed between two consecutive shutdowns. Each cycle consists of an idle period with no processing and a busy period during which the material is processed. We analyse separately the idle period and the busy period. First, we deal with the idle period and define, for $0 \leqslant x \leqslant D$,

$\alpha(D - x) =$ the expected time until the inventory of unprocessed material exceeds the level D, given that at epoch 0 the unprocessed inventory equals x and the plant is closed down,

$\beta(D - x) =$ the expected holding costs incurred until the inventory of unprocessed material exceeds the level D, given that at epoch 0 the unprocessed inventory is x and the plant is closed down.

Clearly, $\alpha(D)$ is the expected length of one idle period and $\beta(D)$ is the expected holding costs incurred during one idle period. Suppose that the inventory is x and the facility is closed down. Then it takes an expected time of $1/\lambda$ before the inventory changes by the arrival of the next batch, and the plant is reopened on account of this batch only if the batch size is larger than $D - x$. Thus, by conditioning on the batch size, it is readily seen that

$$\alpha(D - x) = \frac{1}{\lambda} + \int_0^{D-x} \alpha(D - x - y) f(y) \, dy, \qquad 0 \leqslant x \leqslant D,$$

$$\beta(D - x) = \frac{hx}{\lambda} + \int_0^{D-x} \beta(D - x - y) f(y) \, dy, \qquad 0 \leqslant x \leqslant D.$$

These equations are renewal equations. Denoting by $M(x)$ the renewal function associated with the probability distribution function $F(x)$ and denoting by $m(x)$ the density of $M(x)$, it follows from relation (1.8) that the solutions of the above renewal equations are given by

$$\alpha(D - x) = \frac{1}{\lambda} + \int_0^{D-x} \frac{1}{\lambda} m(y) \, dy = \frac{1}{\lambda} \{1 + M(D - x)\}, \qquad 0 \leqslant x \leqslant D,$$

$$\beta(D - x) = \frac{hx}{\lambda} + \frac{h}{\lambda} \int_0^{D-x} (x + y)m(y)\,\mathrm{d}y$$

$$= \frac{hx}{\lambda}\{1 + M(D - x)\} + \frac{h}{\lambda}(D - x)M(D - x) - \frac{h}{\lambda}\int_0^{D-x} M(y)\,\mathrm{d}y, \qquad 0 \leqslant x \leqslant D,$$

where the latter equality uses partial integration. In particular,

$$\alpha(D) = \frac{1}{\lambda}\{1 + M(D)\}, \qquad \beta(D) = \frac{h}{\lambda}DM(D) - \frac{h}{\lambda}\int_0^D M(y)\,\mathrm{d}y.$$

Next we analyse the busy period. Denote by the random variables τ_B and C_B the length of one busy period and the holding costs incurred during one busy period. A busy period starts when the plant is reopened and ends when the plant becomes empty. Letting X_D be the inventory present when the plant is reopened, we have that $E(\tau_B | X_D = x) = \tau(x)$ and $E(C_B | X_D = x) = c(x)$, where the functions $\tau(x)$ and $c(x)$ have already been determined in example 1.11. Thus, by (1.43),

$$E(\tau_B) = \frac{E(X_D)}{\sigma - \lambda E(Y_1)}, \qquad E(C_B) = \frac{hE(X_D^2)}{2\{(\sigma - \lambda E(Y_1))\}} + \frac{h\lambda E(X_D)E(Y_1^2)}{2\{\sigma - \lambda E(Y_1)\}^2}.$$

Hence for a given D-policy the long-run average costs per unit time equals, with probability 1,

$$g(D) = \frac{(h/\lambda)\{DM(D) - \int_0^D M(y)\,\mathrm{d}y\} + K + E(C_B)}{(1/\lambda)\{1 + M(D)\} + E(\tau_B)}. \tag{1.45}$$

To find $E(X_D)$ and $E(X_D^2)$, note that

$$X_D = D + \gamma_D,$$

where the overshoot variable γ_D is the 'residual life at time D' in a renewal process having the batch sizes $Y_1, Y_2, \ldots$ as interoccurrence times. The moments $E(\gamma_D)$ and $E(\gamma_D^2)$ can be expressed in terms of the renewal function M, but we shall directly invoke asymptotic expansions in order to get a tractable solution for the D-policy. We approximate the first two moments of γ_D as well as the renewal quantities $M(D)$ and $\int_0^D M(y)\,\mathrm{d}y$. Putting, for abbreviation,

$$\mu_k = E(Y_1^k), \qquad k = 1, 2, \ldots,$$

and supposing that D is sufficiently large compared with μ_1, we have by (1.13) that

$$E(\gamma_D) \approx \frac{\mu_2}{2\mu_1}, \qquad E(\gamma_D^2) \approx \frac{\mu_3}{3\mu_1},$$

independently of D. Here it is assumed that $\mu_3 < \infty$. Also, by (1.5) and (1.20),

$$M(D) \approx \frac{D}{\mu_1} + \frac{\mu_2}{2\mu_1^2} - 1, \qquad \int_0^D M(y)\,dy \approx \frac{D^2}{2\mu_1} + D\left(\frac{\mu_2}{2\mu_1^2} - 1\right)$$

$$+ \frac{\mu_2^2}{4\mu_1^3} - \frac{\mu_3}{6\mu_1^2}.$$

Substituting the above asymptotic expansions in (1.45), we find that for a given D-policy the average cost per unit time is approximated by

$$g_{\text{app}}(D) = \left(\frac{D}{\mu_1} + \frac{\mu_2}{2\mu_1^2}\right)^{-1} \left\{ K\left(\frac{\lambda}{\sigma}\right)(\sigma - \lambda\mu_1) \right.$$

$$\left. - \tfrac{1}{2}h\left(\frac{D^2}{\mu_1} + \frac{D\mu_2}{\mu_1^2} + \frac{\mu_2^2}{2\mu_1^3} - \frac{\mu_3}{3\mu_1^2}\right) \right\} + hD + \frac{h\lambda\mu_2}{2(\sigma - \lambda\mu_1)}. \quad (1.46)$$

This approximate cost function is minimal for

$$D_{\text{app}} = \frac{-\mu_2}{2\mu_1} + \frac{\mu_2}{\mu_1}\sqrt{\frac{2K\lambda(\sigma - \lambda\mu_1)\mu_1^3}{h\sigma} + \frac{\mu_3\mu_1}{\mu_2^2} + \frac{1}{3\mu_2^2} - \frac{1}{4}}$$

provided the right side of this formula is well defined and positive; otherwise $D_{\text{app}} = 0$. Using the results of exercise 1.7, we find that for a given D-policy the actual average cost per unit time equals

$$g(D) = \frac{K(\lambda/\sigma)(\sigma - \lambda\mu_1) - h\{D + \int_0^D M(y)\,dy\}}{1 + M(D)} + hD + \frac{h\lambda\mu_2}{2(\sigma - \lambda\mu_1)}.$$

The actual cost function $g(D)$ is minimal for the unique solution D_{opt} of the equation

$$D_{\text{opt}} + \int_0^{D_{\text{opt}}} M(y)\,dy = \frac{K\lambda(\sigma - \lambda\mu_1)}{\sigma h}.$$

In Table 1.2 we give for various batch-size distributions the values D_{app} and D_{opt} together with the following two relative error percentages in costs:

$$\Delta_1 = 100\times\left|\frac{g(D_{\text{app}}) - g(D_{\text{opt}})}{g(D_{\text{opt}})}\right|, \qquad \Delta_2 = 100\times\left|\frac{g_{\text{app}}(D_{\text{app}}) - g(D_{\text{app}})}{g(D_{\text{app}})}\right|.$$

In all examples we assume $\mu_1 = 1$, $\sigma = 1$ and $h = 1$. The arrival rate λ has the two values 0.8 and 0.5 and the setup cost K is varied as 5 and 25. The squared coefficient of variation $c_Y^2 = (\mu_2 - \mu_1^2)/\mu_1^2$ of the batch size has the values $\frac{1}{3}$, $\frac{1}{2}, \frac{3}{4}, 1\frac{1}{2}$ and 3. The value $c_Y^2 = \frac{1}{3}$ corresponds to the E_3 distribution, the values $c_Y^2 = \frac{1}{2}$ and $\frac{3}{4}$ correspond to mixtures of E_1 and E_2 distributions with the same scale parameters, and the values $c_Y^2 = 1\frac{1}{2}$ and 3 correspond to H_2 distributions with balanced means. In addition, for purposes of sensitivity

Table 1.2   Approximate and exact values for the control level D

c_Y^2	D_{app}	D_{opt}	Δ_1	Δ_2	D_{app}	D_{opt}	Δ_1	Δ_2	
	$K = 5$				$K = 25$				
$\frac{1}{3}$	0.710	0.708	0.00	0.06	2.214	2.214	0.00	0.00	$\lambda = 0.8$
$\frac{1}{2}$	0.677	0.680	0.00	0.08	2.155	2.155	0.00	0.00	
$\frac{1}{2}*$	0.664	0.658	0.01	0.17	2.148	2.148	0.00	0.00	
$\frac{3}{4}$	0.629	0.636	0.00	0.18	2.068	2.068	0.00	0.00	
$\frac{3}{4}*$	0.611	0.601	0.02	0.24	2.059	2.059	0.00	0.00	
$1\frac{1}{2}$	0.000	0.391	7.59	14.8	1.545	1.629	0.24	1.50	
3	0.000	0.218	7.09	23.4	0.318	1.049	15.6	20.5	
$\frac{1}{3}$	1.006	1.006	0.00	0.01	2.911	2.911	0.00	0.00	$\lambda = 0.5$
$\frac{1}{2}$	0.964	0.965	0.00	0.04	2.847	2.847	0.00	0.00	
$\frac{1}{2}*$	0.953	0.952	0.00	0.05	2.842	2.842	0.00	0.00	
$\frac{3}{4}$	0.903	0.906	0.00	0.19	2.753	2.753	0.00	0.00	
$\frac{3}{4}*$	0.888	0.885	0.00	0.17	2.746	2.746	0.00	0.00	
$1\frac{1}{2}$	0.271	0.594	12.5	26.7	2.259	2.298	0.13	1.24	
3	0.000	0.338	39.0	70.2	1.142	1.588	11.9	21.7	

analysis, we consider also mixtures of E_1 and E_3 distributions with the same scale parameters for the two values $c_Y^2 = \frac{1}{2}$ and $\frac{3}{4}$ (cf. equations A.13 and A.14 in appendix B). The latter two cases are denoted by an asterisk in the table. The numerical investigations show that D_{app} yields excellent results provided $D_{app} \geqslant \theta\mu_1$ with $\theta = 1$ for $0 < c_Y^2 \leqslant 1$ and $\theta = 1\frac{1}{2}c_Y^2$ otherwise. Also, it appears that the optimal value of D is rather insensitive to more than the first two moments of the batch size when $0 \leqslant c_Y^2 \leqslant 1$.   □

1.6   COMPOUND POISSON PROCESS

We first give a formal definition.

Definition 1.3

A stochastic process $\{X(t), t \geqslant 0\}$ is said to be a compound Poisson process if it can be represented by

$$X(t) = \sum_{i=1}^{N(t)} Y_i, \qquad t \geqslant 0,$$

where $\{N(t), t \geqslant 0\}$ is a Poisson process and $Y_1, Y_2, \ldots$ are independent and identically distributed random variables which are also independent of the process $\{N(t)\}$.

Compound Poisson processes arise in a variety of contexts. As an example, consider an insurance company at which claims arrive according to a Poisson process and the claim sizes are independent and identically distributed random variables, which are also independent of the arrival process. Then the cumulative amount claimed up to time t is a compound Poisson variable. Also, the compound Poisson process has applications in inventory theory. Suppose customers asking for a given product arrive according to a Poisson process. The demands of the customers are independent and identically distributed random variables, which are also independent of the arrival process. Then the cumulative demand up to time t is a compound Poisson variable.

We now discuss the distribution of the compound Poisson variable $X(t)$. The mean and the variance of $X(t)$ follow directly from formulae (A.6) and (A.7) in appendix A and the relations $E(N(t)) = \lambda t$ and $E(N(t)^2) = \lambda^2 t^2 + \lambda t$. Thus

$$E(X(t)) = \lambda t E(Y_1), \qquad \text{var}(X(t)) = \lambda t E(Y_1^2).$$

We give the probability distribution of $X(t)$ for the case in which the random variables $Y_1, Y_2, \ldots$ are non-negative and have a discrete probability distribution

$$\phi(j) = P\{Y_1 = j\}, \qquad j = 0, 1, \ldots$$

Define, for any $t > 0$,

$$r_t(j) = P\{X(t) = j\}, \qquad j = 0, 1, \ldots.$$

Put for abbreviation $\phi_n(j) = P\{Y_1 + \cdots + Y_n = j\}$, $j = 0, 1, \ldots$ and $n = 1, 2, \ldots$. Then, by conditioning on the number of arrivals up to time t, we have, for any $t > 0$,

$$r_t(j) = \sum_{n=0}^{\infty} P\{X(t) = j \mid N(t) = n\} P\{N(t) = n\}$$

$$= \sum_{n=0}^{\infty} \phi_n(j) e^{-\lambda t} \frac{(\lambda t)^n}{n!}, \qquad j = 0, 1, \ldots, \tag{1.47}$$

with the convention $\phi_0(0) = 1$ and $\phi_0(j) = 0$, $j \geqslant 1$. In particular, $r_t(0) = e^{-\lambda t\{1 - \phi(0)\}}$. Although the probabilities $\phi_n(j)$, $j \geqslant 0$, can recursively be computed from the convolution formula

$$\phi_n(j) = \sum_{k=0}^{j} \phi(k)\phi_{n-1}(j - k), \qquad j = 0, 1, \ldots, \qquad n = 1, 2, \ldots,$$

the relation (1.47) can be improved upon for computational purposes. The following efficient recursion scheme applies to the probabilities $r_t(j)$, $j \geqslant 0$. For fixed t, the probabilities $r_t(j), j \geqslant 0$, can recursively be computed from

$$r_t(0) = e^{-\lambda t\{1 - \phi(0)\}} \tag{1.48}$$

$$r_t(j) = \frac{\lambda t}{j} \sum_{k=0}^{j-1} (j - k)\phi(j - k)r_t(k), \qquad j = 1, 2, \ldots.$$

For the special case of $\phi(1) = 1$, (1.48) reduces to the well-known recursion scheme for computing Poisson probabilities. The proof of the recursion scheme (1.48) is instructive since it is based on the useful technique of *generating functions*. This technique is closely related to the Laplace transform method.

For a non-negative random variable N with discrete probability distribution $\{p(j), j = 0, 1, \ldots\}$ the generating function $P(z)$ is defined by

$$P(z) = \sum_{j=0}^{\infty} p(j)z^j, \qquad |z| \leqslant 1. \qquad (1.49)$$

The probability distribution $\{p(j), j = 0, 1, \ldots\}$ can be recovered analytically from the compressed function $P(z)$ by

$$p(j) = \frac{1}{j!} \frac{d^j P(z)}{dz^j}\bigg|_{z=0}, \qquad j = 0, 1, \ldots. \qquad (1.50)$$

It is remarked that the Fast Fourier Transform method provides an effective tool for the numerical computation of the probabilities $p(j)$ when an explicit expression for $P(z)$ is given.

Returning to the compound Poisson model, define for fixed $t > 0$ the generating functions,

$$R(z) = \sum_{j=0}^{\infty} r_t(j)z^j, \qquad Q(z) = \sum_{j=0}^{\infty} \phi(j)z^j, \qquad |z| \leqslant 1.$$

For any $n \geqslant 1$ we have by (1.49) and the independence of $Y_1, \ldots, Y_n$ that the generating function of the convolution probabilities $\phi_n(j), j \geqslant 0$, is given by

$$\sum_{j=0}^{\infty} \phi_n(j)z^j = E(z^{Y_1 + \cdots + Y_n}) = E(z^{Y_1}) \cdots E(z^{Y_n}) = \{Q(z)\}^n.$$

From (1.47) we now obtain, after an interchange of the order of summation,

$$R(z) = \sum_{n=0}^{\infty} e^{-\lambda t} \frac{(\lambda t)^n}{n!} \sum_{j=0}^{\infty} \phi_n(j)z^j = \sum_{n=0}^{\infty} e^{-\lambda t} \frac{(\lambda t)^n}{n!} \{Q(z)\}^n$$

$$= e^{-\lambda t\{1 - Q(z)\}}, \qquad |z| \leqslant 1.$$

By differentiation we get from this relation

$$R'(z) = \lambda t Q'(z)R(z), \qquad |z| \leqslant 1.$$

Next, by applying the Leibniz rule for the differentiation of a product, we find, at $z = 0$,

$$R^{(j)}(0) = \lambda t \sum_{k=0}^{j-1} \binom{j-1}{k} Q^{(j-k)}(0)R^{(k)}(0), \qquad j = 1, 2, \ldots,$$

and so, using (1.50), we get the recursion scheme (1.48).

1.7 APPROXIMATIONS FOR A CONTINUOUS REVIEW (s, S) INVENTORY SYSTEM WITH SERVICE LEVEL CONSTRAINTS

In this section we use renewal theory to obtain easily applicable results for a widely used inventory control model with probabilistic demand. Before doing this, we introduce some concepts from inventory theory. In designing an inventory control system it should be specified how often the inventory position is reviewed. Two possible procedures are continuous review updating the inventory status after each demand transaction and periodic review updating the inventory status only at discrete, usually equally spaced points in time. At the review times opportunities are provided for placing replenishment orders. It depends on the specific situation in practice what type of review procedure should be used, but with the increasing availability of data-processing equipment the continuous review procedure will be applied more and more. A major advantage of continuous review is that, in order to provide the same level of customer service, it requires less safety stock against fluctuations in the demand than does periodic review. Another important characteristic of an inventory control system is what happens when a demand exceeds the current inventory. Two extreme cases for handling this situation are the lost-sales case and the backlog case. In the former case any demand in excess of current inventory is lost, whereas in the latter case any excess demand is backlogged until stock becomes available by the arrival of a replenishment order. In practice, a combination of these two extreme cases is often used. However, in practical inventory applications the lost-sales case and the backlog case are usually quite close to each other when the required level of customer service is sufficiently high. The questions that have to be answered in an inventory control system are 'how much to order' and 'when to order', where the usual objective is the minimization of holding and ordering costs subject to the requirement that a certain grade of customer service is provided. Answers to these questions will be derived for the continuous review inventory system with backlogging of excess demand, but the analysis below needs only slight modifications for the lost-sales case and the periodic review model.

Consider a continuous review inventory system in which the demands for a single product occur at epochs generated by a Poisson process with rate λ. The successive demands for the product are independent, non-negative random variables having a common probability distribution function F with given mean μ_1 and standard deviation σ_1. Also, the successive demands are independent of the Poisson process generating the demand events. It is assumed that the demand distribution function F has a probability density f. Demand exceeding the on-hand inventory is backlogged until it can be satisfied from an incoming replenishment order. A replenishment order can be placed at any time. The lead time of each replenishment order is a positive constant L. The control of the system is based on the *inventory position* defined as

The inventory position = (the inventory on hand) + (total amount on order)
                         − (total backlog).

The inventory position is reviewed just after each demand transaction. The control rule is an (s, S) rule under which the inventory position is ordered up to the level S if at a review the inventory position is at or below the reorder point s and no ordering is done otherwise. Here the parameters s and S satisfy $0 \leqslant s < S$. In the discussion below we assume that the difference $S - s$, being roughly the order quantity, is already given. The purpose is only to determine the reorder point s such that the following service level constraint is satisfied:

The fraction of demand satisfied directly from on-hand inventory $= \beta$, (1.51)

where β with $0 < \beta < 1$ is a prespecified value. The analysis to be given applies equally well to other service measures such as the average number of stockout occurrences per unit time. In practice service level constraints are widely used, especially when shortage costs are difficult to specify. The service level requirement usually influences the optimal order quantity to a minor degree only and has primarily effect on the value of the reorder point. The reorder point s may be seen as the sum of the expected lead-time demand and the safety stock. In the case of probabilistic demand and non-zero lead times, a positive safety stock provides a buffer against fluctuations in the lead-time demand. The amount of safety stock needed depends on the required level of customer service.

The order quantity and the reorder point are often determined separately in practice. The determination of the order quantity is usually based on holding and replenishment cost considerations only, while the determination of the reorder point involves the consideration of the service level requirement and the lead times. Empirical support for this sequential approach is provided in example 3.10 of chapter 3, where we deal with the simultaneous determination of s and S such that the long-run average holding and replenishment costs per unit time are minimized subject to a service level constraint on the fraction of demand to be met directly from the inventory on hand. There the replenishment cost consists of a fixed setup cost of $K > 0$ per replenishment order, while the holding cost is $h > 0$ per unit time for each unit of on-hand inventory. For this particular cost structure we empirically found that in practical applications the optimal value of $S - s$ is fairly independent of the required service level provided the customer service is sufficiently high. It is worth while to point out that for the choice

$$S - s = \sqrt{\frac{2K\lambda\mu_1}{h}} \tag{1.52}$$

the average holding and replenishment costs are often quite close to the minimum average costs. The argument to get the *economic order quantity* formula (1.52) is instructive and generally useful for complex stochastic models. Consider the deterministic equivalent of the stochastic inventory system where the demand occurs at a constant rate of $\lambda\mu_1$ per unit time. In the deterministic inventory model the optimal order quantity is clearly independent of the constant lead time. Hence we can assume zero lead times so that replenishments are done only when the inventory becomes zero. For the control rule under which a replenishment order of size Q is placed each time the inventory becomes zero, the average cost per unit time is easily verified to be given by

$$\frac{K + hQ^2/(2\lambda\mu_1)}{Q/(\lambda\mu_1)}.$$

This average cost function is minimal when Q is given by the economic order quantity formula. Although the resulting value of Q is a practically useful approximation to the parameter $S - s$ in the stochastic inventory system, the deterministic approach predicts poorly the actual average holding costs in the stochastic system. A good approximation to the average holding and ordering costs of an (s, S) rule is discussed in remark 1.1 below.

We now turn to the determination of the reorder point s in the stochastic inventory system such that the service level constraint (1.51) is satisfied when the quantity $S - s$ is given. An exact analysis for the reorder point s is rather complicated and does not lead in general to computationally tractable results. In order to obtain implementable results, we have to compromise between mathematical and practical standpoints. Therefore our analysis for the reorder point s will be approximative. However, this approximate analysis is backed up by sound principles from renewal theory. Also, the approximate analysis has the advantage of being quite flexible so that the analysis applies with minor modifications to variants of the present inventory system such as the lost-sales inventory system and the periodic review inventory system (see remark 1.1 below).

The approximate analysis

To do the approximate analysis, we make the following assumption:

(a) $S - s$ is sufficiently large relative to the average demand between two consecutive reviews, say $S - s \geqslant \theta\mu_1$ with

$$\theta = \begin{cases} 1 & \text{if } \sigma_1^2/\mu_1^2 \leqslant 1 \\ 1\tfrac{1}{2}(\sigma_1^2/\mu_1^2) & \text{if } \sigma_1^2/\mu_1^2 > 1. \end{cases}$$

(b) For the relevant (s, S) rules the on-hand inventory just after arrival of a replenishment order is positive except for a negligible probability.

We now first analyse the service level of a given (s, S) rule with $0 \leqslant s < S$. Define a replenishment cycle as the time elapsed between two consecutive arrivals of replenishment orders. By the theory of regenerative processes, we have that the long-run fraction of demand not satisfied directly from on-hand inventory equals the average amount of demand that goes short per replenishment cycle divided by the average demand per replenishment cycle. Using assumption (b), the average amount of demand that goes short per cycle equals approximately the average shortage present just prior to the arrival of a replenishment order. Thus,

The fraction of demand not met directly from on-hand inventory $\approx \dfrac{A}{B}$,   (1.53)

where

$A =$ the average shortage just prior to the arrival of a replenishment order,

$B =$ the average demand per replenishment cycle.

To determine the ratio in (1.53), we define the following random variables

$\gamma =$ the undershoot of the reorder point s at a review at which a replenishment order is placed,

$\xi =$ the total demand in the lead time of a replenishment order.

Since excess demand is backlogged, every demand is ultimately satisfied and so the average demand per replenishment cycle equals the average replenishment size. Hence

$$B = S - s + E(\gamma).$$ (1.54)

To determine the average shortage A, we tag one of the replenishment orders. We first observe that the inventory position just prior to the placing of the tagged order is distributed as $s - \gamma$. Also, since the lead times of the replenishment orders are constant, the replenishment orders do not cross in time. In other words, any replenishment order placed earlier than the tagged order has arrived when the tagged order comes in, while no order placed later than the tagged order arrives earlier than the tagged order. Hence the on-hand inventory just prior to the arrival of the tagged replenishment order is distributed as $s - \gamma - \xi$. Thus

$$A \approx \int_{s}^{\infty} (x - s)h(x)\,\mathrm{d}x,$$ (1.55)

where

$h(x) =$ the probability density of $\gamma + \xi$.

To elaborate the relations (1.54) and (1.55), we need a tractable approximation for the distribution of the undershoot variable γ. Denoting by $D_1, D_2, \ldots$ the sizes of the successive demands, we have in renewal-theoretic terms that γ is the excess life at 'time' $S - s$ in a renewal process with 'interoccurrence times'

$D_1, D_2, \ldots$. In view of assumption (a) and the asymptotic expansion (1.12) in section 1.1, we make the following approximation:

$$P\{\gamma \leqslant x\} \approx \frac{1}{\mu_1} \int_0^x \{1 - F(y)\} \, dy, \qquad x \geqslant 0,$$

that is for $S - s$ sufficiently large relative to the average review time demand μ_1, the distribution of the undershoot variable γ is approximately independent of the value of $S - s$. Also, by (1.13) in section 1.1,

$$E(\gamma) \approx \frac{\sigma_1^2 + \mu_1^2}{2\mu_1}, \qquad (1.56)$$

which gives an approximation for the quantity B. We next approximate the density $h(x)$ in (1.55). Using the approximation for $P\{\gamma \leqslant x\}$ and the law of total probability, we find

$$P\{\gamma + \xi \leqslant x\} \approx \frac{1}{\mu_1} \int_0^x P\{\xi \leqslant x - y\}\{1 - F(y)\} \, dy$$

$$= \frac{1}{\mu_1} \int_0^x P\{\xi \leqslant u\} \, du$$

$$- \frac{1}{\mu_1} \int_0^x P\{\xi \leqslant u\} F(x - u) \, du, \qquad x \geqslant 0,$$

and so, by differentiation and noting that $F(0) = 0$,

$$h(x) \approx \frac{1}{\mu_1} P\{\xi \leqslant x\} - \frac{1}{\mu_1} \left[\int_0^x P\{\xi \leqslant u\} f(x - u) \, du + P\{\xi \leqslant x\} F(0) \right]$$

$$= \frac{1}{\mu_1} P\{\xi \leqslant x\} - \frac{1}{\mu_1} \int_0^x P\{\xi \leqslant x - y\} f(y) \, dy, \qquad x > 0.$$

Observe now that $\int_0^x P\{\xi \leqslant x - y\} f(y) \, dy$, $x \geqslant 0$, is the probability distribution function of the random variable

$$\eta = \xi + D,$$

where the generic variable D represents the review time demand having probability density $f(x)$. Denoting by $\xi(x)$ and $\eta(x)$ the probability densities of the random variables ξ and η for $x > 0$, we obtain by partial integration that

$$\int_s^\infty (x - s) h(x) \, dx = \frac{1}{2} \int_s^\infty h(x) \, d(x - s)^2 = -\frac{1}{2} \int_s^\infty (x - s)^2 h'(x) \, dx$$

$$= \frac{1}{2\mu_1} \left\{ \int_s^\infty (x - s)^2 \eta(x) \, dx - \int_s^\infty (x - s)^2 \xi(x) \, dx \right\}. \qquad (1.57)$$

Here it is assumed that $\lim_{x \to \infty} x^2 h(x) = 0$, which is satisfied when the demand variable D has a finite third moment. The relations (1.53) to (1.57) together yield that for a given (s, S) rule with $0 \leqslant s < S$ the long-run fraction of demand that is not satisfied directly from the stock on hand is approximately given by

$$\frac{\int_s^\infty (x-s)^2 \eta(x)\, dx - \int_s^\infty (x-s)^2 \xi(x)\, dx}{2\mu_1 \{ S - s + (\sigma_1^2 + \mu_1^2)/2\mu_1 \}}.$$

Next we conclude that under the service level constraint (1.51) and a given value for $S - s$ the reorder point s can be approximated by solving the equation

$$\int_s^\infty (x-s)^2 \eta(x)\, dx - \int_s^\infty (x-s)^2 \xi(x)\, dx = (1 - \beta)2\mu_1 \left(S - s + \frac{\sigma_1^2 + \mu_1^2}{2\mu_1} \right), \quad (1.58)$$

where

$\xi(x) =$ the probability density of the lead-time demand,

$\eta(x) =$ the probability density of the lead-time plus review-time demand.

Remark 1.1   Additional results for the (s, S) inventory system

(a) An examination of the above analysis shows that the result (1.58) also applies when the lead times are stochastic, provided the probability of orders crossing in time is negligible for the relevant (s, S) rules. The effect of the variability of the lead time on the reorder point may be considerable.

(b) For the lost-sales inventory model, we may also apply the equation (1.58) for the calculation of the reorder point s, provided we replace $1 - \beta$ by $(1 - \beta)/\beta$ in the right side of (1.58). To see this, the starting point is again that the long-run fraction of demand that is lost equals the ratio of the expected amount of demand in a cycle that is lost and the total expected amount of demand in a cycle. As before, the expected amount of demand in a cycle that is lost may be approximated by (1.55). The total expected amount of demand in a cycle is now the sum of the expected amount of demand in a cycle that is satisfied and the expected amount of demand in a cycle that is lost. The latter amount is approximated by (1.55), while the former amount equals the average order size and is thus approximated by (1.54). It now follows that for the lost-sales case the reorder point s may be determined approximately from

$$\frac{\int_s^\infty (x-s)h(x)dx}{S - s + E(\gamma) + \int_s^\infty (x-s)h(x)dx} = 1 - \beta,$$

showing that for the lost-sales case equation (1.58) may be used with $1 - \beta$ replaced by $(1 - \beta)/\beta$. Obviously, for β sufficiently close to 1, the backlog model provides a good approximation to the lost-sales model.

(c) For a given (s, S) rule with $S - s$ sufficiently large relative to μ_1, we have the following useful approximations:

The average number of orderings per unit time $\approx \lambda \left(\dfrac{S-s}{\mu_1} + \dfrac{\mu_2}{2\mu_1^2} \right)^{-1}$ (1.59)

and

$$\text{The average on-hand inventory} \approx \left(\frac{S-s}{\mu_1} + \frac{\mu_2}{2\mu_1^2} \right)^{-1}$$

$$\times \left\{ \int_0^S (S-y)\xi(y)\,dy + \left(\frac{S-s}{\mu_1} + \frac{\mu_2}{2\mu_1^2} - 1 \right) \int_0^s (s-y)\xi(y)\,dy \right.$$

$$\left. + \frac{(S-s)^2}{2\mu_1^2} + \left(\frac{\mu_2}{2\mu_1} - 1 \right)(S-s) + \frac{\mu_2^2}{4\mu_1^3} - \frac{\mu_3}{6\mu_1^2} \right\}, \tag{1.60}$$

where μ_k denotes the kth moment of the demand variable D. The latter approximation requires that the probability of a lead-time demand larger than the reorder point s is very small, a requirement which will be satisfied in practical applications with a sufficiently high service level (in deriving equation 1.60 we approximated $\int_0^{S-s} M(y)\Phi_L(S-y)\,dy$ by $\int_0^{S-s} M(y)\,dy$ with $\Phi_L(x)$ denoting the probability distribution function of the lead-time demand and $M(x)$ denoting the renewal function associated with the density of the demand D). The derivation of the above approximations proceeds along similar lines as the derivation of the approximations to the periodic review (s, S) inventory system with zero lead times dealt with in example 1.3. We omit here the details of how to handle the complicating factor of positive lead times in the continuous review (s, S) inventory system (cf. also example 3.10 in section 3.6 of chapter 3). The approximations (1.59) and (1.60) apply both to the backlog model and the lost-sales model.

(d) The approximation analysis for the continuous review (s, S) inventory system applies equally well to the periodic review (s, S) inventory system. In the latter system the (aggregated) demands in successive periods $t = 1, 2, \ldots$ are independent, non-negative random variables having a common probability distribution function F with given mean μ_1 and standard deviation σ_1. Also, the inventory position is only reviewed at the beginning of each Rth period where the review parameter R is a given positive integer. Replenishment opportunities occur only at the review epochs. Denoting by $\mu_R = R\mu_1$ and $\sigma_R = R^{1/2}\sigma_1$ the mean and the standard deviation of the review-time demand, an examination of the above analysis shows that the equation (1.58) for the reorder point s also applies to the periodic review (s, S) inventory system with the review parameter R, provided we replace in the right side of (1.58) the constants μ_1 and σ_1 by μ_R and σ_R. Also, we point out that for the case of $R = 1$ the average on-hand inventory at the beginning of a period just after any delivery may be approximated by (1.60).

The numerical calculation of the reorder point

The equation (1.58) for the reorder point looks simple and its left side is convexly decreasing in s, but still some further work should be done to make this equation suited for routine computations in practice. First, we remark that a discrete version of the relation (1.58) is rather easy to solve. Suppose the demand distribution F is discrete and has $p(j)$ as the probability of a demand of j units, $j = 0, 1, \ldots$. The discrete distributions of the demand variables ξ and η are straightforward to compute. The demand variable ξ has a compound Poisson distribution $\{\xi(j), j \geq 0\}$ to be computed from the recurrence relation (cf. equation 1.48)

$$\xi(0) = e^{-\lambda\{1 - p(0)\}L}, \qquad \xi(j) = \frac{\lambda L}{j} \sum_{k=0}^{j-1} (j - k)p(j - k)\xi(k) \qquad \text{for} \quad j = 1, 2, \ldots.$$

The distribution $\{\eta(j), j \geq 0\}$ can next be computed from

$$\eta(j) = \sum_{k=0}^{j} p(j - k)\xi(k), \qquad j = 0, 1, \ldots.$$

The discrete version of the relation (1.58) requires the computation of the *largest* integer s such that

$$\sum_{k=s}^{\infty} (k - s)^2 \eta(k) - \sum_{k=s}^{\infty} (k - s)^2 \xi(k) \geq (1 - \beta)2\mu_1\left(S - s + \frac{\sigma_1^2 + \mu_1^2}{2\mu_1}\right). \quad (1.61)$$

Note that the infinite sums in (1.61) can be reduced to finite sums. The first sum in the left side of (1.61) is written as $E(\eta^2) - 2sE(\eta) + s^2 - \sum_{k=0}^{s}(k - s)^2\eta(k)$. Similarly, the other infinite sum is handled.

We next discuss solution methods for (1.58) which use only the first two moments of the demand densities $\xi(x)$ and $\eta(x)$. In practice it often happens that only the first two moments of ξ and η are available rather than their complete probability distributions. The first two moments of ξ and η can easily be expressed in the first two moments of the demand D and the lead time L. Using the formulas (A.6) and (A.7) in appendix A and the fact that for the continuous review inventory system with fixed lead time L the number of demand occurrences in the lead time has a Poisson distribution with mean λL, we find, for the mean and the variance of ξ,

$$\mu(\xi) = \lambda L\mu_1, \qquad \sigma^2(\xi) = \lambda L(\sigma_1^2 + \mu_1^2).$$

Also, the mean and variance of η are given by

$$\mu(\eta) = \mu(\xi) + \mu_1, \qquad \sigma^2(\eta) = \sigma^2(\xi) + \sigma_1^2.$$

Many practical inventory applications concern fast-moving items for which the demand densities $\xi(x)$ and $\eta(x)$ can very well be approximated by normal densities. To be specific, consider the case in which the coefficients of variation of ξ and η are both less that 0.5 (say). Note that for a normal distribution with a coefficient of variation less than 0.5 the probability of negative outcomes is negligible for most situations of practical interest. The standard normal distribution function $\Phi(x)$ and the associated probability density $\phi(x)$ are given by

$$\Phi(x) = \frac{1}{\sqrt{2\pi}} \int_{-\infty}^{x} e^{-u^2/2} \, du \qquad \text{and} \qquad \phi(x) = \frac{1}{\sqrt{2\pi}} e^{-x^2/2}.$$

Also, define the standard normal loss integrals $I(x)$ and $J(x)$ by

$$I(x) = \int_{x}^{\infty} (u - x)\phi(u) \, du \qquad \text{and} \qquad J(x) = \int_{x}^{\infty} (u - x)^2 \phi(u) \, du.$$

The following useful relations are readily verified by partial integration,

$$I(x) = \phi(x) - x\{1 - \Phi(x)\} \quad \text{and} \quad J(x) = (1 + x^2)\{1 - \Phi(x)\} - x\phi(x). \quad (1.62)$$

If we fit normal densities to the demand densities $\xi(x)$ and $\eta(x)$ by matching the first two moments, the equation (1.58) reduces to

$$\sigma^2(\eta) J\left(\frac{s - \mu(\eta)}{\sigma(\eta)} \right) - \sigma^2(\xi) J\left(\frac{s - \mu(\xi)}{\sigma(\xi)} \right)$$

$$= (1 - \beta) 2\mu_1 \left(S - s + \frac{\sigma_1^2 + \mu_1^2}{2\mu_1} \right).$$

The standard Newton–Raphson iteration method provides an effective method to solve this equation, since the function evaluations of $J(x)$ and its derivative $-2I(x)$ are very cheap when using the above relations for $I(x)$ and $J(x)$. Numerical procedures for the standard normal distribution function $\Phi(x)$ are widely available. An excellent starting point for the Newton–Raphson iteration is the value s_0 determined by the equation

$$\sigma^2(\eta) J\left(\frac{s - \mu(\eta)}{\sigma(\eta)} \right) = (1 - \beta) 2\mu_1 \left(S - s + \frac{\sigma_1^2 + \mu_1^2}{2\mu_1} \right),$$

from which s_0 may be obtained by a single-pass calculation using the fact that the inverse function of $J(x)$ can be approximated very accurately by the ratio of two polynomials.

The normal distribution cannot be used to fit to a positive demand variable having a coefficient of variation larger than 0.5. However, the gamma, lognormal and Weibull distributions can in principle be fitted to each positive demand variable by matching the first two moments. For each of these fitting distributions the equation (1.58) for the reorder point is rather easy to solve by Newton–Raphson iteration. The left side of (1.58) is written in terms of incomplete gamma functions (cf. appendix B) when gamma or Weibull densities are fitted to $\xi(x)$ and $\eta(x)$, while the left side of (1.58) is written in terms of the standard normal distribution function when lognormal densities are fitted to $\xi(x)$ and $\eta(x)$. In many practical inventory applications the empirical demand distributions can be represented very well by gamma distributions. In these cases the calculation of the reorder point may be simplified further by approximating the gamma density by a generalized Erlangian density by matching the first two or three moments, depending on whether the squared coefficient of variation of the gamma density is between 0 and $\frac{1}{2}$ or is larger than $\frac{1}{2}$ (see in appendix B the densities A.11 and A.15 with the respective specifications A.12 and A.17). For these generalized Erlangian densities the integrals in (1.58) can be expressed into terms of cumulative Poisson probabilities.

A caveat should be issued against the blind application of two-moment approximations. Generally stated, it may be hazardous to use two-moment approximations when the squared coefficients of variation of the demand variables ξ and η get larger (erratic demand), in particular when the required service level β is close to 1. Then the tail behaviour of the demand densities $\xi(x)$ and $\eta(x)$ becomes quite important, as is also expressed by the equation (1.58) in which the quadratic terms $(x - s)^2$ put much emphasis on the tails of these densities.

To illustrate the foregoing, we consider in Table 1.3 a number of examples

Table 1.3   Approximate reorder points and the actual service levels

		$c_D^2 = 0.25$		$c_D^2 = 0.50$		$c_D^2 = 1.00$		$c_D^2 = 2.00$		$c_D^2 = 5.00$	
		s	$\beta(s,S)$	s	$\beta(s,S)$	s	$\beta(s,S)$	s	$\beta(s,S)$	s	$\beta(s,S)$
$\beta = 0.90$	true	57	0.901	59	0.899	64	0.902	73	0.904	99	0.910
	gamma	57	0.901	59	0.899	64	0.902	73	0.904	100	0.912
	logn.	57	0.901	59	0.899	64	0.902	74	0.908	104	0.922
	no(We)	57	0.901	59	0.899	63	0.896	71	0.895	94	0.897
	Erlng	57	0.901	59	0.899	64	0.902	74	0.908	100	0.912
$\beta = 0.95$	true	66	0.951	69	0.950	75	0.949	88	0.952	124	0.956
	gamma	66	0.951	70	0.953	76	0.952	89	0.954	125	0.957
	logn.	66	0.951	70	0.953	77	0.955	91	0.958	135	0.968
	no(We)	65	0.947	68	0.946	73	0.942	84	0.942	115	0.942
	Erlng	66	0.951	70	0.953	76	0.952	90	0.956	124	0.956
$\beta = 0.99$	true	83	0.990	88	0.990	99	0.990	119	0.990	176	0.991
	gamma	84	0.991	90	0.992	101	0.992	122	0.992	180	0.992
	logn.	86	0.993	93	0.994	105	0.994	131	0.995	209	0.997
	no(We)	80	0.987	84	0.985	91	0.983	109	0.983	159	0.984
	Erlng	84	0.991	90	0.992	100	0.991	123	0.992	178	0.991

in which the customer demand D has a negative binomial distribution

$$p(j) = \binom{r+j-1}{j} p^r (1-p)^j, \qquad j = 0, 1, \ldots.$$

The parameters p and r of this distribution are uniquely determined by $p = \mu_1/\sigma_1^2$ and $r = p\mu_1/(1-p)$, provided $\sigma_1^2/\mu_1 > 1$. In all examples we take

$$\mu_1 = 5, \qquad \lambda = 10, \qquad L = 1 \qquad \text{and} \qquad S - s = 50.$$

The squared coefficient of variation c_D^2 of the demand D is varied as 0.25, 0.50, 1, 2 and 5, and the required service level β has the three values 0.90, 0.95 and 0.99. Denoting by c_η^2 and c_ξ^2 the squared coefficients of variation of the demand variables η and ξ, we find for $c_D^2 = 0.25$, 0.50, 1, 2 and 5 that c_η^2 and c_ξ^2 have the respective values

$$c_\eta^2 = 0.105, 0.128, 0.174, 0.264 \text{ and } 0.537$$

and

$$c_\xi^2 = 0.125, 0.150, 0.200, 0.300 \text{ and } 0.600.$$

For this range of values of c_D^2, c_η^2 and c_ξ^2, we give in Table 1.3 for various probability densities fitted to $\eta(x)$ and $\xi(x)$ the approximate values of the reorder point s together with the *actual* values of the associated service levels. The actual values $\beta(s, S)$ of the service levels of the various approximate (s, S) policies are based on the original negative binomial distribution for the demand D and have been computed by an exact method to be discussed in example 3.10 in section 3.6 of chapter 3. The approximation being denoted by 'true' in Table 1.3 results from (1.61) in which the actual demand distributions $\{\xi(k)\}$ and $\{\eta(k)\}$ are used. The probabilities $\xi(k)$ are calculated from the recursion scheme (1.48) in section 1.6 and the probabilities $\eta(k)$ are next calculated from $\eta(k) = \sum_{j=0}^{k} \xi(k-j)p(j)$. Note that in applying the recursion scheme the negative binomial probabilities $p(j)$ may be calculated recursively from

$$p(0) = p^r \qquad \text{and} \qquad p(j) = \frac{(r+j-1)}{j}(1-p)p(j-1), \qquad j = 1, 2, \ldots.$$

The other approximations being denoted by 'gamma', 'logn', 'no (We)' and 'Erlng' in Table 1.3 are obtained by solving the equation (1.58) in which respectively gamma, lognormal, normal (Weibull) and generalized Erlangian densities are fitted to the densities $\eta(x)$ and $\xi(x)$ by matching the first two moments. Here the normal density is used for the cases with $c_D^2 = 0.25$, 0.50 and 1, while the Weibull density is used for the other cases. The following conclusions may be drawn from our numerical investigations. The 'true' approximation to the reorder point performs very well for all values of the parameters. It indeed appears that some care should be exercised in using two-moment approximations, in particular when the required service level β is very close to 1. As a rule of thumb, it can be stated that in practical inventory applications two-moment approximations may be used for relatively smooth demand with $0 \leqslant c_\xi^2$, $c_\eta^2 \leqslant 0.25$ (say). The results in Table 1.3 clearly demonstrate that for erratic demand the tail behaviour of the demand densities $\eta(x)$ and $\xi(x)$ is essential; additional information about the distributional form should then guide the choice of an appropriate fitting distribution. In case of a negative binomial demand D the gamma density apparently approximates quite well the actual densities $\eta(x)$ and $\xi(x)$. Also, a comparison of the gamma and generalized Erlangian approximations to the reorder point suggests that the third moment in addition to the first two moments is a significant improvement when the demand is erratic.

In order to demonstrate the quality of the approximations (1.59) and (1.60) to the average ordering frequency F and the average on-hand inventory I, we give in Table 1.4 the exact and the approximate values of F and I for some of the cases considered in Table 1.3. The approximate values of I were calculated from the following simplification of the approximation (1.60). We replaced the integral

$$\int_0^z (z-y)\xi(y)\,\mathrm{d}y \qquad \text{by} \qquad z - \mu(\xi) + \sigma(\xi)I\left(\frac{z-\mu(\xi)}{\sigma(\xi)}\right) \qquad \text{for} \quad z = s, S,$$

with $I(k)$ being the easily computable normal loss integral in (1.62), and we replaced the third moment μ_3 by $2\mu_2^2/\mu_1 - \mu_2\mu_1$ being the value of μ_3 for the gamma distribution with first two moments μ_1 and μ_2. This approximation to the average on-hand inventory yields good results for both smooth demand and erratic demand.

Finally, Table 1.5 presents some numerical results for the lost-sales model. The numerical results support the claim that the lost-sales inventory model and the backlog inventory model do not differ significantly when the level of customer service is high. Also, the numerical results show that the variability in the lead time may have a considerable effect on the reorder point. For the lead time L we consider the two cases: (a) $P\{L=1\} = 1$ with $E(L) = 1$ and $\sigma^2(L) = 0$, and (b) $P\{L=0.5\} = 0.25$, $P\{L=1\} = 0.5$ and $P\{L=1.5\} = 0.25$ with $E(L) = 1$ and $\sigma^2(L) = 0.125$. It is noted that for the case of a stochastic lead time

Table 1.4   Numerical results for the ordering frequency and the average on-hand inventory

	$c_D^2 = 0.25$					$c_D^2 = 5$			
(s, S)	F_{app}	F_{exa}	I_{app}	I_{exa}	(s, S)	F_{app}	F_{exa}	I_{app}	I_{exa}
$(57, 107)$	0.941	0.950	37.2	34.6	$(99, 149)$	0.769	0.776	80.8	80.9
$(66, 116)$	0.941	0.950	44.1	43.1	$(124, 174)$	0.769	0.776	104.4	105.1
$(83, 133)$	0.941	0.950	59.7	59.8	$(176, 226)$	0.769	0.776	156.0	156.5

Table 1.5 Numerical results for the lost-sales (s, S) inventory system

β	$\sigma^2(L)$	(s, S)	β_{act}	F_{app}	F_{act}	I_{app}	I_{act}
0.90	0	(62, 112)	0.919(3)	0.909	0.847(7)	43.1	44.6(3)
0.95	0	(75, 125)	0.960(3)	0.909	0.881(7)	53.6	54.9(3)
0.99	0	(98, 148)	0.991(1)	0.909	0.908(8)	75.4	76.0(5)
0.90	0.125	(70, 120)	0.924(3)	0.909	0.850(7)	51.0	52.3(5)
0.95	0.125	(85, 135)	0.962(3)	0.909	0.881(7)	63.6	64.5(6)
0.99	0.125	(114, 164)	0.993(1)	0.909	0.910(8)	91.4	92.1(5)

the formulae for $\mu(\xi)$ and $\sigma^2(\xi)$ should be adjusted according to

$$\mu(\xi) = \lambda E(L)\mu_1 \quad \text{and} \quad \sigma^2(\xi) = \lambda E(L)(\sigma_1^2 + \mu_1^2) + \lambda^2 \sigma^2(L)\mu_1^2.$$

In the examples of Table 1.5 the demand D has a negative binomial distribution with $E(D) = 5$ and $c_D^2 = 1$, the arrival rate $\lambda = 10$, the difference $S - s = 50$, and the parameter β is varied as 0.90, 0.95 and 0.99. For the (s, S) policies obtained from the equation (1.58) with $1 - \beta$ replaced by $(1 - \beta)/\beta$, we give in Table 1.5 the actual values β_{act} of the service level and the approximate and actual values of the average ordering frequency F and the average on-hand inventory I. The actual values of these operating characteristics were obtained by computer simulation. In each example 125,000 customer demands were simulated. In the simulation the variance reduction technique of conditional expectations discussed in Carter and Ignall (1975) was used for the service level; the negative binomial distribution for the demand was generated by using the fact that this distribution can be represented as a mixture of Poisson distributions with the gamma density as the mixture density. In Table 1.5 the notation 0.919(3) for the service level is used to denote that the simulated value is 0.919 with [0.916, 0.922] as the 95 per cent confidence interval; similarly, the notations 0.847(7) and 44.6(3) for the average ordering frequency and the average on-hand inventory mean that the simulated values are 0.847 and 44.6 with [0.840, 0.854] and [44.3, 44.9] as the respective 95 per cent confidence intervals. The numerical results show that the approximations are practically useful for the lost-sales model. Also, our numerical investigations indicate that for the case of stochastic lead times the approximation to the reorder point performs reasonably well, provided the probability of orders crossing in time is sufficiently small (say, below 5 per cent); in the examples with $\sigma^2(L) = 0.125$ in Table 1.5 we have, for the respective policies $(s, S) = (70, 120), (85, 135)$ and $(114, 164)$, that the simulated values of the probability of orders crossing in time are 3.6, 3.8 and 4.4 per cent.

1.8 POISSON ARRIVALS SEE TIME AVERAGES

An essential part of the analysis of many queueing and inventory systems with Poisson input is the observation that in any small time interval of the same

length the occurrence of a Poisson event is equally likely so that the distribution of the state of the system just prior to the occurrence of a Poisson event is the same as the state distribution at an arbitrary point in time. In other words, since Poisson events occur completely random in time, the proportion of events occurring when the system is in some state is the same as the proportion of time the system is in that state. This remarkable equality is in general not true when the events are not generated by a Poisson process. To illustrate this, consider a queueing system in which the arrival process of customers is a renewal process with interarrival times always larger than 1 and the service times are equal to 1. In this example every arrival finds the system empty, whereas the system is not always empty of customers. This section explains in more detail the frequently used property 'Poisson arrivals see time averages'. For ease of presentation we use the terminology of Poisson arrivals. However, the results below also apply to Poisson processes in other contexts.

For some specific problem, let the stochastic process $\{X(t), t \geq 0\}$ describe the evolution of the state of a system and let $\{N(t), t \geq 0\}$ be a Poisson arrival process of customers to that system. As examples:

(a) $X(t)$ is the number of customers present at time t in a queueing system.
(b) $X(t)$ gives both the inventory level and the prevailing production rate at time t in a production/inventory problem with a variable production rate.

It is assumed that the Poisson arrival process $\{N(t), t \geq 0\}$ can be seen as an exogenous factor to the system and is not affected by the system itself. More precisely, at each time t the future arrivals occurring after time t are independent of the history of the system up to time t. It is not necessary to specify how the Poisson arrival process $\{N(t)\}$ precisely interacts with the state process $\{X(t)\}$. It will be generally true that

> The long-run fraction of customers who see upon arrival the system in some state = the long-run fraction of time the system is in that state.

This important property is usually expressed as *Poisson arrivals see time averages*. A rigorous proof of this result was given by Wolff (1982) under quite general conditions. We do not give the details of the proof, but only discuss the crucial relation underlying the property of Poisson arrivals see time averages. This relation in itself is of great importance. We assume that the process $\{X(t), t \geq 0\}$ is regenerative, that is there exist time points at which the process probabilistically starts over again (see section 1.3). Letting a cycle be the time elapsed between two consecutive regeneration epochs of the process $\{X(t)\}$, define

$$T = \text{the length of one cycle,}$$
$$N = \text{the number of arrivals during one cycle.}$$

It is assumed that $0 < E(T) < \infty$. Also, for a given set B of states of the process

$\{X(t)\}$, define

T_B = the amount of time the system is in the set B during one cycle,
N_B = the number of arrivals during one cycle who find the system in the set B.

The following remarkable relation was proved in Wolff (1982):

$$E(N_B) = \lambda E(T_B), \qquad (1.63)$$

that is the expected number of arrivals during one cycle who find the system in the set B equals the average arrival rate times the expected amount of time the system spends in the set B during one cycle. This relation is characteristic for the Poisson process and is related to the memoryless property of the Poisson process. In particular, by taking B as the set of all states,

$$E(N) = \lambda E(T).$$

Hence, by (1.63),

$$\frac{E(T_B)}{E(T)} = \frac{E(N_B)}{E(N)}.$$

Now, by the theory of renewal reward processes, $E(T_B)/E(T)$ and $E(N_B)/E(N)$ give the long-run fraction of time the system is in the set B and the long-run fraction of customers who see upon arrival the system in the set B. The latter equality states that these two fractions are the same when the arrival process is a Poisson process.

1.9 ASYMPTOTIC EXPANSIONS FOR RUIN AND WAITING TIME PROBABILITIES

In many applied probability problems asymptotic expansions provide a simple alternative to computationally intractable solutions. A nice example is the ruin probability in risk theory. This probability arises in the following model. Claims arrive at an insurance company according to a Poisson process $\{N(t), t \geqslant 0\}$ with rate λ. The successive claim amounts $S_1, S_2, \ldots$ are positive, independent random variables having a common probability distribution function $B(s)$, and are also independent of the arrival process. In the absence of claims, the company's reserve increases at a constant rate of $\sigma > 0$ per unit time. It is assumed that $\sigma > \lambda E(S)$, that is the average premium received per unit time is larger than the average claim rate. Here the generic variable S denotes the individual claim amount. Denote by the compound Poisson variable

$$X(t) = \sum_{i=1}^{N(t)} S_i$$

the total amount claimed up to time t. If the company's initial reserve is $x > 0$, then the company's total reserve at time t is $x + \sigma t - X(t)$. We say that a ruin occurs at time t if $x + \sigma t - X(t) < 0$. Defining, for each $x \geqslant 0$,

$$q(x) = P\{X(t) > x + \sigma t \quad \text{for some} \quad t \geqslant 0\}, \tag{1.64}$$

then $q(x)$ is the probability that a ruin will ever occur with initial capital x. Since a ruin may occur only at the claim epochs, we can equivalently write

$$q(x) = P\left\{ \sum_{j=1}^{k} S_j - \sigma \sum_{j=1}^{k} \tau_j > x \quad \text{for some} \quad k \geqslant 1 \right\}, \tag{1.65}$$

where $\tau_1, \tau_2, \ldots$ denote the interoccurrence times of successive claims. We are interested in the asymptotic behaviour of $q(x)$ for large x.

The ruin probability $q(x)$ arises in a variety of contexts. As another example consider a production/inventory situation in which demands for a given product arrive according to a Poisson process. The successive demands are independent and identically distributed random variables. On the other hand, inventory replenishments of the product occur at a constant rate of $\sigma > 0$ per unit time. In this context, the ruin probability $q(x)$ represents the probability that a shortage will ever occur when the initial inventory is x.

The ruin probability as waiting-time probability

A less obvious context in which the ruin probability appears is the following standard queueing system with Poisson input and general service times. Customers arrive at a single-server station according to a Poisson process with rate λ. The service or work requirements $S_1, S_2, \ldots$ of the successive customers are independent and identically distributed random variables, and are also independent of the arrival process. A customer who finds upon arrival that the server is idle immediately obtains service; otherwise the customer waits in line. The server works at a constant rate of $\sigma > 0$ per unit time whenever the system is not empty, i.e. the service requirement S of a customer has a processing time of S/σ. The customers are served in order of arrival. Defining the amount $V(t)$ of work in the system at time t as the sum of the remaining service requirements of the customers present at time t, let for $n = 1, 2, \ldots,$

$V_n =$ the amount of work in the system just prior to the arrival epoch of the nth customer.

It will be shown below that

$$q(x) = \lim_{n \to \infty} P\{V_n > x\}, \qquad x \geqslant 0. \tag{1.66}$$

This result has several important consequences. First, letting, for $n = 1, 2, \ldots,$

$D_n =$ the delay in queue of the nth customer,

we have

$$\lim_{n \to \infty} P\{D_n > x\} = q(\sigma x), \qquad x \geqslant 0. \tag{1.67}$$

This result follows by noting that under service in order of arrival the delay in queue of the nth customer equals the amount V_n of work divided by the processing rate σ of the server. Second, using the property Poisson arrivals see time averages, it can be seen that

$$\lim_{t \to \infty} P\{V(t) > x\} = q(x), \qquad x > 0, \tag{1.68}$$

that is $q(x)$ represents also the steady-state distribution of the work in the system at an arbitrary point in time.

A proof of (1.66) proceeds as follows. Denoting by $\tau_1, \tau_2, \ldots$ the interarrival times for the successive customers, it is easily seen that, for $n = 1, 2, \ldots,$

$$V_{n+1} = \begin{cases} V_n + S_n - \sigma \tau_{n+1} & \text{if } V_n + S_n - \sigma \tau_{n+1} \geqslant 0, \\ 0 & \text{if } V_n + S_n - \sigma \tau_{n+1} < 0. \end{cases}$$

Hence, letting $U_n = S_n - \sigma \tau_{n+1}$ for $n \geqslant 1$,

$$V_{n+1} = \max(0, V_n + U_n).$$

Substituting this equation in itself, it follows that

$$V_{n+1} = \max\{0, U_n + \max(0, V_{n-1} + U_{n-1})\} = \max(0, U_n, U_n + U_{n-1} + V_{n-1}).$$

Continuing in this way and using $V_1 = 0$, we find, for each $n \geqslant 1$,

$$V_{n+1} = \max(0, U_n, U_n + U_{n-1}, \ldots, U_n + U_{n-1} + \cdots + U_1).$$

Since the random variables $U_1, U_2, \ldots$ are independent and identically distributed, we have that $(U_n, \ldots, U_1)$ has the same joint distribution as $(U_1, \ldots, U_n)$. Hence, by the duality principle,

$$V_{n+1} = \max\{0, U_1, U_1 + U_2, \ldots, U_1 + \cdots + U_n\}, \qquad n \geqslant 1.$$

Hence, using $P\{V_n > x\} = P\{\sum_{j=1}^{k} U_j > x \text{ for some } 1 \leqslant k \leqslant n - 1\}$,

$$\lim_{n \to \infty} P\{V_n > x\} = P\left\{ \sum_{j=1}^{k} S_j - \sigma \sum_{j=1}^{k} \tau_{j+1} > x \text{ for some } k \geqslant 1 \right\}, \qquad x > 0.$$

This relation and (1.65) verify (1.66).

In the context of the queueing application, $q(0)$ represents the limiting probability that an arrival finds the system not empty. By the property of 'Poisson arrivals see time averages', the fraction of arrivals finding the system not empty equals the fraction of time the system is not empty. Thus, by the relation (1.36),

$$q(0) = \rho, \tag{1.69}$$

where the load factor ρ is defined by

$$\rho = \frac{\lambda E(S)}{\sigma}.$$

We now turn to the determination of the ruin probability $q(x)$. For that purpose, we derive first an integro-differential equation for $q(x)$. For ease of presentation we assume that the probability distribution function $B(x)$ of the claim sizes has a density $b(x)$ at every point x. Fix $x > 0$. To compute $q(x - \Delta x)$ with Δx small, we condition on what happens in the first $\Delta x/\sigma$ time units. In the absence of claims the company's capital grows from $x - \Delta x$ to x. However, since the claims arrive according to a Poisson process with rate λ, a claim occurs in the first $\Delta x/\sigma$ time units with probability $\lambda \Delta x/\sigma + o(\Delta x)$, in which case the company's capital becomes $x - Y$ if Y is the size of that claim. In this case a ruin occurs only if $Y > x$. Thus, we get, for fixed $x > 0$,

$$q(x - \Delta x) = \left(1 - \frac{\lambda \Delta x}{\sigma}\right)q(x) + \frac{\lambda \Delta x}{\sigma} \int_x^\infty b(y)\,\mathrm{d}y + \frac{\lambda \Delta x}{\sigma} \int_0^x q(x - y)b(y)\,\mathrm{d}y + o(\Delta x).$$

Subtracting $q(x)$ from both sides of this equation, dividing by $h = -\Delta x$ and letting $h \to 0$, we obtain the integro-differential equation

$$q'(x) = -\frac{\lambda}{\sigma}\{1 - B(x)\} + \frac{\lambda}{\sigma}q(x) - \frac{\lambda}{\sigma}\int_0^x q(x - y)b(y)\,\mathrm{d}y, \qquad x > 0. \quad (1.70)$$

In the case where the probability density $b(x)$ is a K_2 density, this integro-differential equation can be solved explicitly by using Laplace transforms (see the results A.32 and A.33 in appendix C). In particular, when $b(x)$ is the exponential density,

$$q(x) = \rho e^{-(1 - \rho)x/E(S)}, \qquad x \geqslant 0. \quad (1.71)$$

Also, the equation (1.70) allows for an explicit solution when $b(x)$ corresponds to a deterministic distribution. In case the claim size (service requirement) S is a constant D, the integro-differential equation (1.70) reduces to the differential equation

$$q'(x) = -\frac{\lambda}{\sigma} + \frac{\lambda}{\sigma}q(x), \qquad 0 < x < D,$$

$$= \frac{\lambda}{\sigma}q(x) - \frac{\lambda}{\sigma}q(x - D), \qquad x > D,$$

with the boundary condition $q(0) = \rho$. By induction it is readily verified that the explicit solution of this differential equation is given by

$$q(x) = 1 - (1 - \rho)\sum_{j=0}^k (-1)^j \frac{(\lambda x/\sigma - \rho j)^j}{j!}e^{\lambda x/\sigma - \rho j}, \qquad kD \leqslant x < (k + 1)D,$$

$$k = 0, 1, \ldots. \quad (1.72)$$

In general we can derive from the integro-differential equation (1.70) an asymptotic expansion for $q(x)$ as $x \to \infty$. Therefore we first rewrite (1.70) as a renewal equation. To do so note that, by the differentiation rule for an integral and by partial integration together with $B(0) = 0$,

$$\frac{d}{dx} \int_0^x q(x-y)\{1-B(y)\}\,dy = q(0)\{1-B(x)\} + \int_0^x q'(x-y)\{1-B(y)\}\,dy$$

$$= q(0)\{1-B(x)\} - q(x-y)\{1-B(y)\}\Big|_0^x$$

$$- \int_0^x q(x-y)b(y)\,dy$$

$$= q(x) - \int_0^x q(x-y)b(y)\,dy.$$

Hence the integro-differential equation (1.70) is equivalent to

$$q(x) = q(0) - \frac{\lambda}{\sigma} \int_0^x \{1-B(y)\}\,dy + \frac{\lambda}{\sigma} \int_0^x q(x-y)\{1-B(y)\}\,dy, \qquad x \geq 0. \tag{1.73}$$

Defining the functions

$$a(x) = q(0) - \frac{\lambda}{\sigma} \int_0^x \{1-B(y)\}\,dy \qquad \text{and} \qquad h(x) = \frac{\lambda}{\sigma}\{1-B(x)\}, \qquad x \geq 0,$$

we write (1.73) in the general form

$$q(x) = a(x) + \int_0^x q(x-y)h(y)\,dy, \qquad x \geq 0. \tag{1.74}$$

This equation has the form of a standard renewal equation except that the function $h(x)$, $x \geq 0$, is not a proper probability density. It is true that the function h is non-negative, but

$$\int_0^\infty h(x)\,dx = \frac{\lambda}{\sigma} \int_0^\infty \{1-B(x)\}\,dx = \frac{\lambda E(S)}{\sigma} < 1.$$

Thus h is the density of a distribution whose total mass is less than 1 with a defect of $1 - \rho$. The equation (1.74) is called a *defective renewal equation*. However, such an equation is easily converted to a standard renewal equation provided the distribution function $B(x)$ satisfies some weak regularity condition. Define the number δ as the unique *positive* solution to the equation

$$\frac{\lambda}{\sigma} \int_0^\infty e^{sy}\{1-B(y)\}\,dy = 1. \tag{1.75}$$

Here it is assumed that the complementary distribution function $1 - B(y)$ is bounded by an exponentially fast decreasing function so that the left side of (1.75) is a finite function of s on some interval $(0, s_0)$. This assumption holds for many probability distribution functions of practical interest, but excludes distributions with extremely long tails like the lognormal distribution. The solution of the equation (1.75) is unique and positive since the left side of (1.75) is a continuous and strictly increasing function of s and is less than 1 at $s = 0$. Also, it follows that the number

$$\mu^* = \frac{\lambda}{\sigma} \int_0^\infty y e^{\delta y} \{1 - B(y)\} \, dy \tag{1.76}$$

is finite. By a simple trick the defective renewal equation (1.74) can be converted to a standard renewal equation. Letting

$$h^*(x) = \frac{\lambda}{\sigma} e^{\delta x} \{1 - B(x)\}, \qquad x \geqslant 0,$$

then, by (1.75) and (1.76), $h^*(x)$, $x \geqslant 0$, is a probability density with finite first moment μ^*. Multiplying both sides of the equation (1.74) by $e^{\delta x}$ and defining the functions

$$q^*(x) = e^{\delta x} q(x) \qquad \text{and} \qquad a^*(x) = e^{\delta x} a(x), \qquad x \geqslant 0,$$

we find that the defective renewal equation (1.74) is equivalent to

$$q^*(x) = a^*(x) + \int_0^x q^*(x - y) h^*(y) \, dy, \qquad x \geqslant 0. \tag{1.77}$$

This equation is a standard renewal equation to which we can apply the key renewal theorem. The function $a^*(x)$, $x \geqslant 0$, is non-negative and monotone. Also, using (1.9) and (1.75),

$$\int_0^\infty a(x) \, dx = \int_0^\infty e^{\delta x} \left[\frac{\lambda}{\sigma} \int_x^\infty \{1 - B(y)\} \, dy \right] dx = \frac{\lambda}{\sigma} \int_0^\infty \{1 - B(y)\} \left(\int_0^y e^{\delta x} \, dx \right) dy$$

$$= \frac{1 - \rho}{\delta}.$$

Hence, by applying the key renewal theorem to the renewal equation (1.77) with the function $q^*(x) = e^{\delta x} q(x)$ as unknown, it follows that

$$\lim_{x \to \infty} e^{\delta x} q(x) = \frac{1}{\mu^*} \frac{1 - \rho}{\delta}.$$

This yields

$$q(x) \approx \gamma e^{-\delta x} \qquad \text{for large } x, \tag{1.78}$$

where the decay parameter δ is defined by (1.75) and the amplitude factor γ is given by

$$\gamma = (1 - \rho) \left[\frac{\lambda \delta}{\sigma} \int_0^\infty y e^{\delta y} \{1 - B(y)\} \, dy \right]^{-1}. \tag{1.79}$$

This asymptotic expansion is practically very useful in view of the remarkable finding that already for relatively small values of x the asymptotic estimate in (1.78) turns out to predict quite well the exact value of $q(x)$ when the load factor ρ is not too small.

To illustrate this we introduce x_α as the smallest number for which

$$\left| \frac{q_{\text{asy}}(x) - q(x)}{q(x)} \right| \leqslant \alpha \qquad \text{for all } x > x_\alpha,$$

where $q_{\text{asy}}(x)$ denotes the asymptotic estimate in the right side of (1.78). The degree of acceptability of the relative error induced when using $q_{\text{asy}}(x)$ rather then $q(x)$ will typically depend on the absolute magnitude of the probability $q(x)$. For example asymptotic estimates of 0.11 and 0.012 to exact values of 0.10 and 0.01 have respective error percentages of 10 and 20 per cent but may yet be quite acceptable for practical purposes. In table 1.6 we give for $\alpha = 0.10$ and $\alpha = 0.05$ the numbers x_α for various distributions and several values of ρ. Denoting by c_S^2 the squared coefficient of variation of the claim size (service requirement) S, we consider the deterministic distribution ($c_S^2 = 0$), the E_3 distribution ($c_S^2 = \frac{1}{3}$), the E_2 distribution ($c_S^2 = \frac{1}{2}$), and H_2 distributions for $c_S^2 = 1\frac{1}{2}$, 4 and 25. The H_2 distributions are considered for both the normalization of balanced means and the gamma normalization (cf. appendix B) in order to demonstrate that how quickly $q_{\text{asy}}(x)$ applies may strongly depend on the normalization used; the numbers between brackets in Table 1.6 correspond to the gamma normalization. The reader is referred to subsection 4.2.1 in chapter 4 for a discussion of exact methods for the calculation of the probabilities $q(x)$. Also, in Table 1.7 we give the numerical values of $q(x)$ and $q_{\text{asy}}(x)$ for several values of x when ρ is varied as 0.2, 0.5 and 0.8 and c_S^2 has

Table 1.6   The numbers x_α for the relative error bound α

$\rho \backslash c_S^2$	0	$\frac{1}{3}$	$\frac{1}{2}$	$1\frac{1}{2}$	4	25	
0.20	1.19	0.77	0.84	2.07(0.00)	1.37(0.18)	1.14(0.23)	$\alpha = 0.10$
0.50	0.80	0.19	0.03	0.93(0.00)	1.10(0.01)	1.03(0.11)	
0.80	0.00	0.00	0.00	0.00(0.00)	0.22(0.00)	0.46(0.00)	
0.90	0.00	0.00	0.00	0.00(0.00)	0.00(0.00)	0.00(0.00)	
0.20	2.04	1.06	1.34	2.98(0.18)	1.90(0.40)	1.57(0.43)	$\alpha = 0.05$
0.50	1.07	0.43	0.35	1.86(0.00)	1.72(0.25)	1.54(0.32)	
0.80	0.00	0.00	0.00	0.07(0.00)	0.92(0.00)	1.10(0.06)	
0.90	0.00	0.00	0.00	0.00(0.00)	0.21(0.00)	0.54(0.00)	

Table 1.7   Exact and asymptotic values for $q(x)$

	x	$c_S^2 = 0$		$c_S^2 = \frac{1}{2}$		$c_S^2 = 1\frac{1}{2}$	
		$q(x)$	$q_{asy}(x)$	$q(x)$	$q_{asy}(x)$	$q(x)$	$q_{asy}(x)$
$\rho = 0.2$	0.5	0.11586	0.07755	0.12462	0.14478	0.13667	0.09737
	1	0.02288	0.03007	0.07146	0.07712	0.09669	0.07630
	2	0.00196	0.00210	0.02144	0.02188	0.05234	0.04685
	3	0.00015	0.00015	0.00617	0.00621	0.03025	0.02877
	5	—	—	0.00050	0.00050	0.01095	0.01085
$\rho = 0.5$	0.5	0.35799	0.30673	0.37285	0.38608	0.39390	0.34055
	1	0.17564	0.18817	0.26617	0.26947	0.31629	0.28632
	2	0.05304	0.05356	0.13106	0.13126	0.21186	0.20239
	5	0.00124	0.00124	0.01517	0.01517	0.07179	0.07149
	10	—	—	0.00042	0.00042	0.01262	0.01262
$\rho = 0.8$	0.5	0.70164	0.67119	0.71197	0.71709	0.72705	0.70204
	1	0.55489	0.56312	0.62430	0.62549	0.66522	0.65040
	2	0.36548	0.36601	0.47582	0.47589	0.56345	0.55825
	5	0.10050	0.10050	0.20959	0.20959	0.35322	0.35299
	10	0.01166	0.01166	0.05343	0.05343	0.16444	0.16444

the values 0, $\frac{1}{2}$ and $1\frac{1}{2}$, the latter value corresponding to an H_2 distribution with balanced means. The numerical results indicate that the asymptotic expansion $q_{asy}(x)$ may be used for practical purposes for x already in the range of $E(S)$ to $2E(S)$, provided ρ is not too small (say, $\rho \geqslant 0.2$). Notice that the asymptotic expansion shows the tendency to apply earlier as ρ increases.

To conclude this section we derive an approximation for $q(x)$ that applies for all $x \geqslant 0$. A common technique is to approximate $q(x)$ by a sum of exponential functions whose coefficients are determined by exact results for $q(x)$. Using the asymptotically exponential tail in (1.78), we suggest approximating $q(x)$ by

$$q_{app}(x) = \alpha e^{-\beta x} + \gamma e^{-\delta x} \qquad \text{for all} \quad x \geqslant 0, \qquad (1.80)$$

where the constants α and β are found by matching both the behaviour of $q(x)$ at $x = 0$ and the first moment of the probability distribution function $1 - q(x)$, $x \geqslant 0$. By integrating both sides of (1.73) over x and using (1.9) and (1.69), we find after some manipulations that

$$\int_0^\infty q(x)\,dx = \frac{\lambda E(S^2)}{2\sigma(1 - \rho)}. \qquad (1.81)$$

Note that in the context of the queueing application this formula gives the average delay in queue of a customer. It is now a matter of simple algebra to verify that by matching the equations (1.69) and (1.81) the constants α and β are determined as

$$\alpha = \rho - \gamma, \qquad \beta = \alpha \left\{ \frac{\lambda E(S^2)}{2\sigma(1-\rho)} - \frac{\gamma}{\delta} \right\}^{-1}. \tag{1.82}$$

It should be pointed out that the approximation (1.80) can be applied only when $\beta > \delta$ since otherwise $q_{app}(x)$ is not in accordance with the asymptotically exponential tail in (1.78) as $x \to \infty$. We lack a theoretical proof that the condition $\beta > \delta$ is satisfied under reasonable assumptions. However, extensive numerical experiments indicate that $\beta > \delta$ holds for a wide class of distributions of practical interest, including the deterministic distribution and mixtures of Erlangian distributions. In fact the results (A.32) and (A.33) in appendix C imply that the approximation (1.80) is exact for K_2 distributions. Numerical investigations show that otherwise (1.80) provides an excellent approximation for practical purposes when the load factor ρ is not too small (say, $\rho \geqslant 0.2$). To illustrate this in Table 1.8 we give for several values of x and ρ the exact values $q(x)$, the approximate values $q_{app}(x)$ and the asymptotic values $q_{asy}(x)$ for the E_{10} and E_3 distributions. In all examples we assume $\sigma = 1$ and $E(S) = 1$.

Finally, we remark that for the particular case of a deterministic distribution $B(x)$ with unit mass at $x = D$ the following approximation usually performs somewhat better than the approximation (1.80):

$$\bar{q}_{app}(x) = \begin{cases} \rho e^{-\eta x} & \text{for} \quad 0 \leqslant x \leqslant x_0 \\ \gamma e^{-\delta x} & \text{for} \quad x > x_0, \end{cases}$$

Table 1.8   The values $q(x)$, $q_{app}(x)$ and $q_{asy}(x)$

		Erlang-10			Erlang-3		
	x	$q(x)$	$q_{app}(x)$	$q_{asy}(x)$	$q(x)$	$q_{app}(x)$	$q_{asy}(x)$
$\rho = 0.2$	0.10	0.1838	0.1960	0.3090	0.1839	0.1859	0.2654
	0.25	0.1590	0.1682	0.2222	0.1594	0.1615	0.2106
	0.50	0.1162	0.1125	0.1282	0.1209	0.1212	0.1432
	0.75	0.0755	0.0694	0.0739	0.0882	0.0875	0.0974
	1.00	0.0443	0.0413	0.0427	0.0626	0.0618	0.0663
$\rho = 0.5$	0.10	0.4744	0.4862	0.5659	0.4744	0.4764	0.5332
	0.25	0.4334	0.4425	0.4801	0.4342	0.4361	0.4700
	0.50	0.3586	0.3543	0.3651	0.3664	0.3665	0.3810
	0.75	0.2808	0.2745	0.2776	0.3033	0.3026	0.3088
	1.00	0.2127	0.2102	0.2111	0.2484	0.2476	0.2502
$\rho = 0.8$	0.10	0.7833	0.7890	0.8219	0.7834	0.7844	0.8076
	0.25	0.7557	0.7601	0.7756	0.7562	0.7571	0.7708
	0.50	0.7020	0.6998	0.7042	0.7074	0.7074	0.7131
	0.75	0.6413	0.6381	0.6394	0.6577	0.6573	0.6597
	1.00	0.5812	0.5801	0.5805	0.6097	0.6093	0.6103

when η and x_0 are determined by the requirement that $\bar{q}_{app}(x)$ should be continuous at $x = x_0$ and that $\int_0^\infty \bar{q}_{app}(x)\,dx$ should be equal to $\lambda D^2 / \{2\sigma(1 - \rho)\}$, in agreement with (1.81). Thus η and x_0 are obtained by solving the two non-linear equations

$$\rho e^{-\eta x_0} = \gamma e^{-\delta x_0} \quad \text{and} \quad \frac{\rho}{\eta}\left(1 - e^{-\eta x_0}\right) + \frac{\gamma}{\delta} e^{-\delta x_0} = \frac{\lambda D^2}{2\sigma(1 - \rho)}.$$

It appears from numerical experiments that x_0 is usually close to D.

*1.10 A PRODUCTION/INVENTORY CONTROL MODEL WITH VARIABLE PRODUCTION RATE AND SERVICE LEVEL CONSTRAINTS

An important practical problem in the area of production and inventory control is the coordination of the production rate with the inventory level in order to cope with random fluctuations in the demand. In coordinating the production rate with the inventory level, a suitable compromise is sought between the inventory on hand, the frequency of changes in the production rate and the customer service.

We consider a production/inventory problem in which a certain product is produced on a single machine that operates only intermittently. In production the facility produces continuously at a constant rate being larger than the average demand rate. Thus the replenishments of inventory occur continuously rather than by jumps in discrete-time like in pure inventory problems as discussed in section 1.7; such production/inventory systems are prevalent in continuous processing plants like in the chemical industry. The production facility is turned off when the inventory level becomes sufficiently high, while the facility is again reactivated when the inventory level has dropped sufficiently low. We assume that it takes no setup time to restart production. The demand process for the product is a compound Poisson process. That is customers arrive according to a Poisson process with rate λ and their demands for the single product are independent random variables having a common probability density $b(x)$ with finite second moment. Denoting the demand of a customer by the generic variable D, it is assumed that

$$\sigma > \lambda E(D),$$

where σ denotes the production rate. To complete the specification of the model, we need to say how demand in excess of current inventory is handled. In what follows we deal with both the *backlog model* in which excess demand is backordered until inventory becomes available and the *lost-sales model* in which excess demand is lost. A unifying treatment of these models will be given. In analysing the production/inventory model, we restrict ourselves to the class of intuitively reasonable control rules of the (m, M) type with $0 < m \leqslant M$. Under the (m, M) rule the production is stopped if the inventory level becomes as high as M, while the production is restarted as soon as the inventory level falls below m. The purpose is to choose levels m and M such that a suitable compromise is found between inventories on hand, shortages and changes in the production rate. In practice it is often not possible to specify costs for each of these conflicting alternatives; in particular, shortage costs are usually difficult to assign. In accordance with common practice, we therefore concentrate on service measures such as the average number of stockouts per unit time and the fraction of demand to be met directly from stock on hand.

A practical approach to the solution of the production/inventory problem is a sequential determination of the control parameters $M - m$ and m. First the difference $\Delta = M - m$, indicating the production lot size, is determined by considering only holding and production costs. Next the switching level m is calculated by taking into account the service level constraint. A usual cost structure consists of a fixed setup cost of $K > 0$ for restarting production and a holding cost of $h > 0$ per unit of time for each unit kept in stock. In analogy to the economic order quantity formula (1.52), we may approximate $\Delta = M - m$ by

$$\Delta_{EOQ} = \sqrt{\frac{2K\{\sigma - \lambda E(D)\}\lambda E(D)}{\sigma h}}. \tag{1.83}$$

The approximation shows a good performance in costs for many practical situations. In remark 1.2 below we discuss an improvement of the approximation (1.83). It will turn out that the optimal value of Δ minimizing the long-run average costs subject to some service level constraint is nearly independent of the required service level, provided this level is sufficiently high. This finding is in support of the frequently used approach of a sequential determination of the control parameters $M - m$ and m.

In the discussion to follow we assume that the difference $M - m$ is given and we focus on the calculation of the switching level m satisfying the service level constraint:

$$\text{The fraction of demand to be satisfied directly from stock on hand} = f, \tag{1.84}$$

where f is a prespecified value between 0 and 1. The analysis to be given below can easily be extended to other service measures such as the average number of stockout occurrences per unit time. Also, we remark that in the backlog model the fraction of time the system is out of stock can be seen to be equal to $\lambda E(D)/\sigma$ times the fraction of demand that is not satisfied directly from stock on hand. Further, it is interesting to point out that the following important buffer design problem may be solved as a lost-sales production/inventory model. Suppose the input process to a finite buffer is described by a compound Poisson process, where an input exceeding the remaining capacity of the buffer causes

~~an overflow with loss of the excess amount. The buffer is emptied~~ at a constant rate whenever the buffer content is positive. Then the determination of the minimum buffer capacity in order to ensure that the average number of overflows per unit time is not more than a specific value is equivalent to the determination of the single-switching level m in a lost-sales production/inventory model having $M = m$ and requiring that the average number of stockout occurrences per unit time equals a specific value.

In order to find the switching level m satisfying the service level constraint (1.84), we first analyse the inventory process under a given (m, M) rule. The stochastic process describing jointly the inventory position and the status of the production facility (on or off) is regenerative. This means that the service level associated with the given (m, M) rule may be related to the behaviour of the process during a regeneration cycle. Let a cycle be defined as the time elapsed between two consecutive epochs at which the inventory position reaches the level M and the production facility is turned off. Assuming that such a cycle starts at epoch 0, define

T = the next epoch at which the inventory position reaches the level M and the facility is turned off,

V = the total demand during the cycle $(0, T]$,

S = the amount of demand that will go short during the cycle $(0, T]$.

Then, by the theory of regenerative processes,

The fraction of demand that is not satisfied directly from stock on hand

$$= \frac{E(S)}{E(V)}. \tag{1.85}$$

Using Wald's equation and the memoryless property of the Poisson process, it follows that

$$E(V) = \lambda E(T)E(D). \tag{1.86}$$

To evaluate the service level of the (m, M) rule, we need tractable expressions for the key elements $E(T)$ and $E(S)$. We first derive some basic relations between the backlog model and the lost-sales model. We write $E_B(T)$ and $E_B(S)$ for $E(T)$ and $E(S)$ in the backlog model. Similarly, we use the notation $E_L(T)$ and $E_L(S)$ for the lost-sales model. Also, define the function

$t_0(x)$ = the expected amount of time until the first epoch at which the inventory position drops below the critical level m, given that at epoch 0 the inventory position is $x + m$ and the production facility is off, $x \geqslant 0$.

Note that the function $t_0(x)$ is the same for the backlog model and the lost-sales model. Obviously, $t_0(M - m)$ is the expected amount of time in a cycle during which the production facility is off and thus $t_0(M - m)/E(T)$ gives the long-run fraction of time the production facility is off. We next establish the following basic relations:

$$E_B(T) = \frac{t_0(M - m)}{1 - \lambda E(D)/\sigma},$$ (1.87)

$$E_L(T) = E_B(T) - \frac{E(S_L)}{\sigma - \lambda E(D)},$$ (1.88)

$$E_L(S) = \left\{ 1 - \frac{\lambda E(D)}{\sigma} \right\} E_B(S).$$ (1.89)

To derive these relations, we first observe that

The expected amount of time in a cycle during which the production facility is on $= E(T) - t_0(M - m)$.

For the backlog model,

The total production per cycle $=$ the total demand per cycle,

and so

$$\sigma\{E_B(T) - t_0(M - m)\} = \lambda E_B(T)E(D),$$

implying (1.87). For the lost-sales model,

The total production per cycle $=$ (the total demand per cycle)

$-$ (the lost demand per cycle),

and so

$$\sigma\{E_L(T) - t_0(M - m)\} = \lambda E_L(T)E(D) - E_L(S).$$

This relation and (1.87) imply (1.88). To derive a relation between $E_L(S)$ and $E_B(S)$, define the random variable I for the backorder model as the amount of time in a cycle during which the inventory position is negative. Using the lack of memory of the Poisson arrival process and using the observation that in the backorder model the inventory position will always reach the level 0 when the initial inventory position is negative, it follows after some reflections that

$$E_B(T) = E_L(T) + E(I)$$ (1.90)

and

$$E_B(S) = E_L(S) + \lambda E(I)E(D).$$ (1.91)

Here we use the fundamental relation (1.63) to establish that the expected amount of demand during the time I equals $\lambda E(I)E(D)$. The relations (1.88), (1.90) and (1.91) imply the result (1.89).

It remains to evaluate the quantities $t_0(M - m)$, $E(T)$ and $E(S)$ for the backlog model. In addition to the function $t_0(x)$ defined above, for the backlog model we define

$\gamma(u|x) =$ the probability density of the undershoot U of the critical level m at the next epoch at which the production facility is turned on when presently the inventory level is $x + m$ and the production is off, $x \geqslant 0$,

$t_1(x; M) =$ the expected amount of time until the next epoch at which the inventory position reaches the level M when presently the inventory position is x and the production is on, $x \leqslant M$,

$s(x; M) =$ the expected amount of demand that will go short until the next epoch at which the inventory position reaches the level M when presently the inventory position is x and the production is on, $x \leqslant M$.

Then, by conditioning arguments,

$$E_B(T) = t_0(M - m) + \int_0^\infty t_1(m - u; M)\gamma(u|M - m)\,du, \qquad (1.92)$$

$$E_B(S) = \int_m^\infty \{u - m + s(m - u; M)\}\gamma(u|M - m)\,du$$

$$+ \int_0^m s(m - u; M)\gamma(u|M - m)\,du. \qquad (1.93)$$

To determine $t_0(M - m)$ and $\gamma(u|M - m)$ we distinguish between the two cases of $M - m = 0$ and $M - m > 0$.

Case of $M - m = 0$

We obviously have

$$t_0(M - m) = \frac{1}{\lambda} \qquad (1.94)$$

and

$$\gamma(u|M - m) = b(u), \qquad u \geqslant 0, \qquad (1.95)$$

where $b(u)$ denotes the probability density of the customer demand.

Case of $M - m > 0$

By conditioning on the first demand,

$$t_0(x) = \frac{1}{\lambda} + \int_0^x t_0(x - y)b(y)\,dy, \qquad x \geqslant 0.$$

Using (1.8), it follows that the solution of this renewal equation is given by

$$t_0(x) = \frac{1}{\lambda}\{1 + M(x)\}, \qquad x \geqslant 0,$$

where $M(x)$ is the renewal function associated with the demand density $b(x)$. An application of the asymptotic expansion (1.5) now yields the approximation

$$t_0(M - m) \approx \frac{E(D^2)}{2\lambda E^2(D)}, \tag{1.96}$$

assuming $M - m$ is sufficiently large compared with $E(D)$, say,

$$M - m \geqslant \begin{cases} E(D) & \text{if } 0 < c_D^2 \leqslant 1, \\ 1\frac{1}{2}c_D^2 E(D) & \text{if } c_D^2 > 1, \end{cases} \tag{1.97}$$

The undershoot density $\gamma(u | M - m)$ can be interpreted as the residual life density in a renewal process in which the interoccurrence times are given by the demands of the customers. Thus, assuming $M - m$ satisfies (1.97) and denoting the probability distribution function of the demand D by $B(x)$, we obtain from (1.12) the approximation

$$\gamma(u | M - m) \approx \frac{1}{E(D)}\{1 - B(u)\}, \qquad u \geqslant 0. \tag{1.98}$$

We next turn to the determination of tractable expressions for the functions $t_1(x; M)$ and $s(x; M)$. Here it should be noted that we are considering the backlog model. The function $t_1(x; M)$ is easy to give. We have

$$t_1(x; M) = \frac{M - x}{\sigma - \lambda E(D)}, \qquad x \leqslant M. \tag{1.99}$$

To see this, note the analogy between the inventory process in the backlog model and the inventory process of the model studied in example 1.11. In terms of the latter model $t_1(x; M)$ represents the expected amount of time needed to empty an infinite capacity buffer with initial content $M - x$ when inputs at size D occur according to a Poisson process with rate λ and the buffer is emptied at a constant rate σ. The formula (1.99) now follows from the first result in (1.34). By the same argument, the expected amount of time needed to raise the inventory position to the level 0 when the starting inventory position x is below 0 is given by $-x/\{\sigma - \lambda E(D)\}$. Thus, using the memoryless property of the Poisson process,

$$s(x; M) = \frac{-\lambda E(D)x}{\sigma - \lambda E(D)} + s(0; M), \qquad x \leqslant 0. \tag{1.100}$$

To find $s(x; M)$ for $x \geqslant 0$, we define the auxiliary function for the backlog model as

$s_\infty(x) = $ the expected amount of demand that will go short in the infinite-time interval $(0, \infty)$ when the starting inventory position is x and the production facility is *always* on.

Then it is readily seen that for the backlog model

$$s_\infty(x) = s(x; M) + s_\infty(M), \qquad x \leqslant M. \tag{1.101}$$

As in (1.100) we have

$$s_\infty(x) = \frac{-\lambda E(D)x}{\sigma - \lambda E(D)} + s_\infty(0), \qquad x \leqslant 0. \tag{1.102}$$

The function $s_\infty(x)$ for $x \geqslant 0$ is obtained by deriving an integro-differential equation. Note that $s_\infty(x)$ is finite for all x, since the production rate σ is larger than the average demand rate $\lambda E(D)$, implying that the inventory process will drift to infinity when the production is always on. Similarly to the derivation of the integro-differential equation (1.70), we find by conditioning upon what may happen in a small time $\Delta x/\sigma$ that

$$s_\infty(x - \Delta x) = \frac{\lambda \Delta x}{\sigma} \int_x^\infty \{y - x + s_\infty(x - y)\}b(y)\,\mathrm{d}y + \frac{\lambda \Delta x}{\sigma} \int_0^x s_\infty(x - y)b(y)\,\mathrm{d}y$$

$$+ \left(1 - \frac{\lambda \Delta x}{\sigma}\right)s_\infty(x) + o(\Delta x), \qquad x > 0.$$

Invoking (1.102), we next obtain

$$s'_\infty(x) = \frac{-\lambda}{\sigma - \lambda E(D)} \int_x^\infty (y - x)b(y)\,\mathrm{d}y - \frac{\lambda}{\sigma}s_\infty(0)\{1 - B(x)\} + \frac{\lambda}{\sigma}s_\infty(x)$$

$$- \frac{\lambda}{\sigma} \int_0^x s_\infty(x - y)b(y)\,\mathrm{d}y, \qquad x > 0. \tag{1.103}$$

This integro-differential equation may be studied in a similar way to the integro-differential equation (1.70) in $q(x)$. However, the close resemblance of these two equations suggests that a simple relation should exist between $s_\infty(x)$ and $q(x)$, the latter being the probability that the inventory position will ever be negative when the starting inventory position is x and the production is always on. Indeed, we have the relation

$$s_\infty(x) = \frac{\sigma}{\sigma - \lambda E(D)} \int_x^\infty q(y)\,\mathrm{d}y, \qquad x \geqslant 0. \tag{1.104}$$

This relation may be obtained as follows. From differentiation of (1.103) we find that $u(x) = s'_\infty(x)$ is a solution to the equation

$$u'(x) = \frac{\lambda}{\sigma - \lambda E(D)}\{1 - B(x)\} + \frac{\lambda}{\sigma}u(x) - \frac{\lambda}{\sigma}\int_0^x u(x-y)b(y)\,dy, \qquad x > 0.$$

(1.105)

A comparison of (1.70) and (1.105) yields that $u(x) = \sigma q(x)/\{\sigma - \lambda E(D)\}$ is also a solution to (1.105). The integro-differential equation (1.105) must have a unique solution, since this equation can be rewritten as a renewal equation (as done before for equation 1.70) and a renewal equation is known to have a unique solution. Thus we find

$$s'_\infty(x) = \frac{\sigma}{\sigma - \lambda E(D)}q(x).$$

From this relation and the fact that $s_\infty(x) \to 0$ as $x \to \infty$ (why?), we get the desired result (1.104). In section 1.9 we derived a simple and accurate approximation to $q(x)$. Substituting this approximation into (1.104) and using (1.101), we obtain the tractable approximation

$$s(x; M) \approx \sum_{i=1}^2 c_i(e^{-d_ix} - e^{-d_iM}), \qquad 0 \leqslant x \leqslant M,$$

(1.106)

with

$$c_1 = \frac{\alpha\sigma}{\beta\{\sigma - \lambda E(D)\}}, \qquad c_2 = \frac{\gamma\sigma}{\delta\{\sigma - \lambda E(D)\}}, \qquad d_1 = \beta \quad \text{and} \quad d_2 = \delta,$$

where α, β, γ and δ are defined by (1.75), (1.76) and (1.80) with S replaced by D. This result requires the mild assumption that the demand density $b(x)$ tends at least exponentially fast to 0 as $x \to \infty$. The approximation (1.106) is exact when the demand D has a K_2 distribution.

Summarizing, we have found tractable expressions for the ingredients of the key elements $E(T)$ and $E(S)$ determining the service level for a given (m, M) rule. Under the service level constraint that the fraction of demand to be met directly from stock on hand equals f, we have for the backlog model that the switching level m is obtained by solving the equation (see equations 1.85 and 1.86),

$$\frac{E_B(S)}{E_B(T)} = (1 - f)\lambda E(D).$$

(1.107)

Using the relations (1.88) and (1.89), it is readily verified that the corresponding equation for the lost-sales model is related to the equation (1.107) by

$$\frac{E_L(S)}{E_L(T)} = (1 - f)\lambda E(D) \Leftrightarrow \frac{E_B(S)}{E_B(T)} = \frac{\sigma}{\sigma - \lambda E(D)f}(1 - f)\lambda E(D).$$

(1.108)

This relation shows that for the production/inventory system with continuously occurring inventory replenishments the solution for the lost-sales model differs significantly from the solution for the backlog model when f is close to 1, as opposed to the pure inventory system with inventory replenishments occurring in discrete time (see section 1.7).

Defining the constant c_f by

$$c_f = \begin{cases} 1 & \text{for the backlog model,} \\[2mm] \dfrac{\sigma}{\sigma - \lambda E(D)f} & \text{for the lost-sales model,} \end{cases}$$

and denoting by Δ the given value of $M - m$, we obtain after some algebra from (1.92) to (1.100) and (1.106) to (1.108) the following ultimate results.

Case of $\Delta = 0$

An approximation to the switching level m is obtained by solving the equation

$$\left\{ \frac{1}{\lambda} + \frac{\lambda E(D)}{\sigma - \lambda E(D)} \right\}^{-1} \left[\frac{\sigma}{\sigma - \lambda E(D)} \int_m^\infty (u - m)b(u)\,du + \{1 - B(m)\} \sum_{i=1}^2 c_i(1 - e^{-d_i m}) \right.$$

$$\left. + \sum_{i=1}^2 c_i e^{-d_i m} \int_0^m (e^{d_i u} - 1)b(u)\,du \right] = (1 - f)c_f \lambda E(D). \tag{1.109}$$

The resulting approximation is exact when the demand has a K_2 distribution.

Case of $\Delta > 0$

An approximation for the switching level m is obtained by solving the equation

$$\left[\frac{E(D^2)}{2\lambda E^2(D)} + \frac{\Delta}{\sigma - \lambda E(D)} + \frac{E(D^2)}{2E(D)\{\sigma - \lambda E(D)\}} \right]^{-1} \left(\frac{\sigma}{2E(D)\{\sigma - \lambda E(D)\}} \right.$$

$$\times \int_m^\infty (u - m)^2 b(u)\,du + \frac{1}{E(D)} \left[\int_m^\infty \{1 - B(u)\}\,du \right] \sum_{i=1}^2 c_i(1 - e^{-d_i(\Delta + m)})$$

$$\left. + \frac{1}{E(D)} \sum_{i=1}^2 c_i e^{-d_i m} \int_0^m (e^{d_i u} - e^{-d_i \Delta})\{1 - B(u)\}\,du \right)$$

$$= (1 - f)c_f \lambda E(D), \tag{1.110}$$

assuming Δ satisfies (1.97). The resulting approximation is exact when the demand is exponentially distributed.

The equations (1.109) and (1.110) can be solved numerically by using a standard procedure for the computation of the zero of an equation. The integrals in the above equations are rather easily evaluated for demand distributions of practical interest such as generalized Erlangian distributions.

Remark 1.2

In this remark we briefly discuss an approximation to the average switching and holding costs per unit time when using an (m, M) rule. We assume a fixed switching cost of $K > 0$ for restarting production and a holding cost of $h > 0$ per unit of time for each unit kept in stock. The average switching cost per unit time obviously equals $K/E(T)$, while the average holding cost per unit time is equal to the expected holding cost incurred during a cycle divided by the expected length $E(T)$ of a cycle. Some reflections show that the expected holding cost incurred during a cycle is the same for the backlog model and the lost-sales model. However, the average costs in these models differ because the average cycle lengths are not the same. The expected holding cost incurred during a cycle may be calculated in a more or less analogous way as the expected shortage $E(S)$ was calculated. We omit here the details and give only the ultimate results due to De Kok (1986). Denote by $g(m, \Delta)$ the long-run average switching and holding costs per unit time when using an (m, M) rule. Assuming the service level of the (m, M) rule is sufficiently high, the average costs $g(m, \Delta)$ may be approximated by

$$
g(m, \Delta) \approx \left[\frac{\sigma\{\Delta + E(U)\}}{\lambda E(D)\{\sigma - \lambda E(D)\}} - c_1(c_2 - e^{-\delta\Delta})e^{-\Delta m} \right]^{-1}
$$
$$
\times \left\{ K + h\left(\frac{\sigma[\frac{1}{2}\Delta^2 - \frac{1}{2}E(U^2) + \{\Delta + E(U)\}m]}{\lambda E(D)\{\sigma - \lambda E(D)\}} + \frac{(c_2 - e^{-\delta\Delta})}{\sigma\delta^3\mu^*}e^{-\delta m} \right.\right.
$$
$$
\left.\left. + \frac{1}{2}\lambda E(D^2)\{\Delta + E(U)\}\left[\frac{1}{\{\lambda E(D)\}^2} - \frac{1}{\{\sigma - \lambda E(D)\}^2} \right] \right)\right\}, \qquad (1.111)
$$

where δ and μ^* are given by (1.75) and (1.76) and the constant c_1 equals

$$
c_1 = \begin{cases} 0 & \text{for the backlog model,} \\[2mm] \dfrac{1}{\sigma\delta^2\mu^*} & \text{for the lost-sales model.} \end{cases}
$$

To specify the first two moments $E(U)$ and $E(U^2)$ of the undershoot variable U and the constant c_2, we distinguish between the cases $\Delta = 0$ and $\Delta > 0$.

Case of $\Delta = 0$

Then

$$
E(U) = E(D), \qquad E(U^2) = E(D^2) \qquad \text{and} \qquad c_2 = 1 + \frac{\sigma\delta}{\lambda}.
$$

Case of $\Delta > 0$

Then

$$E(U) = \frac{E(D^2)}{2E(D)}, \qquad E(U^2) = \frac{E(D^3)}{3E(D)} \qquad \text{and} \qquad c_2 = \frac{\sigma}{\lambda E(D)},$$

provided Δ satisfies (1.97). For the case of $\Delta > 0$, the optimal value of Δ minimizing the average costs subject to the service level constraint (1.84) may be approximated as the positive solution to the equation

$$h\Delta\{\Delta + E(U)\} + \frac{h\{ + E(U)\}^2 e^{-\delta\Delta}}{\sigma/\{\lambda E(D)\} - e^{-\delta\Delta}} - \tfrac{1}{2}h\{\Delta^2 - E(U^2)\}$$

$$-\frac{h}{\delta}\{\Delta + E(U)\} = K\frac{\lambda E(D)}{\sigma}\{\sigma - \lambda E(D)\} \qquad (1.112)$$

independent of the required service level f, provided f is close enough to 1 (say $f \geqslant 0.90$). The above equation for Δ applies both to the backlog model and the lost-sales model. Also, it can be shown that this equation does not depend on the specific service measure considered.

We conclude this section with a validation of the various approximations. First, in Table 1.9 we give the switching level m and the associated service level for a number of numerical examples for both the backlog model and the lost-sales model and for either of the cases $\Delta = 0$ and $\Delta > 0$. In the latter case Δ is

Table 1.9   Some numerical results for the service level

c_D^2	σ	f	Backlog case (m, M)	f_{gamma}	f_{logn}	Lost-sales case (m, M)	f_{gamma}	f_{logn}
$\frac{1}{3}$	1.25	0.95	(9.75, 9.75)	0.953(4)	0.951(5)	(5.17, 5.17)	0.951(1)	0.951(2)
$\frac{1}{3}$	1.25	0.99	(14.92, 14.92)	0.988(3)	0.989(3)	(9.88, 9.88)	0.990(1)	0.990(1)
$\frac{1}{3}$	2.00	0.95	(3.74, 3.74)	0.951(2)	0.949(2)	(2.97, 2.97)	0.951(1)	0.949(1)
$\frac{1}{3}$	2.00	0.99	(5.66, 5.66)	0.991(1)	0.988(2)	(4.84, 4.84)	0.991(1)	0.989(1)
$\frac{4}{5}$	1.25	0.95	(13.41, 13.41)	0.955(6)	0.946(6)	(7.04, 7.04)	0.951(2)	0.949(2)
$\frac{4}{5}$	1.25	0.99	(20.59, 20.59)	0.989(3)	0.988(4)	(13.59, 13.59)	0.990(1)	0.989(1)
$\frac{4}{5}$	2.00	0.95	(5.32, 5.32)	0.950(3)	0.940(3)	(4.19, 4.19)	0.951(1)	0.946(2)
$\frac{4}{5}$	2.00	0.99	(8.14, 8.14)	0.990(1)	0.984(2)	(6.94, 6.94)	0.990(1)	0.986(1)
$\frac{1}{3}$	1.25	0.95	(8.05, 11.21)	0.952(4)	0.943(6)	(3.46, 6.63)	0.951(1)	0.949(2)
$\frac{1}{3}$	1.25	0.99	(13.22, 16.38)	0.991(3)	0.993(2)	(8.17, 11.34)	0.990(1)	0.991(1)
$\frac{1}{3}$	2.00	0.95	(1.87, 6.87)	0.949(2)	0.946(3)	(1.11, 6.11)	0.949(1)	0.949(1)
$\frac{1}{3}$	2.00	0.99	(3.79, 8.79)	0.990(1)	0.988(2)	(2.98, 7.98)	0.990(1)	0.989(1)
$\frac{4}{5}$	1.25	0.95	(11.60, 14.76)	0.953(7)	0.942(8)	(5.23, 8.39)	0.951(1)	0.948(2)
$\frac{4}{5}$	1.25	0.99	(18.78, 21.94)	0.991(3)	0.991(4)	(11.77, 14.93)	0.990(1)	0.989(1)
$\frac{4}{5}$	2.00	0.95	(3.14, 8.14)	0.951(2)	0.943(3)	(2.01, 7.01)	0.951(1)	0.949(2)
$\frac{4}{5}$	2.00	0.99	(5.96, 10.96)	0.990(2)	0.983(2)	(4.76, 9.76)	0.990(1)	0.986(1)

predetermined according to (1.83) with $K = 25$ and $h = 1$. In all examples we take $\lambda = 1$ and $E(D) = 1$. The production rate σ has the two values 1.25 and 2.00, the required service level f is varied as 0.95 and 0.99, and the squared coefficient of variation c_D^2 of the demand D has the two values $\frac{1}{3}$ and $\frac{4}{5}$. Assuming a gamma distribution for the demand D, the switching level m is computed from (1.109) and (1.110) for the respective cases of $\Delta = 0$ and $\Delta > 0$. The actual service levels of the approximate (m, M) rules are determined by computer simulation. In each example 250,000 customer demands were simulated and the notation 0.952(4) is used to denote that the simulated value is 0.952 with [0.948, 0.956] as the 95 per cent confidence interval. In order to study the effect of the shape of the underlying demand density on the service level, the actual service level of the (m, M) rule is not only simulated for gamma demand but also for lognormal demand with the same first two moments; the respective service levels are indicated by f_{gamma} and f_{logn} in Table 1.9. The lognormal density is a good testing density in sensitivity analysis because of its long tail. The following conclusions may be drawn from our numerical investigations. The approximate (m, M) rule shows an excellent performance with respect to the required service level. Also, the actual service level of an (m, M) rule is fairly insensitive to more than the first two moments of the demand D, provided the demand density satisfies a reasonable shape constraint and c_D^2 is not too large, say $0 \leqslant c_D^2 \leqslant 1$. The latter finding suggests that mixtures of E_{k-1} and E_k densities with the same scale parameters may be used for doing the calculations in the practically important case of relatively smooth demand D with $0 \leqslant c_D^2 \leqslant 1$.

Second, in Table 1.10 we give numerical results showing that the economic order quantity formula for Δ performs very well in costs and that the formula (1.111) gives an excellent prediction to the actual average holding and switching costs. In the examples $\lambda = 1$, $E(D) = 1$, $K = 25$ and $h = 1$, the required service level f has the two values 0.90 and 0.95, the production rate σ is varied as 1.25 and 2.00, and the squared coefficient of variation c_D^2 of the demand D has the two values 0.5 and 2. The demand D has a gamma distribution. The values Δ_{EOQ} and Δ^* are calculated from the relations (1.83) and (1.112). Table 1.10 gives these

Table 1.10 Some numerical results for the average costs

c_D^2	σ	f	Δ_{EOQ}	m	g_{pre}	g_{act}	Δ^*	m	g_{pre}	g_{act}
0.5	1.25	0.90	3.16	6.77	7.24	7.27(0.09)	5.36	5.90	7.03	7.09(0.08)
0.5	1.25	0.95	3.16	9.31	9.63	9.67(0.10)	5.36	8.43	9.42	9.37(0.09)
0.5	2.00	0.90	5.00	1.35	5.68	5.67(0.02)	5.72	1.19	5.64	5.65(0.02)
0.5	2.00	0.95	5.00	2.31	6.61	6.59(0.02)	5.72	2.16	6.57	6.56(0.02)
2.0	1.25	0.90	3.16	15.52	13.25	13.32(0.20)	6.32	14.02	13.01	13.08(0.19)
2.0	1.25	0.95	3.16	20.85	18.27	18.43(0.25)	6.32	19.35	18.03	18.01(0.25)
2.0	2.00	0.90	5.00	4.48	8.10	8.12(0.03)	6.59	3.92	8.00	8.03(0.03)
2.0	2.00	0.95	5.00	6.71	10.25	10.24(0.05)	6.59	6.15	10.15	10.16(0.04)

two values of Δ together with the associated switching levels m obtained from (1.110) and the predicted and actual values g_{pre} and g_{act} of the average holding and switching costs. The value g_{pre} is calculated from (1.111) and the value g_{act} is obtained by computer simulation of 250,000 customer demands, where the notation 7.27(0.09) for g_{act} means that the simulated value is 7.27 with [7.18, 7.36] as the 95 per cent confidence interval. Table 1.10 concerns the backlog model and the case of $\Delta > 0$; however, the formula (1.111) for the average costs yields excellent results as well for the lost-sales model and the case of $\Delta = 0$.

EXERCISES

1.1 A machine is subject to breakdowns. Each breakdown requires an exponentially distributed repair time with mean $1/\mu$. The running time of the machine until the next breakdown is exponentially distributed with mean $1/\lambda$. The repair times and the running times are assumed to be independent of each other. The machine is in working condition at time 0. Determine an asymptotic expansion for the expected number of breakdowns up to time t.

1.2 An item deteriorates gradually in time by aging. The failure rate of an item having age x is $\alpha(x) = \alpha x$ for some constant $\alpha > 0$, that is an item with age x will fail in the next Δt time units with probability $\alpha(x)\Delta t$ for small Δt. The item is replaced upon failure or upon reaching the age T, whichever occurs first. Determine an asymptotic expansion for the expected number of replacements up to time t.

1.3 Verify for the Erlang-r distribution with mean r/μ that the associated renewal function $M(t)$ can be computed from the rapidly converging series

$$M(t) = \sum_{n=1}^{\infty} \left\{ 1 - \sum_{j=0}^{nr-1} e^{-\mu t} \frac{(\mu t)^j}{j!} \right\}, \qquad t \geq 0.$$

(*Hint*: an Erlang-r distributed random variable can be represented as the sum of r independently exponentially distributed random variables.)

1.4 (a) Verify for the probability density $f(t) = p_1 \mu_1 e^{-\mu_1 t} + p_2 \mu_2 e^{-\mu_2 t}$ that the associated renewal function is given by

$$M(t) = \frac{t}{E(X_1)} + \left\{ \frac{E(X_1^2)}{2E^2(X_1)} - 1 \right\}(1 - e^{-(p_1 \mu_1 + p_2 \mu_2)t}), \qquad t \geq 0.$$

(b) Verify for the probability density $f(t) = p\mu e^{-\mu t} + (1-p)\mu^2 t e^{-\mu t}$ that the associated renewal function is given by

$$M(t) = \frac{t}{E(X_1)} + \left\{ \frac{E(X_1^2)}{2E^2(X_1)} - 1 \right\}(1 - e^{-\mu(2-p)t}), \qquad t \geq 0.$$

(*Hint*: take the Laplace transform of both sides of the renewal equation for $M(t)$.)

1.5 Consider a renewal process in which the interoccurrence times between renewals have an Erlang-r distribution. Verify that the limiting distribution of the excess life has a probability density of the form $\sum_{j=1}^{r} p_j f_j(x)$ where $p_j = 1/r$ for all j and $f_j(x)$ is the density of an Erlang-j distribution. Could you give a heuristic explanation of this result?

1.6 For a renewal process let $M_2(t) = E(N(t)^2)$ be the second moment of the number of renewals up to time t. Verify that $M_2(t)$ satisfies the renewal equation

$$M_2(t) = 2M(t) - F(t) + \int_0^t M_2(t-x)f(x)\,dx, \qquad t \geqslant 0,$$

where $f(x)$ is the density of the interoccurrence times of the renewals. *Note:* using this equation it can be shown that

$$\lim_{t \to \infty} \left\{ \text{var}\{N(t)\} - \frac{\mu_2 - \mu_1^2}{\mu_1^3} t - \left(\frac{5\mu_2^2}{4\mu_1^4} - \frac{2\mu_3}{3\mu_1^3} - \frac{\mu_2}{2\mu_1^2} \right) \right\} = 0,$$

where μ_k denotes the kth moment of the density f (see Heyman and Sobel, 1982).

1.7 Verify for the residual life variable γ_t discussed in section 1.1 that its first two moments are given by

$$E(\gamma_t) = E(X_1)\{1 + M(t)\} - t, \qquad\qquad\qquad t > 0,$$

and

$$E(\gamma_t^2) = E(X_1^2)\{1 + M(t)\} - 2E(X_1)\left\{ t + \int_0^t M(x)\,dx \right\} + t^2, \quad t > 0.$$

1.8 An item is replaced every T time units and upon failure. The lifetimes of the successive items are independent positive random variables having a common probability density $f(x)$. A fixed cost of $c_1 > 0$ is incurred for each planned replacement and a fixed cost of $c_2 > c_1$ for each failure replacement. Determine an asymptotic expansion for the long-run average cost per unit time. For the special case in which the lifetimes have an Erlang-2 distribution, determine the optimal value of T for the block-replacement rule.

1.9 Consider a service facility that is alternately on and off. The on and off times are mutually independent random variables and have respective probability densities $f(x)$ and $g(x)$. Customers arrive at the facility according to a Poisson process with rate λ. Supposing that an on period begins at epoch 0, show that the probability that the first arrival finds upon arrival the facility on is given by

$$\left\{ \frac{1}{\lambda} - \frac{1}{\lambda} \int_0^\infty e^{-\lambda t} f(t)\,dt \right\} \left\{ 1 - \int_0^\infty e^{-\lambda t} f(t)\,dt \int_0^\infty e^{-\lambda t} g(t)\,dt \right\}^{-1}.$$

(*Hint:* defining $p(t)$ as the probability that the facility will be on at time t, establish a renewal equation for $p(t)$ and use Laplace transform results given in appendix C.)

1.10 At a production facility orders arrive according to a renewal process with a mean interarrival time $1/\mu$. A production is started only when N orders have accumulated. The production time is negligible. A fixed cost of $K > 0$ is incurred for each production setup and holding costs at the rate hj with $h > 0$ are incurred when j orders are waiting to be processed. Verify that the value of N for which the long-run average cost per unit time is minimal is one of the two integers nearest to $\sqrt{2K\mu/h}$.

1.11 For the periodic review (s, S) inventory model dealt with in example 1.3, prove that the long-run fraction of weeks at the beginning of which the inventory position is at least y just after any ordering is equal to $\{1 + M(S - y)\}/\{1 + M(S - s)\}$ for $s \leqslant y \leqslant S$. (*Hint*: taking y fixed and using an appropriate cost structure, set up a renewal equation similar to the one in example 1.3.)

1.12 Suppose a non-stationary Poisson process with a time-dependent arrival rate $\lambda(t)$, that is the probability of an arrival to occur in $(t, t + \Delta t]$ is $\lambda(t)\Delta t + o(\Delta t)$ for small Δt. Denoting by $N(t)$ the number of arrivals in $(0, t]$ and assuming that $\lambda(t)$ is (piecewise) continuous in t, verify that $N(t + s) - N(t)$ is Poisson distributed with mean $a(t + s) - a(t)$ for all $t, s > 0$, where $a(t) = \int_0^t \lambda(x)\,dx$. (*Hint*: for t fixed set up a differential equation in $p_j(s) = P\{N(t + s) - N(t) = j\}$ by considering $p_j(s + \Delta s)$ for small Δs.)

1.13 Passengers arrive at a bus stop according to a Poisson process with rate λ. Buses depart from the stop according to a renewal process with interdeparture time A. Using renewal–reward processes, prove that the average waiting time of a passenger equals $E(A^2)/2E(A)$. Could you give a heuristic explanation of why this answer is the same as the average residual life in a renewal process?

1.14 Assume a Poisson process $\{N(t), t \geqslant 0\}$ with interarrival times $X_1, X_2, \ldots$. Verify that for fixed $t > 0$ and $n \geqslant 1$,

$$E(X_k | N(t) = n) = \frac{t}{n + 1} \qquad \text{for} \qquad 1 \leqslant k \leqslant n.$$

(*Hint*: evaluate $P\{\sum_{j=1}^k X_j > x | N(t) = n\}$.)

1.15 A tour boat that makes trips through the canals of Amsterdam leaves every T minutes from the Central Station. Potential customers for a trip arrive at the Central Station according to a Poisson process with rate λ. A potential customer waits until the next departure with probability $e^{-\mu t}$ if upon arrival of the customer the time until the next departure is t minutes. A fixed cost of $K > 0$ is incurred for each trip of a tour boat and a reward of $r > 0$ is earned for each customer joining a trip. Show that the value of T maximizing the long-run average reward per unit time is the unique solution to the equation $e^{-\mu T}(r\lambda T + r\lambda/\mu) = r\lambda/\mu - K$ provided $r\lambda/\mu > K$.

1.16 In an inventory system for a single product the demand process for that product is a Poisson process with rate λ. Each demand which cannot be satisfied directly from inventory on hand is lost. Opportunities to replenish the inventory on hand occur according to a Poisson process with rate μ. For technical reasons a replenishment can only be made when the inventory equals zero. Each replenishment order consists of Q units. The lead time of a replenishment order is negligible. There is a fixed cost of $K > 0$ per replenishment order, a holding cost of h per unit inventory per unit time and a lost-sales cost of π per unit demand lost. Determine the long-run average cost per unit time. Also, determine the fraction of demand which is lost.

1.17 At a processor jobs arrive according to a Poisson process with rate λ. A job is only accepted for execution when the processor is idle. The processing times of the accepted jobs are independent random variables distributed as the generic variable S. Show that the long-run fraction of time that the processor is idle and the long-run fraction of jobs that is rejected are both equal to $1/\{1 + \lambda E(S)\}$.

1.18 Potential building projects of types $1, .., N$ are offered to a constructor according to independent Poisson processes with respective rates $\lambda_1, ..., \lambda_N$. Each offer should be either accepted or rejected upon its occurrence. Here the constructor is only able to handle one project at a time. An accepted project of type j occupies the constructor for a random time τ_j and yields a random payoff ξ_j. The successive offers are assumed to be independent of each other. The constructor accepts only jobs of type $j \in A$ whenever he is idle, where A is a given subset of $\{1, ..., N\}$. Verify that the average payoff per unit time and the fraction of time the constructor is idle are given by

$$\frac{\sum_{j \in A} \lambda_j E(\xi_j)}{1 + \sum_{j \in A} \lambda_j E(\tau_j)} \quad \text{and} \quad \frac{1}{1 + \sum_{j \in A} \lambda_j E(\tau_j)}$$

(This problem is taken from Keilson, 1969.)

1.19 A production process in a factory yields as a derivative waste which is temporarily stored on the factory site. The amount of waste that is produced in a week has an exponential distribution with mean $1/\mu$. Opportunities to remove the waste from the factory site occur at the end of each week only. The following control rule is used. The total amount of waste present is removed when the amount present is larger than z. There is a fixed cost of $K > 0$ for removing the waste. Also, there is a holding cost of $h > 0$ per unit waste per week where the holding costs in each week are charged against the amount of waste present at the end of that week. Verify that the critical level z for which the long-run average cost per week is minimal is the solution to the quadratic equation $\frac{1}{2}h\mu^2 z^2 + h\mu z - \mu K = 0$.

1.20 A production machine gradually deteriorates in time. The machine has N possible working conditions $1, \ldots, N$ which describe increasing degrees of deterioration. Here working condition 1 represents a new system and working condition N represents a failed system. In each working condition i with $1 \leqslant i < N$ the system stays during an exponentially distributed time with mean $1/\mu$. The changes to the working condition cannot be observed except for a failure which is detected immediately. The machine is replaced by a new one upon failure or upon having worked during a time T, whichever occurs first. Each planned replacement involves a fixed cost of $J_1 > 0$, whereas a replacement because of a failure involves a fixed cost of $J_2 > 0$. The replacement time is negligible in both cases. Also, the system incurs an operating cost of $a_i > 0$ for each time unit the system is operating in working condition i. Use the result of exercise 1.14 to verify that the long-run average cost per unit time is given by

$$\left\{ T \sum_{k=0}^{N-1} p_k + \frac{N}{\mu}\left(1 - \sum_{k=0}^{N} p_k\right) \right\}^{-1}$$

$$\times \left\{ J_1 + (J_2 - J_1) \sum_{k=0}^{N-1} p_k + \sum_{k=0}^{N-1} p_k \sum_{i=1}^{k+1} a_i \frac{T}{k+1} + \sum_{k=N}^{\infty} p_k \sum_{i=1}^{N} a_i \frac{T}{k+1} \right\},$$

where $p_k = e^{-\mu T}(\mu T)^k/k!$, $k \geqslant 0$. (This problem is motivated by one in Luss, 1976.)

1.21 Consider the continuous-review inventory model dealt with in section 1.7. Verify for the backlog case that under an (s, S) rule the probability of a stockout occurrence during a cycle may be approximated by

$$\int_s^\infty (x - s)\eta(x)\,dx - \int_s^\infty (x - s)\xi(x)\,dx.$$

Here a stockout is said to occur when the on-hand inventory drops from a positive level to the zero level. Next argue that for the service level constraint requiring that the average number of stockout occurrences is not more than a specific value γ, the reorder point s may be determined approximately from

$$\int_s^\infty (x - s)\eta(x)\,dx - \int_s^\infty (x - s)\xi(x)\,dx = \frac{\gamma}{\lambda\mu_1}\left(S - s + \frac{\sigma_1^2 + \mu_1^2}{2\mu_1}\right).$$

1.22 In a periodic review inventory system the demands for a single product in the successive periods are independent random variables having a common probability density with mean μ_1 and standard deviation σ_1. Excess demand is backlogged. The inventory position is reviewed every R periods with R a given positive integer. At each review the inventory position is ordered up to the level S. In other words, at each review a replenishment order equal to the size of the preceding review time demand is placed. The lead time of a replenishment order is constant and equals periods with L a non-negative integer. For the service level constraint requiring that the fraction of demand to be satisfied directly from stock on hand equals a specific value β, show that the order-up-to-level S may be determined approximately from the equation

$$\int_S^\infty (x - S)\eta(x)\,\mathrm{d}x = (1 - \beta)R\mu_1,$$

where $\eta(x)$ is the probability density of the lead time plus review time demand. In case the demand density $\eta(x)$ is approximated by a normal density, verify that the above equation reduces to the easily solvable equation

$$\sigma(\eta)I(k) = (1 - \beta)R\mu_1$$

when using the representation $S = \mu(\eta) + k\sigma(\eta)$ with $k\sigma(\eta)$ denoting the safety stock. Here $\mu(\eta) = (R + L)\mu_1$ and $\sigma(\eta) = \sqrt{R + L}\,\sigma_1$, while the normal loss integral $I(k)$ is given in formula (1.62) in section 1.7.

1.23 A chemical is produced by a continuous processing plant at a constant rate of σ per unit time. The product has a fixed lifetime of m such that it perishes after m time units, and must be outdated. Demand requests for the product occur according to a Poisson process with rate λ, where the request size has a probability density $b(x)$ with mean μ_1. It is assumed that $\sigma > \lambda\mu_1$. The inventory is issued oldest first, and any demand in excess of current inventory is lost. Demonstrate how results for the lost-sales production/inventory model controlled by an (m, m) rule may be used to determine approximations to operating characteristics such as the expected outdates per unit time and the expected shortages per unit time. (This problem is based on Graves, 1982.)

BIBLIOGRAPHIC NOTES

The very readable monograph of Cox (1962) contributed much to the popularization of renewal theory. A good account of renewal theory can also be found in the texts of Heyman and Sobel (1982), Karlin and Taylor (1975) and Ross (1983). A basic paper on renewal theory and regenerative processes is that of Smith (1958), a paper which recognized the usefulness of renewal–reward processes in the analysis of applied probability problems. Also, the book of Ross (1970) was influential in promoting the application of renewal–reward processes. In section 1.3 the examples 1.6 and 1.7 dealing with the application of renewal theory in maintenance and reliability are adapted from Taylor (1975) and Barlow and Proschan (1975). In section 1.5 the example 1.12 is taken from the paper of Yadin and Naor (1963) which initiated the study of simple control rules for queueing systems. The example 1.13 is based on Heyman (1977) and Levy and Yechiali (1975), while the example 1.14 is adapted from Tijms (1976). In section 1.6 the recursion scheme for the calculation of compound Poisson probabilities is due to Adelson (1966). In section 1.7 the analysis of (s, S) inventory systems with service level constraints follows Tijms and Groenevelt (1984); see also the excellent text of Silver and Peterson (1985) for further work on inventory models. In section 1.8 the discussion of the fundamental property that Poisson arrivals see time averages is based on Wolff (1982). In section 1.9 the powerful technique of deriving asymptotic estimates for ruin and waiting time probabilities comes from Feller (1971). Section 1.10 dealing with production/inventory models with service level constraints is based on De Kok, Tijms and Van Der Duyn Schouten (1984, 1985); see also De Kok (1985) and Doshi, Van Der Duyn Schouten and Talman (1978).

REFERENCES

Adelson, R. M. (1966). 'Compound Poisson distributions', *Operat. Res. Quart.*, **17**, 73–75.

Barlow, R. E., and Proschan, F. (1975). *Statistical Theory of Reliability and Life Testing*, Holt, Rinehart and Winston, New York.

Carter, G., and Ignall, E. J. (1975). 'Virtual measures: a variance reduction technique for simulation', *Management Sci.*, **21**, 607–616.

Cox, D. R. (1962). *Renewal Theory*, Methuen, London.

De Kok, A. G. (1986). *Production-Inventory Control Models, Approximations and Algorithms*, CWI Tract No. 22, CWI, Amsterdam.

De Kok, A. G., Tijms, H. C., and Van Der Duyn Schouten, F. A. (1984). 'Approximations for the single-product production-inventory problem with compound Poisson demand and service level constraints', *Adv. Appl. Prob.*, **16**, 378–401.

De Kok, A. G., Tijms, H. C., and Van Der Duyn Schouten, F. A. (1985). 'Inventory levels to stop and restart a single machine producing one product', *European J. Operat. Res.*, **20**, 239–247.

Delves, L. M., and Walsh, J. (1974). *Numerical Solution of Integral Equations.*, Clarendon Press, Oxford.

Doshi, B. T., Van Der Duyn Schouten, F. A., and Talman, A. J. J. (1978). 'A production-inventory control model with a mixture of back-orders and lost-sales', *Management Sci.*, **24**, 1078–1086.

Feller, W. (1971). *An Introduction to Probability Theory and Its Applications*, Vol. II, 2nd ed., Wiley, New York.

Graves, S. C. (1982). 'Application of queueing theory to continuous perishable inventory systems', *Management Sci.*, **28**, 400–404.

Heyman, D. P. (1977). 'The T-policy for the M/G/1 queue', *Management Sci.*, **23**, 775–778.

Heyman, D. P., and Sobel, M. J. (1982). *Stochastic Models in Operations Research*, Vol. I: *Stochastic Processes and Operating Characteristics*, McGraw-Hill, New York.

Karlin, S., and Taylor, H. M. (1975). *A First Course in Stochastic Processes*, 2nd ed., Academic Press, New York.

Keilson, J. (1969). 'A simple algorithm for contract acceptance', *Opsearch*, **7**, 157–166.

Levy, Y., and Yechiali, U. (1975). 'Utilization of idle time in an M/G/1 queueing system', *Management Sci.*, **22**, 202–211.

Luss, H. (1976). 'Maintenance policies when deterioration can be observed by inspections', *Operat. Res.*, **24**, 359–366.

Ross, S. M. (1970). *Applied Probability Models with Optimization Applications*, Holden-Day, San Francisco.

Ross, S. M. (1983). *Stochastic Processes*, Wiley, New York.

Ross, S. M. (1985). *Introduction to Probability Models*, 3rd ed., Academic Press, New York.

Silver, E. A., and Peterson, R. (1985). *Decision Systems for Inventory Management and Production Planning*, 2nd ed., Wiley, New York.

Smith, W. L. (1958). 'Renewal theory and its ramifications', *J. Roy. Statist. Soc. B.*, **20**, 243–302.

Taylor, H. M. (1975). 'Optimal replacement under additive damage and other failure models', *Naval Res. Logist. Quart.*, **22**, 1–18.

Tijms, H. C. (1976). 'Optimal control of the workload in an M/G/1 queueing system with removable server', *Math. Operationsforsch. u. Statist.*, **7**, 933–943.

Tijms, H. C., and Groenevelt, H. (1984). 'Simple approximations for the reorder point in periodic review and continuous review (s, S) inventory systems with service level constraints', *European J. Operat. Res.*, **17**, 175–190.

Wolff, R. W. (1982). 'Poisson arrivals see time averages', *Operat. Res.*, **30**, 223–231.

Yadin, M., and Naor, P. (1963). 'Queueing systems with removable service station'. *Operat. Res. Quart.*, **14**, 393–405.

CHAPTER 2

Markov chains with operations research and computer science applications

2.0 INTRODUCTION

The notion of what is nowadays called a Markov chain was devised by the Russian mathematician A. A. Markov when at the beginning of this century he investigated the alternation of vowels and consonants in Poeshkin's poem *Onegin*. Therefore he developed a probability model in which the outcomes of successive trials are allowed to be dependent on each other such that each trial depends only on its immediate predecessor. This model, being the simplest generalization of the probability model of independent trials, appeared to give an excellent description of the alternation of vowels and consonants and enabled Markov to calculate a very accurate estimate of the frequency at which consonants occur in the above-mentioned poem of Poeshkin.

The Markov model is no exception to the rule that simple models are often the most useful models for analysing practical problems. A Markov process allows us to model the uncertainty in many real-world systems that evolve dynamically in time. The basic concepts of a Markov process are those of a *state* of a system and a state *transition*. In specific applications the modelling 'art' is to find an adequate state description such that the associated stochastic process has indeed the Markovian property that the knowledge of the present state is sufficient to predict the future stochastic behaviour of the process. The theory of Markov processes has applications to a wide variety of fields including biology, economics and physics among others, but in this chapter we primarily deal with applications to operations research and computer science. As in the previous chapter, the emphasis in this chapter is again on basic results and modelling aspects. In order not to distract the reader's attention from the main ideas, we present only heuristic derivations providing insights, in the safe knowledge that there are many excellent texts in which the interested reader may find detailed, rigorous proofs. Rather than concentrating on proofs, we discuss a wide variety of practical problems that can be solved by Markov chain analysis, where in each problem much attention is paid to an appropriate identification of the state and the state transition. The student cannot be urged enough to try the problems at the end of this chapter in order to acquire skills to model new situations. The modelling is often more difficult than the mathematics!

In section 2.1 we consider discrete-time Markov chains in which state transitions can occur only at given, discrete points in time, and we discuss a number of basic results. Most of these results concern the analysis of the long-run behaviour of the process, in particular the calculation of the long-run frequencies at which the various states occur. In section 2.2 we give an application of the discrete-time Markov chain model to insurance, inventory, queueing and buffer design.

In sections 2.3 and 2.4 we consider the continuous-time analogue of the discrete-time Markov chain. In the continuous-time Markov chain the holding times in the various states are exponentially distributed rather than deterministic, but the state transitions are again governed by a Markov chain. Equivalently, a continuous-time Markov chain can be represented by infinitesimal transition rates (generalizing the Poisson process). This representation allows us to work with the flowrate equation technique. This technique is easy to visualize by a graphical representation and is widely used by practitioners. In section 2.3 we focus on the calculation of the state probabilities by using the flow-equating technique, but we also pay some attention to the powerful uniformization technique for analysing the transient behaviour of the process (including the calculation of the first-passage time distributions). In section 2.4 we model and analyse several practical problems ranging from a cash balance problem via a reliability design problem to a resource allocation problem in data communication. In several of these problems it will be seen that the measures of system performance are (fairly) insensitive to the distributional form of the underlying random phenomena and require only their average values. In section 2.5 we provide a general insight into the insensitivity phenomenon and discuss some stochastic networks having an insensitive product-form solution for the state probabilities.

2.1 DISCRETE-TIME MARKOV CHAINS

A discrete-time Markov chain is a stochastic process which is the simplest generalization of a sequence of independent random variables. A Markov chain is a random sequence in which the dependency of the successive events goes back only one unit in time. In other words, the future probabilistic behaviour of the process depends only on the present state of the process and is not influenced by its past history. This is called the *Markovian* property. Through the very simple structure of Markov chains, these processes are extremely useful in a wide variety of practical probability problems.

Formally, let $\{X_n, n = 0, 1 \ldots\}$ be a sequence of random variables with a *discrete* state space I. We interpret the random variable X_n as the state of a system at time n. The set of possible values of the process is finite or countably infinite.

Definition 2.1

The stochastic process $\{X_n, n = 0, 1, \ldots\}$ is called a discrete-time Markov chain if, for each $n = 0, 1, \ldots$,

$$P\{X_{n+1} = i_{n+1} | X_0 = i_0, \ldots, X_n = i_n\} = P\{X_{n+1} = i_{n+1} | X_n = i_n\} \quad (2.1)$$

for all possible values of $i_0, \ldots, i_{n+1}$.

In the sequel we consider only Markov chains with time-homogeneous transition probabilities, that is we assume that

$$P\{X_{n+1} = j | X_n = i\} = p_{ij}, \qquad i, j \in I,$$

independently of the time parameter n. The probabilities p_{ij} are called the *one-step transition probabilities* and satisfy

$$p_{ij} \geqslant 0, \qquad i, j \in I, \qquad \text{and} \qquad \sum_{j \in I} p_{ij} = 1, \qquad i \in I.$$

The Markov chain $\{X_n, n = 0, 1, \ldots\}$ is completely determined by the probability distribution of the initial state X_0 and the one-step transition probabilities p_{ij}.

Example 2.1   Operating characteristics for an (s, S) inventory system

Suppose a single-product inventory system in which the demands for the product in the successive weeks $n = 0, 1, \ldots$ are independent and identically distributed random variables. The probability of a demand for j units during a week is given by $\varphi(j), j = 0, 1, \ldots$, where the first moment of the weekly demand is finite and positive. It is assumed that any demand in excess of current inventory is lost. The inventory position is reviewed only at the beginning of each week and at each review there is an opportunity for placing a replenishment order. It is supposed that the delivery of each replenishment order is instantaneous. The inventory level is controlled by an (s, S) rule with $0 \leqslant s < S$. Under this rule the inventory position is ordered up to the level S if at a review the inventory position is at or below the reorder point s and no ordering is done otherwise.

Defining, for $n = 0, 1, \ldots$,

$X_n =$ the inventory position at the beginning of the nth week just after delivery of a replenishment order (if any),

the process $\{X_n, n = 0, 1, \ldots\}$ is a discrete-time Markov chain with state space $I = \{s + 1, \ldots, S\}$. The Markovian property (2.1) is an immediate consequence of the independence of the weekly demands and the form of the control rule used. Specifically, denoting by D_n the demand to occur during the nth week, we have

$$X_{n+1} = \begin{cases} X_n - D_n & \text{if } X_n - D_n > s, \\ S & \text{if } X_n - D_n \leqslant s. \end{cases}$$

Alternatively, we could have defined as a Markov chain the process describing the inventory position at the beginning of a week just before any ordering. It will be seen later that the Markov chain $\{X_n\}$ describing the inventory position just after any ordering is more convenient from a computational point of view.

The one-step transition probabilities p_{ij} of the Markov chain $\{X_n\}$ are easily found. We distinguish between the cases of $j \neq S$ and $j = S$. For the starting state i with $s + 1 \leqslant i < S$, we have

$$p_{ij} = \begin{bmatrix} P\{\text{weekly demand} = i - j\} = \varphi(i - j), & s + 1 \leqslant j \leqslant i, \\ P\{\text{weekly demand} \geqslant i - s\} = \sum_{k \geqslant i-s} \varphi(k), & j = S, \\ 0, & \text{otherwise.} \end{bmatrix} \quad (2.2)$$

Similarly, we find for the starting state $i = S$,

$$p_{Sj} = \begin{bmatrix} \varphi(S - j), & s + 1 \leqslant j \leqslant S - 1, \\ \varphi(0) + \sum_{k \geqslant S-s} \varphi(k), & j = S. \end{bmatrix} \quad (2.3) \qquad \square$$

The following example illustrates the powerful technique of *embedded Markov chains*. Many stochastic processes can be analysed by using properly chosen embedded stochastic processes that are discrete-time Markov chains.

Example 2.2 *A single-server queueing system with general interarrival times and exponential service times*

Consider a single-server station at which customers arrive according to a renewal process with interarrival times $A_1, A_2, \ldots$. A customer who finds upon arrival that the server is idle enters service immediately; otherwise the customer waits in line. The service times of the successive customers are independent random variables having a common exponential distribution with mean $1/\mu$, and are also independent of the arrival process. It is assumed that the average service time is smaller than the average interarrival time. A customer leaves the system upon service completion. This queueing system is usually abbreviated as the GI/M/1 queue.

Define $X(t)$ as the number of customers in the system at time t. The stochastic process $\{X(t), t \geqslant 0\}$ does not possess the Markovian property that the future behaviour of the process only depends on its present state. Clearly, to predict the future behaviour of the process, the knowledge of the number of customers present does not suffice in general but the knowledge of the time elapsed since the last arrival is required too. Note that, by the memoryless property of the exponential distribution, the elapsed service time of the service in progress (if any) is not relevant. Actually by the lack of memory of the exponential services, we can find an embedded Markov chain for the continuous-time process $\{X(t)\}$. Consider the process at the epochs when customers arrive. At these epochs the time elapsed since the last arrival is known and equals zero. Thus, letting, for $n = 1, 2, \ldots$,

X_n = the number of customers present just prior to the nth arrival epoch

(with $X_0 = 0$, by convention), the embedded stochastic process $\{X_n, n = 0, 1, \ldots\}$ is a discrete-time Markov chain. This Markov chain has the countably infinite state space $I = \{0, 1, \ldots\}$. A relation between the limiting distributions of the discrete-time process $\{X_n\}$ and the continuous-time process $\{X(t)\}$ will be discussed in section 4.1 of chapter 4.

Denoting by C_{n+1} the number of customers served during the interarrival time A_{n+1} between the nth and $(n+1)$th customer, we have that $X_{n+1} = X_n + 1 - C_{n+1}$. Clearly, the distribution of C_{n+1} depends on X_n. This distribution follows by observing that as long as the server is busy the number of service completions is described by a Poisson process with rate μ, since the service times are independent and have a common exponential distribution with mean $1/\mu$. To give the one-step transition probabilities of the Markov chain $\{X_n\}$, we distinguish between the cases $j \neq 0$ and $j = 0$. We assume for ease that the common probability distribution function of the interarrival times has a probability density g. Using the law of total probability (see appendix A) with the interarrival time as the conditioning variable, we obtain, for $i \geqslant 0$ and $1 \leqslant j \leqslant i + 1$,

$$p_{ij} = P\{X_{n+1} = j \mid X_n = i\}$$

$$= \int_0^\infty P\{i + 1 - j \text{ service completions occur during } A_{n+1} \mid A_{n+1} = t\} g(t) \, dt$$

$$= \int_0^\infty e^{-\mu t} \frac{(\mu t)^{i+1-j}}{(i+1-j)!} g(t) \, dt. \tag{2.4}$$

The probability p_{i0} is easiest to compute from $p_{i0} = 1 - \sum_{j=1}^{i+1} p_{ij}$, $i \geqslant 0$. $\qquad \square$

The long-run behaviour of the Markov chain

The analysis below focuses on the long-run behaviour of the Markov chain $\{X_n\}$ when the process is considered over an *infinite* number of transitions in the future. One might expect that the influence of the initial state of the process will fade out ultimately. However, this is in general not true. As an illustration, suppose that the state space has two non-empty disjoint sets each having the closedness property that the process stays for ever in the set of states once it is in that set. Then the initial state will typically be reflected in the long-run behaviour of the process. In most practical situations, however, the Markov chain has some regeneration state with the property that this state will be reached with probability 1 from any other state. Then the effect of the initial state will fade out ultimately. From now on we make the following assumption.

Assumption 2.1

The Markov chain $\{X_n, n = 0, 1, \ldots\}$ has some regeneration state r (say) such that

$$E(N \mid X_0 = i) < \infty \qquad \text{for all } i \in I,$$

where N is the first time beyond epoch 0 that the process makes a transition into state r.

Under this assumption we discuss the long-run behaviour of the Markov chain. For that purpose, define for each $n = 1, 2, \ldots$ the n-step transition probabilities

$$p_{ij}^{(n)} = P\{X_n = j \mid X_0 = i\}, \qquad i, j \in I,$$

where $p_{ij}^{(1)} = p_{ij}$. That is $p_{ij}^{(n)}$ is the probability that n time units later the process will be in state j, given that the present state is i.

In general, $\lim_{n \to \infty} p_{ij}^{(n)}$ need not exist. To see this, consider the trivial Markov chain with two states $i = 1, 2$ and one-step transition probabilities $p_{12} = p_{21} = 1$ and $p_{11} = p_{22} = 0$. Then the probabilities $p_{ij}^{(n)}$, $n \geq 1$, alternate between 0 and 1, and so $\lim_{n \to \infty} p_{ij}^{(n)}$ does not exist. The periodicity of this Markov chain is the reason that the n-step transition probabilities have no ordinary limits. Here a Markov chain is called *periodic* if the state space can be divided into a finite number of non-empty disjoint sets $I_1, \ldots, I_P$ with $P \geq 2$ such that each one-step transition from a state in the set I_k is into a state of the set I_{k+1} with $I_{P+1} = I_1$. A Markov chain which is not periodic is called *aperiodic*.

Under the above assumption 2.1 on the Markov chain $\{X_n\}$, we have that

$$\lim_{n \to \infty} \frac{1}{n} \sum_{k=1}^{n} p_{ij}^{(k)} = \pi_j \qquad \text{(say)} \tag{2.5}$$

exists for all $i, j \in I$ and is independent of the initial state i. Also,

$$\pi_j \geq 0, \qquad j \in I, \qquad \text{and} \qquad \sum_{j \in I} \pi_j = 1. \tag{2.6}$$

In case the Markov chain is *aperiodic*, then in addition

$$\lim_{n \to \infty} p_{ij}^{(n)} = \pi_j \qquad \text{for all} \qquad i, j \in I. \tag{2.7}$$

A proof of these statements can be found in Chung (1967). The probabilities π_j, called the *limiting* or *steady-state* probabilities, have also the following remarkable interpretation. For each $j \in I$, it holds with probability 1 that

$$\text{The long-run fraction of time the process is in state } j = \pi_j, \qquad (2.8)$$

independently of the initial state $X_0 = i$. To prove this result, we invoke the theory of renewal–reward processes. Fix state $j \in I$ and superimpose the following reward structure on the Markov chain. A reward 1 is earned each time the process makes a transition to state j; otherwise no rewards are earned. Then the long-run average reward per unit time equals the long-run fraction of time the process is in state j. In view of assumption 2.1 it is easily seen that the long-run average reward per unit time for the initial state $X_0 = i$ is the same as for the initial state $X_0 = r$, and thus the left side of (2.8) is independent of the initial state $X_0 = i$. Suppose now that the initial state is $X_0 = r$. Define a cycle as the time elapsed between two consecutive transitions to state r. Using assumption 2.1 and the Markovian property, it follows that both the lengths of the successive cycles and the total rewards earned during the cycles are independent and identically distributed random variables with finite expectations. Thus, letting Z_n be the cumulative reward earned up to time n, we have by the relations (1.14) and (1.15) in section 1.2 of chapter 1 that the average reward per unit time satisfies

$$\lim_{n \to \infty} \frac{Z_n}{n} = \lim_{n \to \infty} \frac{E(Z_n)}{n} \qquad \text{with probability } 1. \qquad (2.9)$$

To evaluate the right side of (2.9), we represent Z_n as

$$Z_n = \sum_{k=1}^{n} \delta_j(X_k),$$

where the function δ_j is defined by

$$\delta_j(i) = \begin{cases} 1 & \text{if } i = j, \\ 0 & \text{if } i \neq j. \end{cases}$$

Using $E(\delta_j(X_k)) = P\{X_k = j \mid X_0 = r\} = p_{rj}^{(k)}$, it follows that

$$\lim_{n \to \infty} \frac{E(Z_n)}{n} = \lim_{n \to \infty} \frac{1}{n} \sum_{k=1}^{n} p_{rj}^{(k)} = \pi_j,$$

which completes the verification of (2.8). Actually, the relation (2.8) is a special case of a more general result given in Chung (1967). To state this result, consider a non-negative function $f(i)$, $i \in I$, and suppose that a reward $f(i)$ is earned each time the process makes a transition to state $i \in I$. In addition to assumption 2.1, we make the following assumption.

Assumption 2.2

The expectation of the total rewards earned up to the first time beyond epoch 0 at which the process makes a transition into state r is finite for each initial state $X_0 = i$.

This assumption follows automatically from assumption 2.1 in case the reward function f is bounded. Then, we have for the long-run average reward per unit time that

$$\lim_{n \to \infty} \frac{1}{n} \sum_{k=1}^{n} f(X_k) = \sum_{j \in I} f(j)\pi_j \qquad \text{with probability} \quad 1, \qquad (2.10)$$

independently of the initial state $X_0 = i$. This useful result is a consequence of the ergodic theorem 2 on p. 92 in Chung (1967) and the assumptions 2.1 and 2.2.

It is needless to say that the quantities π_j, $j \in I$, are extremely important for practical applications. We state without proof that the numbers π_j, $j \in I$, are the *unique* non-negative solution to the following system of linear equations:

$$x_j = \sum_{k \in I} x_k p_{kj}, \qquad j \in I, \qquad (2.11)$$

$$\sum_{j \in I} x_j = 1. \qquad (2.12)$$

For a proof see Chung (1967). The equations (2.11) are called the *balance* or *equilibrium* equations and the equation (2.12) is called the *normalizing* equation. In fact it holds that a solution to only the balance equations (2.11) is uniquely determined up to a multiplicative constant to be found from the normalizing equation. A convenient and effective method for calculating the numerical solution to the linear equations (2.11) to (2.12) is the successive overrelaxation method described in appendix D.

We give some heuristic explanations of the balance equations (2.11). For that purpose, we first derive the *Chapman–Kolmogoroff forward equations* to compute the n-step transition probabilities. These equations are given by

$$p_{ij}^{(n+1)} = \sum_{k \in I} p_{ik}^{(n)} p_{kj}, \qquad n = 1, 2, \ldots; \quad i, j \in I. \qquad (2.13)$$

The probability distribution of the state that occurs a time $n + 1$ later can be computed by conditioning on the state that occurs a time n from now on. Using the law of total probability (see Appendix A) and the Markovian property (2.1), it follows that

$$P\{X_{n+1} = j \mid X_0 = i\} = \sum_{k \in I} P\{X_{n+1} = j \mid X_0 = i, X_n = k\} P\{X_n = k \mid X_0 = i\}$$

$$= \sum_{k \in I} P\{X_{n+1} = j \mid X_n = k\} P\{X_n = k \mid X_0 = i\},$$

which verifies (2.13). We are now in a position to show that the numbers π_j

satisfy the balance equations. Summing both sides of (2.13) over $n = 1, \ldots, t$ and dividing by t, we obtain after an interchange of the order of summation

$$\frac{1}{t} \sum_{n=1}^{t} p_{ij}^{(n+1)} = \sum_{k \in I} p_{kj} \frac{1}{t} \sum_{n=1}^{t} p_{ik}^{(n)}.$$

Letting $t \to \infty$, interchanging the order of limit and summation and using (2.5), we obtain the balance equations (2.11). We omit the (easy) proof that the interchange operation is allowed (for a finite state space I this is obvious).

Alternatively, the balance equations (2.11) can heuristically be argued by using the principle of conservation of flow of probability. By (2.8),

The long-run fraction of transitions that occur from state $j = \pi_j$.

Also,

The long-run fraction of transitions that occur into state j

$= \sum_{k \in I}$ the long-run fraction of transitions that occur from k to j

$= \sum_{k \in I} p_{kj} \times$ (the long-run fraction of transitions that occur from k)

$= \sum_{k \in I} p_{kj} \pi_k.$

Since the long-run transition rate from state j must be equal to the long-run transition rate into state j, we get the balance equations (2.11) for the steady-state probabilities π_j.

We now summarize the several interpretations for the limiting probabilities π_j. The extremely useful result (2.8) states that if we are to consider the process over a very large number of future transitions, then we can predict beforehand with almost certainty that the fraction of transitions occurring to state j will be π_j. In case the Markov chain is aperiodic, we have by (2.7) that π_j can also be interpreted as the probability that the process will be found in state j at an arbitrary point in time in the far distant future. This interpretation provides the following heuristic argument to memorize the equilibrium equations (2.11). The distribution of the state at time ∞ is computed by conditioning upon the state at time $\infty - 1$. Under the condition that the system is in state k at time $\infty - 1$, and this occurs with probability π_k, the system is in state j at time ∞ with probability p_{kj}. Thus, using the law of total probability, this heuristic reasoning leads to

$$\pi_j = P\{X_\infty = j\} = \sum_{k \in I} P\{X_\infty = j \mid X_{\infty-1} = k\} P\{X_{\infty-1} = k\}$$

$$= \sum_{k \in I} p_{kj} \pi_k.$$

2.2 APPLICATIONS OF DISCRETE-TIME MARKOV CHAINS

In this section we give some applications of discrete-time Markov chains to a variety of applied probability problems in operations research and computer science.

Example 2.1   (continued) Operating characteristics for an (s, S) inventory system

Operating characteristics of interest for the lost-sales (s, S) inventory system include the long-run frequency of ordering and the long-run fraction of demand that is lost. These operating characteristics can be expressed in terms of the steady-state probabilities π_j, $s + 1 \leqslant j \leqslant S$, of the Markov chain $\{X_n\}$ describing the inventory position at the beginning of a week just after any ordering. Note that this Markov chain satisfies assumption 2.1 since state S can be reached from any other state. To give the average ordering frequency, we first observe that the probability of placing a replenishment order in the next week equals $\sum_{k \geqslant j-s} \varphi(k)$ when the current inventory is j just after any ordering. Thus, using that π_j gives the long-run fraction of weeks having an on-hand inventory of j just after any ordering, we find

$$\text{The average ordering frequency} = \sum_{j=s+1}^{S} \pi_j \sum_{k \geqslant j-s} \varphi(k).$$

Also, noting that the expected amount of demand that will be lost during the coming week equals $\sum_{k > j}(k - j)\varphi(k)$ when the current inventory equals j just after any ordering, we obtain

$$\text{The average amount of demand that is lost per week} = \sum_{j=s+1}^{S} \pi_j \sum_{k > j}(k - j)\varphi(k).$$

Next the long-run fraction of demand that is lost follows as the ratio of the average amount of demand that is lost per week and the average weekly demand. Similarly, the reader may verify that the long-run fraction of weeks in which a stockout occurs equals $\sum_{j=s+1}^{S} \pi_j \sum_{k > j} \varphi(k)$.

It remains to specify the state probabilities π_j, $s + 1 \leqslant j \leqslant S$. In order to set up the balance equations for these probabilities, it is recommended that the heuristic but suggestive formula $P\{X_\infty = j\} = \sum_k P\{X_\infty = j \mid X_{\infty - 1} = k\} P\{X_{\infty - 1} = k\}$ is used to reason directly rather than to substitute formally the expressions for the one-step probabilities into the general form (2.11). We find the balance equations

$$\pi_j = \sum_{k=j}^{S} \varphi(k - j)\pi_k, \qquad\qquad s + 1 \leqslant j \leqslant S - 1,$$

$$\pi_j = \varphi(0)\pi_S + \sum_{k=s+1}^{S} \left\{ \sum_{l \geqslant k-s} \varphi(l) \right\} \pi_k, \qquad j = S,$$

together with the normalization equation

$$\sum_{j=s+1}^{S} \pi_j = 1.$$

This system of linear equations can recursively be solved by using the fact that a solution to only the balance equations is determined uniquely up to a multiplicative constant. Starting with $\bar{\pi}_S = 1$ (say), we recursively compute $\bar{\pi}_S \to \bar{\pi}_{S-1} \to \cdots \to \bar{\pi}_{s+1}$ from the balance equations. Next, applying the normalization equation, we obtain the desired probabilities π_j from $\pi_j = \bar{\pi}_j / \sum_{k=s+1}^{S} \bar{\pi}_k$, $s + 1 \leqslant j \leqslant S$. We finally remark that a recursive solution of the balance equations would not have been possible in case we had chosen, as the Markov chain, the process describing the inventory position at the beginning of a week just before any ordering. $\qquad \square$

Example 2.2 *(continued) A single-server queueing system with general interarrival times and exponential service times*

For the GI/M/1 queueing system we shall determine the probability distribution of the delay in queue of an arbitrary customer. Here it is assumed that customers are served in order of arrival. For the delay distribution we need the steady-state probabilities π_j, $j \geqslant 0$, of the Markov chain $\{X_n\}$ describing the number of customers present just prior to arrival epochs. We note that this Markov chain satisfies assumption 2.1 and is aperiodic. Define W_q as the delay in queue of an arbitrary customer to arrive when the system has been in operation for a very long time. Since π_j estimates the probability that j other customers will be found present by a customer to arrive in the far distant future, it follows by the law of total probability that

$$P\{W_q > t\} = \sum_{j=1}^{\infty} P\{\text{a customer is delayed more than a time } t \,|\, \text{the customer}$$
$$\text{finds upon arrival } j \text{ other customers present}\} \pi_j.$$

If an arriving customer finds the server busy, then by the lack of memory of the exponential distribution, the remaining service time of the service in progress has the same exponential distribution as the original service time. Hence the delay in queue of a customer finding upon arrival j other customers present is distributed as the sum of j independent exponentials with the same means $1/\mu$ and thus has an Erlang-j distribution. Using (1.22) in section 1.4 of chapter 1, it now follows that

$$P\{W_q > t\} = \sum_{j=1}^{\infty} \pi_j \sum_{k=0}^{j-1} e^{-\mu t} \frac{(\mu t)^k}{k!}, \qquad t \geqslant 0. \tag{2.14}$$

The steady-state probabilities π_j, $j \geqslant 0$, are easily determined. Using the one-step transition probabilities p_{ij} given by (2.4), we obtain, from (2.11) and (2.12),

$$\pi_j = \sum_{k=j-1}^{\infty} \pi_k \int_0^{\infty} e^{-\mu t} \frac{(\mu t)^{k+1-j}}{(k+1-j)!} g(t)\, dt, \qquad j \geqslant 1, \qquad (2.15)$$

$$\sum_{j=0}^{\infty} \pi_j = 1. \qquad (2.16)$$

We have omitted the redundant equation for π_0. By probabilistic arguments it can be shown that $\pi_j = c\omega^j$, $j \geqslant 0$, for some constants $c, \omega > 0$. We do not give these arguments, but just substitute a solution of this geometric form in (2.15). This yields, after some algebra,

$$c\omega^j = c \int_0^{\infty} e^{-\mu t} \omega^{j-1} e^{\omega \mu t} g(t)\, dt, \qquad j \geqslant 1.$$

Hence

$$\omega = \int_0^{\infty} e^{-\mu t (1-\omega)} g(t)\, dt. \qquad (2.17)$$

It is not difficult to show that the equation (2.17) has a unique solution ω with $0 < \omega < 1$. This equation can numerically be solved by the standard Newton–Raphson method. Using (2.16), it follows that

$$c = 1 - \omega.$$

Since the equations (2.15) and (2.16) determine uniquely the steady-state probabilities, we have verified that

$$\pi_j = (1 - \omega)\omega^j, \qquad j = 0, 1, \ldots. \qquad (2.18)$$

Finally, using (2.18), we can simplify (2.14) to

$$P\{W_q > t\} = \omega e^{-\mu t (1-\omega)}, \qquad t \geqslant 0. \qquad (2.19)$$

As an application of the GI/M/1 queueing model, we consider an inventory system which periodically receives a scheduled delivery of some product being the output of a manufacturing process. The replenishments of the product occur at a constant rate of one unit every T time units. The demand process for the product is a Poisson process with a demand rate of μ items per unit time. Any demand in excess of the current inventory is lost. We are interested in the long-run average inventory on hand.

Some reflections show that the inventory process describing the number of items in stock may be identified with the process describing the number of customers present in a GI/M/1 queueing system with deterministic arrivals. The delivery of an item from the manufacturing process is to be interpreted as an arrival of a customer. The items in stock are to be interpreted as customers awaiting service completion where the service time of the item at the top of the stockpile is given by the time until the next occurrence of a demand. This identification is possible due to the assumption of excess demand being lost and the lack of memory of the Poisson demand process (why?). The long-run average inventory on hand is now obtained from the relations

$$\text{The average time an item is kept in stock} = \sum_{i=0}^{\infty} \frac{i+1}{\mu} \pi_i$$

and

The average inventory on hand = (the average replenishment rate of items) × (the average time an item is kept in stock).

Here the average replenishment rate of items is equal to $1/T$. The first relation is obvious by noting that π_i represents the probability of having i items in stock just before a replenishment is to occur and by using the lack of memory of the Poisson demand process. To see the second relation, suppose that a holding cost at a rate of 1 per unit time is incurred for each item kept in stock. Then the average holding cost per unit time is just the average inventory on hand. On the other hand, the average holding cost per unit time equals the average number of items delivered per unit time multiplied by the average holding cost incurred per item. The latter cost is obviously equal to the average time an item is kept in stock, and so we obtain the last relation stated above. Thus we find

$$\text{The average inventory on hand} = \frac{1}{T\mu(1-\omega)},$$

with ω being the unique positive solution to $\omega = e^{-\mu T(1-\omega)}$. □

Example 2.3   A vehicle insurance problem

A transport firm has effected an insurance contract for a fleet of vehicles. The premium payment is due at the beginning of each year. There are four possible premium classes with a premium payment of P_i in class i where $P_{i+1} < P_i$, $i = 1, 2, 3$. The size of the premium depends on the previous premium and the claim history during the past year. In case no damage is claimed in the past year and the previous premium is P_i, the next premium payment is P_{i+1} (with $P_5 = P_4$, by convention); otherwise the highest premium P_1 is due. Since the insurance contract is for a whole fleet of vehicles, the transport firm has obtained the option to decide only at the end of the year whether the accumulated damage during that year should be claimed or not. In case a claim is made, the insurance company compensates the accumulated damage minus an own risk which amounts to r_i for premium class i. The total damages in the successive years are independent random variables having a common probability distribution function F with density f.

We are interested in the claim limits which minimize the yearly cost for the transport firm. Under a claim rule $(\alpha_1, \ldots, \alpha_4)$ the transport firm claims only damages larger than α_i when the current premium class is i. To find the optimal claim limits, we first derive for a given claim rule $(\alpha_1, \ldots, \alpha_4)$ the average cost per year and next minimize this cost function with respect to the parameters $\alpha_1, \ldots, \alpha_4$. Consider a given claim rule $(\alpha_1, \ldots, \alpha_4)$ where $\alpha_i > r_i$, $i = 1, \ldots, 4$. The average cost per year can be obtained by considering the Markov chain which describes the evolution of the premium class for the transport firm. Letting

$X_n = $ the premium class for the transport firm in the nth year,

the stochastic process $\{X_n\}$ is a Markov chain with four possible states $i = 1, \ldots, 4$. The one-step transition probabilities p_{ij} are easily found. A one-step transition from states i to state 1 occurs only if at the end of the present year a damage is claimed; otherwise a transition from state i to state $i + 1$ occurs with state $5 \equiv$ state 4. Since for premium class i only cumulative damages larger than α_i are claimed, it follows that

$$p_{i1} = 1 - F(\alpha_i), \qquad i = 1, \ldots, 4,$$
$$p_{i,i+1} = F(\alpha_i), \qquad i = 1, 2, 3 \quad \text{and} \quad p_{44} = F(\alpha_4).$$

The other one-step transition probabilities p_{ij} are equal to zero. By (2.11) and (2.12) the steady-state probabilities π_j, $1 \leqslant j \leqslant 4$, of the Markov chain $\{X_n\}$ are the unique non-negative solution to the following linear equations:

$$\pi_4 = F(\alpha_3)\pi_3 + F(\alpha_4)\pi_4,$$
$$\pi_3 = F(\alpha_2)\pi_2,$$
$$\pi_2 = F(\alpha_1)\pi_1,$$
$$\pi_1 = \{1 - F(\alpha_1)\}\pi_1 + \{1 - F(\alpha_2)\}\pi_2 + \{1 - F(\alpha_3)\}\pi_3 + \{1 - F(\alpha_4)\}\pi_4,$$

together with the normalizing equation $\pi_1 + \pi_2 + \pi_3 + \pi_4 = 1$. These linear equations can be solved recursively. Taking π_4 as the parameter, we recursively compute π_3, π_2 and π_1 from the first three equations and next apply the normalizing equation. Then we determine the expected costs incurred during a year in which premium P_j is paid. Denoting these costs by $c(j)$, we have by (2.8) that the long-run average cost per year equals, with probability 1,

$$g(\alpha_1, \ldots, \alpha_4) = \sum_{j=1}^{4} c(j)\pi(j).$$

The one-year costs $c(j)$ consist of the premium P_j and any damages not compensated that year by the insurance company. By conditioning on the

Table 2.1   The optimal claim limits and the minimal
average cost

	Gamma			lognormal		
	$c_D^2 = 1$	$c_D^2 = 4$	$c_D^2 = 25$	$c_D^2 = 1$	$c_D^2 = 4$	$c_D^2 = 25$
α_1^*	5908	6008	6280	6015	6065	6174
α_2^*	7800	7908	8236	7931	7983	8112
α_3^*	8595	8702	9007	8717	8769	8890
α_4^*	8345	8452	8757	8467	8519	8640
g^*	9058	7698	6030	9174	8318	7357

cumulative damage in a year, it follows that

$$c(j) = P_j + \int_0^{\alpha_j} s f(s)\,ds + r_j \int_{\alpha_j}^{\infty} f(s)\,ds.$$

The optimal claim limits follow by minimizing the function $g(\alpha_1, \ldots, \alpha_4)$ with respect to the parameters $\alpha_1, \ldots, \alpha_4$. Efficient numerical procedures are widely available to minimize a function of several variables. In Table 2.1 we give for a number of examples the optimal claim limits $\alpha_1^*, \ldots, \alpha_4^*$ together with the minimal average cost g^*. In all examples we take

$$P_1 = 10,000, \qquad P_2 = 7500, \qquad P_3 = 6000, \qquad P_4 = 5000,$$
$$r_1 = 1500, \qquad r_2 = 1000, \qquad r_3 = 750, \qquad r_4 = 500.$$

The average damage size is 5000 in each example, but the squared coefficient of variation of the damage size D is varied as $c_D^2 = 1$, 4 and 25. To see the effect of the shape of the probability density of the damage size on the claim limits, we take the gamma distribution and the lognormal distribution, both having the same first two moments. In particular, the minimal average cost becomes increasingly sensitive to the distributional form of the damage size D when c_D^2 gets larger.

Example 2.4   Buffer design for a communication system with multiple transmission channels

Consider a communication system with a finite capacity buffer and multiple transmission channels. Messages arrive at the buffer according to a Poisson process with rate λ. The messages are first stored in the buffer which has only capacity for N messages. Each message which finds upon arrival that the buffer is full is lost and does not influence the system. At fixed clock times $t = 0, 1, \ldots$ messages are taken out from the buffer and are synchronously transmitted. Each transmission channel can transmit only one message at a time. The transmission time is constant and equals one time unit for each message. There are c transmission channels and so at most c messages can synchronously be transmitted. The transmission of a message can only start at a clock time so that a message which finds upon arrival in the buffer that a transmission line is idle has to wait until a subsequent clock time.

We are interested in the long-run fraction of arrivals which find upon arrival that the buffer is full and thus overflow. In practice one typically wishes to design the buffer size in such a way that the overflow probability is extremely small (say, on the order of 10^{-6}).

To find the overflow probability for a given buffer size N with $N \geqslant c$, we define, for each $n = 0, 1, \ldots,$

$X_n =$ the number of messages in the buffer at clock time n
just prior to transmission.

Using the lack of memory of the Poisson arrival process of the messages, it follows that the process $\{X_n\}$ is a discrete-time Markov chain with state space $I = \{0, 1, \ldots, N\}$. Denoting by A_n the number of message that will arrive at the buffer during the period between the clock times n and $n + 1$, we have

$$X_{n+1} = \begin{cases} \min(A_n, N) & \text{if } X_n < c \\ \min(X_n - c + A_n, N) & \text{if } X_n \geqslant c. \end{cases}$$

Since the messages arrive according to a Poisson process with rate λ, the probability of k arrivals during the transmission time of one time unit is given by

$$a_k = e^{-\lambda} \frac{\lambda^k}{k!}, \qquad k = 0, 1, \ldots.$$

Denoting by p_{ij} the one-step transition probabilities of the Markov chain $\{X_n\}$ we have, for $0 \leqslant i < c$,

$$p_{ij} = \begin{cases} a_j, & 0 \leqslant j < N, \\ \sum_{k=N}^{\infty} a_k, & j = N, \end{cases}$$

while, for $c \leqslant i \leqslant N$,

$$p_{ij} = \begin{cases} a_{j-i+c} & i - c \leqslant j < N, \\ \sum_{k=N-i+c}^{\infty} a_k, & j = N. \end{cases}$$

The steady-state probabilities π_j, $0 \leqslant j \leqslant N$, of the Markov chain $\{X_n\}$ are the unique solution of the linear equations

$$\pi_j = a_j \sum_{k=0}^{c-1} \pi_k + \sum_{k=0}^{j} \pi_{k+c} a_{j-k}, \qquad\qquad 0 \leqslant j \leqslant N-c, \qquad (2.20)$$

$$\pi_j = a_j \sum_{k=0}^{c-1} \pi_k + \sum_{k=0}^{N-c} \pi_{k+c} a_{j-k}, \qquad\qquad N-c < j \leqslant N-1, \; (2.21)$$

$$\pi_N = \left(\sum_{j=N}^{\infty} a_j \right) \sum_{k=0}^{c-1} \pi_k + \sum_{k=0}^{N-c} \pi_{k+c} \left(\sum_{j=N-k}^{\infty} a_j \right), \qquad (2.22)$$

$$\sum_{j=0}^{N} \pi_j = 1. \qquad\qquad\qquad (2.23)$$

A simple and accurate method to solve these linear equations is the iterative method of successive overrelaxation described in appendix D. In the computations the infinite sums $\sum_{j=m}^{\infty} a_j$ in (2.22) are replaced by $1 - \sum_{j=0}^{m-1} a_j$, while the Poisson probabilities a_k are recursively computed.

We are now in a position to give the overflow probability or the long-run fraction of messages which overflow. Therefore observe that

The fraction of messages which overflow

$$= \frac{\text{(the average number of overflows per unit time)}}{\text{(the average number of messages arriving per unit time)}}.$$

Clearly, the average arrival rate of messages is λ. The average number of overflows per unit time can be derived in several ways. It is instructive to give both a rather straightforward derivation and a more subtle one. First, given that presently the buffer contains j messages just prior to transmission, it follows with the notation $j_c = \max(j-c, 0)$ that

The expected number of overflows in the next transmission time unit

$$= \sum_{k=N-j_c}^{\infty} (k-N+j_c)a_k = \lambda - N + j_c + \sum_{k=0}^{N-j_c} (N-j_c-k)a_k.$$

Hence, by (2.10), we have that, with probability 1,

The average number of overflows per unit time

$$= \sum_{j=0}^{N} \pi_j \left\{ \lambda - N + j_c + \sum_{k=0}^{N-j_c} (N-j_c-k)a_k \right\}.$$

Alternatively, since each message is either transmitted or overflows,

The average number of arrivals per unit time
= (the average number of messages served per unit time)
+ (the average number of overflows per unit time).

Since each transmission channel can serve only one message at a time,

The average number of messages served per unit time
= the average number of busy transmission channels.

Noting that π_j gives the fraction of clock times at which j messages are present, we have that, with probability 1,

The average number of busy channels $= \sum_{j=0}^{c-1} j\pi_j + c \sum_{j=c}^{N} \pi_j.$

The above relations together yield the fraction of messages which overflow

$$\prod_{\text{over}} = \frac{1}{\lambda}\left(\lambda - \sum_{j=0}^{c-1} j\pi_j - c \sum_{j=c}^{N} \pi_j \right). \tag{2.24}$$

In addition we find the following identity for the average number of overflows per unit time

$$\sum_{j=0}^{N} \pi_j \left\{ \lambda - N + j_c + \sum_{k=0}^{N-j_c} (N - j_c - k)a_k \right\} = \lambda - \left(\sum_{j=0}^{c-1} j\pi_j + c \sum_{j=c}^{N} \pi_j \right).$$

This identity could be used as an extra accuracy check when solving the linear equations (2.20) to (2.23). As an illustration we give in Table 2.2 for several values of $\rho = \lambda/c$ and c the minimal buffer size N needed to achieve an overflow probability less than 10^{-6}. In view of this small overflow probability, we have taken the extremely small accuracy number $\varepsilon = 10^{-9}$ in the stopping criterion of the successive overrelaxation method. Note from the results in Table 2.2 that N sharply increases for ρ close to 1.

We conclude this example with approximations for the operating characteristics $E(L_q)$ and $E(W_q)$ which are defined as the average number of messages in the buffer and the average delay in the buffer of a transmitted message. Assuming that for the given buffer size N the associated overflow probability is extremely small, we approximately have

$$E(L_q) \approx \sum_{j=c}^{N} (j - c)\pi_j + \frac{\lambda}{2}. \tag{2.25}$$

Table 2.2 The minimal buffer size for an overflow probability of 10^{-6}

$c\backslash\rho$	0.2	0.5	0.7	0.8	0.9
1	6	11	19	29	56
2	7	12	20	30	57
5	9	14	22	32	59
10	11	18	26	36	63
25	25	30	39	49	76

Next $E(W_q)$ can be determined from the exact relation

$$E(L_q) = \lambda(1 - \textstyle\prod_{\text{over}})E(W_q). \tag{2.26}$$

To argue (2.25), we approximate the time-average number of messages in the buffer by one-half the sum of the average number of messages in the buffer at the beginning and at the end of a transmission interval. The average number of messages left in the buffer at the beginning of a transmission interval equals the first term in the right side of (2.25). Also, since the overflow probability is negligible, approximately λ messages are added to the buffer content during the transmission interval. To verify (2.26) it is helpful to interpret $E(L_q)$ and $E(W_q)$ in terms of average costs by superimposing the following cost structure on the system. Suppose that for each message *entering* the buffer the system incurs a cost at a constant rate 1 per unit time as long as that message stays in the buffer. Then the system incurs a cost at rate j whenever j messages are in the buffer. Hence the average cost incurred by the system per unit time is $E(L_q)$, while the average cost incurred per entering message is $E(W_q)$. On the other hand,

    The average cost incurred by the system per unit time
      = (the average number of messages entering the buffer per unit time)
        × (the average cost incurred per entering message).   (2.27)

Since the fraction of messages entering the buffer is $1 - \prod_{\text{over}}$, it can be seen that the average arrival rate of entering messages equals $\lambda(1 - \prod_{\text{over}})$, yielding the desired result (2.26). The formula (2.26) is a special case of the fundamental law of Little that applies to almost all practical queueing systems (cf. also section 4.1 in chapter 4). The above reasoning used to find the relation between $E(L_q)$ and $E(W_a)$ is quite flexible and has many other interesting applications (see Ross, 1983, 1985).   ☐

2.3 CONTINUOUS-TIME MARKOV CHAINS

So far we have considered Markov processes in which the changes of the state only occurred at fixed times $t = 0, 1, \dots$. However, in numerous practical situations changes of the state may occur at each point of time. One of the most appropriate models to analyse such situations is the continuous-time Markov chain model. In this model the times between successive transitions are exponentially distributed, while the succession of states, disregarding how long it takes between transitions, is described by a discrete-time Markov chain. A wide variety of applied probability problems can be modelled as a continuous-time Markov chain by a proper state description.

We give a loose definition of a continuous-time Markov chain. Let $\{X(t), t \geq 0\}$ be a continuous-time stochastic process with discrete state space I. Suppose that this process has the following properties:

(a) If the process moves to state i, it stays in state i during an exponentially distributed time with mean $1/v_i$ independently of how the process reached state i and how long it took to get there.

(b) If the process leaves state i, it moves to state j with probability p_{ij} $(j \neq i)$ independently of the duration of the stay in state i. The transition probabilities p_{ij} satisfy

$$\sum_{j \neq i} p_{ij} = 1 \qquad \text{for all} \qquad i \in I.$$

(c) The rates v_i, $i \in I$, are bounded.

A stochastic process $\{X(t), t \geq 0\}$ with these properties is a continuous-time Markov chain. Note the convention that a transition from a state is always to a *different* state. In view of the assumptions that the amount of time the process stays in some state is exponentially distributed and that the next state visited is independent of the duration of that stay, the process has the *Markovian property* that the future behaviour of the process depends only on the present state and is independent of the past history. Alternatively, the continuous-time Markov chain can be defined by requiring the Markovian property

$$\begin{aligned} P\{X(t+s) = j \,|\, X(s) = i, \quad X(u) = x(u), \quad 0 \leq u < s\} \\ = P\{X(t+s) = j \,|\, X(s) = i\} \end{aligned} \qquad (2.28)$$

for all $s, t \geq 0$, i, j, $x(u) \in I$, and assuming that the transition rates are time-homogeneous and bounded.

We now specify the transition rates of the continuous-time Markov chain $\{X(t)\}$. Suppose that the present state is i and consider the next Δt time units with Δt small. Then, using the characterization (1.28) of the exponential distribution and noting that the probability of two or more state transitions within a time Δt is negligibly small compared with Δt when $\Delta t \to 0$, it follows that

$$P\{X(t + \Delta t) = i \,|\, X(t) = i\} = 1 - v_i \Delta t + o(\Delta t), \qquad (2.29)$$

where $o(\Delta t)$ is some term which is negligibly small compared with Δt when $\Delta t \to 0$. Hence the probability that the process leaves state i in the next Δt time units is $v_i \Delta t + o(\Delta t)$. If the process leaves state i, it moves to state j with probability $p_{ij}(j \neq i)$, independently of the duration of the stay in state i. Thus, for each $i \in I$,

$$P\{X(t + \Delta t) = j \,|\, X(t) = i\} = \lambda_{ij} \Delta t + o(\Delta t), \qquad j \neq i, \qquad (2.30)$$

where

$$\lambda_{ij} = v_i p_{ij}, \qquad j \neq i. \qquad (2.31)$$

The values λ_{ij}, $i, j \in I$, $j \neq i$, are called the *transition rates* of the continuous-time Markov chain. In fact the continuous-time Markov chain is completely determined by the distribution of the initial state and the transition rates λ_{ij}, $i, j \in I$, $j \neq i$. Note that the sojourn-time rates v_i, $i \in I$, follow from

$$v_i = \sum_{j \neq i} \lambda_{ij}, \tag{2.32}$$

while the one-step transition probabilities p_{ij}, $j \neq i$, next follow from (2.31). In solving specific problems, it suffices to specify the transition rates λ_{ij}. We emphasize that the λ_{ij}'s are not probabilities but infinitesimal rates. It is true that $\lambda_{ij}\Delta t$ for small Δt can be interpreted as a transition probability.

We now give some examples of continuous-time Markov chains.

Example 2.5   A single-server queueing system with Poisson arrivals and Erlangian services.

Suppose a single-server station at which customers arrive according to a Poisson process with rate λ. The service times of the customers are independent and identically distributed random variables, and are also independent of the arrival process. The common probability distribution function of the service times has an Erlang-r density

$$f(x) = \mu^r \frac{x^{r-1}}{(r-1)!} e^{-\mu x}.$$

It is assumed that the average service time r/μ is less than the average interarrival time $1/\lambda$. A customer who finds upon arrival that the server is idle, enters service immediately; otherwise the customer waits in queue. The above queueing system is usually abbreviated as the $M/E_r/1$ queue. For the special case of exponential service ($r = 1$) the notation $M/M/1$ queue is used.

We consider first the special case of the $M/M/1$ queue. Letting, for each $t \geqslant 0$,

$$X(t) = \text{the number of customers present at time } t,$$

the process $\{X(t), t \geqslant 0\}$ is a continuous-time Markov chain with state space $I = \{0, 1, \ldots\}$. To show that $\{X(t)\}$ is indeed a Markov process it is easiest to verify the Markovian property (2.28). Since the Poisson arrival process and the exponential service times are both memoryless, it follows that the present state of the process is only relevant to predict the future behaviour of the process and so the Markovian property (2.28) is satisfied. The transition rates λ_{ij} of the process $\{X(t)\}$ are easily determined. Clearly, from state 0 the only possible transition is to state 1, while from state $i \geqslant 1$ the only possible transitions are to the states $i - 1$ and $i + 1$. By the representations (1.32) and (1.27) the probability of an arrival during any small time interval of length Δt is $\lambda \Delta t + o(\Delta t)$ and the probability that a service in progress is completed within the next Δt time units is $\mu \Delta t + o(\Delta t)$. Notice that the probability of both an arrival and a service completion within a time Δt is of the order of magnitude $(\Delta t)^2$ and therefore negligible. It now follows that

$$P\{X(t + \Delta t) = i + 1 \,|\, X(t) = i\} = \lambda \Delta t + o(\Delta t), \qquad i \geqslant 0,$$
$$P\{X(t + \Delta t) = i - 1 \,|\, X(t) = i\} = \mu \Delta t + o(\Delta t), \qquad i \geqslant 1.$$

Hence,

$$\lambda_{ij} = \begin{cases} \lambda, & j = i + 1, & i \geqslant 0 \\ \mu, & j = i - 1, & i \geqslant 1, \\ 0, & \text{otherwise.} \end{cases} \tag{2.33}$$

Next we consider the $M/E_r/1$ queue. In the case $r > 1$, the process describing the number of customers present is no longer Markovian. The future behaviour of the process does not depend only on the number of customers present but depends in addition on the elapsed service time of the service in progress (if any). Nevertheless, using an ingenious method developed by the queueing theory pioneer A. K. Erlang early in this century, the above process can be analysed by a continuous-time Markov chain. The *method of phases* makes it possible to analyse a wide variety of practical probability problems in which the underlying distributions are not necessarily exponential but can be represented as sums or mixtures of exponentials. The powerful method of phases decomposes the distribution concerned into a set of exponential distributions and next exploits the memoryless property of the exponential distribution. For the specific problem of the $M/E_r/1$ queue, note that the Erlangian service time S of a customer can be represented as $S = Y_1 + \cdots + Y_r$, where $Y_1, \ldots, Y_r$ are independent random variables having a common exponential distribution with mean $1/\mu$. The method of phases proceeds now as follows. Imagine that the service time of each customer consists of r independent phases (or tasks) each having an exponential distribution with mean $1/\mu$. If a customer enters service, its associated r phases are sequentially served one phase at a time. The customer leaves after the last phase is completed. Define now, for each $t \geqslant 0$,

$X(t) = $ the number of service phases yet to be completed at time t,

then the process $\{X(t), t \geqslant 0\}$ is a continuous-time Markov chain with state space $I = \{0, 1, \ldots\}$. The Markovian property (2.28) follows from the fact that the Poisson arrival process of the customers (batches of phases) and the service times of the phases are both memoryless. The Markov process $\{X(t)\}$ determines uniquely the number of customers present. Since the service time of each customer consists of exactly r phases, we find that for any $t > 0$ the following relation holds:

$$\text{The number of customers present at time } t = j \tag{2.34}$$

when $X(t)$ satisfies $(j - 1)r < X(t) \leqslant jr$ for some non-negative integer j. The

Markov process $\{X(t)\}$ can only make transitions from state i to the states $i-1$ and $i+r$, where a transition from i to $i-1$ corresponds to a service completion of one phase and a transition from i to $i+r$ corresponds to the arrival of r new phases. By the same arguments as above,

$$P\{X(t+\Delta t)=i+r\,|\,X(t)=i\} = \lambda\,\Delta t + o(\Delta t), \qquad i \geqslant 0,$$
$$P\{X(t+\Delta t)=i-1\,|\,X(t)=i\} = \mu\,\Delta t + o(\Delta t), \qquad i \geqslant 1,$$

and so

$$\lambda_{ij} = \begin{cases} \lambda, & j=i+r, \quad i \geqslant 0, \\ \mu, & j=i-1, \quad i \geqslant 1, \\ 0, & \text{otherwise.} \end{cases} \tag{2.35}$$

$\square$

Example 2.6   Performance measures for response areas for emergency units

Consider a city with two emergency units that cooperate in responding to some type of alarms arising in that city. The alarms arrive at a central dispatcher who sends exactly one unit to each alarm. The two units in the city only respond to the alarms within that city. For the given dispatch strategy the city is divided into two districts. The emergency unit i is the first-due unit for response area i. This means that an alarm arriving when both units are available is served by unit 1 if the alarm is in district 1 and is served by unit 2 otherwise, while an alarm arriving when only one of the units is available is served by the available one. In case both units are not available, the alarm is settled by some unit from outside the city.

Suppose that for the given design of the two response areas the alarms from the areas 1 and 2 arise according to independent Poisson processes with respective rates λ_1 and λ_2. The service times of the alarms are independent random variables, where the time needed to serve an alarm from district j by unit i has an exponential distribution with mean $1/\mu_{ij}$. Here the service times include travel times.

Letting for $i=1, 2$,

$$X_i(t) = \begin{cases} 0 & \text{if unit } i \text{ is free at time } t, \\ 1 & \text{if unit } i \text{ is servicing an alarm from district 1 at time } t, \\ 2 & \text{if unit } i \text{ is servicing an alarm from district 2 at time } t, \end{cases}$$

the two-dimensional stochastic process $\{X(t)=(X_1(t), X_2(t))\}$ is a continuous-time Markov chain with nine possible states. The Markovian property (2.28) follows directly from the fact that the Poisson process generating the alarms and the exponentially distributed service times both have the memoryless property. The transition rates of the process $\{X(t)\}$ are easily determined. For instance, a transition from state $(0,0)$ to state $(0,2)$ occurs only if an alarm arises from area 2 and a transition from state $(2, 1)$ to $(0, 1)$ occurs only if unit 1 completes its service of an alarm from area 2. Using the representations (1.32) and (1.27) in section 1.4 of chapter 1 and noting that the probability of two or more events within a small time interval of length Δt is $o(\Delta t)$, it follows for small Δt that

$$P\{X(t + \Delta t) = (0, 2) \mid X(t) = (0, 0)\} = \lambda_2 \, \Delta t + o(\Delta t)$$

and

$$P\{X(t + \Delta t) = (0, 1) \mid X(t) = (2, 1)\} = \mu_{12} \, \Delta t + o(\Delta t).$$

In this way, we find

$$\lambda_{(0,0)(1,0)} = \lambda_1, \quad \lambda_{(0,0)(0,2)} = \lambda_2, \quad \lambda_{(j,0)(j,k)} = \lambda_{(0,j)(k,j)} = \lambda_k \quad \text{for } j, k = 1, 2,$$

$$\lambda_{(j,0)(0,0)} = \lambda_{(j,j)(0,j)} = \mu_{1j}, \quad \lambda_{(0,j)(0,0)} = \lambda_{(j,j)(j,0)} = \mu_{2j} \quad \text{for } j = 1, 2,$$

$$\lambda_{(1,2)(0,2)} = \mu_{11}, \quad \lambda_{(1,2)(1,0)} = \mu_{22}, \quad \lambda_{(2,1)(0,1)} = \mu_{12}, \quad \lambda_{(2,1)(2,0)} = \mu_{21},$$

while the other transition rates are zero.

The long-run behaviour of the continuous-time Markov chain

We next proceed with the study of the long-run behaviour of the continuous-time Markov chain $\{X(t)\}$. In most practical situations the process $\{X(t)\}$ will possess a regeneration state that can be reached from each initial state with probability 1. Then the effect of the initial state will ultimately fade out. We now make the following assumption.

Assumption 2.3

The continuous-time Markov chain $\{X(t)\}$ has a regeneration state r (say) such that

$$E(T \mid X(0) = i) < \infty \qquad \text{for all } i \in I, \tag{2.36}$$

where T is the first time beyond epoch 0 that the process makes a transition into state r.

Under this assumption, we can invoke from the theory of regenerative processes the following results. For each state $j \in I$,

$$\lim_{t \to \infty} P\{X(t) = j \mid X_0 = i\} = p_j \text{ (say) exists}, \tag{2.37}$$

independently of the initial state i. In addition,

$$p_j \geqslant 0 \qquad \text{and} \qquad \sum_{j \in I} p_j = 1. \tag{2.38}$$

Roughly speaking, we may interpret p_j as the probability that an outside observer entering the system when it has been in operation for a very long time will find the system in state j. Also, we have the useful interpretation that the long-run fraction of time the system will spend in state j equals p_j with probability 1. The probabilities p_j are called the *limiting* or *steady-state* probabilities.

We next discuss the computation of the limiting probabilities p_j. It can be shown that these probabilities are the unique non-negative solution to the following system of linear equations:

$$v_j x_j = \sum_{k \neq j} \lambda_{kj} x_k, \tag{2.39}$$

$$\sum_{j \in I} x_j = 1. \tag{2.40}$$

A proof can be found in Chung (1967). The equations (2.39) are known as the *balance* or *equilibrium* equations and the equation (2.40) is called the *normalizing* equation. A solution to only the balance equations is uniquely determined up to a multiplicative constant to be found from the normalizing equation. A convenient numerical method for solving the linear equations (2.39) and (2.40) is the successive overrelaxation method given in appendix D.

To provide insights, we discuss two different derivations of the balance equations for the probabilities p_j. The first derivation is based on the useful technique of *Kolmogoroff's forward differential equations*. For each $t \geq 0$, let

$$p_{ij}(t) = P\{X(t) = j \mid X(0) = i\}, \qquad i, j \in I.$$

To compute the probability distribution of the state that occurs a time $t + \Delta t$ later, we condition on the state that occurs a time t from now on. Using the Markovian property of the process $\{X(t)\}$ and using (2.29) and (2.30), it follows from the law of total probability that

$$
\begin{aligned}
p_{ij}(t + \Delta t) &= P\{X(t + \Delta t) = j \mid X(0) = i\} \\
&= \sum_{k \in I} P\{X(t + \Delta t) = j \mid X(0) = i, X(t) = k\} P\{X(t) = k \mid X(0) = i\} \\
&= \sum_{k \in I} P\{X(t + \Delta t) = j \mid X(t) = k\} p_{ik}(t) \\
&= \sum_{k \neq j} \lambda_{kj} \Delta t \, p_{ik}(t) + (1 - v_j \Delta t) p_{ij}(t) + o(\Delta t).
\end{aligned}
$$

Hence

$$\frac{p_{ij}(t + \Delta t) - p_{ij}(t)}{\Delta t} = \sum_{k \neq j} \lambda_{kj} p_{ik}(t) - v_j p_{ij}(t) + \frac{o(\Delta t)}{\Delta t}.$$

Now, letting $\Delta t \to 0$ and interchanging the order of limit and summation, we obtain, for all $t > 0$ and $i, j \in I$,

$$p'_{ij}(t) = \sum_{k \neq j} \lambda_{kj} p_{ik}(t) - v_j p_{ij}(t). \tag{2.41}$$

This system of differential equations is known as Kolmogoroff's forward differential equations. We state without proof that the above limiting operations are allowed by the boundedness of the rates v_i, $i \in I$. Next, we let $t \to \infty$ in (2.41). It can be shown that $p'_{ij}(t)$ converges to zero and that the limit and summation in the right side of (2.41) can be interchanged. Thus we find

$$0 = \sum_{k \neq j} \lambda_{kj} p_k - v_j p_j, \qquad j \in I,$$

showing that the limiting probabilities p_j, $j \in I$, satisfy the balance equations (2.39).

A heuristic way to see these balance equations is based on the principle of the conservation of flow of probability. Recalling that p_j is the long-run fraction of time the process is in state j and $v_j = \sum_{l \neq j} \lambda_{jl}$ is the rate at which the process leaves state j when in that state, it will intuitively be clear that

The average number of transitions out of state j per unit time $= v_j p_j$.

Also, noting that λ_{kj}, $k \neq j$, is the rate at which transitions into state j occur when in state k, we have

The average number of transitions into state j per unit time

$= \sum_{k \neq j}$ (the average number of transitions from k to j per unit time)

$= \sum_{k \neq j} \lambda_{kj} \times$ (the long-run fraction of time the process is in state k)

$= \sum_{k \neq j} \lambda_{kj} p_k$.

Hence, the balance equations (2.39) for the limiting probabilities p_j represent the intuitively obvious identity,

The rate at which the process leaves state j

$=$ the rate at which the process enters state j.   (2.42)

More generally, the balance principle applies to any set A with $A \neq I$:

The rate at which the process leaves a set A of states

$=$ the rate at which the process enters that set of states.   (2.43)

This balance principle can be translated into the formula

$$\sum_{j \in A} p_j \sum_{k \notin A} \lambda_{jk} = \sum_{k \notin A} p_k \sum_{j \in A} \lambda_{kj}.$$ (2.44)

It is recommended to memorize the balance principle (2.43) rather than the technical formula (2.44). Alternatively, this formula could directly be obtained from (2.39) by summing both sides of (2.39) over $j \in A$ and by performing some manipulations with the sums involved. In particular, the balance principle (2.43) with a properly chosen set A is very useful in case the possible states can be numbered as $i = 0, 1, \ldots, N$ with $N \leqslant \infty$ such that from each state i only transitions can be made to state $i - 1$ or higher numbered states. Then the steady-state probabilities p_j, $j \geqslant 0$, can recursively be computed. Assuming that

$$I = \{0, 1, \ldots, N\} \qquad \text{and} \qquad \lambda_{ij} = 0 \qquad \text{for } j \leqslant i - 2, \ i \geqslant 2,$$

it follows from the balance principle (2.43) with $A = \{i, \ldots, N\}$, $i \geqslant 1$, that

$$\lambda_{i,i-1} p_i = \sum_{k=0}^{i-1} p_k \sum_{j=i}^{N} \lambda_{kj}, \qquad i = 1, 2, \ldots .$$

Thus, initializing $\bar{p}_0 = 1$ (say), we recursively compute from this relation $\bar{p}_1, \bar{p}_2, \ldots, \bar{p}_N$. Next, using the normalizing equation (2.40), we obtain the desired probabilities p_j from $p_j = \bar{p}_j / \sum_{k=0}^{N} \bar{p}_k$. The above recursion equation may greatly simplify the computations in 'exponential' queueing systems in which service completions occur singly. In the sequel we will encounter several examples of such queueing systems.

Continuous-time Markov chains with a reward structure

In applications one often encounters continuous-time Markov chains on which a reward structure of the following form is superimposed. A reward at rate $h(j)$ is earned whenever the process is in state j and moreover a lump reward $K(j)$ is earned each time the process makes a transition into state j. These reward functions are assumed to be finite and non-negative. In addition to assumption 2.3, we make the following assumption.

Assumption 2.4

The expectation of the total rewards earned up to the first time beyond epoch 0 at which the process makes a transition into state r is finite for each initial state $X_0 = i$.

Define

$$Z(t) = \text{the cumulative reward earned up to time } t, \qquad t \geqslant 0.$$

Now, we have for the long-run average reward per unit time that, with probability 1,

$$\lim_{t \to \infty} \frac{Z(t)}{t} = \sum_{j \in I} h(j) p_j + \sum_{j \in I} K(j) \sum_{k \neq j} p_k \lambda_{kj}, \qquad (2.45)$$

independently of the initial state $X_0 = i$. We state this result without proof and refer to Chung (1967). Intuitively, the first term in the right side of (2.45) can be seen by noting that p_j gives the long-run fraction of time the process is in state j, while the second term can be seen by noting that $p_k \lambda_{kj} (k \neq j)$ is the average number of transitions from state k to state j per unit time and so $\sum_{k \neq j} p_k \lambda_{kj}$ is the average number of transitions into state j per unit time.

A uniformization technique for continuous-time Markov chains

The foregoing discussion mainly concentrated on the calculation of the limiting distribution of the continuous-time Markov chain. To conclude this section, we discuss a powerful uniformization technique for calculating time-dependent state probabilities and first-passage time distributions. The uniformization technique transforms the original continuous-time Markov chain with non-identical transition times into an equivalent continuous-time Markov process in which the transition epochs are generated by a Poisson process at a *uniform* rate and the transitions from state to state are described by a (discrete-time) Markov chain that allows for *fictitious* transitions from a state to itself.

In the continuous-time Markov chain $\{X(t)\}$ being characterized by the infinitesimal transition rates $\lambda_{ij} (j \neq i)$, the amount of time the process stays in each state i is exponentially distributed with mean $1/v_i$, where $v_i = \sum_{j \neq i} \lambda_{ij}$, and if the process leaves state i it moves to state $j (\neq i)$ with probability $p_{ij} = \lambda_{ij}/v_i$. Notice the convention of no self-transitions. The fact that the exponentially distributed holding times in the various states will usually have different means obstructs in general a direct derivation of a computationally tractable, closed-form expression for the time-dependent state probabilities

$$p_{ij}(t) = P\{X(t) = j \mid X(0) = i\} \qquad \text{for } t > 0 \quad \text{and} \quad i, j \in I.$$

Nevertheless, as an alternative to the linear differential equations (2.41), the probabilities $p_{ij}(t)$ may be calculated from a tractable, closed-form equation. To obtain such an equation, let v be any finite number such that

$$v \geqslant v_i \qquad \text{for all } i \in I,$$

and consider the following continuous-time Markov process $\{\bar{X}(t), t \geqslant 0\}$ with state space I. The process $\{\bar{X}(t)\}$ makes state transitions at epochs generated by a Poisson process with rate v and the state transitions are governed by a (discrete-time) Markov chain with one-step transition probabilities

$$\bar{p}_{ij} = \begin{cases} (v_i/v) p_{ij} & \text{for } j \neq i, \\ 1 - v_i/v & \text{for } j = i. \end{cases} \tag{2.46}$$

In terms of the original continuous-time Markov chain $\{X(t)\}$, we can consider the transitions from state i to occur at a uniform rate of v rather than at a state-dependent rate of v_i, but only a fraction v_i/v are real transitions out of state i and the remainder are fictitious transitions that leave the process in state i. Using the lack of memory of the exponential distribution, it can be seen that the Markov process $\{\bar{X}(t)\}$ with a uniform transition rate is probabilistically identical to the original Markov process $\{X(t)\}$ with state-dependent transition rates. Thus, letting $\bar{p}_{ij}(t) = P\{\bar{X}(t) = j | \bar{X}(0) = i\}$, we have

$$\bar{p}_{ij}(t) = p_{ij}(t) \quad \text{for} \quad t > 0 \quad \text{and} \quad i,j \in I.$$

Let now $(\bar{p}_{ij}^{(n)})$, $i,j \in I$, be the n-step transition probabilities of the discrete-time Markov chain with the one-step transition probabilities $(\bar{p}_{ij})$, $i,j \in I$. Using the fact that the probability of exactly n state transitions of the process $\{\bar{X}(t)\}$ during a given time t equals the Poisson probability $e^{-vt}(vt)^n/n!$, it follows by conditioning that

$$\bar{p}_{ij}(t) = \sum_{n=0}^{\infty} \bar{p}_{ij}^{(n)} e^{-vt} \frac{(vt)^n}{n!} \quad \text{for} \quad t > 0 \quad \text{and} \quad i,j \in I, \qquad (2.47)$$

with the convention $\bar{p}_{ij}^{(0)} = 1$ for $j = i$ and $\bar{p}_{ij}^{(0)} = 0$ otherwise. Now, for any fixed $t > 0$ and $i \in I$, the desired probabilities $p_{ij}(t), j \in I$, can be calculated by applying the following recursion scheme. Letting $\varphi_j(n) = \bar{p}_{ij}^{(n)} e^{-vt}(vt)^n/n!$ and using the recurrence relation (2.13) for the probabilities $\bar{p}_{ij}^{(n)}$, it follows that for fixed $t > 0$ and $i \in I$ the probabilities

$$p_{ij}(t) = \sum_{n=0}^{\infty} \varphi_j(n), \qquad j \in I,$$

can be calculated by applying the recursion scheme

$$\varphi_j(n) = \frac{vt}{n} \sum_{k \in I} \varphi_k(n-1) \bar{p}_{kj} \quad \text{for} \quad n \geqslant 1 \quad \text{and} \quad j \in I, \qquad (2.48)$$

starting with $\varphi_j(0) = e^{-vt}$ for $j = i$ and $\varphi_j(0) = 0$ otherwise.

Finally, we remark that the uniformization technique may equally well be applied for the calculation of the distribution of the first-passage time until some state (or set of states) is reached (cf. also example 2.9 in the next section).

2.4   APPLICATIONS OF CONTINUOUS-TIME MARKOV CHAINS

In this section we give a number of applications of the continuous-time Markov chain model. These applications concern a variety of practical problems and illustrate the wide applicability of the continuous-time Markov chain model.

Example 2.5 (continued)   A single-server queueing system with Poisson arrivals and Erlangian services

The steady-state distribution of the number of customers in the system and the delay in queue of an arbitrary customer can efficiently be computed for the $M/E_r/1$ queueing system. Therefore we have to solve the balance equations for the continuous-time Markov chain $\{X(t)\}$ describing the number of uncompleted service phases in the system.

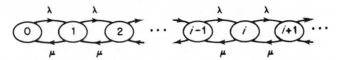

Figure 2.1 The transition rate diagram for the uncompleted service phases

In general, to write down the balance equations, it may be helpful to use a *transition rate diagram*. The nodes of the diagram represent the states and the arrows in the diagram give the possible state transitions. An arrow from node i to node j, is only drawn when the transition rate λ_{ij} is positive, in which case the arrow is labelled with the value λ_{ij}.

We first discuss the special case of the M/M/1 queue ($r = 1$). The transition rates λ_{ij} of the process $\{X(t)\}$ are given by (2.33). In Figure 2.1 we display these rates in the transition rate diagram. Using the balance principle 'rate out of state i = rate into state i', we obtain for the steady-state probabilities p_i of the process $\{X(t)\}$ the balance equations

$$\lambda p_0 = \mu p_1,$$
$$(\lambda + \mu)p_i = \lambda p_{i-1} + \mu p_{i+1}, \qquad i \geqslant 1.$$

These equations can easily be rewritten as

$$\mu p_i = \lambda p_{i-1}, \qquad i \geqslant 1. \tag{2.49}$$

Actually, the equations (2.49) could directly be obtained from the balance principle (2.43) with $A = \{i, i+1, \ldots\}$, $i \geqslant 1$. Iterating (2.49), we find

$$p_i = \left(\frac{\lambda}{\mu}\right)^i p_0, \qquad i \geqslant 1. \tag{2.50}$$

Using the normalizing equation $\sum_{i=0}^{\infty} p_i = 1$, we have

$$p_0 = 1 - \frac{\lambda}{\mu}.$$

Next we consider the M/E$_r$/1 queueing system with $r \geqslant 1$. The transition rates λ_{ij} of the process $\{X(t)\}$ are given by (2.35). We display these rates in Figure 2.2. The only possible transitions out of state i are to state $i - 1$ or higher numbered states. Thus we can recursively compute the steady-state probabilities f_j(say) of the continuous-time Markov chain $\{X(t)\}$. Using the general balance principle 'rate

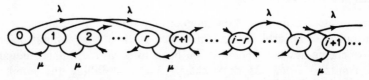

Figure 2.2 The transition rate diagram for the uncompleted service phases

out of the set of states $A = \{i, i+1, \ldots\}$ = rate into that set of states', we obtain

$$\mu f_i = \lambda \sum_{k=i-r}^{i-1} f_k, \qquad i \geqslant 1, \tag{2.51}$$

with the convention that $f_k = 0$ for $k < 0$. From this recursion relation we successively compute $f_1, f_2, \ldots$ starting with f_0 as the parameter. As a matter of fact, we have for f_0 the explicit expression

$$f_0 = 1 - \lambda E(S),$$

where $E(S) = r/\mu$ denotes the mean service time. This result follows from the equation (1.36) with $\sigma = 1$ in section 1.4 of chapter 1 by noting that $1 - f_0$ represents the fraction of time the server is busy. The load factor $\lambda E(S)$ is called the server utilization.

We are now in a position to compute the steady-state distributions of both the number of customers present and the delay in queue of a customer.

It is assumed that the customers are served in order of arrival. Define

$$p_j^* = \lim_{t \to \infty} P\{j \text{ customers are present at time } t\}, \qquad j = 0, 1, \ldots,$$

and

$$\bar{W}_q(x) = \lim_{n \to \infty} P\{\text{the delay in queue of the } n\text{th customer}$$

$$\text{exceeds } x\}, \qquad x \geqslant 0.$$

It can be shown from the theory of regenerative processes that these limits exist and are independent of the initial state. By (2.34), we have

$$p_0^* = f_0, \qquad p_j^* = \sum_{k=(j-1)r+1}^{jr} f_k, \qquad j \geqslant 1. \tag{2.52}$$

To find the complementary distribution function $\bar{W}_q(x)$, we need the fundamental property that 'Poisson arrivals see time averages'. In the case of a Poisson arrival process of customers the customer-average state probabilities are equal to the time-average state probabilities, that is (see section 1.8 of chapter 1)

The fraction of customers who find upon arrival j uncompleted phases present
   = the fraction of time that j uncompleted phases are present.

In other words, Poisson arrivals see the state of the system as it is at a completely random point in time. Thus f_j estimates the probability that a customer to arrive in the far distant future will find j uncompleted service phases in the system. Hence, by the law of total probability,

$$\bar{W}_q(x) = \sum_{j=1}^{\infty} P\{\text{a customer has to wait more than a time } x \,|\, \text{the customer}$$

$$\text{finds upon arrival } j \text{ uncompleted phases present}\} f_j.$$

Using the lack of memory of the exponential distribution, it follows that the delay in queue of a customer finding upon arrival j uncompleted phases present is distributed as the sum of j independent exponentials with the same means $1/\mu$ and thus has an Erlang-j distribution. Hence, using (1.22) in section 1.4 of chapter 1,

$$\overline{W}_q(x) = \sum_{j=1}^{\infty} f_j \sum_{k=0}^{j-1} e^{-\mu x} \frac{(\mu x)^k}{k!}, \qquad x \geq 0.$$

For the special case of the M/M/1 queue ($r = 1$), we have by (2.50) that this expression can be simplified to

$$\overline{W}_q(x) = \frac{\lambda}{\mu} e^{-(\mu - \lambda)x}, \qquad x \geq 0. \tag{2.53}$$

This implies that for the M/M/1 queue the average delay in queue per customer equals

$$E(W_q) = \frac{\lambda}{\mu(\mu - \lambda)} \tag{2.54}$$

This formula is a special case of the Pollaczek–Khintchine formula for the M/G/1 queue. This single-server queueing system has Poisson arrivals but general service times. In the M/G/1 queue the average delay per customer equals

$$E(W_q) = \frac{\lambda \{E(S)\}^2}{2\{1 - \lambda E(S)\}} (1 + c_S^2), \tag{2.55}$$

where $E(S)$ and c_S^2 denote the mean and the squared coefficient of variation of the service time S of a customer. The Pollaczek–Khintchine formula (2.55) is easily proved by using results obtained earlier. Therefore define the work in the system at time t as the sum of the remaining service times of the customers in the system at time t. Since there is a single server and customers are served in order of arrival, the delay in queue of a customer equals the work in the system as seen by that customer upon arrival. Using the property that 'Poisson arrivals see time averages', it follows that for the M/G/1 queue the average delay $E(W_q)$ equals the time-average work in the system. Formula (1.35) with $\sigma = 1$ in section 1.4 of chapter 1 gives the average amount of work in the system. The desired result (2.55) next follows from (1.35) by noting that $1 + c_S^2 = E(S^2)/E^2(S)$.

It is remarkable that the average delay in the M/G/1 queue depends on the service time distribution only through the first two moments. The formula (2.55) specifies how the average delay increases as the variance of the service time increases while the arrival rate and the average service time are kept fixed. In particular, the average delay for constant service is one-half the average delay for exponential service. Also, the formula (2.55) shows that in the case of variability in the arrival process or service times one should not try to equalize the arrival rate and the service rate, since otherwise the average delay in queue of a customer becomes infinitely large. More specifically, denoting by ρ the server utilization

$\lambda E(S)$, the slope of increase of $E(W_q)$ as a function of ρ is proportional to $(1 - \rho)^{-2}$, as follows by differentiation of $E(W_q)$. As an illustration of this qualitative result, a small increase in the arrival rate λ when the load $\rho = 0.9$ causes an increase in the average delay 25 times greater than it would cause when the load $\rho = 0.5$. This non-intuitive finding shows the danger of designing a system with a high utilization level (say, $\rho \geqslant 0.8$), since then a small increase in the traffic input may cause dramatic degradation in system performance. $\square$

Example 2.6 (continued) *Performance measures for response areas for emergency units*

For the given design of the two response areas, performance measures of interest are the fraction of alarms that are lost and the fraction of time that each unit is busy. These performance measures can easily be expressed in terms of the steady-state probabilities of the continuous-time Markov chain $\{X(t)\}$. To do so, we denote these steady-state probabilities by $p(j, k)$ for $0 \leqslant j, k \leqslant 2$. Since $p(j, k)$ represents the long-run fraction of time the system is in state (j, k), it follows that

$$\text{The fraction of time that unit 1 is busy} = \sum_{j=1}^{2} \sum_{k=0}^{2} p(j, k).$$

A similar expression applies for the fraction of time that unit 2 is busy. To find the fraction of alarms from district $l(= 1, 2)$ that are lost, we use the property that 'Poisson arrivals see time averages'. That is the long-run fraction of alarms from district l that find upon occurrence the system in state (j, k) equals the long-run fraction of time the system is in state (j, k). Thus, for each district $l = 1, 2$,

$$\text{The fraction of alarms from district } l \text{ that are lost} = \sum_{j=1}^{2} \sum_{k=1}^{2} p(j, k).$$

It remains to calculate the steady-state probabilities $p(j, k)$. To do so, it is helpful to display in Figure 2.3 the transition rates of the continuous-time Markov chain $\{X(t)\}$. Next, by applying the balance principle 'rate out of a state = rate into that state', we obtain the balance equations

$$(\lambda_1 + \lambda_2)p(0, 0) = \mu_{21}p(0, 1) + \mu_{22}p(0, 2) + \mu_{11}p(1, 0) + \mu_{12}p(2, 0),$$
$$(\lambda_1 + \lambda_2 + \mu_{21})p(0, 1) = \mu_{11}p(1, 1) + \mu_{12}p(2, 1),$$
$$(\lambda_1 + \lambda_2 + \mu_{22})p(0, 2) = \lambda_2 p(0, 0) + \mu_{11}p(1, 2) + \mu_{12}p(2, 2),$$
$$(\lambda_1 + \lambda_2 + \mu_{11})p(1, 0) = \lambda_1 p(0, 0) + \mu_{21}p(1, 1) + \mu_{22}p(1, 2),$$
$$(\mu_{11} + \mu_{21})p(1, 1) = \lambda_1 p(0, 1) + \lambda_1 p(1, 0),$$
$$(\mu_{11} + \mu_{22})p(1, 2) = \lambda_1 p(0, 2) + \lambda_2 p(1, 0),$$
$$(\lambda_1 + \lambda_2 + \mu_{12})p(2, 0) = \mu_{21}p(2, 1) + \mu_{22}p(2, 2),$$
$$(\mu_{12} + \mu_{21})p(2, 1) = \lambda_2 p(0, 1) + \lambda_1 p(2, 0),$$
$$(\mu_{12} + \mu_{22})p(2, 2) = \lambda_2 p(0, 2) + \lambda_2 p(2, 0).$$

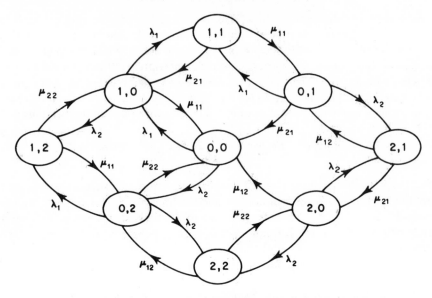

Figure 2.3 The state transition rate diagram for the emergency units

Also, we have the normalizing equation

$$\sum_{j=0}^{2} \sum_{k=0}^{2} p(j,k) = 1.$$

These linear equations have to be solved by some numerical procedure.

In the above analysis the assumption of exponentially distributed service times was made because it simplifies the analysis. In practice, however, this assumption will not be satisfied, since the service times include setup times so that they will nearly always exceed some positive threshold value. Therefore it is important to know whether the results obtained for exponentially distributed service times are approximately correct for the actual service-time distributions. It was shown in Wolff and Wrightson (1976) that for the two-server loss model with generally distributed service times the above performance measures depend on the service-time distributions only through their means when the service times do not depend on the unit that serves. In the case of service times depending on the unit that serves, this insensitivity property is no longer exactly true but numerical experiments indicate that then the dependence on the distributional forms of the service times is quite weak (cf. also exercise 2.17). Thus for practical purposes the calculations can be done under the simplifying assumption of exponentially distributed service times. In the sequel it will be seen that the robustness of the present model is characteristic for many stochastic service systems in which *no* queueing is allowed. It is good to point out that opposite to stochastic service systems with losses are stochastic service systems in which requests for service, finding upon arrival that all of the servers are busy, are delayed until a free server becomes available; in delay systems it is usually not possible to calculate performance measures by using the service time only through its first moment, as is demonstrated, for example, by the Pollaczek–Khintchine formula (2.55) for the delay system dealt with in example 2.5, cf. the discussion in section 2.5. □

Example 2.7   A cash balance problem

Consider a mutual fund that keeps cash in a bank account for withdrawals of customers redeeming their shares and for deposits of customers buying new shares. The cash outflows and the cash inflows can be described by independent compound Poisson processes. The withdrawals occur according to a Poisson process with rate λ_1 and the magnitude of each withdrawal has a discrete probability distribution $\{\phi_1(j), j = 1, 2, \ldots\}$, while the deposits occur according to a Poisson process with rate λ_2 and each deposit has a discrete probability distribution $\{\phi_2(j), j = 1, 2, \ldots\}$. At each point of time the fund may decide to increase the cash balance by selling stocks at a cost of c_1 per dollar value of the stocks or to decrease the cash balance by buying stocks at a cost of c_2 per dollar value. The amount of time involved with a change of the cash balance is negligible. Also, since the fund's cash in the bank account could have been invested elsewhere, the fund incurs a positive holding cost at rate hk whenever the cash balance is $k \geqslant 0$. The fund controls the cash balance by the following two-critical-numbers rule. The cash balance is decreased to the level b if it exceeds this level, is increased to the level a if it drops below this level and is left unchanged otherwise. The control parameters a and b are positive integers with $0 < a < b$. We are interested in the long-run average cost per unit time for this control rule.

To derive the average cost for the given control rule, define, for each $t \geqslant 0$,

$X(t) =$ the amount of cash in the bank account at time t.

Here we agree that $X(t)$ represents the cash balance just after any transaction that may occur at time t. Hence the possible states of the process $\{X(t)\}$ are the integers $a, a + 1, \ldots, b$. The process $\{X(t)\}$ is a continuous-time Markov chain due to the fact that the withdrawals and deposits occur according to independent Poisson processes. We next determine the transition rate diagram. In accordance with our convention for continuous-time Markov chains that a transition from a state is always to a different state, we disregard in the transition rate diagram the possible self-transitions into the states a and b. However, these self-transitions will properly be taken into account when evaluating the average cost. It is readily verified that

$$P\{X(t + \Delta t) = i + j | X(t) = i\} = \lambda_2 \phi_2(j)\Delta t + o(\Delta t), \qquad 1 \leqslant j < b - i, a \leqslant i < b,$$

$$P\{X(t + \Delta t) = b | X(t) = i\} = \left\{ \lambda_2 \sum_{k \geqslant b - i} \phi_2(k) \right\} \Delta t + o(\Delta t), \qquad a \leqslant i < b,$$

$$P\{X(t + \Delta t) = i - j | X(t) = i\} = \lambda_1 \phi_1(j)\Delta t + o(\Delta t), \qquad 1 \leqslant j < i - a, a < i \leqslant b,$$

$$P\{X(t + \Delta t) = a | X(t) = i\} = \left\{ \lambda_1 \sum_{k \geqslant i - a} \phi_1(k) \right\} \Delta t + o(\Delta t), \qquad a < i \leqslant b,$$

while the other transition probabilities are zero. In Figure 2.4 we display the transition rates in the transition rate diagram. Using the balance principle 'rate out of state i = rate into state i', we find for the steady-state probabilities p_i, $a \leqslant i \leqslant b$, of the process $\{X(t)\}$ the balance equations

$$(\lambda_1 + \lambda_2)p_j = \sum_{i=a}^{j-1} \lambda_2 \phi_2(j - i)p_i + \sum_{k=j+1}^{b} \lambda_1 \phi_1(k - j)p_k, \qquad a < j < b,$$

$$\lambda_2 p_a = \sum_{k=a+1}^{b} \left\{ \lambda_1 \sum_{l \geqslant k - a} \phi_1(l) \right\} p_k,$$

$$\lambda_1 p_b = \sum_{k=a}^{b-1} \left\{ \lambda_2 \sum_{l \geqslant b - k} \phi_2(l) \right\} p_k.$$

Also, we have the normalizing equation

$$\sum_{k=a}^{b} p_k = 1.$$

A simple and accurate method to solve the above linear equations is the successive overrelaxation method described in appendix D.

Once the steady-state probabilities have been computed, we can determine the average cost of the given control rule. Since p_i gives the fraction of time the system is in state i and since the withdrawals of cash occur according to a Poisson process, it follows by using the property that 'Poisson arrivals see time

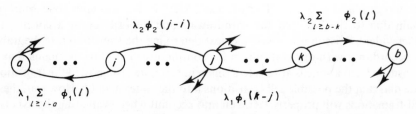

Figure 2.4 The transition rate diagram for the cash balance problem

averages' that, for each $a \leqslant i \leqslant b$,

The average number of times that per unit time the cash balance drops from i to $j = \lambda_1 \phi_1(i-j)p_i$, $j = i-1, i-2, \ldots$.

As a consequence,

The average number of times that per unit time stocks are sold

$$= \sum_{i=a}^{b} \sum_{j \leqslant a-1} \lambda_1 \phi_1(i-j)p_i = \lambda_1 \sum_{i=a}^{b} p_i \left\{ 1 - \sum_{k=0}^{i-a} \phi_1(k) \right\},$$

with the convention $\phi_1(0) = 0$. Also, letting $\beta_1 = \sum_j j\phi_1(j)$, the average size of a withdrawal,

The average selling costs of stocks per unit time

$$= c_1 \sum_{i=a}^{b} p_i \sum_{j \leqslant a-1} (a-j)\lambda_1 \phi_1(i-j)$$

$$= \lambda_1 c_1 \sum_{i=a}^{b} p_i \left\{ a - i + \beta_1 + \sum_{k=0}^{i-a} (i-a-k)\phi_1(k) \right\}. \qquad (2.56)$$

By similar arguments, we find, with the notation $\phi_2(0) = 0$ and $\beta_2 = \sum_j j\phi_2(j)$,

The average number of times that per unit time stocks are bought

$$= \lambda_2 \sum_{i=a}^{b} p_i \left\{ 1 - \sum_{k=0}^{b-i} \phi_2(k) \right\},$$

The average buying costs of stocks per unit time

$$= \lambda_2 c_2 \sum_{i=a}^{b} p_i \left\{ i - b + \beta_2 + \sum_{k=0}^{b-i} (b-i-k)\phi_2(k) \right\}. \qquad (2.57)$$

Further,

The average holding cost of cash per unit time $= h \sum_{i=a}^{b} ip_i$. $\qquad (2.58)$

Together (2.56), (2.57) and (2.58) give the average cost per unit time under the two-critical-numbers control rule.   □

Example 2.8   Performance analysis for an unloader at a container terminal

Consider a container terminal with a finite number of N trailers which bring loads of containers from ships to a single unloader. The unloader can serve only one trailer at a time and the unloading time per trailer has an exponential distribution with mean $1/\mu_s$. A trailer leaves when it is unloaded and returns at the unloader with a next load of containers after an exponentially distributed trip time with mean $1/\lambda$. However, after a trailer is unloaded, the unloader needs an extra finishing time for the unloaded containers before the unloader is available to unload a next trailer. This finishing time has an exponential distribution with mean $1/\mu_f$. The unloading times, the trip times and the finishing times are assumed to be independent of each other.

We are interested in performance measures such as the average number of trailers that are unloaded per unit time and the probability that an arbitrary trailer has to wait. To find these performance measures, define, for each $t \geqslant 0$,

$X_1(t) = $ the number of trailers present at the unloader at time t,

$$X_2(t) = \begin{cases} 0 & \text{if the unloader is available for unloading at time } t, \\ 1 & \text{otherwise.} \end{cases}$$

Clearly, the process $\{X(t) = (X_1(t), X_2(t)), t \geqslant 0\}$ is a continuous-time Markov chain with state space $I = \{(i,j) | i = 0, \ldots, N, \ j = 0, 1\}$. The transition rate diagram in Figure 2.5 is easily verified by using the fact that the arrival rate of trailers at the unloader equals the sum of $N - i$ individual rates λ when i trailers are already present at the unloader.

We denote the steady-state probabilities of the process $\{X(t)\}$ by $p(i,j)$. Using the balance principle 'rate out of a state = rate into that state', we obtain the balance equations

$$N\lambda p(0,0) = \mu_f p(0,1),$$
$$\{\mu_s + (N-i)\lambda\}p(i,0) = (N-i+1)\lambda p(i-1,0) + \mu_f p(i,1), \qquad 1 \leqslant i \leqslant N,$$
$$\{\mu_f + (N-i)\lambda\}p(i,1) = (N-i+1)\lambda p(i-1,1) + \mu_s p(i+1,0), \qquad 0 \leqslant i \leqslant N,$$

with $p(-1,1) = p(N+1,0) = 0$ by convention. Also, we have the normalizing equation

$$\sum_{i=0}^{N} \{p(i,0) + p(i,1)\} = 1.$$

A closer look at the balance equations reveals that they can recursively

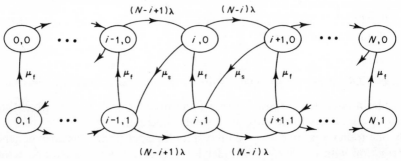

Figure 2.5 The transition rate diagram for the process $\{X(t)\}$

be solved as follows. Starting with $\bar{p}(0,0) = 1$, we successively compute $\bar{p}(0,0) \to \bar{p}(0,1) \to \bar{p}(1,0) \to \cdots \to \bar{p}(N-1,1) \to \bar{p}(N,0) \to \bar{p}(N,1)$. Next, by the normalizing equation, the desired probabilities $p(i,j)$ follow from $p(i,j) = \bar{p}(i,j)/\sum_{i,j}\bar{p}(i,j)$. We are now in a position to specify the following performance measures:

P_I = the fraction of time the unloader is idle,

$E(L_q)$ = the average number of trailers waiting to be unloaded,

λ_T = the average number of trailers unloaded per unit time,

$E(W_q)$ = the average delay in queue of a trailer,

Π_W = the probability that an arbitrary trailer has to wait.

Obviously,

$$P_I = p(0,0)$$

and

$$E(L_q) = \sum_{i=1}^{N} (i-1)p(i,0) + \sum_{i=1}^{N} ip(i,1).$$

To find the throughput λ_T, we first observe that in the long run the average number of trailers unloaded per unit time must be equal to the average number of trailers arriving at the unloader per unit time. Hence

λ_T = the average number of arrivals at the unloader per unit time.

The average arrival rate of trailers at the unloader when i trailers are there is equal to $(N-i)\lambda$. Since $p(i,0) + p(i,1)$ gives the fraction of time that i trailers are at the unloader, we have

$$\lambda_T = \sum_{i=0}^{N} (N-i)\lambda\{p(i,0) + p(i,1)\}.$$

As a matter of fact, this result uses the property that 'Poisson arrivals see time averages'. Note that the arrival process of trailers at the unloader when i trailers are there can be described by a 'state-dependent' Poisson process with rate $(N-i)\lambda$. This observation and the basic relation (1.63) in section 1.8 of chapter 1 can be used to provide a rigorous proof of the above result for λ_T. We omit further details. The throughput λ_T can also be computed in a different way. To do so, note that during the time the unloader is unloading trailers the process of departures of trailers can be described by a Poisson process with rate μ_s. The long-run fraction of time the unloader is unloading trailers equals $\sum_{i=1}^{N} p(i,0)$. Thus the throughput λ_T can alternatively be computed as

$$\lambda_T = \mu_s \sum_{i=1}^{N} p(i,0).$$

Hence we have the identity

$$\sum_{i=0}^{N} (N-i)\lambda\{p(i,0)+p(i,1)\} = \mu_s \sum_{i=1}^{N} p(i,0).$$

This identity could be used as an extra accuracy check when solving the linear equations for the steady-state probabilities.

The average delay $E(W_q)$ can be obtained from the relation

$$E(L_q) = \lambda_T E(W_q).$$

This relation can be derived by repeating the arguments used to obtain the similar relation (2.26) and by noting that the arrival rate of trailers at the unloader is given by λ_T(cf. also the discussion of Little's formula in section 4.1 of chapter 4).

To determine the delay probability Π_W, we observe that

The fraction of trailers delayed at the unloader
= the average number of trailers arriving per unit time and finding
the unloader busy/the average number of trailers arriving per unit time.

Thus

$$\Pi_W = \frac{1}{\lambda_T}\left\{ \sum_{i=1}^{N}(N-i)\lambda p(i,0) + \sum_{i=0}^{N}(N-i)\lambda p(i,1) \right\} = 1 - \frac{N\lambda p(0,0)}{\lambda_T}.$$

In the above discussion we have made simplifying assumptions about the probability distributions of the trip time, the unloading time and the finishing time in order to keep the analysis tractable. In reality, exponential distributions cannot always be expected to occur. In particular, the assumption of exponentially distributed trip times is rather unrealistic. Therefore we have investigated the effect of the simplifying assumptions on the performance measures. Using the method of phases discussed in example 2.5, it is possible to extend the above analysis by an appropriately defined continuous-time Markov chain when each of the distributions involved corresponds to either a sum or a mixture of exponentials. The reader is asked to do this in exercise 2.18. In Table 2.3 we

Table 2.3 Numerical results for the performance measures

(c_T^2, c_U^2)	P_{I}	λ_{T}	$E(W_q)$	Π_W
$(0.5, 0.5)$	0.3714	0.180	2.68	0.5761
$(1, 0.5)$	0.3727	0.179	2.80	0.5841
$(4, 0.5)$	0.3751	0.179	3.01	0.5978
$(0.5, 1)$	0.3780	0.178	3.27	0.5680
$(1, 1)$	0.3786	0.178	3.32	0.5736
$(4, 1)$	0.3795	0.177	3.41	0.5835

give for some numerical examples the values of the performance measures $P_1, \lambda_T, E(W_q)$ and Π_W for various distributions of the trip time and unloading time when the means are kept fixed. We assume the numerical data

$$N = 10, \qquad E(T) = 50, \qquad E(U) = 3 \qquad \text{and} \qquad E(F) = 0.5,$$

where the generic variables T, U and F denote the trip time, the unloading time and the finishing time. The squared coefficient of variation c_T^2 of the trip time is varied as 0.5, 1 and 4 with $c_T^2 = 0.5$ corresponding to the E_2 distribution, $c_T^2 = 1$ to the exponential distribution and $c_T^2 = 4$ to an H_2 distribution with the gamma normalization (A.17) in appendix C. The squared coefficient of variation c_U^2 of the unloading time has the two values 0.5 and 1 with $c_U^2 = 0.5$ corresponding to the E_2 distribution and $c_U^2 = 1$ to the exponential distribution. In all examples the finishing time F has an exponential distribution. It appears from our numerical investigations that the above performance measures are fairly insensitive to more than the means of the various distributions so that practical analysis may be based on the simplifying assumption of exponential distributions. This empirical finding is typical for many practical queueing systems with *finite-source* input.   □

The next example discusses among others useful methods for computing first-passage time distributions in continuous-time Markov chains.

Example 2.9   Reliability of a one-unit operating system with multiple standby units

A computer system uses one operating unit but has built in redundancy in the form of $N \geqslant 1$ standby units. The operating unit has an exponential time with mean $1/\lambda$ to failure. If the operating unit fails it is replaced immediately by a standby unit, if available. Each failed unit enters repair immediately and is again available after an exponentially distributed repair time with mean $1/\mu$. There are ample repair facilities so that any number of units can be in repair simultaneously. The repairs of the various units occur independently of each other, and also the repair times are independent of the operating times. It is assumed that a unit in standby status cannot fail. The system is down only if all $N + 1$ units are down. We are interested in both the long-run fraction of time the system is down and the system reliability defined as the probability that the system will not go down during $[0, t]$.

To analyse the system, define, for each $t \geqslant 0$,

$$X(t) = \text{the number of units in repair at time } t.$$

The process $\{X(t), \ t \geqslant 0\}$ is a continuous-time Markov chain with states $0, 1, \ldots, N + 1$. The transition rate diagram is displayed in Figure 2.6. The average availability of the system follows from the steady-state probabilities

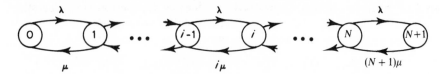

Figure 2.6 The transition rate diagram for the number of units down

$\{p_i\}$ of the process $\{X(t)\}$. Since the system is down only if all $N + 1$ units are in repair, we have

The fraction of time the system is down $= p_{N+1}$.

Using the balance principle 'rate out of the set of states $\{i, \ldots, N + 1\}$ = rate into that set of states', we obtain

$$i\mu p_i = \lambda p_{i-1}, \qquad 1 \leqslant i \leqslant N + 1.$$

This recurrence relation and the normalization equation $\sum_{k=0}^{N+1} p_k = 1$ together yield

$$p_i = \frac{(\lambda/\mu)^i/i!}{\sum_{k=0}^{N+1} (\lambda/\mu)^k/k!}, \qquad 0 \leqslant i \leqslant N + 1. \tag{2.59}$$

In particular,

$$p_{N+1} = \frac{(\lambda/\mu)^{N+1}/(N+1)!}{\sum_{k=0}^{N+1} (\lambda/\mu)^k/k!}. \tag{2.60}$$

The truncated Poisson distribution (2.59) is known as *Erlang's loss distribution* and the formula (2.60) is often called *Erlang's loss formula*. The distribution (2.59) also appears in a telephone trunking problem with Poisson input and c lines where incoming calls that find all lines occupied are lost (cf. also section 4.3 in chapter 4). The one-unit operating problem is in fact identical with the telephone trunking problem, as is readily seen by identifying the number of units in repair with the number of lines occupied and by using the lack of memory of the Poisson process. Invoking a famous insensitivity result from teletraffic theory for loss systems, we have that the limiting distribution of the number of units in repair is also given by formula (2.59) when the repair time has a general probability distribution with mean $1/\mu$, so that only the first moment of the repair time is required. Another context in which this insensitivity result is useful is an inventory model for slow-moving items dealt with in example 4.6 of section 4.4 of chapter 4.

Methods for the computation of the system reliability

In order to calculate the system reliability, which is defined as the probability that the system will not go down during the next t time units when the present state has all units in a good condition, we parametrize with regard to the starting state. Denoting by the first-passage time T the first epoch at which the process $\{X(t)\}$ enters the state $N + 1$ and thus the system goes down, we define, for any initial state $i \neq N + 1$,

$$Q_i(t) = P\{T > t \mid X(0) = i\}, \qquad t \geqslant 0.$$

In particular, $Q_0(t)$ is the desired system reliability.

We first discuss the powerful method of *Kolmogoroff's backward differential equations* for calculating the first-passage time distributions. This method derives a simultaneous system of linear differential equations for the probabilities $Q_i(t)$, $0 \leqslant i \leqslant N$ and $t > 0$. Fix the initial state $i \neq N + 1$ and $t > 0$. To find the probability that at time $t + \Delta t$ the process has still not visited state $N + 1$, we go backwards in time and condition on the state at time Δt. Under the condition that at time Δt the process is in state j, the probability that the process will not visit state $N + 1$ in the next t time units is $Q_j(t)$. The only possible states at time Δt are the states $i - 1$, $i + 1$ and i when $X(0) = i$. Hence, using the law of total probability, we find

$$Q_i(t + \Delta t) = i\mu\Delta t Q_{i-1}(t) + \lambda\Delta t Q_{i+1}(t) + (1 - i\mu\Delta t - \lambda\Delta t)Q_i(\Delta t) + o(\Delta t),$$

with $Q_{-1}(t) = Q_{N+1}(t) = 0$ for $t \geqslant 0$ by convention. Subtracting $Q_i(t)$ from both sides of the above relation, dividing by Δt and letting $\Delta t \to 0$, we obtain the system of linear differential equations

$$\begin{aligned} Q_i'(t) = i\mu Q_{i-1}(t) + \lambda Q_{i+1}(t) - (i\mu + \lambda)Q_i(t), \\ 0 \leqslant i \leqslant N \qquad \text{and} \qquad t > 0. \end{aligned} \tag{2.61}$$

This system has the boundary condition

$$Q_i(0) = 1, \qquad 0 \leqslant i \leqslant N. \tag{2.62}$$

In general the above system of linear differential equations can only be solved numerically. The solving of linear differential equations with constant coefficients is a well-studied problem in numerical analysis and effective Runge–Kutta methods are known (cf. Grassmann, 1977, and Van As, 1984). It is noteworthy that the tail probabilities $Q_i(t)$ can be shown to decrease exponentially fast to zero as $t \to \infty$.

Using (2.61) and (2.62), it is a simple matter to derive a system of linear equations that uniquely determines the expected first-passage times to state $N + 1$. Denoting by m_i, $0 \leqslant i \leqslant N$, the first moment of the time until the first entry into state $N + 1$ when the initial state is i, we have by (1.9) in section 1.1 of chapter 1 that

$$m_i = \int_0^\infty Q_i(t)\,\mathrm{d}t, \qquad 0 \leqslant i \leqslant N.$$

Integrating both sides of (2.61) over t and using $\int_0^t Q_i'(x)\,\mathrm{d}x = Q_i(t) - Q_i(0) \to -1$

as $t \to \infty$, we find that the numbers m_i satisfy the linear equations

$$i\mu m_{i-1} + \lambda m_{i+1} - (i\mu + \lambda)m_i = -1, \qquad 0 \leqslant i \leqslant N,$$

with $m_{-1} = m_{N+1} = 0$, by convention. Similarly, the higher moments of the first-passage times can be calculated. Letting $m_i^{(k)}$ be the kth moment of the time until the first passage to state $N + 1$ when the initial state is i, it is left to the reader to verify that

$$i\mu m_{i-1}^{(k)} + \lambda m_{i+1}^{(k)} - (i\mu + \lambda)m_i^{(k)} = -km_i^{(k-1)}, \qquad 0 \leqslant i \leqslant N, \qquad (2.63)$$

with $m_{-1}^{(k)} = m_{N+1}^{(k)} = 0$, by convention. Note that $m_i^{(1)} = m_i$. Thus the moments $m_i^{(k)}$, $0 \leqslant i \leqslant N$, can be computed by solving successively k systems of linear equations. The moments of the first-passage times provide a method to approximate the first-passage distributions $Q_i(t)$ by sums of exponentials (see exercise 2.28).

For the particular case of $N = 1$ standby unit, the linear differential equations (2.61) can be solved explicitly. We only state the solution for $Q_0(t)$. By taking the Laplace transforms of both sides of the differential equations (2.61) with $N = 1$, we can determine explicitly the Laplace transform of $Q_0(t)$ from the resulting two linear equations. Next, by inversion of the Laplace transform, it can be seen that

$$Q_0(t) = e^{-(\lambda + \mu/2)t} \left\{ \cosh\left(\frac{t}{2}\sqrt{4\lambda\mu + \mu^2}\right) + \frac{2\lambda + \mu}{\sqrt{4\lambda\mu + \mu^2}} \right.$$

$$\left. \times \sinh\left(\frac{t}{2}\sqrt{4\lambda\mu + \mu^2}\right) \right\}. \qquad t \geqslant 0, \qquad (2.64)$$

where $\cosh(z) = \frac{1}{2}(e^z + e^{-z})$ and $\sinh(z) = \frac{1}{2}(e^z - e^{-z})$. Also, using (2.63) with $N = 1$, we obtain

$$m_0 = \frac{2}{\lambda} + \frac{\mu}{\lambda^2} \quad \text{and} \quad m_0^{(2)} = \frac{(\lambda + \mu)(6\lambda + 2\mu)}{\lambda^4}.$$

Although effective numerical methods exist for solving the linear differential equations (2.61), it may be useful to discuss an alternative method for calculating the system reliability. Such an alternative method is provided by the uniformization technique discussed at the end of section 2.2. Recall that the continuous-time Markov chain allows for the representation that the holding time in each state i is exponentially distributed with mean v_i^{-1} and that at the end of the stay in state i the process jumps to state $j(\neq i)$ with probability p_{ij}. In the problem under consideration,

$$v_i = \begin{cases} i\mu + \lambda, & 0 \leqslant i \leqslant N, \\ (N+1)\mu, & i = N+1, \end{cases}$$

and

$$
p_{ij} = \begin{cases}
\lambda/(i\mu + \lambda), & j = i+1, \quad 0 \leqslant i \leqslant N, \\
i\mu/(i\mu + \lambda), & j = i-1, \quad 0 \leqslant i \leqslant N, \\
1, & j = N, \quad i = N+1, \\
0, & \text{otherwise.}
\end{cases}
$$

For the calculation of the first-passage time until reaching state $N + 1$, we can equivalently consider the following Markov process $\{\bar{X}(t)\}$ in which state $N + 1$ is made absorbing and state transitions occur at a uniform rate of ξ, with ξ chosen as

$$
\xi = \max_{0 \leqslant i \leqslant N} v_i.
$$

The transition epochs of the process $\{\bar{X}(t)\}$ are generated by a Poisson process with rate ξ and the state transitions are described by a (discrete-time) Markov chain with one-step transition probabilities

$$
\bar{p}_{ij} = \begin{cases}
(v_i/\xi)p_{ij}, & j \neq i, \quad 0 \leqslant i \leqslant N, \\
1 - v_i/\xi, & j = i, \quad 0 \leqslant i \leqslant N \\
1, & j = N+1, \quad i = N+1, \\
0, & \text{otherwise.}
\end{cases} \tag{2.65}
$$

Thus, by allowing fictitious transitions from a state to itself, we uniformize the rate at which the process makes state transitions. Also, it is convenient to make state $N + 1$ absorbing when we are interested only in the behaviour of the process until it reaches state $N + 1$ for the first time. Using the lack of memory of the exponential distribution, it can be seen that the first-passage time distribution function $1 - Q_i(t)$ for the original continuous-time Markov chain $\{X(t)\}$ can be calculated as the probability distribution function of the time until the Markov process $\{\bar{X}(t)\}$ is absorbed in state $N + 1$ when the initial state is i. Denoting by $(\bar{p}_{ij}^{(n)})$, $0 \leqslant i, j \leqslant N + 1$, the n-step transition probabilities of the discrete-time Markov chain with one-step transition probabilities $(\bar{p}_{ij})$ and using the fact that state $N + 1$ is absorbing, we have that $\sum_{l \neq N+1} \bar{p}_{il}^{(n)}$ represents the probability that this Markov chain will not visit state $N + 1$ during the first n transitions when the starting state is i. Hence, by conditioning on the number of transitions of the process $\{\bar{X}(t)\}$ up to time t, we find that

$$
Q_i(t) = \sum_{n=0}^{\infty} \sum_{l \neq N+1} \bar{p}_{il}^{(n)} e^{-\xi t} \frac{(\xi t)^n}{n!}, \qquad t > 0 \quad \text{and} \quad 0 \leqslant i \leqslant N,
$$

with the convention $\bar{p}_{il}^{(0)} = 1$ for $l = i$ and $\bar{p}_{il}^{(0)} = 0$ otherwise. For any fixed $t > 0$, the following recursion scheme can be applied for calculating $Q_i(t)$ for $i = 0, \ldots, N$. Letting, for fixed $t > 0$,

$$\xi_i(n) = \sum_{l \neq N+1} \bar{p}_{il}^{(n)} e^{-\xi t} \frac{(\xi t)^n}{n!}, \qquad n \geqslant 1 \quad \text{and} \quad 0 \leqslant i \leqslant N,$$

and using the backward recurrence equation (verify!)

$$\sum_{l \neq N+1} \bar{p}_{il}^{(n)} = \sum_{k \neq N+1} \bar{p}_{ik} \sum_{l \neq N+1} \bar{p}_{kl}^{(n-1)}, \qquad n \geqslant 1 \quad \text{and} \quad 0 \leqslant i \leqslant N,$$

it readily follows that $Q_i(t) = \sum_{n=0}^{\infty} \xi_i(n)$ is obtained by applying the recursion scheme

$$\xi_i(n) = \frac{\xi t}{n} \sum_{k=0}^{N} \bar{p}_{ik} \xi_k(n-1), \qquad n \geqslant 1 \quad \text{and} \quad 0 \leqslant i \leqslant N,$$

starting with $\xi_i(0) = e^{-\xi t}$ for $0 \leqslant i \leqslant N$. Here we remark that in the case ξt is such a large number that exponent underflow occurs when evaluating $e^{-\xi t}$, this numerical difficulty may be eliminated as follows. For some appropriate number ξ_0 with $0 < \xi_0 < \xi$, we split off the term $e^{-\xi_0 t}$ from $e^{-\xi t}$, initiate the recursion scheme with $\tilde{\xi}_i(0) = e^{-(\xi-\xi_0)t}$ and multiply by $e^{-\xi_0 t}$ at the end of the recursive calculations.

Numerical considerations

To conclude this section, we present some numerical results for the system reliability and discuss the effect of the distributional forms of the lifetimes and repair times on this performance measure. Therefore the lifetime L of a unit is assumed to have a Weibull distribution with mean

$$E(L) = 1,$$

where the squared coefficient of variation c_L^2 is varied as 0.5, 1 and 2. Note that for $c_L^2 = 1$ the Weibull distribution reduces to the exponential distribution. The repair time R of a unit has either an exponential distribution or a deterministic distribution. Table 2.4 deals with the $N = 1$ standby unit and gives for various

Table 2.4   The effect of the distributional forms on the system reliability

		$E(T)$	c_T	$Q_0(2)$	$Q_0(5)$	$Q_0(10)$
$c_L^2 = 1$	exp	12.0	0.99	0.851	0.662	0.435
	det	11.6(0.3)	1.00(0.02)	0.840(0.007)	0.650(0.009)	0.424(0.010)
$c_L^2 = 0.5$	exp	27.1(0.5)	0.99(0.02)	0.942(0.005)	0.841(0.007)	0.699(0.009)
	det	32.5(0.6)	0.99(0.02)	0.947(0.005)	0.863(0.007)	0.737(0.009)
$c_L^2 = 2$	exp	6.7(0.2)	1.05(0.02)	0.708(0.009)	0.462(0.010)	0.232(0.009)
	det	6.0(0.2)	1.06(0.02)	0.682(0.009)	0.426(0.010)	0.196(0.008)

cases the values of the mean $E(T)$, the coefficient of variation c_T and the tail probabilities $Q_0(t) = P\{T > t\}$ for $t = 2$, 5 and 10. In all cases of Table 2.4 the average repair time is taken as $E(R) = 1/10$. The results corresponding to the cases of exponential and deterministic repair times are denoted by 'exp' and 'det' in the table. For the cases in which the lifetimes and repair times are not both exponentially distributed, we have used computer simulation to obtain the values of the performance measures. In each case the simulated values and the corresponding 95 per cent confidence intervals are based on 10,000 independent runs; the notation 11.6(3) denotes that the simulated value is 11.6 with [11.3, 11.9] as the 95 per cent confidence interval. The numerical results in Table 2.4 show that the distributional form of the lifetime has a much larger effect on the system reliability than the distributional form of the repair time; for the important case of exponentially distributed lifetimes the system reliability is fairly insensitive to the distributional form of the repair time. In Table 2.5 we give some numerical results indicating the effect of an extra standby unit on the system reliability. Assuming that both the lifetimes and the repair times are exponentially distributed, the number N of standby units is varied as 1, 2 and 3. The average lifetime $E(L)$ is taken equal to 1 and the average repair time $E(R)$ has the two values $\frac{1}{5}$ and $\frac{1}{10}$. The numerical results in Table 2.5 show that the effect of an extra standby unit is considerably larger than one would intuitively expect. Another remarkable finding is that in the cases considered the coefficient of variation c_T is very close to 1, suggesting that the first-passage time T is approximately exponentially distributed. Indeed, our numerical investigations indicate that

$$P\{T > t\} \approx e^{-t/E(T)} \qquad \text{for all} \quad t \geqslant 0 \qquad (2.66)$$

is an excellent approximation when the average failure rate $1/E(L)$ is sufficiently small compared with the average repair rate $1/E(R)$, as will be the case in most practical applications. This empirical finding is in agreement with theoretical results in Keilson (1979) and Solovyev (1971) stating that under general conditions the time until the first occurrence of a 'rare' event in a regenerative stochastic process is approximately exponentially distributed. The importance of

Table 2.5   The effect of an extra standby unit on the system reliability

	$E(R) = 1/5$					$E(R) = 1/10$			
	$E(T)$	c_T	$Q_0(5)$	$Q_0(10)$	$Q_0(25)$	$E(T)$	c_T	$Q_0(5)$	$Q_0(10)$	$Q_0(25)$
$N = 1$	7	0.98	0.493	0.238	0.027	12	0.99	0.662	0.435	0.124
$N = 2$	68	1.00	0.932	0.866	0.694	233	1.00	0.979	0.959	0.899
$N = 3$	984	1.00	0.995	0.990	0.975	6864	1.00	0.999	0.999	0.996

a result such as (2.66) is that it provides qualitative insight. It is noted that for the particular case of the $N = 1$ standby unit the key quantity $E(T)$ can easily be calculated from an analytical expression; see the reliability problem at the end of appendix A.

As a final remark, the above discussion shows that numbers are often indispensable for gaining system understanding, which is ultimately the primary purpose of stochastic modelling. Here both analytical methods and computer simulation might be useful for that purpose. □

Example 2.10 Satellite capacity allocation to two competing user classes

Messages from two sources are sent to a satellite communication system which has s circuits for handling the submitted messages. The two sources 1 and 2 consist of M_1 and M_2 users respectively. Each user of type j generates messages for the satellite according to a Poisson process with rate λ_j whenever the user has no message in service at the satellite, and generates no new messages otherwise. The users act independently of each other. The handling time of a message of type j is exponentially distributed with mean $1/\mu_j$. Each circuit is able to handle a message of any type but can transmit only one message at a time. No queueing is allowed in the satellite system so that an arriving message gets immediate access to a free circuit when that message is admitted to the system. The following acceptance/rejection rule is used for the messages submitted by the two competing user classes. Messages of type 1 are always accepted whenever not all of the s circuits are occupied, whereas messages of type 2 are only accepted when less than L messages of type 2 are being processed and not all of the circuits are occupied. In this example, being an application of the general problem of dynamic allocation of limited resources to competing users with random requests for service, we are interested in the value of the control parameter L that minimizes a weighted sum of the rejection rates of the messages of the types 1 and 2 with c_1 and c_2 as the respective weights.

Assuming a given value of the control parameter L, the stochastic process $\{(X_1(t), X_2(t)), t \geqslant 0\}$ defined by

$X_j(t)$ = the number of messages of type j in the satellite system at time t,

is a continuous-time Markov chain with $I = \{(i_1, i_2) | 0 \leqslant i_1 + i_2 \leqslant s, 0 \leqslant i_2 \leqslant L\}$ as state space. Denoting by $R_j(L)$ the average number of messages of type j that is rejected per unit time, the design criterion is given by

$$R(L) = c_1 R_1(L) + c_2 R_2(L)$$

To evaluate this design criterion, note that messages of type 1 [2] arrive at the satellite system according to a Poisson process with the state-dependent rate $(M_1 - i_1)\lambda_1 [(M_2 - i_2)\lambda_2]$ whenever the system is in state (i_1, i_2). Then, letting $p(i_1, i_2)$ denote the steady-state probability that the process is in state

(i_1, i_2), it follows that

$$R_1(L) = \sum_{i_2=0}^{L} \{M_1 - (s - i_2)\} \lambda_1 p(s - i_2, i_2)$$

and

$$R_2(L) = \sum_{i_1=0}^{s-L} (M_2 - L) \lambda_2 p(i_1, L) + \sum_{i_2=0}^{L-1} (M_2 - i_2) \lambda_2 p(s - i_2, i_2).$$

Other interesting operating characteristics are the average throughput $T(L)$ and the average server occupancy $O(L)$, which are defined as the average number of messages served per unit time and the fraction of time a circuit is occupied. Since messages of type j are completed at an exponential rate of $i_j \mu_j$ whenever i_j messages of type j are in the system and since $p(i_1, i_2)$ represents the long-run fraction of time the system is in state (i_1, i_2), it follows that the average throughput equals

$$T(L) = \sum_{i_1, i_2} (i_1 \mu_1 + i_2 \mu_2) p(i_1, i_2).$$

Also, noting that the average server occupancy equals $(1/s)$ times the average number of occupied circuits, it follows that the average server occupancy equals

$$O(L) = \frac{1}{s} \sum_{i_1, i_2} (i_1 + i_2) p(i_1, i_2).$$

It remains to evaluate the steady-state probabilities $p(i_1, i_2)$. The familiar balance principle 'rate out of a state = rate into that state' yields for any state (i_1, i_2),

$$\{i_1 \mu_1 + i_2 \mu_2 + (M_1 - i_1) \lambda_1 + (M_2 - i_2) \lambda_2 \iota(L - i_2)\} p(i_1, i_2)$$
$$= (M_1 - i_1 + 1) \lambda_1 p(i_1 - 1, i_2) + (M_2 - i_2 + 1) \lambda_2 p(i_1, i_2 - 1)$$
$$+ (i_1 + 1) \mu_1 p(i_1 + 1, i_2) + (i_2 + 1) \mu_2 p(i_1, i_2 + 1),$$

when $i_1 + i_2 < s$; otherwise the left side of this equation should be changed as $(i_1 \mu_1 + i_2 \mu_2) p(i_1, i_2)$. Here $\iota(x) = 1$ for $x \geqslant 1$ and $\iota(x) = 0$ otherwise, and $p(i_1, i_2) = 0$ for the infeasible states (i_1, i_2). These linear equations with

$$\sum_{i_1, i_2} p(i_1, i_2) = 1$$

enable us to calculate the steady-state probabilities $p(i_1, i_2)$.

As an illustration, we consider the following numerical data

$$s = 10, \qquad M_1 = 10, \qquad M_2 = 10 \qquad \lambda_1 = 3, \qquad \lambda_2 = 1, \qquad \mu_1 = 4,$$
$$\mu_2 = 1, \qquad c_1 = 1, \qquad \text{and} \qquad c_2 = 1.$$

In Table 2.6 we give the values of the average rejection rate $R(L)$, the average throughput $T(L)$ and the average server occupancy $O(L)$ for several values of the

Table 2.6 The performance measures $R(L)$, $T(L)$ and $O(L)$.

L	$R(L)$	$T(L)$	$O(L)$
4	4.456	19.833	0.746
5	4.387	19.790	0.789
6	4.653	19.587	0.811

control parameter L. It turns out from the calculations that the design criterion $R(L)$ assumes its minimum value for the L-policy with $L = 5$.

The following remarks are in order with regard to the above model. First, in reality transmission times are not always expected to be exponentially distributed. To find out the effect of the distributional form of the transmission times on the performance measures, we numerically analysed the case of generalized Erlangian distributed transmission times by the method of phases. These additional numerical investigations indicated that under any L-policy the state probabilities and thus the performance measures depend on the transmission times of the types 1 and 2 only through their respective means $1/\mu_1$ and $1/\mu_2$. This empirical finding is not surprising since many stochastic service systems in which no queueing is possible have the property that the state probabilities are insensitive to the form of the service-time distributions and depend only on their means. Indeed, for the present model controlled by an L-policy this insensitivity property can be proved theoretically by using the general concept of local balance. For the special case of the L-policy with $L = s$ the insensitivity result is quite known and is sometimes referred to as the generalized Engset formula (cf. Cohen, 1957). Also, for the present finite-source population model with blocking, it can be shown that the state probabilities are also correct when the think time of a user of type j until sending a new request is generally distributed with mean $1/\lambda_j$. A proof of the insensitivity results can be found in section 2.5.

Second, the above discussion limited itself to the easily implementable L-policy, but other control rules are conceivable. The question of how to compute the optimal control rule among the class of all possible control rules will be addressed to in the next chapter dealing with Markovian decision problems (cf. example 3.7 and exercise 3.12 in chapter 3). It should be noted that the best L-policy is in general not optimal among the class of all possible control rules. However, we found that using the best L-policy rather than the overall optimal control rule often leads to only a small deviation from the optimal value of the design criterion. For example, in the case of the above numerical data the minimum rejection rate of the messages equals 4.246, being only 3.2 per cent less than the rejection rate of 4.387 of the best L-policy. The minimum rejection rate is achieved by the following control rule. Each arriving message of type 1 is accepted whenever not all of the circuits are occupied, whereas a message of type 2 finding upon arrival that i messages of type 1 are present, is accepted only when less than L_i messages of type 2 are in service and not all of the circuits are occupied; for the above data $L_0 = L_1 = 6$, $L_2 = L_3 = L_4 = 5$, $L_5 = 4$, $L_6 = L_7 = 3$, $L_8 = 2$, $L_9 = 1$ and $L_{10} = 0$. It is remarkable that for this more general control rule the above-mentioned insensitivity property is no longer exactly true, but numerical investigations indicate that the dependence on the distributional form of the service times is quite weak. For example, taking an Erlang-2 distribution for the service times of the requests of type 1 with the other data kept the same, the average rejection rate for the above L_i-policy equals 4.272, as opposed to 4.246 for the case of exponential services.   □

*2.5   STOCHASTIC NETWORKS AND INSENSITIVITY

Queueing network models are a useful analysis tool in a wide variety of areas such as computer performance evaluation, communication network design, and production planning in flexible manufacturing. Generally speaking, a network of queues is a collection of service nodes with jobs (users) moving between the nodes and making random requests for service at the nodes. What follows is restricted to *closed* networks of queues, that is, the system contains a fixed number of circulating jobs. The analysis to be given can easily be extended to *open* networks in which jobs may enter and leave the system.

In this section we present some basic results for queueing networks. We will investigate the steady-state probabilities of the number of jobs at the various nodes in the network. The analysis will be based on an appropriate continuous-time Markov chain representation of the network. The already frequently used balance principle (2.42) and the corresponding equations (2.39) will therefore be exploited further. It will appear that a more detailed notion of the balance principle leads to the so called *product-form solution* for the state probabilities and guarantees that this solution is also valid for non-exponential service requirements, the latter result being known as the *insensitivity property*. For the case of exponential services and a single class of jobs, the product-form solution is intricately related to the *node-local-balance* condition requiring that for any node l the rate out of a state due to a change at node l is equal to the rate into that state due to a change at node l. Node-local-balance can sometimes be interpreted as a more refined form of local balance when special queue disciplines are used. Insensitivity of the stochastic network is guaranteed only by the stronger condition of *job-local-balance* requiring that for any marked job the rate out of a state due to that job is equal to the rate into that state due to that same job. Insensitivity is therefore often coupled with the product-form solution for the state probabilities.

In what follows, the product-form solution and insensitivity will be demonstrated both for a rather general non-blocking network and for a two-node closed network with blocking. The latter model encompasses, among others, the stochastic service model studied in example 2.10. In the non-blocking network model any number of jobs can be simultaneously present at any node and the routing of a job is independent of the configuration of the jobs over the various nodes. In the blocking model the actual routing of a job becomes state dependent.

Before presenting the analysis of the network models, the following remark is in order. In the foregoing treatment of continuous-time Markov chains the convention was followed to define the infinitesimal transition rates in such a way that transitions from a state to itself were excluded. In the applications to be discussed below self-transitions may naturally occur. In setting up the balance equations, however, the self-transitions may just as well be included as not since they contribute equally much to the rate out of a state as to the rate into that state.

A closed non-blocking network model

To introduce the product-form solution which was first found in the pioneering work of Jackson (1963), we first consider the following simple model. Suppose a network of N nodes numbered as $1,\ldots,N$, where each node has a single server. At each node there is no bound on the number of jobs that can be simultaneously present at that node. A fixed number of M identical jobs move around in the network. The service requirement of a job at node i is exponentially distributed with mean $1/\mu_i$. All service requirements are assumed to be independent of each other. The queue discipline at each node is first-come–first-served (FIFO) and the server works at a unity rate. An incoming job at node i finding the server busy joins the end of the queue at node i and waits until its turn at service comes. When a job has been served at node i it moves to node j with probability p_{ij}, where $\sum_{j=1}^{N} p_{ij} = 1$ for all i. Here it is allowed that $p_{ii} > 0$. It is assumed that the Markov matrix (p_{ij}) is irreducible, that is, for any i and j there is an integer $n \geqslant 1$ such that the n-step transition probability $p_{ij}^{(n)}$ is positive. In words, a particular job starting at any node can reach any other node.

The irreducibility assumption guarantees that a unique positive solution $\{\lambda_i, 1 \leqslant i \leqslant N\}$ exists to the equilibrium equations

$$\lambda_j = \sum_{i=1}^{N} \lambda_i p_{ij}, \qquad j = 1,\ldots,N \tag{2.67}$$

together with the normalization equation $\sum_{j=1}^{N} \lambda_j = 1$, see (2.11) and (2.12). It is noted that the steady-state probabilities λ_i of the routing matrix can easily be shown to be proportional to the average arrival rates of the jobs into the nodes i $(i = 1,\ldots,N)$.

This network model can be analysed as a continuous-time Markov chain with states $\bar{n} = (n_1, n_2,\ldots,n_N)$, where n_i denotes the number of jobs present at node i and $n_1 + \cdots + n_N = M$. For this continuous-time Markov chain, denote by $p(\bar{n})$ the limiting probability of state $\bar{n}$. Then Jackson (1963) suggested the product-form solution

$$p(\bar{n}) = c \prod_{j=1}^{N} \left(\frac{\lambda_j}{\mu_j}\right)^{n_j}, \qquad (2.68)$$

where c is a normalizing constant such that the sum of all state probabilities $p(\bar{n})$ equals 1. The usual way to prove (2.68) is to verify that the corresponding balance equations (2.39) are satisfied by (2.68), using that the solution to (2.39) is uniquely determined up to a multiplicative constant. For any state $\bar{n}$, let $\bar{n}_{ij}$ denote the state which is equal to state $\bar{n}$ except for that n_i is increased by 1 and n_j is decreased by 1. Then the balance equations (2.39) adjusted for self-transitions read for the present model as

$$p(\bar{n}) \sum_{j:n_j>0} \mu_j = \sum_{j:n_j>0} \left\{ \sum_{i=1}^{N} p(\bar{n}_{ij})\mu_i p_{ij} \right\}. \qquad (2.69)$$

Following Whittle (1967), we now make the following important observation. The equations (2.69) are certainly satisfied by (2.68) when it can be verified that (2.68) satisfies for each j with $n_j > 0$ the detailed balance equation

$$p(\bar{n})\mu_j = \sum_{i=1}^{N} p(\bar{n}_{ij})\mu_i p_{ij}. \qquad (2.70)$$

Rather than expressing global balance like (2.69), the equation (2.70) represents local balance for node j alone. For each node j, the equation (2.70) states that

> The rate out of a state due to a change at node j
> = the rate into that state due to a change at node j.   (2.71)

This property of *node-local-balance* is in general not satisfied in a stochastic network, but can indeed be proved in the present model. To do so, note that from the product-form solution (2.68) we would conclude that

$$p(\bar{n}_{ij}) = p(\bar{n})\frac{\lambda_i \mu_j}{\lambda_j \mu_i}.$$

A direct substitution of this equation into the equation (2.70) leads after a cancellation of common terms to $\lambda_j = \sum_{i=1}^{N} \lambda_i p_{ij}$, which is indeed true by (2.67). We have thus proved the product-form formula by showing that the node-local-balance property is satisfied in the present model.

Remark 2.1.

A product-form solution of the form (2.68) applies also to open queueing networks. Suppose the same network model as above except for (i) jobs arrive from outside the network to each node i according to a Poisson process with rate γ_i, and (ii) the routing matrix $P = (p_{ij})$, $i, j = 1, \ldots, N$ is defective, that is, $\sum_{j=1}^{N} p_{ij} \leq 1$ for each node i, and with probability $1 - \sum_{j=1}^{N} p_{ij}$ a job leaves the system after being served at node i. It is assumed that the n-fold matrix product $P^n \to 0$ as $n \to \infty$, that is, each newly incoming job will eventually leave the network. Then, by the same arguments as above, it follows that the limiting probability of having n_j jobs at node $j(j = 1, \ldots, N)$ at an arbitrary time is again given by (2.68), where the n_j's can now assume any non-negative integer value so that the normalization constant $c = \prod_{j=1}^{N} (1 - \lambda_j/\mu_j)$. Here it should be assumed $\lambda_j/\mu_j < 1$ for all j, where the λ_j's are now the unique non-negative solution to the linear equations

$$\lambda_j = \gamma_j + \sum_{i=1}^{N} \lambda_i p_{ij}, \qquad j = 1, \ldots, N.$$

Noting that at each node the average input equals the average throughput, the reader can easily verify that λ_i can be interpreted as the total arrival rate of jobs into node i.

Returning to the closed network model, a first question is: for which other exponential networks does a product-form solution apply. A second question is whether the product-form solution is also valid for non-exponential services. The answer to the second question is in the negative when service is in order of arrival and thus jobs may have to wait before their service can start. However, for certain other queue disciplines having the property that immediate service is provided to any newly incoming job, it can be established that the stochastic network is insensitive, that is, the state probabilities depend on the service requirements only through their means.

Before discussing further the above questions, we first develop a useful notation which describes a variety of queue disciplines. Imagine, for the moment, an isolated service node. The queue discipline prescribes how the jobs in the queue at the node are served. It is convenient to view the queue as a set of ordered places, each of which may be occupied by one job. Here the places $1, \ldots, x$ are occupied when x jobs are present at the node. We consider now the following parametrization of queue disciplines. A node has a queue discipline (f, ϕ, δ) if the following happens when x jobs are present at the node:

(i) A total amount of service is provided at rate $f(x)$ (service capacity).
(ii) A fraction $\phi(i|x)$ of this service amount is given to the job at the ith place $(i = 1, \ldots, x)$; when the job at the ith place leaves the queue, then the jobs at the places $i+1, \ldots, x$ are shifted to the places $i, \ldots, x-1$ respectively.
(iii) A newly arriving job at the node will be assigned to place $i(i = 1, \ldots, x+1)$ with probability $\delta(i|x)$; when the arriving job is assigned to place i with $i \neq x + 1$, then the jobs previously at the places $i, \ldots, x$ are shifted to the places $i + 1, \ldots, x + 1$.

It is assumed that $f(x) > 0$ if $x > 0$ and that

$$\sum_{i=1}^{x} \phi(i|x) = \sum_{i=1}^{x} \delta(i|x-1) = 1. \tag{2.72}$$

As illustrated by the following examples the parametrization (f, ϕ, δ) covers a wide class of queue disciplines.

(a) Imagine a multi-server node with $s < \infty$ identical servers each of which can handle no more than one job at a time, and the first-come–first-served queue discipline. Then take

$$f(x) = \min(x, s), \quad \phi(i|x) = 1/\min(x, s) \quad \text{for } 1 \leqslant i \leqslant \min(x, s), \ \delta(x+1|x) = 1.$$

(b) Imagine a single-server node and the last-come–first-served preemptive-resume queue discipline. Under this queue discipline the total service effort is given to the job that last arrived and the service of a job returning to the server after having been interrupted is continued from the point at which it was left off. Then take

$$f(x) = 1, \qquad \phi(1|x) = 1 \qquad \text{and} \qquad \delta(1|x) = 1.$$

(c) Imagine a node with the processor-sharing queue discipline. Under this discipline all jobs in the queue receive simultaneous service where all jobs are sharing the service capacity equally. Then, for given service capacity function $f(x)$, we can take

$$\phi(i|x) = \frac{1}{x} \text{ for } 1 \leqslant i \leqslant x \quad \text{and} \quad \delta(i|x) = \frac{1}{x+1} \text{ for } 1 \leqslant i \leqslant x+1.$$

(d) Imagine a node with an infinite number of identical servers where a free server is immediately assigned to each arriving job. Then we can take

$$f(x) = x, \quad \phi(i|x) = \frac{1}{x} \text{ for } 1 \leqslant i \leqslant x$$

and

$$\delta(i|x) = \frac{1}{x+1} \text{ for } 1 \leqslant i \leqslant x+1.$$

It is important to observe that each of the queue disciplines in the examples (b)–(d) has the property that

$$\delta(i|x-1) = \phi(i|x) \qquad \text{for all } i, x. \tag{2.73}$$

Also, note that these queue disciplines have the feature that service is provided immediately to any newly arriving job. The first-come–first-served queue discipline for the delay system in example (a) is lacking this feature and does not satisfy the property (2.73).

Following Kelly (1979), a queue discipline (f, ϕ, δ) satisfying (2.73) is called *symmetric*. It will appear that in non-blocking networks symmetric queue disciplines guarantee insensitivity. Before showing this, it is helpful to extend the product-form solution (2.68) to a rather general closed queueing network with different types of jobs and exponential services. In this extension the concept of node-local-balance need be strengthened to job-local-balance. It is the latter concept that is the key to insensitivity.

Consider a network of N nodes numbered as $1,\ldots,N$, where at each node l we assume a queue discipline (f_l, ϕ_l, δ_l) defined as above. At each node any number of jobs can be simultaneously present. A fixed number of M (different) jobs move around in the network. The jobs are numbered as $1,\ldots,M$. The service requirement of job k at node l is exponentially distributed with mean $1/\mu_l(k)$ for all k and l. All service requirements are assumed to be independent of each other. When job k has been served at node i it moves to node j with probability $p_{ij}(k)$. For each k, it is assumed that the routing matrix $(p_{ij}(k))$ is irreducible. This assumption implies that for each k there exists a unique solution $\{\lambda_j(k), 1 \leqslant j \leqslant N\}$ to the equilibrium equations

$$\lambda_j(k) = \sum_{i=1}^{N} \lambda_i(k)p_{ij}(k), \qquad j = 1,\ldots,N \tag{2.74}$$

together with the normalization equation $\sum_{j=1}^{N} \lambda_j(k) = 1$.

Let $\{r_{li}; i = 1,\ldots,n_l, l = 1,\ldots,N\}$ denote the state for which n_l is the number of jobs present at node l and r_{li} is the job-number of the job at the ith place in the queue at node l. Then the above network can be analysed as a continuous-time Markov chain with such states. Denote by $p(\cdot)$ the limiting distribution of this continuous-time Markov chain. First, under the assumption that the queue discipline (f_l, ϕ_l, δ_l) at node l is symmetric for each $l = 1,\ldots,N$, it will be shown that the state probabilities are given by

$$p(\{r_{li}; i = 1,\ldots,n_l, l = 1,\ldots,N\}) = c \prod_{l=1}^{N} \left[\prod_{i=1}^{n_l} \frac{\lambda_l(r_{li})}{\mu_l(r_{li})} \frac{1}{f_l(i)} \right], \tag{2.75}$$

where c is a normalizing constant. The proof that these probabilities indeed satisfy the global balance equations (2.39) is given by showing that the following detailed balance equations hold in the present model. For each job k,

$$\begin{array}{c} \text{the rate out of a state due to job } k \\ = \text{rate into that state due to job } k. \end{array} \tag{2.76}$$

Following the terminology of Hordijk and Van Dijk (1983), a network satisfying (2.76) is said to have the *job-local-balance* property. Note that job-local-balance implies node-local-balance and global balance. To prove that the product-form solution (2.75) satisfies job-local-balance, we fix the job with job-number k (say). Consider state $\{r_{li}; i = 1,\ldots,n_l, l = 1,\ldots,N\}$ and assume that for this state job k occupies the ith place in the queue at node l. For ease of notation, we denote by $[l,i]$ the state $\{r_{li}; i = 1,\ldots,n_l, l = 1,\ldots,N\}$ and denote by $[j,t]$ the state which only differs from state $[l, i]$ in that job k has moved to place t in the queue at node j and the places of the other jobs are only changed according to the shift protocol given in the definition of the queue discipline (f, ϕ, δ). Also, let $p[j, t]$ denote the probability given by (2.75). Then, the rate out of state $[l, i]$ due to job k is given by

$$p[l, i] f_l(n_l) \phi_l(i | n_l) \mu_l(k). \tag{2.77}$$

The rate into state $[l, i]$ due to job k is given by

$$\sum_{\substack{j \neq l \\ t = 1, \ldots, n_j + 1}} p[j, t] f_j(n_j + 1) \phi_j(t | n_j + 1) \mu_j(k) p_{jl}(k) \delta_l(i | n_l - 1)$$

$$+ \sum_{t = 1, \ldots, n_l} p[l, t] f_l(n_l) \phi_l(t | n_l) \mu_l(k) p_{ll}(k) \delta_l(i | n_l - 1). \tag{2.78}$$

To verify that (2.77) is equal to (2.78) for the product-form solution (2.75), note that from (2.75) we would conclude

$$p[j, t] = p[l, i] \frac{\lambda_j(k) \mu_l(k) f_l(n_l)}{\lambda_l(k) \mu_j(k) f_j(n_j + 1)} \qquad \text{for} \quad j \neq l, t = 1, \ldots, n_j + 1,$$

and $p[l, t] = p[l, i]$ for $t = 1, \ldots, n_l$. Substituting these equations into (2.78) and using the symmetry assumption $\delta_l(i | n_l - 1) = \phi_l(i | n_l)$ together with (2.72), we find that (2.78) equals

$$p[l, i] f_l(n_l) \phi_l(i | n_l) \mu_l(k) \sum_{j=1}^{N} p_{jl}(k) \frac{\lambda_j(k)}{\lambda_l(k)}.$$

The latter expression equals (2.77), by (2.74). We have now proved that the product-form solution (2.75) has the job-local-balance property and thus satisfies the global balance equations (2.39). In the above proof it is not only essential that the queue discipline at each node is symmetric but as well that the queue discipline at each node depends only on the number of jobs present at that node.

The product-form solution (2.75) has the following corollary. Suppose that there is only a single class of jobs, where each job has the same service and routing characteristics so that $\mu_l(k) \equiv \mu_l$ and $\lambda_l(k) \equiv \lambda_l$. Then it follows from (2.75) that the limiting probability of having n_l jobs at node l ($l = 1, \ldots, N$) at an arbitrary time is given by

$$p(n_1, \ldots, n_N) = cM! \prod_{l=1}^{N} \left(\frac{\lambda_l}{\mu_l} \right)^{n_l} \left\{ \prod_{i=1}^{n_l} \frac{1}{f_l(i)} \right\}. \tag{2.79}$$

This result is obtained under the assumption of symmetric queue disciplines. However, as opposed to (2.75), the product-form solution (2.79) is also true without this symmetry assumption provided all services are exponential. This can be argued as follows. By the memoryless property of the exponential distribution, the rate at which jobs leave node j when n_j jobs are there is always

$f_j(n_j)\mu_j$ regardless of which queue discipline (f_j, ϕ_j, δ_j) is used. Thus the state distribution of the total number of jobs present at the various nodes is not affected when we assume that at each node the last-come–first-served preemptive-resume queue discipline is used where the total service effort is given to the job which last arrived. The latter queue discipline is symmetric so that the product-form solution (2.79) holds. Another way to obtain (2.79) is to proceed along the same lines as in the derivation of (2.68).

We return to the product-form solution (2.75) which was established for exponential services under the assumption of symmetric queue disciplines. It will be shown that this product-form solution is also valid for non-exponential services with the same means $1/\mu_l(k)$ as before, provided the crucial assumption of symmetric queue disciplines is maintained. The proof is again based on the verification of job-local-balance for an appropriately defined continuous-time Markov chain that takes also account of the residual life times of the services in progress. Actually, the proof will be given only for a special class of service requirement distributions, namely finite mixtures of Erlangian distributions with the same scale parameters. However, any service requirement distribution can be approximated arbitrarily closely by a member of this class (cf. also section 4.1 in Chapter 4) and thus in view of general convergence results it suffices to verify (2.75) for this special class of distributions. The advantage of the use of mixtures of Erlangian distributions with the same scale parameters is that we can easily define a supplemented continuous-time Markov chain that also describes the residual service requirements.

A mixture of Erlangian distributions with the same scale parameters has a probability density of the form

$$b(x) = \sum_{j=1}^{m} q(j)v^j \frac{x^{j-1}}{(j-1)!} e^{-vx}, \qquad x \geqslant 0, \tag{2.80}$$

where $q(j) \geqslant 0$ and $\sum_j q(j) = 1$. The following interpretation can be given to a service requirement density of this form. With probability $q(j)$ the service requirement is the sum of j independent exponential phases each with the same mean $1/v$, cf. also appendix B. Denoting by $1/\mu = \sum_j (j/v)q(j)$ the mean of the density (2.80), we define the probabilities

$$q^e(r) = \frac{\mu}{v} \sum_{j=r}^{m} q(j), \qquad r = 1, \ldots, m. \tag{2.81}$$

Using the equilibrium excess distribution (1.12) in section 1.1 of Chapter 1, it is readily seen that $q^e(r)$ can be interpreted roughly as the probability that an arbitrary service requirement when inspected at a randomly chosen moment has still to go through r exponential phases until completed.

Suppose now that for all k and l the service requirement of job k at node l has a probability density of the form (2.80) with parameters $m = m_l(k)$, $v = v_l(k)$ and $q(j) = q_l(j;k)$, $1 \leqslant j \leqslant m_l(k)$, and mean $1/\mu_l(k)$. Let $q_l^e(r;k)$ denote the corresponding probability (2.81). To analyse the network model by a continuous-time Markov chain, we have to supplement the states with information about the residual service requirements of the jobs so that the process becomes Markovian. Let state $\{(r_{li}, a_{li}),\ i = 1, \ldots, n_l,\ l = 1, \ldots, N\}$ correspond to the situation that n_l jobs are present at node l and the job at the ith place in the queue at node l has job-number r_{li} and has still to go through a_{li} exponential service phases. Denote by $p(\cdot)$ the limiting distribution of the so defined finite-state, continuous-time Markov chain. Then the following result holds

$$p(\{(r_{li}, a_{li});\ i = 1, \ldots, n_l,\ l = 1, \ldots, N\}) = c \prod_{l=1}^{N} \left[\prod_{i=1}^{n_l} \frac{\lambda_l(r_{li})}{\mu_l(r_{li})} \frac{q_l^e(a_{li}; r_{li})}{f_l(i)} \right], \qquad (2.82)$$

where c is the normalizing constant. The desired product-form solution (2.75) follows directly from (2.82) by summing over all a_{li} and using that $\sum_a q_l^e(a; k) = 1$.

To prove the result (2.82), we proceed along the same lines as in the derivation of (2.75) for the exponential case and verify the job-local-balance property (2.76). To do so, fix job k and state $\{(r_{li}, a_{li});\ i = 1, \ldots, n_l, l = 1, \ldots, N\}$. For ease of notation we denote this state by $[l, i, a]$ when for this state job k has the ith place at node l and has still to go through $a = a_{li}$ exponential phases. Also, let $[j, t, b]$ denote the state which only differs from state $[l, i, a]$ in that job k has moved to place t at node j at which it has b residual phases of service, where the other jobs have only changed their places according to the shift-protocol given in the definition of the queue discipline (f, ϕ, δ). Denote by $p[j, t, b]$ the probability given by (2.82). Then the rate out of state $[l, i, a]$ due to job k is given by

$$p[l, i, a] f_l(n_l) \phi_l(i \mid n_l) v_l(k), \qquad (2.83)$$

while the rate into state $[l, i, a]$ due to job k is given by

$$\sum_{\substack{j \neq l \\ t = 1, \ldots, n_j + 1}} p[j, t, 1] f_j(n_j + 1) \phi_j(t \mid n_j + 1) v_j(k) p_{jl}(k) \delta_l(i \mid n_l - 1) q_l(a; k)$$

$$+ \sum_{t = 1, \ldots, n_l} p[l, t, 1] f_l(n_l) \phi_l(t \mid n_l) v_l(k) p_{ll}(k) \delta_l(i \mid n_l - 1) q_l(a; k)$$

$$+ p[l, i, a + 1] f_l(n_l) \phi_l(i \mid n_l) v_l(k). \qquad (2.84)$$

In the same way as before, it is next verified that the expressions (2.83) and (2.84) are equal to each other for the product-form solution (2.82). This verification uses the symmetry condition $\phi_l(i \mid n_l) = \delta_l(i \mid n_l - 1)$ together with (2.72), the equation (2.74) and the identity $q_l^e(a; k) = q_l^e(a + 1; k) + q_l(a; k) \mu_l(k) / v_l(k)$ with $q_l^e(1; k) = \mu_l(k) / v_l(k)$. The technical details are left to the reader.

We emphasize again that job-local-balance was the key to insensitivity. In the non-blocking model with queue disciplines (f_l, ϕ_l, δ_l) the symmetry of these disciplines is not only sufficient but also necessary for insensitivity, see Hordijk and Van Dijk (1983). This explains that delay systems with first-come–first-served queue disciplines cannot be insensitive.

Additional discussion on closed queueing networks without blocking can be found in subsection 4.2.2 of Chapter 4.

A closed two-node network with blocking

In establishing job-local-balance and thus insensitivity for the non-blocking model, it was important that the routing of each job was independent of the configuration of the jobs over the various nodes. In networks with blocking the actual routing becomes state dependent and thus strong conditions on the routing matrix have to be imposed in order to guarantee insensitivity. The essential features of such routing conditions are reflected in the following simple blocking model, which, however, is general enough to cover a number of practical applications.

Consider a closed network model with two nodes in cyclic order. If a job has completed service at one of the nodes it requests for service at the other node. A fixed number of M jobs move around in the network. There are p types of jobs, where M_h of the jobs are of type h $(h = 1, \ldots, p)$ with $M_1 + \cdots + M_p = M$. The following blocking protocol is assumed.

Blocking protocol

If a job of type h arrives at node l when n_l jobs are already present at node l including n_l^h jobs of type h, then the arriving job of type h is accepted at node l with a probability of the form

$$A_l(n_l)A_l^h(n_l^h), \qquad l = 1, 2 \quad \text{and} \quad h = 1, \ldots, p. \tag{2.85}$$

If a service request is rejected, the blocked job remains at (or returns to) its place at the preceding node as just before its service completion and has to undergo a complete new service at that node.

In most applications the functions $A_l(\cdot)$ and $A_l^h(\cdot)$ assume only the values 1 and 0. The service requirement of a job of type h at node l has a general probability distribution with mean $1/\mu_l(h)$. All service requirements are assumed to be independent of each other. At each node l a queue discipline of the form (f_l, ϕ_l, δ_l) is followed. Here (f_l, ϕ_l, δ_l) is defined as before with the only modification that the placing function δ_l now refers only to jobs that are accepted at node l. It is assumed that the queue discipline is symmetric at both nodes. In many applications of the network model with blocking we have at each node a symmetric queue discipline of the form

$$f_l(x) = x \quad \text{and} \quad \phi_l(i|x) = \delta_l(i|x-1) = 1/x \qquad \text{for all } i, x, \tag{2.86}$$

corresponding to the case that each accepted job at a node is immediately provided with a free server. We give two examples of closed two-node systems with finite-source input. It is noted that such systems with a very large number of sources behave approximately like open systems with Poisson input.

(i) The generalized Engset model with ideal grading

Suppose a stochastic service system with s servers and M sources with $M \geqslant s$, where each source can generate requests for service by one of the servers. A request for service hunts over K randomly chosen servers for a free server, where K is a given integer with $1 \leqslant K \leqslant s$. In case the request finds a free server among the chosen K servers, it is provided immediately a free server and its source becomes busy; otherwise the request is lost and its source starts a new think time until it generates a new request. A source cannot generate any new requests during the time it has a request in service. Upon completion of a service request, its source becomes idle and starts again a think time until the generation of a next request. There are p types of sources, where M_h sources are of type h $(h = 1,\ldots,p)$ with $M_1 + \cdots + M_p = M$. Each source type may have its own characteristics for the think time and the service requirements of the requests generated by that source type. All think times and service requirements are assumed to be independent of each other.

This stochastic service system can be modelled as a closed network with two nodes 1 and 2 in cyclic order and a fixed number of M jobs. Node 1 has M available places and generates jobs for node 2 that has only s available places. Node 1 is in fact an infinite-server node and node 2 is the only node at which blocking can occur. The blocking characteristics are determined by

$$A_1(n_1) = A_1^h(n_1^h) \equiv 1 \qquad \text{for } h = 1,\ldots,p,$$

and

$$A_2(n_2) = 1 - \binom{n_2}{K} \Big/ \binom{s}{K}, \qquad A_2^h(n_2^h) \equiv 1 \qquad \text{for } h = 1,\ldots,p,$$

with the convention $\binom{n}{m} = 0$ for $n < m$. For the particular case of $K = s$, we have $A_2(n_2) = 1$ for $n_2 < s$ and $A_2(n_2) = 0$ otherwise. At both nodes we have a symmetric queue discipline (f_l, ϕ_l, δ_l) that can be represented by (2.86).

(ii) The resource allocation problem of example 2.10 with the L-policy

This problem can be modelled as a cyclic closed network with two nodes 1 and 2. Node 1 is an infinite-server node, while node 2 has s servers and is the only node at which blocking can occur. A fixed number of jobs move around in the network. There are two types of jobs, where M_1 jobs are of type 1 and M_2 jobs of type 2. Each job type h has its own service characteristics at node l with $1/\mu_1(h) = 1/\lambda_h$ and $1/\mu_2(h) = 1/\mu_h$ $(h = 1, 2)$. Jobs of type 1 are accepted at node 2 only when not all of the servers are busy, while the acceptance of jobs of type 2 at node 2 is controlled by the L-policy. For this problem the blocking characteristics are determined by

$$A_1(n_1) = A_1^h(n_1^h) \equiv 1 \quad \text{for } h = 1, 2,$$

and

$$A_2(n_2) = \begin{cases} 1 & \text{for } n_2 < s, \\ 0 & \text{otherwise,} \end{cases} \quad A_2^1(n_2^1) \equiv 1,$$

$$A_2^2(n_2^2) = \begin{cases} 1 & \text{for } n_2^2 < L, \\ 0 & \text{otherwise.} \end{cases}$$

We now resume our discussion on the two-node closed network model. Let us say that at time t the network is in state $\bar{n} = (n_l^h)$ when at time t there are n_l^h jobs of type h present at node $l(h = 1, \ldots, p; \ l = 1, 2)$, where $n_1^h + n_2^h = M_h$ for all h. For any state $\bar{n} = (n_l^h)$, let

$$n_l = \sum_{h=1}^{p} n_l^h$$

denote the total number of jobs present at node $l(l = 1, 2)$. Then, under the assumption of symmetric queue disciplines, it will be shown that the limiting probability that the network is in state $\bar{n}$ at an arbitrary time is given by

$$p(\bar{n}) = c' \prod_{l=1}^{2} \left[\left\{ \sum_{i=1}^{n_l} \frac{A_l(i-1)}{f_l(i)} \right\} \left\{ \prod_{h=1}^{p} \prod_{j=1}^{n_l^h} \left(\frac{A_l^h(j-1)}{\mu_l(h)} \right) \right\} \right], \quad (2.87)$$

where c' is the normalization constant. This insensitive product-form solution involves the service requirements only through their means. It is remarked that for the particular case of exponential services the result (2.87) is also valid for non-symmetric queue disciplines provided that the mean service times $1/\mu_l(h)$ are independent of the job type h at each node l with a non-symmetric queue discipline (f_l, ϕ_l, δ_l). This can be seen by the same arguments as below (2.79).

To prove the insensitivity result (2.87), we proceed in a similar way as in the proof of (2.75). We show (2.87) only for the case that each service requirement distribution is a mixture of Erlangian distributions with the same scale parameters. Next, by deep continuity arguments, the result (2.87) can be established for generally distributed service requirements with means $1/\mu_l(h)$. Hence, suppose that the service requirement of a job of type h at node l has a probability density of the form (2.80) with parameters $m = m_l(h)$, $v = v_l(h)$ and $q(j) = q_l(j; h)$, $1 \leqslant j \leqslant m_l(h)$. The corresponding residual life probabilities (2.81) are denoted by $q_l^c(r; h)$. Then, using the same notation as in the previous non-blocking model, we define a continuous-time Markov chain with states $\{(r_{li}, a_{li}); \ i = 1, \ldots, n_l, \ l = 1, 2\}$. Also, let h_{li} denote the type of the job with job-number r_{li} and denote by n_l^h the number of jobs with $h_{li} = h$. It will be shown that the limiting distribution of this continuous-time Markov chain is given by

$$p(\{(r_{li}, a_{li}); i = 1, \ldots, n_l, l = 1, 2\})$$

$$= c \prod_{l=1}^{2} \left[\left\{ \prod_{i=1}^{n_l} \frac{A_l(i-1)}{f_l(i)} q_l^e(a_{li}; h_{li}) \right\} \left\{ \prod_{h=1}^{p} \sum_{j=1}^{n_l^h} \left(\frac{A_l^h(j-1)}{\mu_l(h)} \right) \right\} \right], \quad (2.88)$$

where c is the normalization constant. The desired result (2.87) follows easily from (2.88) by summing over all a_{li} and using that $\sum_a q_l^e(a; h) = 1$. We verify (2.88) by proving that the present model satisfies job-local-balance. To do so, fix an arbitrary job (say, job k) and consider a state $\{(r_{li}, a_{li}); i = 1, \ldots, n_l, l = 1, 2\}$. As before, we abbreviate this state by $[l, i, a]$ when $r_{li} = k$ and $a_{li} = a$. Also, we define state $[j, t, b]$ as before and denote by $p[j, t, b]$ the corresponding probability (2.88). Let $h = h_{li}$ denote the type of job k. Noting that the network is symmetric in $l = 1, 2$, it is no restriction to assume $l = 1$. Then the rate out of state $[1, i, a]$ due to job k is given by

$$p[1, i, a] f_1(n_1) \phi_1(i | n_1) v_1(h). \quad (2.89)$$

The rate into state $[1, i, a]$ due to job k is given by

$$\sum_{t=1}^{n_2+1} p[2, t, 1] f_2(n_2 + 1) \phi_2(t | n_2 + 1) v_2(h) A_1(n_1 - 1) A_1^h(n_1^h - 1)$$

$$\times \delta_1(i | n_1 - 1) q_1(a; h) + p[1, i, 1] f_1(n_1) \phi_1(i | n_1) v_1(h) \quad (2.90)$$

$$\times \{1 - A_2(n_2) A_2^h(n_2^h)\} q_1(a; h) + p[1, i, a + 1] f_1(n_1) \phi_1(i | n_1) v_1(h).$$

To verify that the expressions (2.89) and (2.90) are equal to each other for the product-form solution (2.88), we first note that (2.88) would imply

$$p[1, i, 1] = p[1, i, a] q_1^e(1; h) / q_1^e(a; h),$$

$$p[1, i, a + 1] = p[1, i, a] q_1^e(a + 1; h) / q_1^e(a; h),$$

$$p[2, t, 1] = p[1, i, a] \frac{f_1(n_1) q_2^e(1; h)}{f_2(n_2 + 1) q_1^e(a; h)} \frac{\mu_1(h) A_2(n_2) A_2^h(n_2^h)}{\mu_2(h) A_1(n_1 - 1) A_1^h(n_1^h - 1)}.$$

Here it is noted that the denominator of the last expression is positive since ~~otherwise the state $[1, i, a]$ would not have been accessible.~~ Substituting these equations in (2.90) and using the symmetry condition $\phi_1(i | n_1) = \delta_1(i | n_1 - 1)$ and $q_l^e(1; h) = \mu_l(h)/v_l(h)$, the equality of both expressions (2.89) and (2.90) follows in view of the identity

$$\frac{1}{q_1^e(a; h)} \left[\frac{\mu_1(h)}{v_1(h)} A_2(n_2) A_2^h(n_2^h) q_1(a; h) \sum_{t=1}^{n_2+1} \phi_2(t | n_2 + 1) \right.$$

$$\left. + \frac{\mu_1(h)}{v_1(h)} \{1 - A_2(n_2) A_2^h(n_2^h)\} q_1(a; h) + q_1^e(a + 1; h) \right] = 1. \qquad \square$$

Finally, the following remark is in order. In the blocking model the most important probabilities are the blocking probabilities rather than the time-average probabilities (2.87). The blocking probability of a job of type h at node l is defined as the probability that a job of type h is blocked upon arrival at node l when coming from the other node. The blocking probabilities depend on the service requirements only through their means. This insensitivity property cannot be concluded directly from the result (2.87), but can be proved by using the more detailed result (2.88) in conjunction with the following two observations. First, the blocking probability of a job of type h at node $l = 2$ (say) equals the average number of visits of a job of type h from node 1 to node 2 per unit time that are blocked divided by the average number of visits of a job of type h from node 1 to node 2 per unit time. Second, supposing that job k is of type h, a blocking of job k at node $l = 2$ can happen only at the moments at which a transition occurs due to a service completion of job k in some state $\{(r_{li}, a_{li}); i = 1, \ldots, n_l, l = 1, 2\}$ with $r_{1i} = k$ and $a_{1i} = 1$ for some i. The rate at which such transitions occur from this state is $p(\{(r_{li}, a_{li}); i = 1, \ldots, n_l, l = 1, 2\}) f_1(n_1)\phi_1(i|n_1)v_1(h)$, where a fraction $1 - A_2(n_2)A_2^h(n_2^h)$ of these transitions leads to a blocking of job k at node $l = 2$. Using these two observations and the product-form solution (2.88), it is a matter of technical manipulations to verify that the blocking probability of a job of type h at node l has indeed the insensitivity property.

EXERCISES

2.1 To meet the demand for water in a region, it is decided that a dam should be built in a river. For the dam capacity there is an option of either 2 units or 3 units. The probability distribution of the number of units of water that flow into the dam during each week is given by $p_0 = \frac{1}{8}$, $p_1 = \frac{1}{4}$, $p_2 = \frac{1}{2}$ and $p_3 = \frac{1}{8}$. If the supply of water by the river exceeds the remaining capacity of the dam, the excess water is lost. The demand for water is 2 units in each week where it is assumed that the water has to be provided at the beginning of each week. If the dam does not contain enough water to satisfy the demand, the shortage will be fulfilled at a cost of 10 per unit shortage. The weekly depreciation costs of the dam have the values of 2 and $2\frac{1}{2}$ for the respective capacities of 2 and 3 units. Determine for which dam capacity the long-run average costs per week are minimal. (*Answer*: capacity 3 with average weekly costs of 6.56.)

2.2 Consider an aperiodic discrete-time Markov chain. Supposed one is asked to predict the present state of the system once the system has undergone many transitions. Give an estimate of the probability to find the system in state j in the case where

(a) no information is available about the past history of the system,

(b) the information is available that the system visited state i five transitions before.

2.3 A factory has a buffer with a capacity of $4\,\text{m}^3$ for temporarily storing waste produced by the factory. Each week the factory produces $k\,\text{m}^3$ waste with a probability p_k, where $p_0 = \frac{1}{8}$, $p_1 = \frac{1}{2}$, $p_2 = \frac{1}{4}$ and $p_3 = \frac{1}{8}$. If the amount of waste produced in a week exceeds the remaining capacity of the buffer, the excess is specially removed at a cost of 30 per m^3. At the end of each week there is a regular opportunity to remove waste from the storage buffer at a fixed cost of 25 and a variable cost of 5 per m^3. The following policy is used. If at the end of the week the storage buffer contains more than $2\,\text{m}^3$ the buffer is emptied; otherwise no waste is removed. Determine the long-run average cost per week. (*Answer*: 17.77.)

2.4 A machine has two critical parts that are subject to failure. The machine can continue to operate if one part has failed. Only in the case where both parts are no longer intact does a repair need to be done. A repair takes one day and after a repair both parts are intact again. At the beginning of each day the machine is examined to determine whether a repair is required or not. If at the beginning of a day both parts are intact, then at the end of the day exactly one part will be intact with probability 0.25 and both parts will be intact with probability 0.5. In case only one part is intact at the beginning of a day, this part will have failed at the end of the day with probability 0.5. Each repair costs 50, while for each day the machine is running a reward of 100 is earned. Determine the long-run average reward per day. (*Answer*: 87.5.)

2.5 Consider the discrete-time Markov chain studied in section 2.1. Verify that for each set A of states with $A \neq I$,

$$\sum_{j \in A} \pi_j \sum_{k \notin A} p_{jk} = \sum_{i \notin A} \pi_i \sum_{j \in A} p_{ij}.$$

Also, derive this relation by a direct probabilistic argument. Discuss the computational advantage of the above relation in the case $I = \{0, 1, \ldots, N\}$ and $p_{ij} = 0$ for $j \leqslant i - 2$. (*Hint*: take $A = \{i, \ldots, N\}$.)

2.6 Suppose an inventory system in which replenishments occur at a constant rate of Q items every T time units. The demand for the item is assumed to be a Poisson process with rate μ. Any demand occurring while the system is out of stock is lost. Set up a Markov chain analysis for the computation of the average on-hand inventory. Give a direct argument showing that the fraction of demand that is lost equals $1 - Q/T\mu$. Here it is assumed that $T\mu > Q$.

2.7 Consider a periodic review inventory system in which every period a replenishment order is placed for the demand that occurred in the last period. The lead time of a replenishment order is either $\frac{1}{4}$ period or $1\frac{1}{2}$ period. If the lead time of the current replenishment order is α periods, then the lead time of the next replenishment order will also be α periods with probability $q(\alpha)$ for $\alpha = \frac{1}{4}$, $1\frac{1}{2}$. For the numerical data $q(\frac{1}{4}) = 0.8$ and $q(1\frac{1}{2}) = 0.5$, calculate the long-run fraction of replenishment orders that cross in time. (*Answer*: 1/7.)

2.8 At the beginning of each day a crucial piece of an electronic equipment is inspected and then classified as being in one of the working conditions $i = 1,\ldots,N$. Here the working condition i is better than the working condition $i + 1$. If the working condition is $i = N$ the piece must be replaced by a new one and such an enforced replacement takes two days. If the working condition is i with $i < N$ there is a choice between replacing preventively the piece by a new one and letting the piece operate for the present day. A preventive replacement takes one day. A new piece has working condition $i = 1$. A piece whose present working condition is i has the next day the working condition j with known probability q_{ij} where $q_{ij} = 0$ for $j < i$. The following replacement rule is used. The current piece is only replaced by a new one when its working condition is greater than the critical value m where m is a given integer with $1 \leqslant m < N$. Determine the fraction of days that the equipment is inoperative and determine the fraction of replacements occurring in the failure state N.

2.9 At the beginning of each day a batch of containers arrives at a stackyard having ample capacity to store any number of containers. The batch size has a discrete probability distribution $\{q_k, k \geqslant 1\}$. Each container stays an exponentially distributed time at the stackyard, where the holding times of the various containers are independent of each other. Show how to calculate the mean and the standard deviation of the number of containers present at the stackyard at the beginning of each day. Write a computer program and solve the case in which the batch size is geometrically distributed with mean 2 and the exponentially distributed holding time of a container has a mean of 5 days. (*Answer*: mean = 11.03 and standard deviation = 3.32.)

2.10 Suppose that a conveyer belt is running at a uniform speed and transporting items on individual carriers equally spaced along the conveyer. There are N work stations placed in order along the conveyer. Each time unit an item for processing arrives and is handled by the first work station that is idle. Any station can process only one item at a time and has no storage capacity. An item that finds all of the N work stations busy is lost. The processing time of an item at station i has an Erlang-r_i distribution with mean m_i, $i = 1,\ldots,N$. The work stations are numbered $i = 1,\ldots,N$ according to the order they are placed along the conveyer. Give a Markov chain analysis aimed at the computation of the loss probability. For the model with $N = 2$ work stations, solve for the two cases:

(a) the processing times at the stations 1 and 2 are exponentially distributed with respective means $m_1 = 0.75$ and $m_2 = 1.25$ (*answer*: 0.0467);
(b) the processing times at the stations 1 and 2 are Erlang-3 distributed with respective means $m_1 = 0.75$ and $m_2 = 1.25$ (*answer*: 0.0133).

(This problem is motivated by Gregory and Litton, 1975.)

2.11 At a telephone exchange calls arrive according to a Poisson process with rate λ. The calls are first put in an infinite capacity buffer before they can be processed further. The buffer is periodically scanned every T time units and only at those scanning epochs calls in the buffer are allocated to free servers. There are c servers and each server can handle only one call at a time. The service times of the calls are independent random variables having a common exponential distribution with mean $1/\mu$. Set up the equilibrium equations of a suitably chosen discrete-time Markov chain in order to determine the average number of calls in the buffer. Also, relate the average delay in the buffer per call to the average number of calls in the buffer.

2.12 At a single-server station potential customers arrive according to a Poisson process with rate λ. A customer joins the system with probability $1/(n + 1)$ when he finds upon arrival n other customers present. The service times of the customers are independent random variables having a common exponential distribution with mean $1/\mu$. Verify that the limiting distribution of the number of customers in the system is a Poisson distribution with mean λ/μ. Also, determine the average number of customers per unit time who actually join the system.

2.13 Messages arrive at a communication channel according to a Poisson process with rate λ. The message length is exponentially distributed with mean $1/\mu$. An arriving message finding that the line is idle is provided with service immediately; otherwise the message waits until access to the line can be given. The communication line is only able to submit one message at a time, but has available two possible transmission rates σ_1 and σ_2 with $0 < \sigma_1 < \sigma_2$. Thus the transmission time of a message is exponentially distributed with mean $1/\sigma_i\mu$ when the transmission rate σ_i is used. It is assumed that $\lambda/\sigma_2\mu < 1$. At any time the transmission line may switch from one rate to the other. The transmission rate is controlled by a single-critical-number rule. The transmission rate σ_1 is used whenever less than R messages are present, while otherwise the faster transmission rate σ_2 is used. The following costs are involved. There is a holding cost at rate hj whenever j messages are in the system. An operating cost at rate $r_i > 0$ is incurred when the line is transmitting a message using rate σ_i, while an operating cost at rate $r_0 \geqslant 0$ is incurred when the line is idle. Derive a recursion scheme for the computation of the limiting distribution of the number of messages present and give an expression for the long-run average cost per unit time. Write a computer program for the calculation of the value of R which minimizes the average cost and solve for the numerical data $\lambda = 0.8$, $\mu = 1$, $\sigma_1 = 1$, $\sigma_2 = 1.5$, $h = 1$, $r_0 = 0$, $r_1 = 5$ and $r_2 = 25$. (*Answer*: $R = 10$ and the minimal average cost is 7.66.)

2.14 In an inventory system for a single product the depletion of stock is due to demand and deterioration. The demand process for the product is a Poisson process with rate λ. The lifetime of each unit product is exponentially distributed with mean $1/\mu$. The stock control is exercised as follows. Each time the stock drops to zero an order for Q units is placed. The lead time of each order is negligible. Determine the average stock and the average number of orders placed per unit time.

2.15 Assume a single processor where jobs arrive according to a Poisson process with rate λ. The processor can handle only one job at a time and has a random processing time S per job where $E(S)$ and $E(S^2)$ are finite with $\lambda E(S) < 1$. One considers replacing this centralized system by a decentralized system with n slower processors. In this latter system the jobs are divided equally among the n processors, while the processing time per job becomes n times as large as before. Verify that the effect of decentralization is that the average delay in queue per job becomes n times larger than in the centralized system.

2.16 In a hierarchical communication system users arrive at the top level according to a Poisson process with rate λ. Each user while at the top level generates tasks for the processor at the second level according to a Poisson process with rate ϕ. A user stays at the top level during an exponentially distributed time with mean $1/\mu$. The processor can handle only one task at a time. The processing time of a task is exponentially distributed with mean $1/\eta$. Show that the limiting distribution of the number of users present at the top level has a Poisson distribution with mean λ/μ and that the average throughput of the processor equals $\phi\lambda/\mu$. Set up the balance equations for the joint distribution of the number of users at the top level and the number of tasks at the processor.

2.17 Consider example 2.6 and denote by S_{ij} the time needed to serve an alarm for district j by unit i. Show how to calculate the performance measures π_L = the fraction of alarms that is lost and P_i = the fraction of time that unit i is busy, when

(a) S_{ij} is distributed as the sum of two independent exponentials for all i, j;
(b) S_{ij} has an H_2 distribution for all i, j.

Letting m_{ij} and c_{ij}^2 denote the mean and the squared coefficient of variation of S_{ij}, assume the numerical data $\lambda_1 = 0.25$, $\lambda_2 = 0.25$, $m_{11} = 0.75$, $m_{12} = 1.25$, $m_{21} = 1.25$ and $m_{22} = 1$. Write a computer program to verify the following numerical results:

(i) $\pi_L = 0.0704$, $P_1 = 0.1302$, $P_2 = 0.1622$ when $c_{ij}^2 = \frac{1}{2}$ for all i, j;
(ii) $\pi_L = 0.0708$, $P_1 = 0.1296$, $P_2 = 0.1615$ when $c_{ij}^2 = 1$ for all i, j;
(iii) $\pi_L = 0.0718$, $P_1 = 0.1282$, $P_2 = 0.1602$ when $c_{ij}^2 = 4$ for all i, j.

Here the values $c_{ij}^2 = \frac{1}{2}$, 1 and 4 correspond to the E_2 distribution, the exponential distribution and the H_2 distribution with balanced means.

2.18 Consider example 2.8 and assume now that the unloading time is distributed as the sum of two independent exponentials. Show how to calculate the various performance measures when

(a) the trip time has an H_2 distribution;
(b) the trip time has an E_2 distribution.

2.19 Consider example 2.8 and suppose now that there are two unloaders each having its own queue. An arriving trailer joins the unloader with the shortest queue and randomly picks an unloader when both queues are equally long. No jockeying between the queues is possible. Extend the analysis in example 2.8.

2.20 An assembly line for a certain product has two stations in series. Each station has only room for a single unit of the product. If the assembly of a unit is completed at station 1, it is forwarded immediately to station 2 provided station 2 is idle; otherwise the unit remains in station 1 until station 2 becomes free. Units for assembly arrive at station 1 according to a Poisson process with rate λ, but a newly arriving unit is only accepted by station 1 when no other unit is present in station 1. Each unit rejected is handled elsewhere. The assembly times at the stations 1 and 2 are exponentially distributed with respective means $1/\mu_1$ and $1/\mu_2$. Show how to calculate the fraction of units that are not accepted by station 1. Also, use Little's formula to determine the average time spent by an accepted unit in the assembly line.

2.21 Suppose a single unloader system at which trains arrive which bring coal from various mines. There are N trains involved in the coal transport. The coal unloader can handle only one train at a time and the unloading time per train has an exponential distribution with mean $1/\mu$. The unloader is subject to breakdowns when the unloader is in operation. The operating time of an unloader has an exponential distribution with mean $1/\eta$ and the time to repair a broken unloader is exponentially distributed with mean $1/\xi$. The unloading of the train that is in service when the unloader breaks down is resumed as soon as the repair of the unloader is completed. An unloaded train returns to the mines for another trainload of coal. The time for a train to complete a trip from the unloader to the mines and back is assumed to have an exponential distribution with mean $1/\lambda$. Show how to calculate performance measures such as the average number of trains at the unloader (L), the average time a train has to wait until unloading is completed (W), and the average number of trains unloaded per unit time (T). Write a computer program and solve for the numerical data $N = 5$, $\mu = 2.5$, $\eta = 0.10$, $\xi = 0.35$ and $\lambda = 0.2$. (*Answer:* $L = 0.765$, $W = 0.903$ and $T = 0.847$.) Extend your analysis in order to investigate the effect of the simplifying assumption of exponentially distributed trip times on the performance measures. (This problem is based on Chelst, Tilles and Pipis, 1981.)

2.22 Suppose we wish to determine the capacity of a stack yard at which containers arrive according to a Poisson process with a rate of $\lambda = 1$ per hour.

A container finding upon arrival that the yard is full is brought elsewhere. The time that a container is stored in the yard is exponentially distributed with a mean of $1/\mu = 10$ hours. Determine the required capacity of the yard so that no more than 1 per cent of the arriving containers find the yard full. (*Answer*: capacity 19.) How does the answer change when the time that a container is stored in the yard is uniformly distributed between 5 and 15 hours?

2.23 Long-parkers and short-parkers arrive at a parking place for cars according to independent Poisson processes with respective rates of $\lambda_1 = 4$ and $\lambda_2 = 6$ per hour. The parking place has room for $N = 10$ cars. Each arriving car which finds all places occupied goes elsewhere. The parking time of long-parkers is uniformly distributed between 1 and 2 hours, while the parking time of short-parkers has a uniform distribution between 20 and 60 minutes. Calculate the probability that upon arrival a car finds all parking places occupied. (*Answer*: 0.215.)

2.24 Suppose a conveyor system at which items for processing arrive according to a Poisson process with rate λ. The service or processing requirements of the items are independent random variables having a common exponential distribution with mean $1/\mu$. The conveyor system has two work stations 1 and 2 that are placed according to this order along the conveyor. Work station i consists of s_i identical service channels, each having a constant processing rate of σ_i $(i = 1, 2)$, that is an item processed at work station i has an average processing time of $1/\sigma_i\mu$. Both work stations have no storage capacity and each service channel can handle only one item at a time. An arriving item is processed by the first work station in which a service channel is free and is lost when no service channel is available at either of the stations. Show how to calculate the fraction of items that is lost and solve for the numerical data $\lambda = 10$, $\mu = 1$, $\sigma_1 = 2$, $\sigma_2 = 1.5$, $s_1 = 5$ and $s_2 = 5$. (*Answer*: 0.0306.) Verify experimentally that the loss probability is nearly insensitive to the distributional form of the service requirement (e.g. compute the loss probability 0.0316 for the above numerical data when the service requirement has an H_2 distribution with balanced means and a squared coefficient of variation of 4).

2.25 Suppose a stochastic service system with Poisson arrivals at rate λ and two different groups of servers, where each arriving customer requires simultaneously a server from both groups. An arrival not finding that both groups have a free server is lost and has no further influence on the system. The ith group consists of s_i identical servers $(i = 1, 2)$ and each server can handle only one customer at a time. An entering customer occupies the two assigned servers from the groups 1 and 2 during independently exponentially distributed times with respective means $1/\mu_1$ and $1/\mu_2$. Show how to calculate the loss probability and solve for the numerical data $\lambda = 1$, $1/\mu_1 = 2$, $1/\mu_2 = 5$, $s_1 = 5$ and $s_2 = 10$. (*Answer*: 0.0464.) Verify experimentally that the loss probability is nearly insensitive to the distributional form of the service times (e.g. compute the loss probability 0.0470 for the above data when the service time in group 1 has an E_2 distribution and the service time in group 2 has an H_2 distribution with balanced means and a squared coefficient of variation of 4).

2.26 Consider the following stochastic service model having applications in both message storage and data multiplexing problems. Customers of the types 1 and 2 arrive at a shared resource with c units according to independent Poisson processes with respective rates λ_1 and λ_2. A customer of type i requires b_i resource units and is rejected anyway when less than b_i units are free. An accepted customer of type i has an exponentially distributed residency time with mean $1/\mu_i$ during which all of the b_i assigned resource units are kept occupied so that the b_i units are relinquished simultaneously when the customer departs. Suppose the complete sharing policy is followed under which an arriving customer of type i is rejected only if less than b_i resource units are free. Show how to calculate the fraction of customers of type i, $i = 1, 2$, that is lost. Solve for the numerical data $c = 50$, $b_1 = 3$, $b_2 = 2$, $\lambda_1 = 10$, $\lambda_2 = 5$, $1/\mu_1 = 1$ and $1/\mu_2 = 1$. (*Answer*: the respective loss probabilities of type 1 and type 2 customers are 0.0863 and 0.0551.) Verify (experimentally) that in the present model the state probabilities are insensitive to the distributional form of residency times. (This problem is taken from Kaufman, 1981.)

2.27 Suppose a production facility has M operating machines and a buffer of B standby machines. Machines in operation are subject to breakdowns. The running times of the operating machines are independent of each other and have a common exponential distribution with mean $1/\lambda$. An operating machine that breaks down is replaced by a standby machine if one is available. A failed machine immediately enters repair. There are ample repair facilities so that any number of machines can be repaired simultaneously. The repair time of a failed machine is assumed to have an exponential distribution with mean $1/\mu$. For given values of μ, λ and M, demonstrate how to calculate the minimum buffer size B in order to achieve that the long-run fraction of time, that less than M machines are operating, is no more than a specific value β. Do you expect the answer to depend on the specific form of the repair-time distribution? Prove your guess by using the general results in section 2.5.

2.28 Consider example 2.9. The moments method approximates the system reliability $Q_0(t)$ by

$$Q_0(t) \approx \sum_{j=1}^{K} a_j e^{-b_j t}, \qquad t \geqslant 0,$$

by using the boundary condition $Q_0(0) = 1$ and by matching a sufficient number of methods. In general the moments method with K fixed is not guaranteed to be feasible, since the determining algebraic equations for the unknowns a_j and

b_j should have a solution such that the coefficients b_j are positive and the resulting sum of exponentials is a positive function of t. The moments method with $K = 1$ always works with $a_1 = 1$ and $b_1 = m_0$. Verify numerically the performance of the moments approximation with $K = 2$ for the case of $N = 1$ standby unit.

2.29 An operating system has $r + s$ identical units where r units must be operating and s units are in preoperation (warm standby). A unit in operation has a constant failure rate of λ, while a unit in preoperation has a constant failure rate of β with $\beta < \lambda$. Failed units enter a repair facility which is able to repair simultaneously at most c units. The repair time of a failed unit has an exponential distribution with mean $1/\mu$. An operating unit that fails is replaced immediately by a unit from the warm standby if one is available. The operating system goes down when less than r units are in operation. Show how to calculate the probability distribution function of the time until the system goes down for the first time when all of the $r + s$ units are in a good condition at time 0.

2.30 In a computer system jobs arrive according to a Poisson process at rate λ_0 when the system is operating and at rate λ_1 when the system is down. The processing times of the jobs have an Erlang-r distribution with mean r/μ. The system can handle only one job at a time and provides service only when it is not down. If the processing of a job is interrupted by a breakdown of the system, the processing of that job is continued as soon as the system is again operating with no loss of service time already provided. The duration of an operating period is exponentially distributed with mean $1/\eta$ and the down period of a failed system has an Erlang-s distribution with mean s/ξ. State a condition under which no infinitely long queue of jobs will ultimately build up and give an explicit expression for the average number of jobs processed per unit time. Next show how to calculate the average number of jobs in the system and the average time spent by a job in the system. (This problem is based on Shogan, 1979.)

2.31 Suppose a communication system with c transmission channels at which messages arrive according to a Poisson process with rate λ. Each message that finds upon arrival all of the c channels busy is lost; otherwise the message is randomly assigned to one of the free channels. The transmission length of an accepted message has an exponential distribution with mean $1/\mu$. However, each separate channel is subject to a randomly changing environment that influences the transmission rate of the channel. Independently of each other the channels alternate between periods of good condition and periods of bad condition. These alternating periods are independent of each other and have exponential distributions with respective means $1/\gamma_g$ and $1/\gamma_b$. The transmission rate of a channel being in a good (bad) condition is $\sigma_g(\sigma_b)$. Set up the balance equations for the calculation of the fraction of messages that is rejected. Noting that $\sigma = (\sigma_b \gamma_g + \sigma_g \gamma_b)/(\gamma_g + \gamma_b)$ is the average transmission rate used by a channel, make some numerical comparisons with the case of a fixed transmission rate σ.

2.32 Messages arrive at a node in a communication network according to a Poisson process with rate λ. Each arriving message is temporarily stored in an infinite capacity buffer until it can be transmitted. The messages have to be routed over one of two communication lines each with a different transmission time. The transmission time over the ith communication line is exponentially distributed with mean $1/\mu_i (i = 1, 2)$, where $1/\mu_1 < 1/\mu_2$ and $\mu_1 + \mu_2 > \lambda$. The faster communication line is always available for service, but the slower line will be used only when the number of messages in the buffer exceeds some critical level. Each line is only able to handle one message at a time and provides non-preemptive service. With the goal of minimizing the average sojourn time (including transmission time) of a message in the system, the following control rule with switching level L is used. The slower line is turned on for transmitting a message when the number of messages in the system exceeds the level L and is turned off again when it completes a transmission and the number of messages left behind is at or below L. Show how to calculate the average sojourn time of a message in the system. (This problem is taken from Lin and Kumar, 1984.)

2.33 Two communication lines in a packet switching network share a finite storage space for incoming messages. Messages of types 1 and 2 arrive at the storage area according to two independent Poisson processes with respective rates λ_1 and λ_2. A message of type j is destined for communication line j and its transmission time is exponentially distributed with mean $1/\mu_j$, $j = 1, 2$. A communication line is only able to transmit one message at a time. The storage space consists of M buffers. Each message requires exactly one buffer and occupies the buffer until its transmission time has been completed. A number of N_j buffers is always reserved for messages of type j and a number of N_0 buffers is to be used by messages of both types, where $N_0 + N_1 + N_2 = M$. An arriving message of type j enters the system only when the storage space is not full and less than $N_0 + N_1$ buffers are occupied by messages of type j; otherwise the message is rejected. Discuss how to calculate the optimal values of N_0, N_1 and N_2 when the goal is to minimize the total rejection rate of both types of messages. Write a computer program and solve for the numerical data $M = 15$, $\lambda_1 = \lambda_2 = 1$ and $\mu_1 = \mu_2 = 1$ (*Answer*: $N_0 = 7$ and $N_1 = N_2 = 4$ with a total rejection rate of 0.207). (This problem is based on Kamoun and Kleinrock, 1980.)

2.34 In a packet switching network with two priority classes, 1-packets and 2-packets arrive at a central node according to independent Poisson processes with respective rates λ_1 and λ_2. The node has a finite capacity buffer consisting of N places with the stipulation that L places should be reserved for 1-packets so that a 2-packet is only admitted to the system when less than $N-L$ places are occupied. Packets which cannot enter the system are lost and have no further influence on the system. An arriving packet occupies one buffer place until service completion. The packets are handled one at a time by a processor, where the handling times of 1-packets and 2-packets are exponentially distributed with respective means $1/\mu_1$ and $1/\mu_2$. A 1-packet is always handled before a 2-packet but cannot preempt a 2-packet being in service. Show how to calculate the loss probability and the average throughput of i-packets, for $i = 1, 2$. (This problem is adapted from Van As, 1984.)

BIBLIOGRAPHIC NOTES

Many good textbooks on stochastic processes are available and most of them treat the topic of Markov chains. Our favourite books include Cox and Miller (1965), Karlin and Taylor (1975) and Ross (1983), each offering an excellent introduction to Markov chain theory. A very fundamental treatment of Markov chains can be found in the book of Chung (1967). The books of Bartholomew (1973) and Bartlett (1978) provide interesting applications of Markov chains to the fields of social sciences and natural sciences. The concept of the embedded Markov chain and its application in example 2.2. are due to Kendall (1953). The powerful technique of equating the flow out of an (aggregated) state to the flow into that state seems to have a long history in teletraffic analysis. The method of phases using fictitious stages with exponentially distributed lifetimes has its origin in the pioneering work of Erlang on stochastic processes in the early 1900s and is discussed in full generality in Cox and Miller (1965); an extension of this method is the supplementary variable technique for constructing Markovian processes in continuous time when the lifetime variables have continuous, non-exponential distributions; see also the book of Kosten (1973). The uniformization technique for continuous-time Markov chains already goes back to Jensen (1953) and is quite useful for both analytical and computational purposes; see also Grassmann (1977), Gross and Miller (1984) and Keilson (1979). Many applications of Markov chains are to be found in operations research and computer science. The buffer design problem in example 2.4 is based on Chu (1970). The resource allocation problem for urban emergency services in example 2.6 is taken from Carter, Chaiken and Ignall (1972); see also Jarvis (1985) and Larson (1974) for related work. The cash balance problem in example 2.7 is of a similar nature as an inventory problem with disposals and returns studied in Heyman (1978). Example 2.8 dealing with unloading of trailers at a stackyard is based on practical work of Van Hee (1984). The resource allocation problem for communication networks in example 2.10 is adapted from Foschini, Gopinath and Hayes (1981). Section 2.5 on stochastic networks and insensitivity follows Hordijk and Van Dijk (1982) and Van Dijk and Tijms (1986). See also Hordijk and Van Dijk (1983), Kelly (1979) and references therein.

REFERENCES

Bartholomew, D. J. (1973). *Stochastic Models for Social Processes*, 2nd ed., Wiley, New York.

Bartlett, M. S. (1978). *An Introduction to Stochastic Processes*, 3rd ed., Cambridge University Press, Cambridge.

Carter, G., Chaiken, J. M., and Ignall, E. J. (1972). 'Response areas for two emergency units', *Operat. Res.*, **20**, 571–594.

Chelst, K., Tilles, A. Z., and Pipis, J. S. (1981). 'A coal unloader: a finite queueing system with breakdowns', *Interfaces*, **11**, no. 5, 12–24.

Chu, W. W. (1970). 'Buffer behaviour for Poisson arrivals and multiple synchronous constant output', *IEEE Trans. Comput.*, **19**, 530–534.

Chung, K. L. (1967), *Markov Chains with Stationary Transition Probabilities*, 2nd ed., Springer–Verlag, Berlin.

Cohen, J. W. (1957). 'The generalized Engset formula', *Philips Telecomm. Review*, **18**, 158–170.

Cox, D. R., and Miller, H. D. (1965). *The Theory of Stochastic Processes*, Chapman and Hall, London.

Foschini, G. J., Gopinath, B., and Hayes, J. F. (1981). 'Optimum allocation of servers to two types of competing customers', *IEEE Trans. Commun.*, **29**, 1051–1055.

Grassmann, W. K. (1977). 'Transient solutions in Markovian queueing systems', *Comput. Ops. Res.*, **4**, 47–53.

Gregory, G., and Litton, C. D. (1975). 'A conveyor model with exponential service times', *Int. J. Prod. Res.*, **13**, 1–7.

Gross, D., and Miller, D. R. (1984). 'The randomization technique as a modelling tool and solution procedure for transient Markov processes', *Operat. Res.*, **32**, 343–361.

Heyman, D. P. (1978). 'Return policies for an inventory system with positive and negative demands', *Naval Res. Logist. Quart.*, **25**, 581–596.

Hordijk, A., and Van Dijk, N. (1982). 'Stationary probabilities for networks of queues', in *Applied Probability—Computer Science, The Interface, Vol. II* (Eds. R. L. Disney and T. J. Ott), pp. 423–451, Birkhauser, Boston.

Hordijk, A., and Van Dijk, N. (1983). 'Networks of queues', in *Modelling and Performance Evaluation Methodology*, (Eds. F. Baccelli and G. Fayolle), Lecture Notes in Control and Information Sciences, No. 60, pp. 158–205. Springer–Verlag, Berlin.

Jackson, J. R. (1963). 'Jobshop-like queueing systems', *Management Sci.*, **10**, 131–142.

Jarvis, J. P. (1985). 'Approximating the equilibrium behaviour of multi-server loss systems', *Management Sci.*, **31**, 235–239.

Jensen, A. (1953). 'Markov chains as an aid in the study of Markoff processes', *Skand. Aktuarietidskr.*, **36**, 87–91.

Kamoun, F., and Kleinrock, L. (1980). 'Analysis of a shared finite storage in a computer network node environment under general traffic conditions', *IEEE Trans. Commun.*, **28**, 992–1003.

Karlin, S., and Taylor, H. M. (1975). *A First Course in Stochastic Processes*, 2nd ed., Academic Press, New York.

Kaufman, J. S. (1981). 'Blocking in a shared resource environment', *IEEE Trans. Commun.*, **29**, 1474–1481.

Keilson, J. (1979). *Markov Chain Models—Rarity and Exponentiality*, Springer–Verlag, Berlin.

Kelly, F. P. (1979). *Reversibility and Stochastic Networks*, Wiley, New York.

Kendall, D. G. (1953). 'Stochastic processes occurring in the theory of queues and their analysis by the method of the embedded Markov chain', *Ann. Math. Statist.*, **24**, 338–354.

Kosten, L. (1973). *Stochastic Theory of Service Systems*, Pergamon Press, London.

Larsen, R. C. (1974). 'A hypercube queueing model for facility location in urban emergency services', *Comput. Ops. Res.*, **1**, 67–95.

Lin, W., and Kumar, P. (1984). 'Optimal control of a queueing system with two heterogeneous servers', *IEEE Trans. Automat. Contr.*, **29**, 696–703.

Ross, S. M. (1983). *Stochastic Processes*, Wiley, New York.

Ross, S. M. (1985). *Introduction to Probability Models*, 3rd ed., Academic Press, New York.

Shogan, A. W. (1979). 'A single server with arrival rate dependent on server breakdowns', *Naval Res. Logist. Quart.*, **26**, 487–497.

Solovyev, A. D. (1971). 'Asymptotic behaviour of the time of first occurrence of a rare event in a regenerating process', *Engineering Cybernetics*, **9**, 1038–1048.

Van As, H. (1984). 'Congestion control in packet switching networks by a dynamic foreground–background storage strategy', in *2nd International Symposium on the Performance of Computer-Communication Systems* (Eds. H. Rudin and W. Bux), pp. 433–448, North-Holland, Amsterdam.

Van Dijk, N., and Tijms, H. C. (1986). 'Insensitivity in two-node blocking network models with applications', in *Teletraffic Analysis and Computer Performance Evaluation* (Eds. O. J. Boxma, J. W. Cohen and H. C. Tijms), North-Holland, Amsterdam.

Van Hee (1984). 'Models underlying decision support systems for port terminal planning', *Wissenschaftliche Zeitschrift Technische Hochschule Leipzig*, **8**, 161–170.

Whittle, P. (1967). 'Nonlinear migration process', *Bull. Inst. Statist.*, **42**, 642–647.

Wolff, R. W., and Wrightson, C. W. (1976). 'An extension of Erlang's loss formula', *J. Appl. Prob.*, **13**, 628–632.

CHAPTER 3

Markovian decision processes and their applications

3.0 INTRODUCTION

In the previous chapter we saw that in the analysis of many operational systems the concepts of a state of a system and state transition are of basic importance. For dynamic systems with a *given* probabilistic law of motion, the simple Markov model is often appropriate. However, in many situations with uncertainty and dynamism the state transitions can be controlled by taking a sequence of actions. The Markov decision model is a versatile and powerful tool to analyse probabilistic sequential decision processes with an infinite planning horizon. This model is an outgrowth of the Markov model and dynamic programming. The latter concept, being developed by Bellman in the early 1950s, is a computational approach for analysing sequential decision processes with a finite planning horizon. The basic ideas of dynamic programming are states, principle of optimality and functional equations. In fact dynamic programming is a recursion procedure for calculating optimal value functions from a functional equation. This functional equation is obtained from the principle of optimality, stating that an optimal policy has the property that whatever the initial state and initial decision are, the remaining decisions must constitute an optimal policy with regard to the state resulting from the first transition—a principle being always valid when the number of states and the number of actions are finite. At much the same time as Bellman (1957) popularized dynamic programming, Howard (1960) used basic principles from Markov chain theory and dynamic programming to develop a policy-iteration algorithm for solving probabilistic sequential decision processes with an infinite planning horizon. In the two decades following the pioneering work of Bellman and Howard, the theory of Markov decision processes has expanded at a fast rate and a powerful technology was developed. However, in that period relatively little effort was put in applying the quite useful Markov decision model to practical problems. The Markov decision model has many potential applications in inventory control, maintenance, resource allocation and computer science, among others. It is to be believed that the Markov decision model will see many significant applications when this versatile model becomes more familiar to engineers, operations research analysts and computer science people and others. To this purpose, we focus in this chapter on the basic concepts and algorithms from Markov decision theory and illustrate the wide applicability of the Markov decision model to a variety of problems. In our discussion we confine ourselves to the optimality criterion of the long-run average cost per unit time. For most applications of Markov decision theory this criterion is believed to be more appropriate than the alternative criterion of the total expected discounted costs.

This chapter is organized as follows. In section 3.1 we discuss the basic concepts of the discrete-time Markov decision model in which the times between the decision epochs are constant. The policy-iteration algorithm is given in section 3.2. In section 3.3 we present a linear programming formulation of the Markov decision model and discuss how to handle probabilistic constraints on the long-run state–action frequencies. The policy-iteration algorithm and the linear programming formulation, being related to each other, require both the solving of a system of simultaneous linear equations in each iteration step. In section 3.4 we discuss the alternative method of value-iteration which avoids the computationally burdensome solving of systems of linear equations and involves only recursive computations. The value-iteration method endowed with quickly converging lower and upper bounds on the minimal costs is usually the most effective method for solving Markov decision problems with a large number of states. Also, in section 3.4 we give a modified value-iteration method with a dynamic relaxation factor to speed up convergence. The semi-Markov decision model is the subject of section 3.5. In this model the times between the decision epochs are not constant but random. The algorithms for the discrete-time Markov decision model will be extended to the semi-Markov decision model. A data transformation is given by which a semi-Markov decision model can be transformed into an equivalent discrete-time Markov decision model so that value-iteration applies as well to the semi-Markov model. Also, for semi-Markov decision processes with exponentially distributed transition times, the effectiveness of value-iteration may be further enhanced by introducing fictitious decision epochs with the purpose of creating sparse matrices of transition probabilities. In the final section 3.6 we show how the flexibility of policy-iteration together with an embedding technique can be used to develop tailor-made algorithms for practical applications having a specific structure.

3.1 DISCRETE-TIME MARKOV DECISION PROCESSES

In section 2.1 of chapter 2 we have considered a dynamic system that evolves over time according to a *fixed* probabilistic law of motion satisfying the Markovian assumption. This assumption states that the next state to be visited depends only on the present state of the system. In this chapter we deal with a dynamic system evolving over time where the probabilistic law of motion can be controlled by taking decisions. Also, costs are incurred (or rewards are earned) as a consequence of the decisions that are sequentially made when the system evolves over time. An *infinite planning horizon* is assumed and the goal is to find a control rule which minimizes the *long-run average cost per unit time.*

A typical example of a controlled dynamic system is an inventory system for a given product where the state is inventory position and is periodically reviewed and the demands for the product between consecutive reviews are independent and identically distributed random variables. The decisions taken at the review times consist of ordering a certain amount of the product depending on the inventory position. The economic consequences of the decisions are reflected in ordering, inventory and shortage costs.

We now introduce the Markov decision model. Consider a dynamic system which is reviewed at equidistant points of time $t = 0, 1, \ldots$. At each review the system is classified into one of a possible number of states and subsequently a decision has to be made. The set of possible states is denoted by I. For each state $i \in I$, a set $A(i)$ of decisions or actions is given. The state space I and the action sets $A(i)$ are assumed to be *finite*. The economic consequences of the decisions taken at the review times (decision epochs) are reflected in costs. This controlled dynamic system is called a *discrete-time Markov decision model* when the following Markovian properties are satisfied. If at a decision epoch the action a is chosen in state i, then regardless of the past history of the system, the following happens:

(a) An immediate cost $c_i(a)$ is incurred.
(b) At the next decision epoch the system will be in state j with probability $p_{ij}(a)$ where

$$\sum_{j \in I} p_{ij}(a) = 1, \qquad i \in I.$$

Note that the one-step costs $c_i(a)$ and the one-step transition probabilities $p_{ij}(a)$ are assumed to be time-homogeneous. In specific problems the 'immediate' costs $c_i(a)$ will often represent the expected cost incurred until the next decision epoch when action a is chosen in state i. Also, it should be emphasized that the choice of the state space and of the action sets usually depends on the cost structure of the specific problem considered. For example, in a production/inventory problem involving a fixed setup cost for restarting production after an idle period, the state description should include a state variable indicating whether the production facility is on or off. Many practical control problems can be modelled as a Markov decision process by choosing appropriately the state space and action sets. Before we develop the required theory for the average cost criterion, we give two examples.

Example 3.1   A maintenance problem for a deteriorating piece of equipment

At the beginning of each day a piece of equipment is inspected to reveal its actual working condition. The equipment can be found in one of the working conditions $i = 1, \ldots, N$. Here the working condition i is better than the working condition $i + 1$. The equipment deteriorates in time. If the present working condition is i and no repair is done, then at the beginning of the next day the equipment has working condition j with probability q_{ij}. It is assumed that the equipment cannot improve on its own, that is $q_{ij} = 0$ for $j < i$, so that $\sum_{j \geqslant i} q_{ij} = 1$. The working condition $i = N$ represents a malfunction that requires a repair taking two days. For the intermediate states i with $1 < i < N$ there is a choice between preventively repairing the equipment and letting the equipment operate for the present day. A preventive repair takes only one day. A repaired system has the working condition $i = 1$. We wish to determine a maintenance rule which minimizes the long-run fraction of time the system is in repair.

This problem can be put in the framework of a discrete-time Markov decision model. To do this, note first that, by assuming a cost of 1 for each day the system is in repair, the long-run average cost per day represents the long-run fraction of days the system is in repair. Also, since a repair for working condition N takes two days and in the discrete-time Markov decision model the state of the system has to be defined at the beginning of each day, we need an auxiliary state for the situation in which a repair is in progress. Thus the set of possible states of the system is chosen as

$$I = \{1, 2, \ldots, N, N + 1\}.$$

Here state i with $1 \leqslant i \leqslant N$ corresponds to the situation in which an inspection reveals working condition i, while state $N + 1$ corresponds to the situation in which a repair is in progress already for one day. Denoting the two possible actions by

$$a = \begin{cases} 1 & \text{if the system is repaired,} \\ 0 & \text{otherwise,} \end{cases}$$

the set of possible actions in state i is chosen as

$$A(1) = \{0\}, \qquad A(i) = \{0, 1\} \quad \text{for } 1 < i < N, \qquad A(N) = A(N + 1) = \{1\}.$$

It is readily verified that the one-step transition probabilities $p_{ij}(a)$ are given by

$$\begin{aligned} p_{i1}(1) &= 1 && \text{for } 1 < i < N, \\ p_{N,N+1}(1) &= 1, \\ p_{N+1,1}(1) &= 1, \\ p_{ij}(0) &= q_{ij} && \text{for } 1 \leqslant i < N \text{ and } j \geqslant i, \\ p_{ij}(a) &= 0 && \text{otherwise.} \end{aligned}$$

Further, the one-step costs $c_i(a)$ are given by

$$c_i(1) = 1 \qquad \text{and} \qquad c_i(0) = 0. \qquad \qquad \square$$

Example 3.2   Optimal control for an electricity plant

An electricity plant has two generators $j = 1$ and 2 for generating electricity. The required amount of electricity fluctuates during the day. The 24 hours of a day are divided into six consecutive periods of 4 hours each. The amount of electricity required in period k is d_k kWh for $k = 1, \ldots, 6$. Also, the generator j has a capacity of generating c_j kWh of electricity per period of 4 hours for $j = 1, 2$. An excess of electricity produced during a period cannot be used for a next period. At the beginning of each period k it has to be decided which generators to use for that period. The following costs are involved. An operating cost of r_j is incurred for each period in which generator j is used. Also, a setup cost of S_j is incurred each time generator j is turned on after having been idle for some time. We wish to determine a control rule which minimizes the long-run average cost per day.

This problem can be modelled as a discrete-time Markov decision process in which a decision has to be made at the beginning of each period of 4 hours. Note that in this process the state transitions are deterministic. In view of the fluctuating demand for electricity and the fixed cost for turning an idle generator on, we choose as state space

$$I = \{(k, m) \mid k = 1, \ldots, 6, \quad m = 1, 2, 3\}.$$

Here the first state variable k indicates the current period k and the second state variable m indicates which generators are currently on, where $m = 1$ (2) means that only generator 1 (2) is on and $m = 3$ means that both generators are on. The possible actions are denoted by

$$a = \begin{cases} 1 & \text{if only generator 1 is used,} \\ 2 & \text{if only generator 2 is used,} \\ 3 & \text{if both generators are used.} \end{cases}$$

Clearly, the action a is only feasible in state (k, m) if the amount of electricity produced under this action is at least as large as d_k.

Noting that the demand process for electricity regenerates itself every 24 hours so that period 1 succeeds period 6, it follows that, for $1 \leqslant k \leqslant 5$,

$$p_{(k,m),s}(a) = \begin{cases} 1 & \text{if } s = (k+1, a), \\ 0 & \text{otherwise,} \end{cases}$$

and

$$p_{(6,m),s}(a) = \begin{cases} 1 & \text{if } s = (1, a), \\ 0 & \text{otherwise.} \end{cases}$$

The one-step costs are given by

$$c_{(k,1)}(1) = r_1, \qquad c_{(k,1)}(2) = S_2 + r_2, \qquad c_{(k,1)}(3) = S_2 + r_1 + r_2,$$
$$c_{(k,2)}(1) = S_1 + r_1, \qquad c_{(k,2)}(2) = r_2, \qquad c_{(k,2)}(3) = S_1 + r_1 + r_2,$$
$$c_{(k,3)}(1) = r_1, \qquad c_{(k,3)}(2) = r_2, \qquad c_{(k,3)}(3) = r_1 + r_2.$$

□

We now introduce some concepts that will be needed in the algorithms to be described in the next sections. A *rule* or *policy* for controlling the system is a prescription for taking actions at each decision epoch. In principle, a control rule may be quite complicated in the sense that the prescribed actions may depend on the whole history of the system. However, in view of the above Markovian assumptions and the fact that the planning horizon is infinitely long, it will intuitively be clear that we need only to consider the so-called *stationary* policies. A stationary policy R is a rule that always prescribes a single action R_i whenever the system is found in state i at a decision epoch. Thus in example 3.1 the rule R prescribing a repair only when the system has a working condition of at least 5 is given by $R_i = 0$ for $1 \leqslant i < 5$ and $R_i = 1$ for $5 \leqslant i \leqslant N + 1$. We refer to Derman (1970) for the non-trivial proof that we may restrict ourselves to the class of stationary policies.

Define, for $n = 0, 1, \ldots,$

X_n = the state of the system at the nth decision epoch.

Under a given stationary policy R, we have

$$P\{X_{n+1} = j \mid X_n = i\} = p_{ij}(R_i),$$

regardless of the past history of the system up to time n. Hence under a given stationary policy R the stochastic process $\{X_n\}$ is a discrete-time Markov chain with one-step transition probabilities $p_{ij}(R_i)$. Also, this Markov chain incurs a cost $c_i(R_i)$ each time the system visits state i. Thus we can invoke the results of section 2.1 to give the long-run average cost per unit time when a given stationary policy is used. Before doing this, we give some notation and a regularity condition for the Markov chains associated with the stationary policies. Assuming that a given stationary policy R is followed, denote the n-step transition probabilities of the Markov chain $\{X_n\}$ by

$$p_{ij}^{(n)}(R) = P\{X_n = j \mid X_0 = i\}, \qquad i,j \in I \quad \text{and} \quad n = 1, 2, \ldots,$$

where $p_{ij}^{(1)}(R) = p_{ij}(R_i)$. Note that, by relation (2.3) in section 2.1 of chapter 2, these transition probabilities satisfy the recursion relation

$$p_{ij}^{(n+1)}(R) = \sum_{k \in I} p_{ik}^{(n)}(R) p_{kj}(R_k), \qquad n = 1, 2, \ldots. \tag{3.1}$$

Let us say that state j can be *reached* from state i under policy R if $p_{ij}^{(n)}(R) > 0$ for some $n \geqslant 1$. In practical applications each stationary policy of practical interest will usually have the *unichain* property that some state exists which can be reached from any other state under that policy. Unless stated otherwise, the following unichain assumption is made in the sequel.

Assumption 3.1

For each stationary policy R, a state r (that may depend on R) exists which can be reached from any other state under policy R.

Using the finiteness of the state space, it is not difficult to prove that assumption 3.1 implies that for each stationary policy R the associated Markov chain $\{X_n\}$ satisfies the assumptions 2.1 and 2.2 in section 2.1 of chapter 2. Thus, for each stationary policy R, we have

$$\pi_j(R) = \lim_{m \to \infty} \frac{1}{m} \sum_{n=1}^{m} p_{ij}^{(n)}(R) \tag{3.2}$$

exists and is independent of the initial state $X_0 = i$. The equilibrium distribution $\{\pi_j(R), j \in I\}$ satisfies the system of linear equations

$$\pi_j(R) = \sum_{k \in I} p_{kj}(R_k)\pi_k(R), \qquad j \in I, \tag{3.3}$$

$$\sum_{j \in I} \pi_j(R) = 1. \tag{3.4}$$

This system of linear equations has a unique solution. Also by the ergodic relation (2.10) in section 2.1, we have, with probability 1,

The long-run average cost per unit time when using rule R

$$= \sum_{j \in I} c_j(R_j)\pi_j(R), \tag{3.5}$$

independently of the initial state. Letting

$$g(R) = \sum_{j \in I} c_j(R_j)\pi_j(R),$$

a stationary policy R^* is said to be *average cost optimal* if $g(R^*) \leqslant g(R)$ for each stationary policy R.

In most applications, it is computationally not feasible to find an average cost optimal policy by computing the associated average cost for each stationary policy separately from (3.3) to (3.5). For example in the case that the state space has N states and each action set consists of two actions, the number of possible stationary policies is 2^N, and this number grows quickly beyond any practical bound. However, an efficient algorithm can be given that constructs a sequence of improved policies until an average cost optimal policy is found.

In improving a given policy R, a key role is played by the so-called *relative values* of the various starting states when policy R is used. The relative values indicate the transient effect of the starting states on the total expected costs under the given policy. In what follows it will be seen that the average cost per unit time and the relative values of the various starting states can be calculated simultaneously by solving a system of linear equations.

Relative values associated with a given policy R

It may be helpful to give first a heuristic discussion of the relative values before presenting a rigorous treatment. The starting point is the obvious relation $\lim_{n\to\infty} V_n(i, R)/n = g(R)$ for all i, where $V_n(i, R)$ denotes the total expected costs over the first n decision epochs when the initial state is i and policy R is used. This relation motivates the heuristic assumption that bias values $v_i(R)$, $i\in I$, exist such that, for each $i\in I$,

$$V_n(i, R) \approx ng(R) + v_i(R) \qquad \text{for } n \text{ large.} \tag{3.6}$$

Note that $v_i(R) - v_j(R) \approx V_n(i, R) - V_n(j, R)$ for n large, so that $v_i(R) - v_j(R)$ measures the difference in total expected costs when starting in state i rather than in state j, given that policy R is followed. Next we heuristically argue that the average cost $g(R)$ and the relative values $v_i(R)$, $i\in I$, satisfy a simultaneous system of linear equations. To do so, note the recursion equation

$$V_n(i, R) = c_i(R_i) + \sum_{j\in I} p_{ij}(R_i)V_{n-1}(j, R), \qquad n \geqslant 1 \quad \text{and} \quad i\in I.$$

This recursion equation follows by a standard conditioning argument. Under the condition that the next state is j, the total expected costs over the remaining next $n-1$ decision epochs is $V_{n-1}(j, R)$. The next state is j with probability $p_{ij}(R_i)$ when the action $a = R_i$ is made in the starting state i. By substituting the asymptotic expansion (3.6) in the recursion equation, we find, after cancelling out common terms,

$$g(R) + v_i(R) \approx c_i(R_i) + \sum_{j\in I} p_{ij}(R_i)v_j(R), \qquad i\in I,$$

yielding the value-determination equations for policy R.

A rigorous way of introducing the relative values associated with a given stationary policy R is to consider the costs incurred until the first return to some regeneration state for policy R. We choose some state r such that for each initial state the Markov chain $\{X_n\}$ associated with policy R will visit state r with probability 1 after finitely many transitions. Thus we can define, for each state $i\in I$,

$T_i(R) = $ the expected time until the first return to state r

when starting in state i and using policy R.

In particular, letting a cycle be the time elapsed between two consecutive visits to the regeneration state r under policy R, we have that $T_r(R)$ is the expected length of a cycle. Also, define, for each $i\in I$,

$K_i(R)$ = the expected costs incurred until the first return to state r
when starting in state i and using policy R.

Here we use the convention that $K_i(R)$ includes the cost incurred when starting in state i but excludes the cost incurred when returning to state r. By the theory of renewal–reward processes, the average cost per unit time equals the expected costs incurred in a cycle divided by the expected length of a cycle and so

$$g(R) = \frac{K_r(R)}{T_r(R)}. \tag{3.7}$$

Next we define the relative values

$$w_i(R) = K_i(R) - g(R)T_i(R), \qquad i \in I. \tag{3.8}$$

Note, as a consequence of (3.7), the normalization

$$w_r(R) = 0. \tag{3.9}$$

In the next theorem we prove that the average cost per unit time and the relative values can be calculated simultaneously by solving a system of linear equations.

Theorem 3.1

Let R be a given stationary policy.

(a) The average cost $g(R)$ and the relative values $w_i(R)$, $i \in I$, satisfy the following system of linear equations in the unknowns g and v_i, $i \in I$:

$$v_i = c_i(R_i) - g + \sum_{j \in I} p_{ij}(R_i)v_j, \qquad i \in I. \tag{3.10}$$

(b) Let the numbers g and v_i, $i \in I$, be any solution to (3.10). Then

$$g = g(R)$$

and, for some constant c,

$$v_i = w_i(R) + c \qquad \text{for all } i \in I.$$

(c) Let s be an arbitrarily chosen state. Then the linear equations (3.10) together with the normalization equation $v_s = 0$ have a unique solution.

Proof. (a) By conditioning on the next state following the initial state i, it can be seen that

$$T_i(R) = 1 + \sum_{\substack{j \in I \\ j \neq r}} p_{ij}(R_i)T_j(R), \qquad i \in I,$$

$$K_i(R) = c_i(R_i) + \sum_{\substack{j \in I \\ j \neq r}} p_{ij}(R_i)K_j(R), \qquad i \in I.$$

Hence,

$$K_i(R) - g(R)T_i(R) = c_i(R_i) - g(R) + \sum_{\substack{j \in I \\ j \neq r}} p_{ij}(R_i)\{K_j(R) - g(R)T_j(R)\}.$$

Next, using (3.9), we find

$$w_i(R) = c_i(R_i) - g(R) + \sum_{j \in I} p_{ij}(R_i)w_j(R), \qquad i \in I.$$

(b) Let $\{g, v_i\}$ be a solution to (3.10). We first verify by induction that the following identity holds for each $m = 1, 2, \ldots$:

$$v_i = \sum_{t=0}^{m-1} \sum_{j \in I} p_{ij}^{(t)}(R)c_j(R_j) - mg + \sum_{j \in I} p_{ij}^{(m)}(R)v_j, \qquad i \in I, \tag{3.11}$$

where $p_{ij}^{(0)}(R) = 1$ for $j = i$ and $p_{ij}^{(0)}(R) = 0$ for $j \neq i$. Clearly, (3.11) is true for $m = 1$. Suppose now that (3.11) is true for $m = n$. Substituting the equations (3.10) in the right side of (3.11) with $m = n$, it follows that, for each $i \in I$,

$$v_i = \sum_{t=0}^{n-1} \sum_{j \in I} p_{ij}^{(t)}(R)c_j(R_j) - ng + \sum_{j \in I} p_{ij}^{(n)}(R)\left\{c_j(R_j) - g + \sum_{k \in I} p_{jk}(R_j)v_k\right\}$$

$$= \sum_{t=0}^{n} \sum_{j \in I} p_{ij}^{(t)}(R)c_j(R_j) - (n+1)g + \sum_{k \in I} \left\{\sum_{j \in I} p_{ij}^{(n)}(R)p_{jk}(R_j)\right\}v_k,$$

where the latter equality involves an interchange of the order of summation. Next, using (3.1), we get (3.11) for $m = n + 1$ which completes the induction step. Put for abbreviation

$$V_m(i, R) = \sum_{t=0}^{m-1} \sum_{j \in I} p_{ij}^{(t)}(R)c_j(R_j). \tag{3.12}$$

Noting that $\sum_{j \in I} p_{ij}^{(t)}(R)c_j(R_j)$ gives the expected cost to be incurred at decision epoch t given that $X_0 = i$ and policy R is used, we have

$V_m(i, R) =$ the expected costs incurred over the first m decision
          epochs given that $X_0 = i$ and policy R is used.

We now write (3.11) in the more convenient form:

$$v_i = V_m(i, R) - mg + \sum_{j \in I} p_{ij}^{(m)}(R)v_j, \qquad i \in I. \tag{3.13}$$

Noting that $V_m(i, R)/m \to g(R)$ as $m \to \infty$ for each i, the result $g = g(R)$ follows by dividing both sides of (3.13) by m and letting $m \to \infty$. To prove the second part of assertion (b), let $\{g, v_i\}$ and $\{g', v_i'\}$ by any two solutions to (3.10). Since $g = g' = g(R)$, it follows from the representation (3.13) that

$$v_i - v_i' = \sum_{j \in I} p_{ij}^{(m)}(R)\{v_j - v_j'\}, \qquad i \in I \quad \text{and} \quad m \geqslant 1. \tag{3.14}$$

By summing both sides of (3.14) over $m = 1, \ldots, n$ and next dividing by n, it follows after an interchange of the order of summation that

$$v_i - v_i' = \sum_{j \in I} \left\{ \frac{1}{n} \sum_{m=1}^{n} p_{ij}^{(m)}(R) \right\} (v_j - v_j'), \qquad i \in I \quad \text{and} \quad n \geq 1.$$

Next, by letting $n \to \infty$ and using (3.2), we obtain

$$v_i - v_i' = \sum_{j \in I} \pi_j(R)(v_j - v_j'), \qquad i \in I.$$

The right side of the equation does not depend on i. Assertion (b) now follows by taking the particular solution $v_i' = w_i(R)$, $i \in I$.

(c) Let $\{g, v_i\}$ be any solution to (3.10). Since $\sum_j p_{ij}(R_i) = 1$ for each $i \in I$, it follows that for any constant γ the numbers g and $v_i' = v_i + \gamma$, $i \in I$, satisfy (3.10) as well. Hence the equations (3.10) together with $v_s = 0$ for some state s have a solution. In view of assertion (b), this solution must be unique. $\qquad \square$

Economic interpretation of the relative values

For any solution $\{g(R), v_i(R)\}$ to the value-determination equations (3.10), the numbers $v_i(R)$, $i \in I$, are called the relative values of the various starting states when policy R is used. An explanation of this nomenclature is the following. Assuming that the Markov chain $\{X_n\}$ associated with rule R is aperiodic, we have, for any two states $i, j \in I$,

$v_i(R) - v_j(R) =$ the difference in total expected costs over an
infinitely long period of time by starting in state i
rather than in state j when using policy R.

In other words, $v_i(R) - v_j(R)$ is the maximum amount that a rational person is willing to pay to start the system in state j rather than in state i when the system is controlled by rule R. This interpretation is an easy consequence of (3.13). Using the assumption that the Markov chain $\{X_n\}$ is aperiodic, we find that $\lim_{m \to \infty} p_{ij}^{(m)}(R)$ exists and is independent of the initial state i. Thus, by (3.13),

$$v_i(R) - v_j(R) = \lim_{m \to \infty} \{V_m(i, R) - V_m(j, R)\},$$

yielding the desired result since $V_m(i, R)$ represents the expected total costs over the first m decision epochs when the initial state is i and policy R is used.

3.2   POLICY-ITERATION ALGORITHM

The relative values associated with a given policy R provide a tool for constructing a new policy $\bar{R}$ whose average cost is no more than that of the current policy R. Before giving an exact discussion of Howard's policy-improvement procedure, we heuristically motivate this procedure. For that purpose, imagine that the system has to be controlled over a very long period with the stipulation that at the first decision epoch it is allowed to choose any feasible action but thereafter policy R must be followed. If at the first decision epoch action a is chosen in state i and thereafter policy R is used, the total expected costs over the first n decision epochs are given by $c_i(a) + \sum_j p_{ij}(a) V_{n-1}(j, R)$. Obviously, we are interested in an action a that minimizes this expression. For n large, we have under the heuristic assumption (3.6) that

$$c_i(a) + \sum_{j \in I} p_{ij}(a) V_{n-1}(j, R) \approx c_i(a) + \sum_{j \in I} p_{ij}(a) v_j(R) + (n-1) g(R),$$

so that we may alternatively minimize the quantity $c_i(a) + \sum_j p_{ij}(a) v_j(R)$. It is to be expected that determining in this way a minimizing action $a = \bar{R}_i$ (say) for each starting state $i \in I$ yields an improved policy $\bar{R}$.

To be precise, we now state the following theorem that will be crucial in what follows.

Theorem 3.2

Let g and v_i, $i \in I$, be given numbers. Suppose that the policy $\bar{R}$ has the property

$$c_i(\bar{R}_i) - g + \sum_{j \in I} p_{ij}(\bar{R}_i) v_j \leq v_i \qquad \text{for each } i \in I. \tag{3.15}$$

Then

$$g(\bar{R}) \leq g. \tag{3.16}$$

The theorem is also true when the inequality signs in (3.15) and (3.16) are reversed.

Proof. We first give an intuitive explanation of the theorem and next we formalize this explanation. Suppose that a control cost of $c_i(a) - g$ is incurred each time the action a is chosen in state i, while a terminal cost of v_j is incurred when the control of the system is stopped and the system is left behind in state j. Then (3.15) states that controlling the system for one step according to rule $\bar{R}$ and stopping next is preferable to stopping directly when the initial state is i. Since this property is true for each initial state, a repeated application of this property yields that controlling the system for m steps according to rule $\bar{R}$ and stopping after that is preferable to stopping directly. Thus, using the notation (3.12) with R replaced by $\bar{R}$, we have, for each initial state $i \in I$,

$$V_m(i, \bar{R}) - mg + \sum_{j \in I} p_{ij}^{(m)}(\bar{R}) v_j \leq v_i \qquad \text{for} \qquad m = 1, 2, \dots. \tag{3.17}$$

Dividing both sides of (3.17) by m and letting $m \to \infty$, it follows that $g(\bar{R}) - g \leq 0$, which was to be proved. It is easy to formalize the above arguments. In the same way as we derived (3.11), we obtain (3.17) by a repeated substitution of the inequality (3.15). Finally, it will be clear from the above proof that the theorem remains true when the inequality signs in (3.15) and (3.16) are reversed. $\square$

In order to improve a given policy R whose average cost $g(R)$ and relative values $v_i(R)$, $i\in I$, have been computed, we apply the above theorem with $g = g(R)$ and $v_i = v_i(R)$, $i\in I$. Thus, by constructing a new policy $\bar{R}$ such that, for each initial state $i\in I$,

$$c_i(\bar{R}_i) - g(R) + \sum_{j\in I} p_{ij}(\bar{R}_i)v_j \leqslant v_i, \tag{3.18}$$

we obtain an improved rule $\bar{R}$ according to $g(\bar{R}) \leqslant g(R)$. In constructing such an improved policy $\bar{R}$ it is important to realize that for each state i *separately* an action $\bar{R}_i$ satisfying (3.18) can be determined. A particular way to find for some state $i\in I$ an action $\bar{R}_i$ satisfying (3.18) is to minimize

$$c_i(a) - g(R) + \sum_{j\in I} p_{ij}(a)v_j(R). \tag{3.19}$$

with respect to $a\in A(i)$. Noting that (3.19) equals $v_i(R)$ for $a = R_i$, it follows that (3.18) is satisfied for the action $\bar{R}_i$ which minimizes (3.19) with respect to $a\in A(i)$. We are now in a position to formulate the following algorithm.

Policy-iteration algorithm

Step 0 (initialization). Choose a stationary policy R.

Step 1 (value-determination step). For the current rule R, compute the unique solution $\{g(R), v_i(R)\}$ to the following system of linear equations:

$$v_i = c_i(R_i) - g + \sum_{j\in I} p_{ij}(R_i)v_j, \qquad i\in I, \tag{3.20}$$

$$v_s = 0, \tag{3.21}$$

where s is an arbitrarily chosen state.

Step 2 (policy-improvement step). For each state $i\in I$, determine an action a_i yielding the minimum in

$$\min_{a\in A(i)} \left\{ c_i(a) - g(R) + \sum_{j\in I} p_{ij}(a)v_j(R) \right\}. \tag{3.22}$$

The new stationary policy $\bar{R}$ is obtained by choosing $\bar{R}_i = a_i$ for all $i\in I$ with the convention that $\bar{R}_i$ is chosen equal to the old action R_i when this action yields the minimum in (3.22).

Step 3 (convergence test). If the new policy $\bar{R}$ equals the old policy R, the algorithm is stopped with policy R. Otherwise, go to step 1 with R replaced by $\bar{R}$.

The policy-iteration algorithm converges after a finite number of iterations. We defer the proof to the appendix at the end of this section.

The policy-iteration algorithm is empirically found to be a remarkably robust algorithm that converges very fast in specific problems. The number of iterations is practically independent of the number of states and of the starting policy, and varies typically between 3 and 15 (say). Also, it can be roughly stated that the average costs of the policies generated by policy-iteration converge at least exponentially fast to the minimum costs.

If the algorithm has converged to a stationary policy R^*, it is easy to prove that policy R^* is average cost optimal. From the policy improvement step, we have, for all $i \in I$,

$$\min_{a \in A(i)} \{c_i(a) - g(R^*) + \sum_{j \in I} p_{ij}(a)v_j(R^*)\} = c_i(R_i^*) - g(R^*) + \sum_{j \in I} p_{ij}(R_i^*)v_j(R^*).$$

$$(3.23)$$

Noting that the right side of (3.23) equals $v_i(R^*)$, it follows from (3.23) that, for an arbitrarily chosen policy R,

$$c_i(R_i) - g(R^*) + \sum_{j \in I} p_{ij}(R_i)v_j(R^*) \geqslant v_i(R^*) \qquad \text{for all} \qquad i \in I.$$

An application of Theorem 3.2 now gives $g(R) \geqslant g(R^*)$, showing that policy R^* is average cost optimal.

Remark 3.1

The finite convergence of the above algorithm is only warranted in the case where assumption 3.1 is satisfied. In specific problems it may happen that for some non-optimal stationary policies the associated Markov chains $\{X_n\}$ lack the unichain property as required in assumption 3.1, while this property is satisfied by the 'relevant' stationary policies. In these situations it could theoretically happen that the above algorithm does not work. However, in practice the algorithm will typically lead to an average cost optimal policy.

We next compute, by the policy-iteration algorithm, average cost optimal policies for the control problems in the examples 3.1 and 3.2.

Example 3.1 (continued) A maintenance problem for a deteriorating piece of equipment

Assume the numerical data given in Table 3.1. The number of possible working conditions equals $N = 5$ and the deterioration probabilities q_{ij} are given in Table 3.1. The policy-iteration algorithm is initialized with the policy which prescribes the repair action $a = 1$ in each state except state 1.

Iteration 1
Step 1. For the current policy $R^{(1)}$ whose actions are

$$R_1^{(1)} = 0, \qquad R_2^{(1)} = R_3^{(1)} = R_4^{(1)} = R_5^{(1)} = R_6^{(1)} = 1,$$

the average cost $g(R^{(1)})$ and the relative values $v_i(R^{(1)})$ can be computed as

Table 3.1   The deterioration probabilities q_{ij}

$i\backslash j$	1	2	3	4	5
1	0.75	0.20	0.05	0	0
2	0	0.50	0.20	0.20	0.10
3	0	0	0.50	0.25	0.25
4	0	0	0	0.30	0.70

the unique solution to the linear equations

$$v_1 = 0 - g + 0.75v_1 + 0.20v_2 + 0.05v_3,$$
$$v_k = 1 - g + v_1, \qquad k = 2, 3, 4,$$
$$v_5 = 1 - g + v_6,$$
$$v_6 = 1 - g + v_1,$$
$$v_6 = 0,$$

where $s = 6$ is taken for the normalizing equation $v_s = 0$. The solution of these linear equations is given by

$$g(R^{(1)}) = 0.2, \qquad v_1(R^{(1)}) = -0.8, \qquad v_2(R^{(1)}) = 0, \qquad v_3(R^{(1)}) = 0,$$
$$v_4(R^{(1)}) = 0, \qquad v_5(R^{(1)}) = 0.8, \qquad v_6(R^{(1)}) = 0.$$

Step 2. The calculations for the policy-improvement step for policy $R^{(1)}$ are displayed in Table 3.2. Note that for state 3 the action $a = 1$ prescribed by the current policy $R^{(1)}$ is one of the minimizing actions and thus the new action in state 3 is taken as $a = 1$. From Table 3.2 we find the new policy $R^{(2)}$ whose actions are

$$R_1^{(2)} = R_2^{(2)} = 0, \qquad R_3^{(2)} = R_4^{(2)} = R_5^{(2)} = R_6^{(2)} = 1.$$

Step 3. The new policy $R^{(2)}$ is different from the previous policy $R^{(1)}$ and hence a next iteration is performed.

Table 3.2   The policy-improvement step for policy $R^{(1)}$

State i	Action a	Test quantity $c_i(a) - g(R^{(1)}) + \sum_{j=1}^{6} p_{ij}(a)v_j(R^{(1)})$	
2	0	$0 - 0.2 + (0.50)(0) + (0.20)(0) + (0.20)(0) + (0.10)(0.8)$	$= -0.12 \leftarrow$
	1	$1 - 0.2 + (1.0)(-0.8)$	$= 0$
3	0	$0 - 0.2 + (0.50)(0) + (0.25)(0) + (0.25)(0.8)$	$= 0$
	1	$1 - 0.2 + (1.0)(-0.8)$	$= 0 \leftarrow$
4	0	$0 - 0.2 + (0.30)(0) + (0.70)(0.8)$	$= 0.36$
	1	$1 - 0.2 + (1.0)(-0.8)$	$= 0 \leftarrow$

Iteration 2

Step 1. For the current policy $R^{(2)}$ the average cost $g(R^{(2)})$ and the relative values $v_i(R^{(2)})$ can be computed as the unique solution to the linear equations

$$v_1 = 0 - g + 0.75v_1 + 0.20v_2 + 0.05v_3,$$
$$v_2 = 0 - g + 0.50v_2 + 0.20v_3 + 0.20v_4 + 0.10v_5,$$
$$v_k = 1 - g + v_1, \qquad k = 3, 4,$$
$$v_5 = 1 - g + v_6,$$
$$v_6 = 1 - g + v_1,$$
$$v_6 = 0.$$

The solution of these linear equations is

$$g(R^{(2)}) = 0.172, \qquad v_1(R^{(2)}) = -0.828, \qquad v_2(R^{(2)}) = -0.178,$$
$$v_3(R^{(2)}) = 0, \qquad v_4(R^{(2)}) = 0, \qquad v_5(R^{(2)}) = 0.828, \qquad v_6(R^{(2)}) = 0.$$

Step 2. The policy-improvement step for policy $R^{(2)}$ is shown in Table 3.3.

Table 3.3   The policy-improvement step for policy $R^{(2)}$

State i	Action a	Test quantity $c_i(a) - g(R^{(2)}) + \sum_{j=1}^{6} p_{ij}(a)v_j(R^{(2)})$		
2	0	$0 - 0.172 + (0.50)(-0.178) + (0.20)(0) + (0.20)(0)$ $+ (0.10)(0.828)$	$= -0.178$	←
	1	$1 - 0.172 + (1.0)(-0.828)$	$= 0$	
3	0	$0 - 0.172 + (0.50)(0) + (0.25)(0) + (0.25)(0.828)$	$= 0.035$	
	1	$1 - 0.172 + (1.0)(-0.828)$	$= 0$	←
4	0	$0 - 0.172 + (0.30)(0) + (0.70)(0.828)$	$= 0.408$	
	1	$1 - 0.172 + (1.0)(-0.828)$	$= 0$	←

From Table 3.3 we find the new policy whose actions are

$$R_1^{(3)} = R_2^{(3)} = 0, \qquad R_3^{(3)} = R_4^{(3)} = R_5^{(3)} = R_6^{(3)} = 1.$$

Step 3. The new policy $R^{(3)}$ is the same as the previous policy $R^{(2)}$ and is thus average cost optimal. The minimum fraction of days the equipment is in repair equals 0.172.   □

In the next example we demonstrate that for the case of *deterministic* state transitions the computational burden in the value-determination step may be reduced considerably in view of the fact that for this special case the average cost and relative values allow for a recursive computation.

Example 3.2 *(continued)* **Optimal control for an electricity plant**

Assume the following numerical data:

$d_1 = 20,$ $\quad d_2 = 40,$ $\quad d_3 = 60,$ $\quad d_4 = 90,$ $\quad d_5 = 70,$ $\quad d_6 = 30,$

$C_1 = 40,$ $\quad C_2 = 60,$

$r_1 = 1000,$ $\quad r_2 = 1100,$ $\quad S_1 = 500,$ $\quad S_2 = 300.$

The policy-iteration algorithm is initialized with the policy $R^{(1)}$ which prescribes to use generator 2 in periods 1, 2 and 3, generators 1 and 2 in periods 4 and 5 and generator 1 in period 6.

Iteration 1

Step 1. To apply the value-determination step for the policy $R^{(1)}$, it is helpful to display in Figure 3.1 the one-step transitions under policy $R^{(1)}$. This state diagram shows that after finitely many transitions the process cycles between the states $(1,1), (2,2), (3,2), (4,2), (5,3)$ and $(6,3)$. These states are said to be recurrent under policy $R^{(1)}$, while the other states are said to be transient under policy $R^{(1)}$. In this example we have determined the recurrent states by inspection. An algorithm to find in a systematic way the recurrent states is given in Fox and Landi (1968). The cost incurred in the recurrent states determine the long-run average cost per period. By the theory of renewal–reward processes, the long-run average cost per period equals the total cost incurred between two consecutive visits to state $(1,1)$ divided by the number of periods between two such visits. Hence,

$$g(R^{(1)}) = \tfrac{1}{6}(1400 + 1100 + 1100 + 2600 + 2100 + 1000) = 1550.$$

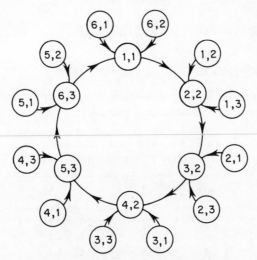

Figure 3.1 The state transition diagram for policy $R^{(1)}$

Once we have computed in this way the average cost per period, it is a simple matter to compute recursively the relative values. This is first done for the recurrent states. Therefore we put the relative value for one of the recurrent states equal to zero, say

$$v_{(1,1)}(R^{(1)}) = 0.$$

Since state $(1,1)$ is reached from state $(6,3)$, we next compute $v_{(6,3)}(R^{(1)})$ from

$$v_{(6,3)}(R^{(1)}) = c_{(6,3)}(1) - g(R^{(1)}) + v_{(1,1)}(R^{(1)}),$$

which yields

$$v_{(6,3)}(R^{(1)}) = -550.$$

Continuing in this way for the other recurrent states, we find successively

$$v_{(5,3)}(R^{(1)}) = 0, \qquad v_{(4,2)}(R^{(1)}) = 1050, \qquad v_{(3,2)}(R^{(1)}) = 600,$$
$$v_{(2,2)}(R^{(1)}) = 150.$$

Next we compute the relative values for the transient states. For instance, we have

$$v_{(6,m)}(R^{(1)}) = c_{(6,m)}(1) - g(R^{(1)}) + v_{(1,1)}(R^{(1)}) \qquad \text{for} \quad m = 1,2,$$

which gives

$$v_{(6,1)}(R^{(1)}) = -550, \qquad v_{(6,2)}(R^{(1)}) = -50.$$

In the same way, we find

$$v_{(1,2)}(R^{(1)}) = -300, \qquad v_{(1,3)}(R^{(1)}) = -300, \qquad v_{(2,1)}(R^{(1)}) = 450,$$
$$v_{(2,3)}(R^{(1)}) = 150,$$
$$v_{(3,1)}(R^{(1)}) = 900, \qquad v_{(3,3)}(R^{(1)}) = 600, \qquad v_{(4,1)}(R^{(1)}) = 850,$$
$$v_{(4,3)}(R^{(1)}) = 550,$$
$$v_{(5,1)}(R^{(1)}) = 300, \qquad v_{(5,2)}(R^{(1)}) = 500.$$

Step 2. The calculations in the policy-improvement step for policy $R^{(1)}$ result in the new policy $R^{(2)}$ shown in Table 3.4

Table 3.4 The actions prescribed by policy $R^{(2)}$ in the states (k, m)

m/k	1	2	3	4	5	6
1	1	1	2	3	3	1
2	2	2	2	3	3	2
3	2	2	2	3	3	2

Table 3.5   The relative values for policy $R^{(2)}$

m/k	1	2	3	4	5	6
1	100	616.67	1133.33	1050	466.67	−416.67
2	0	416.67	833.33	1250	666.67	−416.67
3	0	416.67	833.33	750	166.67	−416.67

Step 3. The new policy $R^{(2)}$ is different from the previous policy $R^{(1)}$ and hence we perform a next iteration.

Iteration 2

Step 1. Under the policy $R^{(2)}$ shown in Table 3.4, the process ultimately cycles between the states (1,2), (2,2), (3,2), (4,2), (5,3) and (6,3). Hence

$$g(R^{(2)}) = \tfrac{1}{6}(1100 + 1100 + 1100 + 2600 + 2100 + 1100) = 1516.67.$$

Analogously as in iteration 1, we compute the relative values for policy $R^{(2)}$ where we put the relative value for state (1,2) equal to zero. These relative values are shown in Table 3.5.

Step 2. The calculations in the policy-improvement step for policy $R^{(2)}$ result in the new policy $R^{(3)}$ which is identical to the previous policy $R^{(2)}$.

Step 3. Since $R^{(3)} = R^{(2)}$ the policy $R^{(2)}$ shown in Table 3.4 is average cost optimal and the minimal average cost per period equals 1516.67. Note for this policy that once the process has left the transient states, both generators are used in the periods 4 and 5 and only generator 2 is used in the other periods.   □

The above example illustrates the key features of the simplified policy evaluation calculations when transitions are deterministic:

(a)   The recurrent states of policy R form a cycle.

(b)   The cost rate $g(R)$ equals the sum of one-step costs in this cycle divided by the number of states in the cycle.

(c)   The relative values for recurrent states can be obtained recursively, in reverse direction to the natural flow around the cycle, after assigning a value 0 to one recurrent state.

(d)   The relative values for transient states are computed first for states which reach the cycle in one step, then for states which reach the cycle in two steps, and so forth.

Appendix.   The finite convergence of the policy-iteration algorithm

Before we prove the finite convergence of the policy-iteration algorithm we introduce some concepts for Markov chains with a finite state space. Consider the Markov chain $\{X_n\}$ associated with a given stationary policy Q. A state i is said to be *transient* under policy Q if a state k exists such that k can be reached from i under Q but i cannot be reached from k under Q. Otherwise, a state i is said to be *recurrent*. Hence for a recurrent state i it holds that if state k can be reached from i then i can be reached from k. It is easy to verify that a transient state cannot be reached from a recurrent state. Thus, denoting by $I(Q)$ the set of states which are recurrent under Q, we have

$$p_{ij}(Q_i) = 0 \quad \text{if} \quad i \in I(Q) \quad \text{and} \quad j \notin I(Q). \tag{3.24}$$

Also, using the finiteness of the state space, it is not difficult to verify that for each transient initial state the Markov chain $\{X_n\}$ enters the set $I(Q)$ with probability 1 after finitely many transitions. In view of (3.24), the process remains forever in the set of recurrent states once it is in that set. Hence, for each $i \in I$,

$$\lim_{n \to \infty} p_{ij}^{(n)}(Q) = 0 \quad \text{if} \quad j \notin I(Q). \tag{3.25}$$

Since the Markov chain $\{X_n\}$ has the unichain property stated in assumption 3.1, it can be shown that any two states in $I(Q)$ can reach each other (see Ross, 1983). We also refer to Ross (1983) for a proof of

$$\pi_i(Q) > 0 \quad \text{if and only if} \quad i \in I(Q). \tag{3.26}$$

We now turn to the convergence proof for the policy-iteration algorithm. We establish a lexicographical ordering for the average cost and the relative values associated with the policies that are generated by the algorithm. For that purpose we need to standardize the relative cost functions since a relative cost function is not uniquely determined. Let us (re)number the possible states as $i = 0, \ldots, N$. In view of the fact that the relative values of a given policy are unique up to an additive constant, the sequence of policies generated by the algorithm does not depend on the particular choice of the relative cost function for a given policy. For each policy Q, we now consider the particular relative cost function $w_i(Q)$ defined by (3.8) where the regeneration state r is chosen as the *largest* state in $I(Q)$.

Let R and $\bar{R}$ be immediate successors in the sequence of policies generated by the algorithm. Suppose that $\bar{R} \neq R$. We assert that either

(a) $$g(\bar{R}) < g(R)$$

or

(b) $g(\bar{R}) = g(R)$ and $w_i(\bar{R}) \leqslant w_i(R)$ for all $i \in I$ with equality for each $i \in I(\bar{R})$ and strict inequality for at least one state $i \notin I(\bar{R})$.

That is each iteration either reduces the cost rate or else reduces the relative value of a transient state. Since the number of possible stationary policies is finite, this assertion implies that the algorithm converges after finitely many iterations. To prove the assertion, the starting point is the relation

$$c_i(\bar{R}_i) - g(R) + \sum_{j\in I} p_{ij}(\bar{R}_i)w_j(R) \leqslant w_i(R), \qquad i\in I, \tag{3.27}$$

with strict inequality only for those states i with $\bar{R}_i \neq R_i$. This relation is an immediate consequence of the construction of policy $\bar{R}$. By Theorem 3.2 and (3.27), we have $g(\bar{R}) \leqslant g(R)$. We now sharpen this result as follows. It holds that $g(\bar{R}) < g(R)$ only if the strict inequality holds in (3.27) for some state i that is recurrent under the new policy $\bar{R}$. To prove this, multiply both sides of (3.27) with $\pi_i(\bar{R})$ and sum over i. Then, using $\pi_i(\bar{R}) \geqslant 0$ for all i with the strict inequality sign only for $i\in I(\bar{R})$, we find after an interchange of the order of summation that

$$\sum_{i\in I} \pi_i(\bar{R})c_i(\bar{R}_i) - g(R) + \sum_{j\in I}\left\{\sum_{i\in I}\pi_i(\bar{R})p_{ij}(\bar{R}_i)\right\}w_j(R) \leqslant \sum_{i\in I}\pi_i(\bar{R})w_i(R)$$

with strict inequality only if there is strict inequality in (3.27) for some $i\in I(\bar{R})$. By (3.3) and (3.5) with R replaced by $\bar{R}$, the latter relation is equivalent to

$$g(\bar{R}) - g(R) + \sum_{j\in I}\pi_j(\bar{R})w_j(R) \leqslant \sum_{i\in I}\pi_i(\bar{R})w_i(R)$$

with strict inequality only if there is strict inequality in (3.27) for some $i\in I(\bar{R})$. This yields the desired sharpening of Theorem 3.2. Thus we find either (a) $g(\bar{R}) < g(R)$ or (b) $g(\bar{R}) = g(R)$ with strict equality in (3.27) for all $i\in I(\bar{R})$. To complete the verification of the above assertion, consider now the latter case (b). Then, by the convention made in the policy-improvement step,

$$\bar{R}_i = R_i, \qquad i\in I(\bar{R}).$$

Using this result, the relation (3.24) with $Q = \bar{R}$ and the fact that any two states in $I(R)$ can reach each other under policy R, it is readily verified that

$$I(R) = I(\bar{R}).$$

For each policy Q the relative values $w_i(Q)$ for $i\in I(Q)$ do not depend on the actions Q_i for $i\notin I(Q)$. This follows easily from definition (3.8) and the fact that the process always stays in the set of recurrent states once it is in that set. Noting that the regeneration state r for the relative cost function is chosen as the largest state in $I(Q)$ so that this particular regeneration state is the same for the policies R and $\bar{R}$, it now follows from the above two equations that

$$w_i(R) = w_i(\bar{R}), \qquad i\in I(\bar{R}). \tag{3.28}$$

The remainder of the proof is now easy. Proceeding in the same way as we derived (3.12), we find by iterating the inequality (3.27) that for $m = 1,2,\ldots,$

$$w_i(R) \geqslant c_i(\bar{R}_i) - g(R) + \sum_{j\in I} p_{ij}(\bar{R}_i)w_j(R)$$

$$\geqslant V_m(i, \bar{R}) - mg(R) + \sum_{j\in I} p_{ij}^{(m)}(\bar{R})w_j(R), \qquad i\in I. \tag{3.29}$$

Recall that in the first one of the above inequalities the strict inequality sign holds for each i with $\bar{R}_i \neq R_i$. Since we are considering the case of $g(\bar{R}) = g(R)$, we obtain by using (3.13) with R replaced by $\bar{R}$ that, for $m = 1, 2, \ldots,$

$$V_m(i, \bar{R}) - mg(R) + \sum_{j \in I} p_{ij}^{(m)}(\bar{R}) w_j(R)$$

$$= w_i(\bar{R}) + \sum_{j \in I} p_{ij}^{(m)}(\bar{R}) \{ w_j(R) - w_j(\bar{R}) \}, \qquad i \in I.$$

Next, by letting $m \to \infty$ and using (3.25) and (3.28), we obtain from this equation and (3.29) that

$$w_i(R) \geq c_i(\bar{R}_i) - g(R) + \sum_{j \in I} p_{ij}(\bar{R}_i) w_j(R) \geq w_i(\bar{R}), \qquad i \in I,$$

where in the first inequality the strict inequality sign holds for each i with $\bar{R}_i \neq R_i$. This completes the proof. $\qquad\qquad\square$

The finite convergence of the policy-iteration algorithm implies that numbers g^* and v_i^*, $i \in I$, exist which satisfy the so-called *average cost optimality equation*

$$v_i^* = \min_{a \in A(i)} \left\{ c_i(a) - g^* + \sum_{j \in I} p_{ij}(a) v_j^* \right\}, \qquad i \in I. \tag{3.30}$$

The constant g^* is uniquely determined as the minimal average cost per unit time, that is

$$g^* = \min_R g(R).$$

Moreover, each stationary policy R^* such that the action R_i^* minimizes the right side of (3.30) for all $i \in I$ is average cost optimal. It is left as an exercise to the reader to verify that these statements are an easy consequence of Theorem 3.2.

*3.3 LINEAR PROGRAMMING FORMULATION

The policy-iteration algorithm solves the average cost optimality equation (3.30) in a finite number of steps by generating a sequence of improving policies. Another convenient way of solving the optimality equation is the application of a linear programming formulation for the average cost case. The linear programming formulation to be given below allows the unichain assumption 3.1 of section 3.1 to be weakened as follows.

Weak unichain assumption

Every average cost optimal stationary policy has the property that some state exists that can be reached from any other state when using that policy.

Define a set C of states to be *closed* under policy R if no transition is possible from a state within C to a state outside C when using policy R. Then the weak unichain assumption requires that every average cost optimal policy has no two disjoint closed sets, but allows the class of transient states to vary from policy to policy and non-optimal policies to have multiple disjoint closed sets. The unichain assumption may be too strong for some applications, e.g. in inventory problems with strictly bounded demands it may be possible to construct stationary policies with disjoint ordering regions such that the levels between which the stock fluctuates remain dependent on the initial level. However, the weak unichain assumption will practically always be satisfied in real-world applications. For the weak unichain case, the minimal average cost per unit time is independent of the initial state and, moreover, the average cost optimality equation (3.30) applies and determines uniquely g^* as the minimal average cost per unit time, (for a proof see Denardo and Fox, 1968). This reference also gives the following linear programming algorithm for the computation of an average cost optimal policy.

Linear programming algorithm

Step 1. Apply the simplex method to compute an optimal basic solution (x_{ia}^*) to the linear program

$$\text{Minimize} \qquad \sum_{i\in I}\sum_{a\in A(i)} c_i(a)x_{ia} \qquad (3.31)$$

subject to

$$\sum_{a\in A(j)} x_{ja} - \sum_{i\in I}\sum_{a\in A(i)} p_{ij}(a)x_{ia} = 0, \qquad j\in I,$$

$$\sum_{i\in I}\sum_{a\in A(i)} x_{ia} = 1, \; x_{ia} \geqslant 0, \quad i\in I \quad \text{and} \quad a\in A(i).$$

Step 2. Start with the non-empty set

$$S := \left\{ i \mid \sum_{a\in A(i)} x_{ia}^* > 0 \right\}$$

and, for any state $i\in S$, set the decision

$$R_i^* := a \text{ for some } a \text{ such that } x_{ia}^* > 0.$$

Step 3. If $S = I$, then the algorithm is stopped with the average cost optimal policy R^*. Otherwise, determine some state $i\notin S$ and action $a\in A(i)$ such that $p_{ij}(a) > 0$ for some $j\in S$, set $R_i^* := a$ and $S := S\cup\{i\}$, and repeat step 3.

The linear program (3.31) can heuristically be explained by interpreting the variables x_{ia} as

x_{ia} = the long-run fraction of decision epochs at which the system
is in state i and action a is made

(or, in the case where each stationary policy induces an aperiodic Markov chain, as the steady-state probability of being in state i and choosing action a). Then the objective of the linear program is the minimization of the long-run average cost per unit time, while the first set of constraints represent the balance equations requiring that for any state $j \in I$ the long-run average number of transitions from state j per unit time must be equal to the long-run average number of transitions into state j per unit time. The last constraint obviously requires that the sum of the fractions x_{ia} must be equal to 1.

Next we sketch a proof that the above linear programming algorithm leads to an average cost optimal policy when the weak unichain assumption is satisfied. Our starting point is the average cost optimality equation (3.30). Since this equation is solvable, the linear inequalities

$$g + v_i - \sum_{j \in I} p_{ij}(a)v_j \leqslant c_i(a), \qquad i \in I \quad \text{and} \quad a \in A(i) \tag{3.32}$$

must have a solution. Next it is readily verified that any solution $\{g, v_i\}$ to these inequalities satisfies $g \leqslant g_i(R)$ for any $i \in I$ and any policy R, where $g_i(R)$ denotes the long-run average cost per unit time under policy R when the initial state is i. The inequalities $g \leqslant g_i(R)$, $i \in I$, follow by a repeated application of the inequalities $g + v_i - \sum_{j \in I} p_{ij}(R_i)v_j \leqslant c_i(R_i)$, $i \in I$; see the proof of Theorem 3.2. Hence we can conclude that for any solution $\{g, v_i\}$ to the linear inequalities (3.32) holds that $g \leqslant g^*$ with g^* being the minimal average cost per unit time. Hence, using the fact that relative values v_i^*, $i \in I$, exist such that $\{g^*, v_i^*\}$ constitutes a solution to (3.32), the linear program

Maximize g $\hspace{6cm}$ (3.33)

subject to

$$g + v_i - \sum_{j \in I} p_{ij}(a)v_j \leqslant c_i(a), \qquad i \in I \quad \text{and} \quad a \in A(i), g, v_i \quad \text{unrestricted}$$

has the minimal average cost g^* as the optimal objective-function value. Next observe that the linear program (3.31) is the dual of the primal linear program (3.33). By the dual theorem of linear programming the primal and dual linear programs have the same optimal objective-function value. Hence the minimal objective-function value of the linear program (3.31) yields the minimal average cost g^*. To show that an optimal basic solution (x_{ia}^*) to the linear program induces an average cost optimal policy, we first prove that the non-empty set

$$S_0 = \left\{ i \mid \sum_{a \in A(i)} x_{ia}^* > 0 \right\}$$

is closed under any stationary policy. The proof proceeds by contradiction. Suppose that $p_{ij}(a) > 0$ for some $i \in S_0$, $j \notin S_0$ and $a \in A(i)$, then it follows from the constraints of the linear program (3.31) that $\sum_{a \in A(j)} x_{ja}^* > 0$, contradicting $j \notin S_0$. By the closedness of the set S_0 under any policy and the assumption that every average cost optimal policy has no two disjoint closed sets, the states $i \notin S_0$ are transient under every average cost optimal policy. This result guarantees that the completion of policy R^* in step 3 of the linear programming algorithm is feasible. It remains to prove that the constructed policy R^* is average cost optimal. To do so, let $\{g^*, v_i^*\}$ be the particular optimal basic solution to the primal linear program (3.33) such that this basic solution is complementary to the optimal basic solution (x_{ia}^*) of the dual linear program (3.31). Then, by the complementary slackness property of linear programming,

$$g^* + v_i^* - \sum_{j \in I} p_{ij}(R_i^*) v_j^* = c_i(R_i^*) \qquad \text{for } i \in S_0.$$

By the construction of policy R^* and the fact that the set S_0 is closed under any policy, we have that the set $I(R^*)$ of recurrent states of policy R^* is contained in the set S_0. Thus, noting that no transition is possible from a recurrent state to a transient state,

$$g^* + v_i^* - \sum_{j \in I(R^*)} p_{ij}(R_i^*) v_i^* = c_i(R_i^*) \qquad \text{for } i \in I(R^*).$$

By iterating these equalities and proceeding in the same way as in the proof of the first part of Theorem 3.1(b), we find that under policy R^* the average cost per unit time equals g^* for each recurrent initial state. Hence, since for any transient initial state the closed set of recurrent states will be reached after finitely many transitions, the average cost per unit time under policy R^* is equal to g^* for each initial state and so policy R^* is average cost optimal. $\qquad\square$

Comparison of linear programming and policy-iteration

In comparing the linear programming and policy-iteration formulations for the Markovian decision problem, the following remarks are in order. The linear programming formulation has the advantage that sophisticated linear programming codes with the additional option of sensitivity analysis are widely available. The policy-iteration formulation usually involves the writing of its own code. The number of iterations required by the simplex method depends heavily on the specific problem considered, whereas the policy-iteration algorithm requires typically only a very small number of iterations regardless of the problem size. In general a statement about the relative efficiency of the two methods seems difficult to make. On the one hand, the policy-iteration algorithm usually requires a much smaller number of iterations than the simplex method. However, on the other hand, the computational burden per iteration is greater for the policy-iteration method than for the simplex method; at each iteration of the former method a simultaneous system of linear equations has to be solved anew, whereas at an iteration of the last method the basic solution of the linear equations is calculated rather easily from the previous basic solution due to the fact that the simplex method changes only one decision per iteration.

It is interesting to note that the policy-iteration algorithm may be interpreted as a modified linear programming algorithm in which pivot operations are performed simultaneously on several variables, rather than a pivot operation on a single variable as in the simplex method. Also, there is considerable flexibility in changing the decisions in an iteration of the policy-iteration algorithm, and it will be seen later in section 3.6 that this flexibility may enable us to formulate very efficient, tailor-made algorithms in applications having a specific structure. In general both policy-iteration and linear programming are unattractive for solving large-scale Markovian decision problems, since both methods require per iteration the solving of a system of linear equations whose size equals the number of states. Another computational method based on successive substitutions will be discussed in the next section; this method is in general the best method for solving large-scale Markov decision problems.

We illustrate the linear programming formulation to the Markovian decision problem dealt with in example 3.1. The specification of the basic elements of the Markovian decision model for this problem is given in section 3.1.

Example 3.1 (continued) A maintenance problem for a deteriorating piece of equipment

The linear programming formulation for this problem is

Minimize $\qquad \sum_{i=2}^{N+1} x_{i1}$

subject to

$$x_{10} - \left(q_{11} x_{10} + \sum_{i=2}^{N-1} x_{i1} + x_{N+1,1} \right) = 0,$$

$$x_{j0} + x_{j1} - \sum_{i=1}^{j} q_{ij} x_{i0} = 0, \qquad 2 \leqslant j \leqslant N-1,$$

$$x_{N1} - \sum_{i=1}^{N-1} q_{iN} x_{i0} = 0,$$

$$x_{N+1,1} - x_{N1} = 0,$$

$$x_{10} + \sum_{i=2}^{N-1} (x_{i0} + x_{i1}) + x_{N1} + x_{N+1,1} = 1,$$

$$x_{10}, x_{i0}, x_{i1}, x_{N1}, x_{N+1,1} \geqslant 0.$$

For the numerical data given in Table 3.1 in section 3.2, the linear program has the optimal basic solution

$$x_{10}^* = 0.5917, \qquad x_{20}^* = 0.2367, \qquad x_{31}^* = 0.0769, \qquad x_{41}^* = 0.0473,$$
$$x_{51}^* = x_{61}^* = 0.0237.$$

This yields an average cost optimal policy $R^* = (0, 0, 1, 1, 1, 1)$ with the minimal average cost $\sum_{i=2}^{N+1} x_{i1}^* = 0.172$, in agreement with the results obtained by the policy-iteration algorithm.   □

Linear programming and probabilistic constraints

The linear programming formulation may often handle conveniently Markovian decision problems with probabilistic constraints. In many practical applications constraints are imposed on certain state frequencies. For example, in inventory problems when shortage costs are difficult to estimate, probabilistic constraints may be placed on the probability of shortage or the fraction of demand that cannot be met directly from stock on hand. Similarly, in a maintenance problem involving a randomly changing state a constraint may be placed on the frequency at which a certain inoperative state occurs.

The following illustrative example taken from Wagner (1975) shows that for control problems with probabilistic constraints it may be optimal to choose the decisions in a random way rather than in a deterministic way. Suppose the daily demand D for some product is described by the probability distribution

$$P\{D = 0\} = P\{D = 1\} = \tfrac{1}{6}, \qquad P\{D = 2\} = \tfrac{2}{3}.$$

The demands on the successive days are independent of each other. At the beginning of each day it has to be decided how much to order of the product. The delivery of any order is instantaneous. The variable ordering cost of each unit is $c > 0$. Any unit that is not sold at the end of the day becomes obsolete and must be discarded. The decision problem is to minimize the average ordering cost per day, subject to the constraint that the fraction of the demand to be met is at least $\tfrac{1}{3}$. This probabilistic constraint is satisfied when using the policy of ordering one unit every day, a policy which has an average cost of c per day. However, this deterministic control rule is not optimal, as can be seen by considering the randomized control rule under which every day no unit is ordered with probability $\tfrac{4}{5}$ and two units are ordered with probability $\tfrac{1}{5}$. Under this randomized rule the probability that the daily demand is met equals $(\tfrac{4}{5})(\tfrac{1}{6}) + (\tfrac{1}{5})(1) = \tfrac{1}{3}$ and the average ordering cost per day equals $(\tfrac{4}{5})(0) + (\tfrac{1}{5})(2c) = (\tfrac{2}{5})c$. It is readily seen that the above randomized rule is optimal.

So far we have considered only stationary policies under which the actions are chosen deterministically. A policy π is called a stationary *randomized* policy when it can be described by a probabilistic distribution $\{\pi_a(i), a \in A(i)\}$ for each state $i \in I$ such that action $a \in A(i)$ is to be chosen with probability $\pi_a(i)$ whenever the process is in state i. Here $\pi_a(i) \geqslant 0$ and $\sum_a \pi_a(i) = 1$. In the case where $\pi_a(i)$ is 0 or 1 for every i and a, the stationary randomized policy π reduces to the familiar stationary policy choosing the actions in a deterministic way. For any policy π, let the state–action frequencies $f_{ia}(\pi)$ be defined by

$f_{ia}(\pi) = $ the long-run fraction of decision epochs at which the process
is in state i and action a is chosen when policy π is used.

Suppose now a Markovian decision problem in which the goal is to minimize the long-run average cost per unit time subject to the following linear constraints on the state–action frequencies,

$$\sum_{i\in I}\sum_{a\in A(i)} \alpha_{ia}^{(l)} f_{ia}(\pi) \leqslant \beta^{(l)} \qquad \text{for} \quad l=1,\dots,L,$$

where $\alpha_{ia}^{(l)}$ and $\beta^{(l)}$ are given constants. Here it is assumed that the constraints allow for a feasible solution. In case the unichain assumption 3.1 of section 3.1 is satisfied, it can be shown that an optimal policy may be obtained by solving the following linear program (see Derman, 1970, and Hordijk and Kallenberg, 1984),

minimize
$$\sum_{i\in I}\sum_{a\in A(i)} c_i(a)x_{ia}$$

subject to

$$\sum_{a\in A(j)} x_{ja} - \sum_{i\in I}\sum_{a\in A(i)} p_{ij}(a)x_{ia} = 0, \qquad j\in I,$$

$$\sum_{i\in I}\sum_{a\in A(i)} x_{ia} = 1,$$

$$\sum_{i\in I}\sum_{a\in A(i)} \alpha_{ia}^{(l)} x_{ia} \leqslant \beta^{(l)}, \qquad l=1,\dots,L,$$

$$x_{ia} \geqslant 0, \quad i\in I \quad \text{and} \quad a\in A(i).$$

Denoting by (x_{ia}^{*}) an optimal basic solution to this linear program and letting the set $S_0 = \{i\,|\,\sum_a x_{ia}^{*} > 0\}$, an optimal stationary randomized policy π^{*} is given by

$$\pi_a^{*}(i) = \begin{cases} x_{ia}^{*}/\sum_a x_{ia}^{*} & a\in A(i) \quad \text{and} \quad i\in S_0, \\ \text{arbitrary,} & \text{otherwise.} \end{cases}$$

Here it is pointed out that the unichain assumption 3.1 is essential for guaranteeing the existence of an optimal stationary randomized policy.

A heuristic approach for handling probabilistic constraints is the Lagrange-multiplier method by which method only stationary (non-randomized) policies are produced. To describe this method, assume a single probabilistic constraint

$$\sum_{i\in I}\sum_{a\in A(i)} \alpha_{ia} f_{ia}(\pi) \leqslant \beta$$

on the state–action frequencies. Here it is assumed that $\alpha_{ia} \geqslant 0$ for all i and a. In the Lagrange multiplier method the above constraint is eliminated by putting it into the criterion function by means of a Lagrange multiplier λ, that is the goal function is changed from $\sum_{i,a} c_i(a)x_{ia}$ to $\sum_{i,a} c_i(a)x_{ia} + \lambda(\sum_{i,a}\alpha_{ia}x_{ia} - \beta)$; the Lagrange multiplier may be interpreted as the cost to each unit that is used from some resource. Thus, for a given value of the Lagrange multiplier $\lambda \geqslant 0$, we consider the unconstrained Markov decision problem with one-step costs

$$c_i^\lambda(a) = c_i(a) + \lambda\alpha_{ia}$$

and one-step transition probabilities $p_{ij}(a)$ as before. Solving this unconstrained Markov decision problem yields an optimal stationary policy $R(\lambda)$ that prescribes always the same action $R_i(\lambda)$ whenever the system is in state i. Let $\beta(\lambda)$ be the constraint level associated with policy $R(\lambda)$, that is

$$\beta(\lambda) = \sum_{i\in I} \alpha_{i,R_i(\lambda)} f_{i,R_i(\lambda)}(R(\lambda)).$$

Note that $f_{i,R_i(\lambda)}(R(\lambda))$ is just the steady-state probability $\pi_i(R(\lambda))$ giving the frequency at which the system visits state i under policy $R(\lambda)$. Since the number of stationary policies is finite, there will be gaps in the range of values of $\beta(\lambda)$ when the Lagrange multiplier λ is varied. Actually, by a standard result in parametric linear programming, there are a finite number of values $\lambda_0 < \lambda_1 < \cdots < \lambda_{m-1} < \lambda_m$ with $\lambda_0 = 0$ and $\lambda_m = \infty$ such that for any $0 \leqslant j \leqslant m-1$ a same optimal stationary policy $R(\lambda) = R(\lambda_j)$ applies for all $\lambda \in [\lambda_j, \lambda_{j+1})$. Also, we have that $\beta(\lambda)$ is piecewise constant and non-increasing in $\lambda \geqslant 0$. The heuristic Lagrangian approach searches now for the stationary policy $R(\lambda_k)$ corresponding to the *first* interval $[\lambda_k, \lambda_{k+1})$ for which

$$\beta(\lambda_k) \leqslant \beta.$$

The gaps in the constraint levels of the stationary policies $R(\lambda)$ explain partly why the average cost of the stationary policy obtained by the Lagrangian approach will in general be larger than the average cost of the stationary randomized policy resulting from the linear programming formulation. Also, it should be pointed out that there is no guarantee that the stationary policy $R(\lambda_k)$ obtained by the Lagrangian approach is the best policy among all stationary policies satisfying the probabilistic constraint, although in most practical situations this may be expected to be the case. In spite of the possible pitfalls of the Lagrangian approach, this approach may be quite useful in practical applications having a specific structure (cf. example 3.10 in section 3.6).

We conclude this section with the following application.

Example 3.3 *Two competing queues with a constraint on the queue size*

A single communication channel is shared for transmitting voice packets and data packets. The channel can transmit only one packet at a time and the transmission time of each packet is the same constant. The time is divided into slots of size corresponding to the transmission time. Transmission of a packet can only start at the beginning of a time slot. Within each time slot voice packets and data packets arrive according to a general input process with $p(i_1, i_2)$ denoting the joint probability of i_1 arrivals of voice packets and i_2 arrivals of data packets, where the arrivals in the successive slots are independent of each other. The voice and data packets have to wait in buffers of respective sizes N_1 and N_2 (the waiting positions are exclusive to any packet in transmission). A packet finding upon arrival that its buffer is full is lost. In assigning the voice and data packets access to the communication channel, it is important that the average queue size of voice packets is kept small so that the conversation they represent should not become incoherent. Therefore the goal is to find an assignment rule that minimizes the long-run average queue size of data packets subject to the constraint that the long-run average queue size of voice packets does not exceed a prespecified level β.

The above control problem can be formulated as a Markov decision problem with a probabilistic constraint. The state of the system is reviewed only at the beginnings of the time slots $t = 1, 2, \ldots$ and is represented by the tuple (i_1, i_2) with $i_1(i_2)$ denoting the number of voice (data) packets in the buffer. Here $0 \leqslant i_1 \leqslant N_1$ and $0 \leqslant i_2 \leqslant N_2$. The set of possible actions consists of the actions $a = 1, 2$. Let us say that action $a = 1$ ($a = 2$) in state (i_1, i_2) prescribes to transmit a voice (data) packet whenever a voice (data) packet is present; otherwise, no packet is transmitted. In the linear programming formulation below it is convenient to use the variables $x_{i_1,0,2}$ and $x_{0,i_2,1}$ for $0 \leqslant i_1 \leqslant N_1$ and $0 \leqslant i_2 \leqslant N_2$ with the convention that

$$x_{i_1,0,2} = 0 \quad \text{for} \quad i_1 \geqslant 1 \quad \text{and} \quad x_{0,i_2,1} = 0 \quad \text{for} \quad i_2 \geqslant 1.$$

It is now left to the reader to verify that the following linear program applies to the above control problem:

Minimize
$$\sum_{i_1,i_2,a} i_2 x_{i_1,i_2,a}$$

subject to

$$x_{i_1,i_2,1} + x_{i_1,i_2,2} - \sum_{\substack{0 \leqslant j_1 \leqslant i_1 \\ 0 \leqslant j_2 \leqslant i_2}} (x_{j_1+1,j_2,1} + x_{j_1,j_2+1,2}) p(i_1 - j_1, i_2 - j_2)$$

$$- (x_{0,0,1} + x_{0,0,2}) p(i_1, i_2) = 0, \qquad 0 \leqslant i_1 \leqslant N_1 - 1, \qquad 0 \leqslant i_2 \leqslant N_2 - 1,$$

$$x_{N_1,i_2,1} + x_{N_1,i_2,2} - \sum_{\substack{0 \leqslant j_1 \leqslant N_1 \\ 0 \leqslant j_2 \leqslant i_2}} x_{j_1+1,j_2,1} \sum_{k \geqslant N_1 - j_1} p(k, i_2 - j_2)$$

$$- \sum_{\substack{0 \leqslant j_1 \leqslant N_1 \\ 0 \leqslant j_2 \leqslant i_2}} x_{j_1,j_2+1,2} \sum_{k \geqslant N_1 - j_1} p(k, i_2 - j_2) - (x_{0,0,1} + x_{0,0,2}) \sum_{k \geqslant N_1} p(k, i_2) = 0,$$

$$0 \leqslant i_2 \leqslant N_2 - 1,$$

$$x_{i_1,N_2,1} + x_{i_1,N_2,2} - \sum_{\substack{0 \leqslant j_1 \leqslant i_1 \\ 0 \leqslant j_2 \leqslant N_2}} x_{j_1+1,j_2,1} \sum_{k \geqslant N_2 - j_2} p(i_1 - j_1, k)$$

$$- \sum_{\substack{0 \leqslant j_1 \leqslant i_1 \\ 0 \leqslant j_2 \leqslant N_2}} x_{j_1,j_2+1,2} \sum_{k \geqslant N_2 - j_2} p(i_1 - j_1, k) - (x_{0,0,1} + x_{0,0,2}) \sum_{k \geqslant N_2} p(i_1, k) = 0,$$

$$0 \leqslant i_1 \leqslant N_1 - 1,$$

$$x_{N_1,N_2,1} + x_{N_1,N_2,2} - \sum_{\substack{0 \leqslant j_1 \leqslant N_1 \\ 0 \leqslant j_2 \leqslant N_2}} x_{j_1+1,j_2,1} \sum_{\substack{k_1 \geqslant N_1 - j_1 \\ k_2 \geqslant N_2 - j_2}} p(k_1, k_2)$$

$$- \sum_{\substack{0 \leqslant j_1 \leqslant N_1 \\ 0 \leqslant j_2 \leqslant N_2}} x_{j_1,j_2+1,2} \sum_{\substack{k_1 \geqslant N_1 - j_1 \\ k_2 \geqslant N_2 - j_2}} p(k_1, k_2) - (x_{0,0,1} + x_{0,0,2}) \sum_{\substack{k_1 \geqslant N_1 \\ k_2 \geqslant N_2}} p(k_1, k_2) = 0,$$

$$\sum_{i_1,i_2,a} x_{i_1,i_2,a} = 1,$$

$$\sum_{i_1,i_2,a} i_1 x_{i_1,i_2,a} \leqslant \beta,$$

$$x_{i_1,i_2,a} \geqslant 0.$$

As an illustration, consider the following numerical data:

$$N_1 = 5, \qquad N_2 = 7, \qquad p(0,0) = 0.36, \qquad p(0,1) = p(1,0) = 0.24,$$
$$p(1,1) = 0.16, \qquad \text{and} \qquad \beta = 0.5.$$

In Table 3.6 we give the optimal stationary randomized policy π^* by the values of the probabilities $\pi^*_{i_1,i_2,1}$ of taking action 1 in state (i_1, i_2). Under this optimal policy the long-run average sizes L_1 and L_2 of voice and data packets have the respective values $0.5 (= \beta)$ and 1.056. Also, in Table 3.6 we give the values of the probabilities $\pi_{i_1,i_2,1}$ corresponding to the stationary (non-randomized) policy obtained by the Lagrangian approach. For this policy, we find $L_1 = 0.463 (< \beta)$ and $L_2 = 1.101$. Comparing the latter result with the optimum, we see that in this case randomization allows for a reduction of about 4 per cent in the average queue size of data packets when a probabilistic constraint is imposed.

Finally, it is interesting to note that in the case of an infinite waiting room for both voice and data packets there exists an optimal randomized policy that is characterized by a single probability π such that $\pi_{i_1,i_2,1} = \pi$ for

Table 3.6   Numerical results

$i_1 \backslash i_2$	Linear program: $L_1 = 0.5$, $L_2 = 1.056$								Lagrangian: $L_1 = 0.463$, $L_2 = 1.101$							
	0	1	2	3	4	5	6	7	0	1	2	3	4	5	6	7
0	1	0	0	0	0	0	0	0	1	0	0	0	0	0	0	0
1	1	1	1	1	0.2606	0	0	0	1	1	1	1	1	0	0	0
2	1	1	1	0	0	0	0	0	1	1	1	0	0	0	0	0
3	1	1	0	0	0	0	0	0	1	1	0	0	0	0	0	0
4	1	0	0	0	0	0	0	0	1	0	0	0	0	0	0	0
5	1	0	0	0	0	0	0	0	1	0	0	0	0	0	0	0

all i_1, i_2 with both $i_1 \geqslant 1$ and $i_2 \geqslant 1$ (see Jain and Ross, 1985). This remarkable result is not true for the case of a finite waiting room (e.g. for the above numerical data the best value of π is 0.8169 with $L_1 = 0.5$ and $L_2 = 1.086$, where the respective rejection probabilities for voice and data packets are 2.6×10^{-6} and 3.6×10^{-4}). A convenient method to calculate the best value of π is to apply Markov chain analysis in conjunction with a numerical method for locating the zero of a function. □

3.4   VALUE-ITERATION ALGORITHM

The policy-iteration algorithm and the linear programming formulation both require that in each iteration a system of linear equations of the same size as the state space is solved. In general, this will be computationally burdensome for a large state space and makes these algorithms computationally unattractive for large-scale Markov decision problems. In this section we discuss an alternative algorithm which avoids solving systems of linear equations but uses instead the recursive solution approach from dynamic programming. This method is the *value-iteration algorithm* which computes recursively a sequence of value functions approximating the minimal average cost per unit time. The value functions provide lower and upper bounds on the minimal average cost and under a certain aperiodicity condition these bounds approximate arbitrarily closely the minimal cost rate. The value-iteration algorithm endowed with these lower and upper bounds is in general the best computational method for solving large-scale Markov decision problems. Moreover, it is easy to write an own computer code of this algorithm for specific problems. Also, it is important to point out that value-iteration may be a convenient method to calculate the average costs in a single Markov chain.

The value-iteration algorithm computes recursively for $n = 1, 2, \ldots$ the value function $V_n(i)$ from

$$V_n(i) = \min_{a \in A(i)} \left\{ c_i(a) + \sum_{j \in I} p_{ij}(a) V_{n-1}(j) \right\}, \qquad i \in I, \qquad (3.34)$$

starting with an arbitrarily chosen function $V_0(i)$, $i \in I$. The quantity $V_n(i)$ can be interpreted as the minimal total expected costs with n periods left to the time horizon when the current state is i and a terminal cost of $V_0(j)$ is incurred when the system ends up at state j (for a proof see Denardo, 1982, and Derman, 1970). This interpretation suggests that for large n the one-step difference $V_n(i) - V_{n-1}(i)$ will come very closely to the minimal average cost per unit time, while the stationary policy whose actions minimize the right side of (3.34) for all i will be very close in costs to the minimal average costs. However, these matters appear to be rather subtle for the average cost criterion due to the effect of possible periodicities in the underlying decision processes. To state this precisely, let, for $n = 1, 2, \ldots,$

$$m_n = \min_{j \in I} \{ V_n(j) - V_{n-1}(j) \} \quad \text{and} \quad M_n = \max_{j \in I} \{ V_n(j) - V_{n-1}(j) \}. \tag{3.35}$$

We first prove the following theorem.

Theorem 3.3

Suppose that the unichain assumption 3.1 holds. For any $n \geq 1$, let $R(n)$ be a stationary policy whose actions minimize the right side of (3.34) for all $i \in I$. Then

$$m_n \leq g^* \leq g(R(n)) \leq M_n, \qquad n \geq 1, \tag{3.36}$$

where $g^* = \min_R g(R)$ denotes the minimal average cost per unit time. Moreover, the sequence $\{ m_n, n \geq 1 \}$ is non-decreasing and the sequence $\{ M_n, n \geq 1 \}$ is non-increasing.

Proof. By the construction of the value function $V_n(i)$ and the rule $R(n)$, we have, for all $i \in I$,

$$V_n(i) \leq c_i(a) + \sum_{j \in I} p_{ij}(a) V_{n-1}(j), \qquad a \in A(i), \tag{3.37}$$

$$V_n(i) = c_i(R_i(n)) + \sum_{j \in I} p_{ij}(R_i(n)) V_{n-1}(j). \tag{3.38}$$

Writing $V_n(i) = V_{n-1}(i) + V_n(i) - V_{n-1}(i)$ and noting that $V_n(i) - V_{n-1}(i) \leq M_n$, it follows from (3.38) that

$$c_i(R_i(n)) + \sum_{j \in I} p_{ij}(R_i(n)) V_{n-1}(j) \leq V_{n-1}(i) + M_n, \qquad i \in I.$$

An application of Theorem 3.2 next yields $g(R(n)) \leq M_n$. To prove $g^* \geq m_n$ we verify that $g(R) \geq m_n$ for each stationary policy R. Letting R be an arbitrarily chosen rule, we have by (3.37) with $a = R_i$ that

$$c_i(R_i) + \sum_{j \in I} p_{ij}(R_i) V_{n-1}(j) \geq V_n(i), \qquad i \in I.$$

Noting that $V_n(i) - V_{n-1}(i) \geq m_n$, we obtain

$$c_i(R_i) + \sum_{j \in I} p_{ij}(R_i) V_{n-1}(j) \geq V_{n-1}(i) + m_n, \qquad i \in I.$$

Hence, by applying Theorem 3.2, $g(R) \geq m_n$. We next verify that $m_k \geq m_{k-1}$ for each $k \geq 2$. Using (3.38) with $n = k$ and (3.37) with $n = k - 1$ and $a = R_i(k)$, we obtain, by subtraction of $V_k(i)$ and $V_{k-1}(i)$,

$$V_k(i) - V_{k-1}(i) \geq \sum_{j \in I} p_{ij}(R_i(k)) \{ V_{k-1}(j) - V_{k-2}(j) \}, \qquad i \in I.$$

This inequality implies

$$V_k(i) - V_{k-1}(i) \geq \min_{j \in I} \{ V_{k-1}(j) - V_{k-2}(j) \} \sum_{j \in I} p_{ij}(R_i(k)), \qquad i \in I,$$

and so $V_k(i) - V_{k-1}(i) \geqslant m_{k-1}$ for all $i \in I$, proving that $m_k \geqslant m_{k-1}$. It is left as an exercise to the reader to verify that $M_k \leqslant M_{k-1}$.   □

Remark 3.2

In the formulation of Theorem 3.3 for ease we have made the assumption that each stationary policy has the unichain property. However, value-iteration applies equally well when it is only required that the minimal average cost g^* per unit time is independent of the initial state. A closer examination of the proof of Theorem 3.3 shows that the following important result then holds. If for an arbitrary vector $x = (x(i), i \in I)$ the new vector Ux is calculated as

$$Ux(i) = \min_{a \in A(i)} \left\{ c_i(a) + \sum_{j \in I} p_{ij}(a)x(j) \right\}, \qquad i \in I,$$

and R is a policy such that the action R_i minimizes the right side of this equation for all $i \in I$, then

$$\min_{j \in I} \{ Ux(j) - x(j) \} \leqslant g^* \leqslant g_i(R) \leqslant \max_{j \in I} \{ Ux(j) - x(j) \}, \qquad i \in I,$$

where $g_i(R)$ denotes the average cost per unit time when the initial state is i and policy R is used (in the case where R has the unichain property, then $g_i(R) = g(R)$ for all i).

In formulating the value-iteration algorithm it is no restriction to assume that each one-step cost $c_i(a)$ is *non-negative*. Otherwise, add a sufficiently large constant to each $c_i(a)$ and note that by this transformation the average cost of any policy changes by the same amount. Then, by choosing $V_0(i)$ such that $0 \leqslant V_0(i) \leqslant \min_a c_i(a)$ for all $i \in I$, we have $V_1(i) \geqslant V_0(i)$ for all i, implying that each term of the non-decreasing sequence $\{ m_n, n \geqslant 1 \}$ is non-negative. Using this we next obtain from Theorem 3.3 that

$$\frac{M_n - m_n}{m_n} \leqslant \varepsilon \text{ implies } 0 \leqslant \frac{g(R_n^*) - g^*}{g^*} \leqslant \varepsilon,$$

that is the average cost of the policy $R(n)$ obtained at the nth iteration cannot deviate more than 100ε per cent from the theoretically minimal average cost when $(M_n - m_n)/m_n \leqslant \varepsilon$. In practical applications one is usually satisfied with a policy whose average cost is sufficiently close to the minimal average cost.

Value-iteration algorithm

Step 0. Choose $V_0(i)$ with $0 \leqslant V_0(i) \leqslant \min_a c_i(a)$ for all $i \in I$. Let $n := 1$.
Step 1. Compute the value function $V_n(i)$, $i \in I$, from

$$V_n(i) = \min_{a \in A(i)} \left\{ c_i(a) + \sum_{j \in I} p_{ij}(a)V_{n-1}(j) \right\}$$

and determine $R(n)$ as a stationary policy whose actions minimize the right side of this equation for all $i \in I$.

Step 2. Compute the bounds

$$m_n = \min_{j \in I} \{V_n(j) - V_{n-1}(j)\} \qquad \text{and} \qquad M_n = \max_{j \in I} \{V_n(j) - V_{n-1}(j)\}.$$

The algorithm is stopped with policy $R(n)$ when $0 \leqslant M_n - m_n \leqslant \varepsilon m_n$ where ε is a prespecified tolerance number (for example $\varepsilon = 0.001$) Otherwise, go to step 3.

Step 3. $n := n + 1$ and go to step 1.

The remaining question is whether the algorithm will be stopped after finitely many iterations. In other words, the question is whether the upper and lower bounds M_n and m_n converge to the same limits. In general M_n and m_n need not have the same limits, as the following example demonstrates. Consider the trivial Markov decision problem with the two states 1 and 2 and a single action a_0, where the one-step costs and the one-step transition probabilities are given by $c_1(a_0) = 1$, $c_2(a_0) = 0$, $p_{12}(a_0) = p_{21}(a_0) = 1$ and $p_{11}(a_0) = p_{22}(a_0) = 0$. Hence the system cycles between the states 1 and 2. It is easily verified that $V_{2k}(1) = V_{2k}(2) = k$, $V_{2k-1}(1) = k$ and $V_{2k-1}(2) = k - 1$ for all $k \geqslant 1$. Hence $m_n = 0$ and $M_n = 1$ for all n, implying that the sequences $\{m_n\}$ and $\{M_n\}$ have different limits. The reason for the oscillating behaviour of $V_n(i) - V_{n-1}(i)$ is the periodicity of the Markov chain describing the state of the system.

Under the assumption that for each stationary policy the associated Markov chain $\{X_n\}$ is *aperiodic*, it holds that $\lim_{n \to \infty} \{V_n(i) - ng^*\}$ exists and is finite for all $i \in I$, implying that

$$\lim_{n \to \infty} m_n = \lim_{n \to \infty} M_n = g^*,$$

provided the minimal average cost g^* per unit of time is independent of the initial state, (for a proof see Bather, 1973, and Schweitzer and Federgruen, 1979). Moreover, the practically important result holds that the bounds converge geometrically fast (see also Zijm, 1985). Actually these convergence results apply also when the aperiodicity of the Markov chains holds only for the average cost optimal policies.

In the case where not every optimal policy induces an aperiodic Markov chain $\{X_n\}$, the periodicity issue can be circumvented by a *perturbation* of the one-step transition probabilities. The perturbation technique is based on the following two observations. First, a Markov chain in which direct transitions from a recurrent state to itself are possible must be aperiodic. Second, the relative frequencies at which the states of a Markov chain are visited do not change when at each transition epoch the Markov chain is allowed only with a constant probability $\tau > 0$ to make a transition according to the original law of motion and is forced to make a self-transition otherwise. In other words, if the one-step transition probabilities p_{ij} of a Markov chain $\{X_n\}$ are perturbed as $\tilde{p}_{ij} = \tau p_{ij}$ for $j \neq i$ and $\tilde{p}_{ii} = \tau p_{ii} + 1 - \tau$ for some constant τ with $0 < \tau < 1$, the perturbed Markov chain $\{\tilde{X}_n\}$ with one-step transition probabilities $\tilde{p}_{ij}$ is aperiodic and has the same equilibrium probabilities as the original Markov chain $\{X_n\}$. The reader may wish to verify the latter assertion by direct substitution into the equilibrium equations (2.11) in section 2.1 of chapter 2. Thus a Markov decision model involving periodicities may be perturbed as follows. Choosing some constant τ with $0 < \tau < 1$, the state space, the action sets, the one-step costs and the one-step transition probabilities of the perturbed Markov decision model are defined by

$$
\begin{aligned}
&\tilde{I} = I, \\
&\tilde{A}(i) = A(i) && \text{for } i \in I, \\
&\tilde{c}_i(a) = c_i(a) && \text{for } i \in \tilde{I} \text{ and } a \in \tilde{A}(i), \\
&\tilde{p}_{ij}(a) = \begin{cases} \tau p_{ij}(a) & \text{for } j \neq i, \quad i \in \tilde{I} \quad \text{and} \quad a \in \tilde{A}(i), \\ \tau p_{ii}(a) + 1 - \tau & \text{for } j = i, \quad i \in \tilde{I} \quad \text{and} \quad a \in \tilde{A}(i). \end{cases}
\end{aligned}
$$

For each stationary policy, the associated Markov chain $\{\tilde{X}_n\}$ in the perturbed model is aperiodic and has the same equilibrium probabilities as the corresponding Markov chain $\{X_n\}$ in the original model. Thus, using (3.5), for each stationary policy, the average cost per unit time in the perturbed model is the same as that in the original model. Therefore the value-iteration algorithm may be applied to the perturbed model in the case where the algorithm is not convergent for the original model. In specific problems involving periodicities the 'optimal' value of τ is usually not clear beforehand; empirical investigations indicate that $\tau = \frac{1}{2}$ is usually a satisfactory choice.

The value-iteration algorithm has in general not the robustness of the policy-iteration algorithm. The number of iterations required by the value-iteration algorithm is typically problem dependent and will usually increase when the number of states becomes larger. Also, the tolerance number ε in the stopping criterion and the starting vector $V_0(i)$, $i \in I$, will affect the number of iterations required.

The convergence rate of the value-iteration algorithm may be accelerated by using a relaxation factor such as in successive overrelaxation for solving a single system of linear equations. Then at the nth iteration a new approximation to the value function $V_n(i)$ is obtained by using both the previous values $V_{n-1}(i)$ and the residuals $V_n(i) - V_{n-1}(i)$. It is possible to select dynamically a relaxation factor and thus avoid the experimental determination of the best value of a fixed relaxation factor. The following modification of the standard value-iteration can be formulated.

Modified value-iteration algorithm with a dynamic relaxation factor

The steps 0, 1 and 2 are as before, while step 3 of the standard value-iteration algorithm is modified as follows.

Step 3(a). Determine states u and v such that

$$V_n(u) - V_{n-1}(u) = m_n \quad \text{and} \quad V_n(v) - V_{n-1}(v) = M_n$$

and compute the relaxation factor

$$\omega = \frac{M_n - m_n}{M_n - m_n + \sum_{j \in I} \{p_{uj}(R_u) - p_{vj}(R_v)\}\{V_n(j) - V_{n-1}(j)\}}. \quad (3.39)$$

Step 3(b). For each $i \in I$, change $V_n(i)$ according to

$$V_n(i) := V_{n-1}(i) + \omega\{V_n(i) - V_{n-1}(i)\}.$$

Step 3(c). $n := n + 1$ and go to step 1.

In the case of a tie when selecting in step 3(a) the state $u(v)$ for which the minimum (maximum) in $m_n(M_n)$ is obtained, the convention is made to choose the minimizing (maximizing) state of the previous iteration when that state is one of the candidates and to choose otherwise the first state achieving the minimum (maximum) in m_n (M_n); thus the ordering of the states may influence the speed of convergence of the modified value-iteration algorithm.

The value of the dynamic relaxation factor ω is motivated as follows. Suppose that at the nth iteration the minimum in the lower bound $m_n = \min_i\{V_n(i) - V_{n-1}(i)\}$ and the maximum in the upper bound $M_n = \max_i\{V_n(i) - V_{n-1}(i)\}$ are achieved by the states u and v. We change the estimate $V_n(i)$ as $\bar{V}_n(i) = V_{n-1}(i) + \omega\{V_n(i) - V_{n-1}(i)\}$ for all i in order to accomplish, at the $(n + 1)$th iteration,

$$c_u(R_u) + \sum_{j \in I} p_{uj}(R_u)\bar{V}_n(j) - \bar{V}_n(u) = c_v(R_v) + \sum_{j \in I} p_{vj}(R_v)\bar{V}_n(j) - \bar{V}_n(v),$$

in the implicit hope that the difference between the new upper and lower bounds M_{n+1} and m_{n+1} will decrease more quickly. Using the relations $m_n = V_n(u) - V_{n-1}(u) = c_u(R_u) + \sum_j p_{uj}(R_u)V_{n-1}(j) - V_{n-1}(u)$ and the similar relations for M_n, it is a matter of simple algebra to verify from the above condition that ω is given by (3.39). We omit the easy proof that always $\omega > 0$ as long as the bounds m_n and M_n have not yet converged.

Numerical experiments indicate that using a dynamic relaxation factor in value-iteration may greatly enhance the speed of convergence of the algorithm (see also the examples 3.7 and 3.8 in section 3.5). The modified value-iteration algorithm is theoretically not guaranteed to converge, but in practice the algorithm will usually work very well. It is important to note that the relaxation factor ω is kept outside the recursion equation (3.34) so that the bounds in (3.36) are not destroyed. Although the bounds in (3.36) apply, it is no longer true that the sequences $\{m_n\}$ and $\{M_n\}$ are monotonic.

To conclude this section, we apply value-iteration to calculate optimal no-claim limits in an insurance problem. The next example shows that periodicities can easily be dealt with when the Markov chains associated with the stationary policies are all known to have the same period, in which case no data transformation is needed.

Example 3.4   Optimal no-claim limits for vehicle insurance

Consider a vehicle insurance which charges reduced premium to motorists who do not make claims over one or more years. When an accident occurs the motorist has the option of either making a claim and thereby perhaps losing a reduction in premium, or paying the costs associated with the accident himself. The decision should be based on a no-claim limit which typically depends on the claim history during the current premium year and the time until the next premium payment. We assume a premium structure for which the premium payment is due at the beginning of each year and the payment depends only on the previous payment and the number of claims made in the past year. The premium structure is as shown in Table 3.7 where for the five possible premiums $\pi(i)$, $1 \leqslant i \leqslant 5$, the condition holds that $\pi(i + 1) < \pi(i)$, $1 \leqslant i \leqslant 4$.

The accidents occur according to a Poisson process at a rate λ per year and the costs associated with an accident have a given probability distribution function F with a positive density f. We wish to determine no-claim limits for which the long-run average cost per year is minimal.

To model this insurance problem as a discrete-time Markov decision process, we assume that the value of λ is such that the probability of more than one accident in a month is negligible. Thus we take a basic unit of one month and take the beginnings of the successive months as the decision epochs. The decisions of whether to claim or not are modelled as follows. Taking decision a at the beginning of a month means that a claim would be made after an

Table 3.7   The premium structure

Current premium	Subsequent premium		
	No claim	One claim	Two or more claims
$\pi(1)$	$\pi(2)$	$\pi(1)$	$\pi(1)$
$\pi(2)$	$\pi(3)$	$\pi(1)$	$\pi(1)$
$\pi(3)$	$\pi(4)$	$\pi(1)$	$\pi(1)$
$\pi(4)$	$\pi(5)$	$\pi(2)$	$\pi(1)$
$\pi(5)$	$\pi(5)$	$\pi(3)$	$\pi(1)$

accident in that month only if the costs associated with the accident exceed a. In view of the premium structure shown in Table 3.7 at the beginning of each month the system can be defined to be in one of the states (t, i) with $t = 0$, $1, \ldots, 11$ and $i = 1, \ldots, 6$, where

t denotes the number of months until the next premium payment,

$i = 1$ means that the next premium is $\pi(1)$,

$i = 2$ means that the next premium is $\pi(2)$ if no more claims are made until the next premium payment,

$i = 3$ means that the next premium is $\pi(3)$ if no more claims are made until the next premium payment,

$i = 4$ means that the next premium is $\pi(4)$ if no more claims are made until the next premium payment,

$i = 5$ means that the last premium was $\pi(4)$ while the next premium is $\pi(5)$ if no more claims are made until the next premium payment,

$i = 6$ means that the last premium was $\pi(5)$ while the next premium is $\pi(5)$ if no more claims are made until the next premium payment.

Next we specify the one-step expected costs and the one-step transition probabilities. Since the accidents occur according to a Poisson process with an average of λ accidents per year and the probability of more than one accident in a month is assumed to be negligible, we take $\lambda/12$ as the probability of one accident in a month and $1 - \lambda/12$ as the probability of no accident in a month. If an accident with associated costs D occurs in a month and action a is followed in that month, then this accident results in a claim only if $D > a$ while for the motorist this accident involves damage costs D in the case $D \leqslant a$ and damage costs 0 in the case $D > a$. Thus the probability of a claim in a month during which action a is followed equals

$$p(a) = \frac{\lambda}{12}\{1 - F(a)\},$$

while for the motorist the expected damage costs in that month are given by

$$c(a) = \frac{\lambda}{12} \int_0^a s f(s) \, ds.$$

It now follows that if action a is chosen in state (t, i) the one-step expected costs are given by $\pi(\min(i, 5)) + c(a)$ for $t = 0$ and by $c(a)$ for $t \neq 0$. Also, for the one-step transition probabilities we distinguish between the states $(0, i)$ and the states (t, i) with $t \neq 0$. In Table 3.8 we list the next premium class $\phi(i)$ which results in the case where a claim is made for the state (t, i) with $t \neq 0$.

If action a is chosen in state (t, i) with $t \neq 0$, then at the beginning of the next month the second state variable equals i with probability $1 - p(a)$ and

Table 3.8　Transition of the premium class

i	1	2	3	4	5	6
$\phi(i)$	1	1	1	1	2	3

equals $\phi(i)$ with probability $p(a)$. Noting the effect of the premium payment in state $(0, i)$ and assuming that the action a is chosen in this state, it is readily verified that at the beginning of the next month the second state variable equals $\min(i + 1, 6)$ with probability $1 - p(a)$ and equals $\phi(\min(i + 1, 6))$ with probability $p(a)$. The way in which the first state variable changes each month is obvious. This variable changes from t to $t - 1$ for $t \geqslant 1$ and from $t = 0$ to $t = 11$.

We have now completely specified the Markov decision model for the insurance problem. The most effective way to compute an average cost optimal policy in this model is to apply the value-iteration algorithm. Define, for $0 \leqslant t \leqslant 11$, $1 \leqslant i \leqslant 6$ and $n \geqslant 1$.

$V_{12n + t}(t, i) =$ the minimal expected total cost if the motorist has still an insurance contract for $t + 12n$ months only and the present state is (t, i).

The following recursion relation applies for $n = 1, 2, \ldots$ (verify!):

$$V_{12n + t}(t, i) = \min_{a}[c(a) + \{1 - p(a)\}V_{12n + t - 1}(t - 1, i)$$
$$+ p(a)V_{12n + t - 1}(t - 1, \phi(i))] \quad \text{for } 1 \leqslant t \leqslant 11, \quad 1 \leqslant i \leqslant 6,$$

$$V_{12n}(0, i) = \min_{a}[\pi(i) + c(a) + \{1 - p(a)\}V_{12(n - 1) + 11}(11, i + 1)$$
$$+ p(a)V_{12(n - 1) + 11}(11, \phi(i + 1))] \quad \text{for } 1 \leqslant i \leqslant 5,$$

$$V_{12n}(0, 6) = \min_{a}[\pi(5) + c(a) + \{1 - p(a)\}V_{12(n - 1) + 11}(11, 6)$$
$$+ p(a)V_{12(n - 1) + 11}(11, \phi(6))],$$

where $V_t(t, i) = 0$ for $0 \leqslant t \leqslant 11$ and $1 \leqslant i \leqslant 6$. We denote by $R_{t, i}(n)$ the action a for which the minimum is assumed in the right side of the recursion relation for $V_{12n + t}(t, i)$. This minimizing action can explicitly be given. To do this, note that the expression between brackets in the recursion relation for $V_{12n + t}(t, i)$ has the derivative $(\lambda/12)f(a)\{a + V_{12n + t - 1}(t - 1, i) - V_{12n + t - 1}(t - 1, \phi(i))\}$ with respect to a. Using $V_{12n + t - 1}(t - 1, i) \leqslant V_{12n + t - 1}(t - 1, j)$ for $i \geqslant j$, it is readily verified that the minimizing action $R_{t, i}(n)$ is given by

$$R_{t, i}(n) = \begin{cases} V_{12n + t - 1}(t - 1, \phi(i)) - V_{12n + t - 1}(t - 1, i), & 1 \leqslant t \leqslant 11, 1 \leqslant i \leqslant 6, \\ V_{12(n - 1) + 11}(11, \phi(i + 1)) - V_{12(n - 1) + 11}(11, i + 1), & t = 0, 1 \leqslant i \leqslant 5, \\ V_{12(n - 1) + 11}(11, \phi(6)) - V_{12(n - 1) + 11}(11, 6), & t = 0, i = 6. \end{cases}$$

Table 3.9   The (nearly) optimal no-claim limits

$c_D^2 = 1,$	Lognormal: $N = 14$, $m_N = 317.3$, $M_N = 317.6$, $g^* \approx 317.5$										
i/t	0	1	2	3	4	5	6	7	8	9	10	11
1	190	0	0	0	0	0	0	0	0	0	0	0
2	355	286	276	267	257	248	239	230	221	213	205	197
3	495	499	485	471	457	444	430	417	405	392	380	368
4	418	662	646	630	615	599	584	569	554	539	524	510
5	291	467	464	460	456	452	447	443	438	433	428	423
6	291	255	260	264	269	272	276	279	282	285	287	289

$c_D^2 = 1,$	Gamma: $N = 15$, $m_N = 306.5$, $M_N = 306.7$, $g^* \approx 306.6$										
i/t	0	1	2	3	4	5	6	7	8	9	10	11
1	200	0	0	0	0	0	0	0	0	0	0	0
2	368	289	280	271	262	254	246	238	230	222	215	207
3	509	502	489	476	463	451	438	426	414	402	391	380
4	423	668	653	637	622	608	593	578	564	550	536	522
5	293	470	466	463	459	455	451	446	442	437	433	428
6	293	257	262	266	271	274	278	281	284	287	289	292

$c_D^2 = 4,$	Lognormal: $N = 14$, $m_N = 289.1$, $M_N = 289.2$, $g^* \approx 289.1$										
i/t	0	1	2	3	4	5	6	7	8	9	10	11
1	217	0	0	0	0	0	0	0	0	0	0	0
2	396	300	292	284	276	268	261	253	246	238	231	224
3	533	511	500	489	478	467	457	446	436	426	416	406
4	418	665	652	640	628	615	603	591	579	568	556	544
5	271	448	445	443	441	438	436	433	430	427	424	421
6	271	237	241	245	248	252	255	258	261	264	266	268

$c_D^2 = 4,$	Gamma: $N = 14$, $m_N = 259.9$, $M_N = 260.0$, $g^* \approx 259.9$										
i/t	0	1	2	3	4	5	6	7	8	9	10	11
1	245	0	0	0	0	0	0	0	0	0	0	0
2	427	305	299	293	287	282	276	271	266	260	255	250
3	563	514	505	497	489	481	473	465	457	450	442	434
4	418	665	656	646	636	627	617	608	599	590	581	572
5	263	440	438	436	434	433	431	429	427	424	422	420
6	263	231	234	238	241	244	247	250	253	255	258	260

$c_D^2 = 4,$	Weibull, $N = 14$, $m_N = 270.5$, $M_N = 270.6$, $g^* \approx 270.6$										
i/t	0	1	2	3	4	5	6	7	8	9	10	11
1	235	0	0	0	0	0	0	0	0	0	0	0
2	416	303	297	290	283	277	271	265	258	252	247	241
3	552	513	504	494	485	476	467	459	450	441	433	424
4	418	666	655	644	634	623	613	602	592	582	572	562
5	266	443	441	439	437	435	433	431	428	426	423	421
6	266	233	237	240	244	247	250	253	256	259	261	264

The Markov decision model of the insurance problem is indeed periodic, but its special periodicity structure is such that each decision process considered only every twelve months is aperiodic. This observation enables us to define upper and lower bounds that both converge to the minimal average twelve-monthly costs. Defining

$$m_n = \min_{1 \leqslant i \leqslant 6} \{V_{12n}(0, i) - V_{12(n-1)}(0, i)\}, \qquad M_n = \max_{1 \leqslant i \leqslant 6} \{V_{12n}(0, i) - V_{12(n-1)}(0, i)\},$$

it can be seen that (3.36) holds where the bounds m_n and M_n both converge to g^*, provided we interpret g^* and $g(R(n))$ as the minimal average twelve-monthly costs and the average twelve-monthly costs of policy $R(n) = (R_{t,i}(n))$ (cf. also Su and Deininger, 1972, dealing with another application of a periodic Markov decision model). Letting N be the first integer $n \geqslant 2$ for which $(M_n - m_n)/m_n \leqslant \varepsilon$, with ε a prespecified tolerance number, the value-iteration algorithm is stopped with the stationary policy $R(N)$. The minimal average yearly cost g^* is approximated by $(m_N + M_N)/2$ and the relative difference percentage of the average yearly cost of policy $R(N)$, and the minimal average yearly cost is at most 100ε per cent.

As an illustration we give some numerical examples in which

$$\pi(1) = 500, \qquad \pi(2) = 375, \qquad \pi(3) = 300, \qquad \pi(4) = 250,$$

$$\pi(5) = 200, \qquad \lambda = 0.5 \quad \text{and} \qquad E(D) = 500,$$

where the random variable D represents the costs associated with an accident. The squared coefficient of variation $c_D^2 = \sigma^2(D)/E^2(D)$ is varied as 1 and 4. In order to show that for larger values of c_D^2 the optimal no-claim limits become increasingly sensitive to more than the first two moments of the costs D, we consider for the costs D a lognormal distribution, a gamma distribution and a Weibull distribution each having the same first two moments (cf. also appendix B). In Table 3.9 we give for the various numerical examples the nearly optimal no-claim limits obtained by the value-iteration algorithm starting with the values $V_0(i) = 0$ for all i and having $\varepsilon = 10^{-3}$ as the tolerance number for the stopping criterion. The results in the table reflect the differences in tail behaviour of the various distributions considered. Also, we give in Table 3.9 the number N of iterations required by the algorithm and the lower and upper bounds m_N and M_N. For the specific insurance problem the number of required iterations turns out to be quite small.   □

3.5 SEMI-MARKOV DECISION PROCESSES

In the previous sections we have considered a decision model in which the decisions can be made only at fixed epochs $t = 0, 1, \ldots$. However, in many optimization problems, the times between consecutive decision epochs are not identical but are random. A possible tool to analyse such problems is the semi-Markov decision model. This model concerns a dynamic system which at random points in time beginning with epoch 0 is observed and classified into one of a possible number of states. The set of possible states is denoted by I. After observing the state of the system, a decision has to be made, and costs are incurred as a consequence of the decision made. For each state $i \in I$, a set $A(i)$ of possible decisions or actions is available. It is assumed that the state space I and the action sets $A(i)$, $i \in I$, are finite. This controlled dynamic system is called a *semi-Markov decision process* when the following Markovian properties are satisfied. If at a decision epoch the action a is chosen in state i, then the time until, and the state at, the next decision epoch depend only on the present state i and the subsequently chosen action a and are thus independent of the past history of the system. Also, the costs incurred until the next decision epoch depend only on the present state and the action chosen in that state. We note that in specific problems the state at the next decision epoch will often depend on the time until that epoch. Also, the costs usually consist of lump costs incurred at discrete points in time and rate costs incurred continuously in time.

The long-run average cost per unit time is taken as the optimality criterion. For this criterion the semi-Markov decision model is in fact determined by the following characteristics:

$p_{ij}(a)$ = the probability that at the next decision epoch the system will be in state j if action a is chosen in the present state i,

$\tau_i(a)$ = the expected time until the next decision epoch if action a is chosen in the present state i,

$c_i(a)$ = the expected costs incurred until the next decision epoch if action a is chosen in the present state i.

In the sequel it is assumed that $\tau_i(a) > 0$ for all $i \in I$ and $a \in A(i)$. We now give two applications of the semi-Markov decision model.

Example 3.5   A production-inventory problem with controllable production rate

Consider a production facility that operates only intermittently to manufacture a single product. The production will be stopped if the inventory is sufficiently high while the production will be restarted when the inventory has dropped sufficiently low. Customers asking for this product arrive according to a Poisson process with rate λ, and the demand of each customer equals one unit. The demand which cannot be satisfied directly from stock on hand is lost. Also, a finite capacity C for the inventory is assumed. In a production run, any desired lot size can be produced. The production time of a lot size of Q units is a random variable T_Q having a probability density $f_Q(t)$. The lot size is added to the inventory at the end of the production run. After the completion of a production run, a new production run is started or the facility is closed down. At each point of time the production can be restarted. The production costs for a lot size of $Q \geqslant 1$ units consist of a fixed setup cost $K \geqslant 0$ and a variable cost of c per unit produced. Also, there is a holding cost of $h > 0$ per unit kept in stock per unit time, and a lost-sales cost of $\pi > 0$ is incurred for each lost demand.

This problem can be modelled as a semi-Markov decision process. The decision epochs are given by the epochs at which a production run is completed and the epochs at which a demand occurs while the facility is shut down. In view of the above cost structure, we choose as state space

$$I = \{0, 1, \ldots, C\},$$

where state i indicates that the inventory on hand equals i units. For each state i, the set $A(i)$ of possible actions is given by

$$A(i) = \{0, 1, \ldots, C - i\}$$

where action $a = 0$ means to start no new production and action $a \geqslant 1$ means to start a new production of a units. To specify the one-step transition probabilities and one-step costs, we introduce the following notation. For each $Q \geqslant 1$, let

$$p_k^Q = \int_0^\infty e^{-\lambda t} \frac{(\lambda t)^k}{k!} f_Q(t) \, dt, \qquad k = 0, 1, \ldots.$$

Noting that the total demand in a time t has a Poisson distribution with mean λt and using the law of total probability with the production time T_Q as the conditioning variable, it follows that p_k^Q is the probability that a total demand of k units occurs during the production time of a lot size Q. Since excess demand is lost, we now obtain, for $1 \leqslant Q \leqslant C - i$,

$$p_{ij}(Q) = \begin{cases} \sum_{k=i}^{\infty} p_k^Q & \text{for } j = Q, \\ p_{Q+i-j}^Q & \text{for } j = Q + 1, \ldots, Q + i, \\ 0 & \text{otherwise,} \end{cases}$$

while

$$p_{ij}(0) = \begin{cases} 1 & \text{for } j = \max(i - 1, 0). \\ 0 & \text{otherwise.} \end{cases}$$

It is immediate that

$$\tau_i(Q) = \begin{cases} E(T_Q) & \text{for } Q = 1, \ldots, C - i, \\ 1/\lambda & \text{for } Q = 0. \end{cases}$$

To give $c_i(Q)$, denote by τ_k the epoch at which the kth demand occurs. Using the lack of memory of the Poisson process and the law of total expectation with the production time T_Q as the conditioning variable, it can be verified that, for $Q \geqslant 1$,

$$c_i(Q) = K + cQ + h \int_0^\infty E\left\{ \sum_{k=1}^i \min(t, \tau_k) \right\} f_Q(t) \, dt$$

$$+ \pi \sum_{k=i}^\infty (k - i) p_k^Q.$$

Since the random variable τ_k has an Erlang-k distribution whose density is given by (1.22) in section 1.4 of chapter 1, it follows with (1.23) that,

$$E\{\min(t, \tau_k)\} = \int_0^t x \lambda^k \frac{x^{k-1}}{(k-1)!} e^{-\lambda x} dx + t \int_t^\infty \lambda^k \frac{x^{k-1}}{(k-1)!} e^{-\lambda x} dx$$

$$= \frac{k}{\lambda} \left\{ 1 - \sum_{j=0}^k e^{-\lambda t} \frac{(\lambda t)^j}{j!} \right\} + t \sum_{j=0}^{k-1} e^{-\lambda t} \frac{(\lambda t)^j}{j!}.$$

Next we find after some simple manipulations that, for $1 \leqslant Q \leqslant C - i$

$$c_i(Q) = K + cQ + h \sum_{k=1}^i \left\{ \frac{k}{\lambda} \left(1 - \sum_{j=0}^k p_j^Q \right) + \sum_{j=0}^{k-1} \frac{j+1}{\lambda} p_{j+1}^Q \right\}$$

$$+ \pi \left\{ \lambda E(T_Q) - i + \sum_{k=0}^i (i - k) p_k^Q \right\}, \qquad i \geqslant 0.$$

Verify that for fixed Q the following recursion relation holds:

$$c_i(Q) = c_{i-1}(Q) + \frac{h}{\lambda} \left\{ i - \sum_{j=0}^{i-1} (i - j) p_j^Q \right\} - \pi \left(1 - \sum_{j=0}^{i-1} p_j^Q \right), \qquad i \geqslant 1,$$

with $c_0(Q) = K + cQ + \pi \lambda E(T_Q)$. It is more efficient to compute $c_i(Q)$, $i \geqslant 0$, from this recursion relation rather than from the above explicit expression. Finally,

$$c_i(0) = \pi, \quad \text{for } i = 0 \quad \text{and} \quad c_i(0) = \frac{hi}{\lambda} \quad \text{for } i \geqslant 1,$$

which completes the specification of the basic elements for the semi-Markov decision model.   □

Example 3.6   Production control for a flexible manufacturing system

Consider a flexible manufacturing facility producing parts, one at a time, for two assembly lines. The time needed to produce one part for assembly line k is exponentially distributed with mean $1/\mu_k$, $k = 1, 2$. Each part produced for line k is put into the buffer for line k. This buffer for line k has space for only N_k parts including the part (if any) in assembly. Each line takes parts one at a time from its buffer as long as the buffer is not empty. At line k, the assembly time for one part is exponentially distributed with mean $1/\lambda_k$, $k = 1, 2$. The production times at the flexible manufacturing facility and the assembly times at the lines are mutually independent. A real-time control for the flexible manufacturing facility is exercised. After each production at this central facility, it must be decided what type of part is to be produced next. Here the system cannot produce for a line whose buffer is full. Also, the system cannot remain idle if not all the buffers are full. The control is based on the full knowledge of the buffer status at both lines. The system incurs a lost-opportunity cost at a rate of γ_k per unit time when line k is idle. The goal is to control the production at the flexible manufacturing facility in such a way that the long-run average cost per unit time is minimal. Note that for this criterion only the ratio γ_2/γ_1 is relevant since we actually try to minimize a weighted sum of the fractions of time that the lines are idle.

This problem can be modelled as a semi-Markov decision process. The decision epochs are given by the epochs at which the flexible manufacturing facility completes a production and the epochs at which the system leaves the blocked state where both buffers are full. The set of possible states is given by

$$I = \{(i_1, i_2) | i_1 = 0, \dots, N_1, i_2 = 0, \dots, N_2\},$$

where state (i_1, i_2) indicates that i_1 parts are present at line 1 and i_2 parts at line 2. Taking as possible actions

$$a = \begin{cases} 0 & \text{if no production is started,} \\ 1 & \text{if a production for line 1 is started,} \\ 2 & \text{if a production for line 2 is started,} \end{cases}$$

the set of possible actions in state (i_1, i_2) is given by

$$A(i_1, i_2) = \begin{cases} \{1, 2\} & \text{if } i_1 < N_1 \text{ and } i_2 < N_2, \\ \{1\} & \text{if } i_1 < N_1 \text{ and } i_2 = N_2, \\ \{2\} & \text{if } i_1 = N_1 \text{ and } i_2 < N_2, \\ \{0\} & \text{if } i_1 = N_1 \text{ and } i_2 = N_2. \end{cases}$$

To derive the one-step transition probabilities, we use the following observation. As long as the buffer for line k is not empty, the times elapsed between completions of parts at line k are independent and exponentially distributed random variables with the same means $1/\lambda_k$ or, equivalently, parts at line k are completed according to a Poisson process with rate λ_k. Hence, when the buffer for line k starts with i parts present, the probability that l of these parts will be completed in the next t time units is given by $e^{-\lambda t}(\lambda t)^l/l!$ when $0 \leqslant l \leqslant i - 1$

and by $1 - \sum_{j=0}^{i-1} e^{-\lambda t} (\lambda t)^j / j!$ when $l = i$. Put for abbreviation

$$A_k(l_1, l_2) = \int_0^\infty e^{-\lambda_1 t} \frac{(\lambda_1 t)^{l_1}}{l_1!} e^{-\lambda_2 t} \frac{(\lambda_2 t)^{l_2}}{l_2!} \mu_k e^{-\mu_k t} \, dt$$

$$= \binom{l_1 + l_2}{l_1} \frac{\lambda_1^{l_1} \lambda_2^{l_2} \mu_k}{(\lambda_1 + \lambda_2 + \mu_k)^{l_1 + l_2 + 1}} \quad \text{for } k = 1, 2 \text{ and } l_1, l_2 \geqslant 0,$$

where the latter equality uses the fact that $e^{-\lambda t} \lambda^l t^{l-1} / (l-1)!$ represents the Erlang-l density and thus integrates to 1 over the interval $(0, \infty)$. Then, by conditioning on the production time at the central facility, we find from the law of total probability that

$$p_{(i_1, i_2)(j_1, j_2)}(1) = \int_0^\infty e^{-\lambda_1 t} \frac{(\lambda_1 t)^{i_1 - j_1 + 1}}{(i_1 - j_1 + 1)!} e^{-\lambda_2 t} \frac{(\lambda_2 t)^{i_2 - j_2}}{(i_2 - j_2)!} \mu_1 e^{-\mu_1 t} \, dt$$

$$= A_1(i_1 - j_1 + 1, i_2 - j_2), \qquad 1 < j_1 \leqslant i_1, \ 1 \leqslant j_2 \leqslant i_2,$$

$$p_{(i_1, i_2)(1, j_2)}(1) = \int_0^\infty \left(1 - \sum_{l=0}^{i_1 - 1} e^{-\lambda_1 t} \frac{(\lambda_1 t)^l}{l!} \right) e^{-\lambda_2 t} \frac{(\lambda_2 t)^{i_2 - j_2}}{(i_2 - j_2)!} \mu_1 e^{-\mu_1 t} \, dt$$

$$= \frac{\lambda_2^{i_2 - j_2} \mu_1}{(\lambda_2 + \mu_1)^{i_2 - j_2 + 1}} - \sum_{l=0}^{i_1 - 1} A_1(l, i_2 - j_2) \qquad 1 \leqslant j_2 \leqslant i_2,$$

$$p_{(i_1, i_2)(j_1, 0)}(1) = \int_0^\infty e^{-\lambda_1 t} \frac{(\lambda_1 t)^{i_1 - j_1 + 1}}{(i_1 - j_1 + 1)!} \left(1 - \sum_{l=0}^{i_2 - 1} e^{-\lambda_2 t} \frac{(\lambda_2 t)^l}{l!} \right) \mu_1 e^{-\mu_1 t} \, dt$$

$$= \frac{\lambda_1^{i_1 - j_1 + 1} \mu_1}{(\lambda_1 + \mu_1)^{i_1 - j_1 + 2}} - \sum_{l=0}^{i_2 - 1} A_1(i_1 - j_1 + 1, l), \qquad 1 < j_1 \leqslant i_1,$$

$$p_{(i_1, i_2)(1, 0)}(1) = 1 - \sum_{(j_1, j_2) \neq (1, 0)} p_{(i_1, i_2)(j_1, j_2)}(1).$$

with the convention $\sum_{j=a}^{b} = 0$ if $b < a$. In the same way the formulas for $p_{(i_1, i_2)(j_1, j_2)}(2)$ follow. Using the second relation in (1.29) in section 1.4 of chapter 1, we find

$$p_{(N_1, N_2)(j_1, j_2)}(0) = \begin{cases} \lambda_1 / (\lambda_1 + \lambda_2) & \text{if } (j_1, j_2) = (N_1 - 1, N_2), \\ \lambda_2 / (\lambda_1 + \lambda_2) & \text{if } (j_1, j_2) = (N_1, N_2 - 1). \end{cases}$$

Next we determine the one-step expected transition times and the one-step expected costs. Obviously, we have

$$\tau_{(i_1, i_2)}(a) = \frac{1}{\mu_a} \quad \text{for } a = 1, 2 \quad \text{and} \quad \tau_{(N_1, N_2)}(0) = \frac{1}{\lambda_1 + \lambda_2},$$

where the latter equation uses the first relation in (1.29). To find the one-step expected costs, note that, for $a = 1, 2$,

$$c_{(i_1, i_2)}(a) = \gamma_1 U_1(i_1, a) + \gamma_2 U_2(i_2, a),$$

where

$U_k(i_k, a) =$ the expected amount of time that line k is idle during the production time of one part for line a when at the beginning of that production time i_k parts are present at line k.

To find $U_k(i_k, a)$, note first that $\{\lambda_k/(\lambda_k + \mu_a)\}^{i_k}$ gives the probability that line k will complete the production of i_k parts before the central facility has completed the production of one part for line a. This follows by using the lack of memory of the exponential distribution and by applying repeatedly the second equation in (1.29) in section 1.4 of chapter 1 (cf. also example 1.8). If the central facility is still producing for line a when line k becomes idle, it takes an exponentially distributed time with mean $1/\mu_a$ until this production is completed. Hence

$$U_k(i_k, a) = \frac{1}{\mu_a}\left(\frac{\lambda_k}{\lambda_k + \mu_a}\right)^{i_k}.$$

This completes the determination of $c_{(i_1,i_2)}(a)$ for $a = 1, 2$. Finally, $c_{(N_1,N_2)}(0) = 0$.

The above formulation as a semi-Markov decision problem may be extended to the general case of $J \geqslant 2$ assembly lines being supplied by a central production facility. However, the computational complexity of the formulation increases dramatically with J, being to some extent due to the denseness of the matrices of transition probabilities. It will be seen later in this section that a simpler and more efficient semi-Markov decision formulation is possible by exploiting the fact that the holding times in the states are exponentially distributed, so that sparse transition matrices are obtained by introducing fictitious decision epochs.   □

We now resume our treatment of the semi-Markov decision model. As before, a stationary policy R is a rule which prescribes the single action $R_i \in A(i)$ whenever the system is observed in state i at a decision epoch. Using the finiteness of the state space, it can be shown that under each stationary policy the number of decisions made in a finite time interval is finite with probability 1. We omit the proof of this result. Also, denoting by X_n the state of the system at the nth decision epoch, it follows that under a stationary policy R the embedded stochastic process $\{X_n\}$ is a discrete-time Markov chain with one-step transition probabilities $p_{ij}(R_i)$. The unichain assumption 3.1 is again made for the Markov chains $\{X_n\}$ associated with the stationary policies. Thus, for each stationary policy R, the Markov chain $\{X_n\}$ has a unique equilibrium probability distribution $\{\pi_j(R), j \in I\}$ satisfying (3.2) to (3.4).

Denoting by

$Z(t) =$ the total costs incurred up to time t,

we have the following result for the long-run average cost per unit time.

Theorem 3.4

Under a stationary policy R,

$$\lim_{t \to \infty} \frac{Z(t)}{t} = \frac{\sum_{j \in I} c_j(R_j) \pi_j(R)}{\sum_{j \in I} \tau_j(R_j) \pi_j(R)} \quad \text{with probability 1,}$$

independently of the initial state.

Proof. We only give a sketch of the proof. Using the theory of renewal reward processes, it is not difficult to verify (see Theorem 3.16 in Ross, 1970):

$$\lim_{t \to \infty} \frac{Z(t)}{t} = \lim_{m \to \infty} \frac{E(\text{costs over the first } m \text{ decision epochs})}{E(\text{time over the first } m \text{ decision epochs})}$$

with probability 1. Denote by τ_n and C_n the time and the costs made between the $(n-1)$th and nth decision epoch (the 0th decision epoch is at epoch 0). Noting that $E(\tau_n | X_0 = i) = \sum_{j \in I} \tau_j(R_j) p_{ij}^{(n-1)}(R)$, it follows after an interchange of the order of summation and using (3.2) that

$$\lim_{m \to \infty} \frac{1}{m} \sum_{n=1}^{m} E(\tau_n | X_0 = i) = \lim_{m \to \infty} \frac{1}{m} \sum_{n=1}^{m} \sum_{j \in I} \tau_j(R_j) p_{ij}^{(n-1)}(R)$$

$$= \sum_{j \in I} \tau_j(R_j) \pi_j(R).$$

Similarly,

$$\lim_{m \to \infty} \frac{1}{m} \sum_{n=1}^{m} E(C_n | X_0 = i) = \sum_{j \in I} c_j(R_j) \pi_j(R).$$

The above relations together imply the desired result. □

Abbreviating the average cost rate of policy R by

$$g(R) = \frac{\sum_{j \in I} c_j(R_j) \pi_j(R)}{\sum_{j \in I} \tau_j(R_j) \pi_j(R)}, \tag{3.40}$$

a stationary policy R^* is said to be average cost optimal if $g(R^*) \leq g(R)$ for each stationary policy R. The algorithms for computing an average cost optimal policy in the discrete-time Markov decision model can easily be extended to the semi-Markov decision model. We first discuss an extension of the policy-iteration algorithm. In the same way as before, we define for a stationary policy R the relative cost function $w_i(R)$, $i \in I$, by (3.8). It is left as an exercise to the reader to verify that Theorems 3.1 and 3.2 require only the following minor modifications for the semi-Markov decision model. Theorems 3.1 and 3.2 remain valid provided we replace g by $g\tau_i(R_i)$ in (3.10) and by $g\tau_i(\bar{R}_i)$ in (3.15). In particular, we have the following important result. Suppose that $g(R)$ and $v_i(R)$,

$i \in I$, are the average cost and the relative values of a stationary policy R. If the stationary policy $\bar{R}$ is constructed such that, for each state $i \in I$,

$$c_i(\bar{R}_i) - g(R)\tau_i(\bar{R}_i) + \sum_{j \in I} p_{ij}(\bar{R}_i)v_j(R) \leqslant v_i(R), \tag{3.41}$$

then

$$g(\bar{R}) \leqslant g(R). \tag{3.42}$$

Actually, by the same arguments as used after (3.27), we obtain that $g(\bar{R}) < g(R)$ if (3.41) holds for each $i \in I$ with strict inequality for at least one state i which is recurrent under $\bar{R}$.

We can now formulate the following algorithm.

The policy-iteration algorithm for the semi-Markov decision model

Step 0 (initialization). Choose a stationary policy R.

Step 1 (value-determination step). For the current rule R, compute the average costs $g(R)$ and the relative values $v_i(R)$, $i \in I$, as the unique solution to the linear equations

$$v_i = c_i(R_i) - g\tau_i(R_i) + \sum_{j \in I} p_{ij}(R_i)v_j, \qquad i \in I, \tag{3.43}$$

$$v_s = 0, \tag{3.44}$$

where s is an arbitrarily chosen state.

Step 2 (policy-improvement step). For each state $i \in I$, determine an action a_i yielding the minimum in

$$\min_{a \in A(i)} \left\{ c_i(a) - g(R)\tau_i(a) + \sum_{j \in I} p_{ij}(a)v_j(R) \right\}. \tag{3.45}$$

The new stationary policy $\bar{R}$ is obtained by choosing $\bar{R}_i = a_i$ for all $i \in I$ with the convention that $\bar{R}_i$ is chosen equal to the old action R_i when this action yields the minimum in (3.45).

Step 3 (convergence test). If the new policy $\bar{R}$ equals the old policy R the algorithm is stopped with policy R. Otherwise, go to step 1 with R replaced by $\bar{R}$.

In the same way as for the discrete-time Markov decision model, it can be shown that under assumption 3.1 the above algorithm converges in a finite number of iterations to an average cost optimal policy. Also, as a consequence of the convergence of the algorithm, there exist numbers g^* and v_i^*, $i \in I$, satisfying the *average cost optimality equation*

$$v_i^* = \min_{a \in A(i)} \left\{ c_i(a) - g^*\tau_i(a) + \sum_{j \in I} p_{ij}(a)v_j^* \right\}, \qquad i \in I. \tag{3.46}$$

The constant g^* is uniquely determined as the minimal average cost per unit time and each stationary policy whose actions minimize the right side of (3.46) for all $i \in I$ is average cost optimal. The proof of these statements is left as an exercise to the reader.

Value-iteration algorithm

For the semi-Markov decision model the formulation of a value-iteration algorithm is not straightforward. A recursion relation for the minimal expected costs over the first n decision epochs does not take into account the non-identical transition times and thus these costs for large n cannot be related to the minimal average cost per unit time. However, by a data transformation, we can convert the semi-Markov decision model into a discrete-time Markov decision model such that for each stationary policy the average costs per unit time are the same in both models. A value-iteration algorithm for the original semi-Markov decision model is then implied by the value-iteration algorithm for the associated discrete-time Markov decision model. The data transformation to be given below may be considered as an extension of the uniformization technique for continuous-time Markov chains discussed in section 2.2 of chapter 2.

We introduce the following *data transformation*. Choose a number τ with

$$0 < \tau \leqslant \min_{i,a} \tau_i(a).$$

Consider now the discrete-time Markov decision model whose state space, action sets, one-step costs and one-step transition probabilities are given by

$$\tilde{I} = I,$$

$$\tilde{A}(i) = A(i), \qquad\qquad i \in \tilde{I},$$

$$\tilde{c}_i(a) = \frac{c_i(a)}{\tau_i(a)}, \qquad\qquad i \in \tilde{I} \text{ and } a \in \tilde{A}(i),$$

$$\tilde{p}_{ij}(a) = \begin{cases} \dfrac{\tau}{\tau_i(a)} p_{ij}(a), & j \neq i, \ i \in \tilde{I} \text{ and } a \in \tilde{A}(i), \\[2ex] \dfrac{\tau}{\tau_i(a)} p_{ij}(a) + \left(1 - \dfrac{\tau}{\tau_i(a)}\right), & j = i, \ i \in \tilde{I} \text{ and } a \in \tilde{A}(i). \end{cases}$$

This discrete-time Markov decision model has the same class of stationary policies as the original semi-Markov decision model. For a given stationary policy R in the discrete-time Markov decision model, denote by $\tilde{\pi}_i(R)$, $i \in I$, the equilibrium probabilities of the associated Markov chain $\{\tilde{X}_n\}$ of states. By direct substitution in the equilibrium equations (2.11) for the Markov chain

$\{\tilde{X}_n\}$, it is readily verified that

$$\tilde{\pi}_i(R) = \gamma \tau_i(R_i)\pi_i(R), \qquad i \in I \text{ with } \gamma = 1/\sum_{j \in I} \tau_j(R_j)\pi_j(R).$$

Hence, for the average cost $\tilde{g}(R)$ of a stationary policy R in the discrete-time Markov decision model, it follows that

$$\tilde{g}(R) = \sum_{j \in I} \tilde{c}_j(R_j)\tilde{\pi}_j(R) = \frac{\sum_{j \in I} c_j(R_j)\pi_j(R)}{\sum_{j \in I} \tau_j(R_j)\pi_j(R)} = g(R).$$

Since for each stationary policy the average costs per unit time are the same in both models, we can solve the discrete-time Markov decision model equally well. In applying the value-iteration algorithm to this model it is no restriction to assume that each $\tilde{c}_i(a) = c_i(a)/\tau_i(a)$ is *non-negative*. Otherwise, add a sufficiently large constant to each $\tilde{c}_i(a)$ and note that by doing so the average cost of each policy changes with the same constant. Also, it is no restriction to assume that in the discrete-time model the Markov chain $\{\tilde{X}_n\}$ is *aperiodic* for each stationary policy, since by choosing τ strictly less than $\min_{i,a} \tau_i(a)$ we have $\tilde{p}_{ii}(a) > 0$ for all i, a and thus the required aperiodicity. An application of the value-iteration algorithm to the discrete-time Markov decision model results in the following algorithm.

Value-iteration algorithm for the semi-Markov decision model

Step 0. Choose $V_0(i)$ such that $0 \leqslant V_0(i) \leqslant \min_a \{c_i(a)/\tau_i(a)\}$ for all i. Let $n := 1$.
Step 1. Compute the function $V_n(i)$, $i \in I$, from

$$V_n(i) = \min_{a \in A(i)} \left[\frac{c_i(a)}{\tau_i(a)} + \frac{\tau}{\tau_i(a)}\sum_{j \in I} p_{ij}(a)V_{n-1}(j) \right.$$

$$\left. + \left\{ 1 - \frac{\tau}{\tau_i(a)} \right\} V_{n-1}(i) \right], \qquad i \in I, \tag{3.47}$$

and determine $R(n)$ as a stationary policy whose actions minimize the right side of (3.47).
Step 2. Compute the bounds

$$m_n = \min_{j \in I}\{V_n(j) - V_{n-1}(j)\}, \qquad M_n = \max_{j \in I}\{V_n(j) - V_{n-1}(j)\}.$$

The algorithm is stopped with policy $R(n)$ when $0 \leqslant (M_n - m_n) \leqslant \varepsilon m_n$, where ε is a prespecified accuracy number. Otherwise, go to step 3.
Step 3. $n := n + 1$ and go to step 1.

In view of the aperiodicity property, the algorithm stops after finitely many iterations with a policy $R(n)$ satisfying

$$0 \leqslant g(R(n)) - g^* \leqslant \varepsilon g^*,$$

where g^* denotes the minimal average cost per unit time. Regarding the choice of τ in the algorithm, it is in general to be recommended taking $\tau = \min_{i,a} \tau_i(a)$ when for this choice the aperiodicity requirement is satisfied; otherwise $\tau = \frac{1}{2} \min_{i,a} \tau_i(a)$ is a reasonable choice. The convergence speed of the above algorithm may be greatly enhanced by using the modified value-iteration algorithm with a dynamic relaxation factor, as discussed in section 3.4.

Linear programming formulation for the semi-Markov decision model

As the last algorithm for the semi-Markov decision model, we give the linear programming formulation under the weak unichain assumption stated in section 3.3. Using the data transformation for converting the semi-Markov decision model into a discrete-time Markov decision model, and using the change of variable $u_{ia} = x_{ia}/\tau_i(a)$, the reader may easily verify that the linear programming algorithm in section 3.3 requires only the following modification to the linear program (3.31):

Minimize
$$\sum_{i \in I} \sum_{a \in A(i)} c_i(a) u_{ia}$$

subject to

$$\sum_{a \in A(j)} u_{ja} - \sum_{i \in I} \sum_{a \in A(i)} p_{ij}(a) u_{ia} = 0, \qquad j \in I,$$

$$\sum_{i \in I} \sum_{a \in A(i)} \tau_i(a) u_{ia} = 1,$$

$$u_{ia} \geqslant 0.$$

We remark that probabilistic constraints such as 'the fraction of time that the system is in some subset I_0 of states should not exceed α' and 'the average frequency of taking some action d in some subset I_0 of states should not exceed β' can easily be dealt with in the linear programming formulation by adding the constraints $\sum_{i \in I_0, a \in A(i)} \tau_i(a) u_{ia} \leqslant \alpha$ and $\sum_{i \in I_0} u_{id} \leqslant \beta$. In the case of probabilistic constraints the average cost optimal policy usually involves randomized decisions (see the discussion in section 3.3).

To conclude this section, we discuss two applications of value-iteration to semi-Markov decision problems. First, we consider the important and general problem of optimal sharing of limited resources between competing users in a random environment. A problem of this type has already been dealt with in example 3.6. In both this example and the next one the usual value-iteration formulation is computationally burdensome because the matrices of transition probabilities are rather dense. In the following it will be shown that for decision problems with exponentially distributed times between consecutive decision epochs, a more efficient value-iteration formulation may be obtained by the simple trick of introducing fictitious decision epochs so that sparse transition matrices are obtained.

Example 3.7 Optimal sharing of memory between processors

In data or computer networks an important problem is the allocation of memory to several types of users. An example of the shared use of memory is the situation in which messages destined for different processors share a common waiting area, where decisions to accept or reject an arriving message are based on the number of waiting messages of each type. Here the goal is to minimize the average rejection rate of messages or equivalently to maximize the average throughput of messages. Note that a policy that rejects certain messages when waiting spaces are available may process more messages on the average than a policy that always accepts a message when the waiting area is not full.

In what follows we consider the decision problem for the special case of $p = 2$ processors. However, at the expense of a somewhat more complicated notation, the methodology applies equally well to the general case of $p \geqslant 2$ processors.

Suppose two processors share a common memory that is able to accommodate a total of M messages. The messages are distinguished by the processor destinations. Messages of the types 1 and 2 destined for the processors 1 and 2 arrive according to independent Poisson processes with respective rates λ_1 and λ_2. When a message arrives a decision to accept or reject that message must be made. A message that is rejected has no further influence on the system. If a message is accepted it stays in the memory until completion of service. The time required to process a message of type i is exponentially distributed with mean $1/\mu_i$. The processor i handles only messages of type i and is able to serve only one message at a time.

The measure of system performance is a weighted sum of the rejection rates of the messages of types 1 and 2 where the respective weights are given by γ_1 and γ_2. Here the rejection rate of messages of type i is defined as the average number of messages of type i that are rejected per unit time. Note that for the important case of $\gamma_1 = \gamma_2 = 1$ the minimization of the average rejection rate is equivalent to the maximization of the average throughput.

The sharing problem will be modelled as a semi-Markov decision problem. In doing so, it is convenient to consider the service completion epochs as fictitious decision epochs in addition to the real decision epochs, being the arrival epochs of messages. The fictitious decision at the service completion epochs is to let the system alone. Note that the inclusion of these fictitious decision epochs does not change the Markovian nature of the decision processes, since the times between state transitions are exponentially distributed and thus have the memoryless property. Below it will become clear that the inclusion of fictitious decision epochs simplifies not only the formulation of the value-iteration algorithm, but, more importantly, reduces as well the computational effort compared with a straightforward formulation as given in example 3.6. Thus we take as state space

$$I = \{(i_1, i_2, k) | i_1, i_2 = 0, 1, \ldots, M; i_1 + i_2 \leqslant M; k = 0, 1, 2\}.$$

Here state (i_1, i_2, k) with $k = 1$ or 2 corresponds to the situation in which a message of type k arrives and finds i_1 messages of type 1 and i_2 messages of type 2 being present in the common waiting area. The auxiliary state $(i_1, i_2, 0)$ corresponds to the situation in which the service of a message is completed and i_1 messages of type 1 and i_2 messages of type 2 are left behind in the waiting area. For the states (i_1, i_2, k) with $k = 1$ or 2 the possible actions are denoted by

$$a = \begin{cases} 0, & \text{reject the arriving message,} \\ 1, & \text{accept the arriving message,} \end{cases}$$

with the stipulation that $a = 0$ is the only feasible decision when $i_1 + i_2 = M$. For notational convenience, we introduce for the states $(i_1, i_2, 0)$ the fictitious decision $a = 0$ which means that the system is left alone. Put for abbreviation,

$$\lambda(i_1, i_2) = \lambda_1 + \lambda_2 + \mu_1 \delta(i_1) + \mu_2 \delta(i_2),$$

where $\delta(0) = 0$ and $\delta(j) = 1$ for $j \geqslant 1$. Using the properties of the exponential distribution in (1.29) in section 1.4 of chapter 1, it is readily seen that the expected time until the next decision epoch when action $a = 0$ is made in state (i_1, i_2, k) is always given by $1/\lambda(i_1, i_2)$, while in that case the next state is one of the four states $(i_1, i_2, 1)$, $(i_1, i_2, 2)$, $(i_1 - 1, i_2, 0)$ and $(i_1, i_2 - 1, 0)$ with respective probabilities $\lambda_1/\lambda(i_1, i_2)$, $\lambda_2/\lambda(i_1, i_2)$, $\mu_1 \delta(i_1)/\lambda(i_1, i_2)$ and $\mu_2 \delta(i_2)/\lambda(i_1, i_2)$. For action $a = 1$ in state $(i_1, i_2, 1)$, the expected time until the next decision epoch is $1/\lambda(i_1 + 1, i_2)$, while the next state is one of the four states $(i_1 + 1, i_2, 1)$, $(i_1 + 1, i_2, 2)$, $(i_1, i_2, 0)$ and $(i_1 + 1, i_2 - 1, 0)$ with respective probabilities $\lambda_1/\lambda(i_1 + 1, i_2)$, $\lambda_2/\lambda(i_1 + 1, i_2)$, $\mu_1/\lambda(i_1 + 1, i_2)$ and $\mu_2 \delta(i_2)/\lambda(i_1 + 1, i_2)$, and similarly for action $a = 1$ in state $(i_1, i_2, 2)$. The one-step expected costs are obvious when the goal is to minimize a weighted sum of the rejection rates of messages of types 1 and 2 with respective weights γ_1 and γ_2. An immediate cost of γ_k is incurred each time action $a = 0$ is made in state (i_1, i_2, k) with $k = 1$ or 2, while a cost of 0 is incurred otherwise.

Now, having specified the basic elements of the semi-Markov decision model, we are in a position to formulate the value-iteration algorithm for the computation of a (nearly) optimal sharing rule. Therefore, according to the data transformation, we choose

$$\tau = \frac{1}{\lambda_1 + \lambda_2 + \mu_1 + \mu_2}.$$

Using the above specifications the value-iteration scheme (3.47) becomes quite simple for the specific problem of sharing resources. Note that the expressions for the one-step transition times $\tau_s(a)$ and the one-step transition probabilities $p_{st}(a)$ have a common denominator and so the ratio $p_{st}(a)/\tau_s(a)$ required in the value-iteration algorithm has a very simple form. More importantly, although the above semi-Markov formulation using fictitious decision epochs requires the extra states $(i_1, i_2, 0)$, the number of additions and multiplications per iteration in the associated value-iteration algorithm is quadratic in the number M of waiting places rather than of the order M^4 as in the case of a straightforward semi-Markov formulation as used in example 3.6. The reason of this reduction in computational burden is the fact that the former formulation has sparse transition matrices and the latter formulation has dense transition matrices. Numerical experiments indicate that the respective numbers of iterations for the two value-iteration formulations do not differ greatly so that there is a considerable overall reduction in computational effort when using the formulation with the fictitious decision epochs. This reduction in computational effort gets larger as M increases.

In specifying the value-iteration scheme (3.47), we have obviously to distinguish between the auxiliary states $(i_1, i_2, 0)$ and the other states. For the states $(i_1, i_2, 0)$ we have

$$V_n(i_1, i_2, 0) = \tau\lambda_1 V_{n-1}(i_1, i_2, 1) + \tau\lambda_2 V_{n-1}(i_1, i_2, 2) + \tau\mu_1 V_{n-1}(i_1 - 1, i_2, 0)$$
$$+ \tau\mu_2 V_{n-1}(i_1, i_2 - 1, 0) + \{1 - \tau\lambda(i_1, i_2)\} V_{n-1}(i_1, i_2, 0),$$

with the convention $V_{n-1}(i_1, i_2, 0) = 0$ when i_1 or i_2 is negative. For the states $(i_1, i_2, 1)$ we find

$$V_n(i_1, i_2, 1) = \min [\gamma_1 \lambda(i_1, i_2) + \tau\lambda_1 V_{n-1}(i_1, i_2, 1) + \tau\lambda_2 V_{n-1}(i_1, i_2, 2)$$
$$+ \tau\mu_1 V_{n-1}(i_1 - 1, i_2, 0) + \tau\mu_2 V_{n-1}(i_1, i_2 - 1, 0)$$
$$+ \{1 - \tau\lambda(i_1, i_2)\} V_{n-1}(i_1, i_2, 1),$$
$$\tau\lambda_1 V_{n-1}(i_1 + 1, i_2, 1) + \tau\lambda_2 V_{n-1}(i_1 + 1, i_2, 2)$$
$$+ \tau\mu_1 V_{n-1}(i_1, i_2, 0) + \tau\mu_2 V_{n-1}(i_1 + 1, i_2 - 1, 0)$$
$$+ \{1 - \tau\lambda(i_1 + 1, i_2)\} V_{n-1}(i_1, i_2, 1)],$$

provided we put $V_{n-1}(i_1, i_2, 1) = V_{n-1}(i_1, i_2, 2) = \infty$ when $i_1 + i_2 = M + 1$ in order to exclude the unfeasible decision $a = 1$ in the states $(i_1, i_2, 1)$ with $i_1 + i_2 = M$. A similar expression applies to $V_n(i_1, i_2, 2)$. We omit this obvious expression for reasons of space. This completes the specification of the main step of the value-iteration algorithm. The other steps of the algorithm are self-explanatory.

We found by experimentation that the speed of convergence of the value-iteration algorithm developed for the sharing problem is greatly enhanced by using the relaxation factor (3.39). The specification of (3.39) for the sharing problem follows by noting that for the function $w(s) = V_n(s) - V_{n-1}(s)$,

$$\sum_{t \in I} \tilde{p}_{st}(a)w(t) = \tau\lambda_1 w(i_1, i_2, 1) + \tau\lambda_2 w(i_1, i_2, 2) + \tau\mu_1 w(i_1 - 1, i_2, 0)$$
$$+ \tau\mu_2 w(i_1, i_2 - 1, 0) + \{1 - \tau\lambda(i_1, i_2)\}w(i_1, i_2, k)$$

when $s = (i_1, i_2, k)$ and $a = 0$, while

$$\sum_{t \in I} \tilde{p}_{st}(a)w(t) = \tau\lambda_1 w(i_1 + 1, i_2, 1) + \tau\lambda_2 w(i_1 + 1, i_2, 2) + \tau\mu_1 w(i_1, i_2, 0)$$
$$+ \tau\mu_2 w(i_1 + 1, i_2 - 1, 0) + \{1 - \tau\lambda(i_1 + 1, i_2)\}w(i_1, i_2, 1)$$

when $s = (i_1, i_2, 1)$ and $a = 1$, where the latter equation with an interchange of the roles of types 1 and 2 applies also when $s = (i_1, i_2, 2)$ and $a = 1$.

Before presenting some numerical results, we wish to point out the following useful extension of the above results. The value-iteration algorithm for the sharing problem needs only a minor adjustment when service completion of messages of type k occur at a general rate of $\mu_k(i_1, i_2)$ whenever i_l messages of type l for $l = 1, 2$ are in the waiting area. The present problem deals with the special case of $\mu_k(i_1, i_2) = \mu_k$ for $i_k \geq 1$ and $\mu_k(i_1, i_2) = 0$ for $i_k = 0$. The value-iteration formulation for the general case follows by an appropriate replacement of μ_k by $\mu_k(i_1, i_2)$. A service completion rate function $\mu_k(i_1, i_2)$ enables us to cover not only the sharing problem with multiple servers but also the following other problem of optimal allocation. Suppose that a limited number of servers must be allocated dynamically to different types of competing customers. There are $J \geq 2$ independent customer or user classes and each user class j generates requests for service according to a Poisson process with rate λ_j. The service requirement of a request of type j is exponentially distributed with mean $1/\mu_j$. There are M identical servers who can handle requests of any type with the restriction that a server can handle only one request at a time. No queueing is allowed so that an arriving request gets immediate access to a free server when that request is accepted. The decision to accept or reject a request for service is based on the type of that request and the number of servers occupied by each type of users. The goal is to minimize a weighted sum of the rejection rates of the user types. This problem arises, for example, in parallel processing in computer systems when M identical computer processor units are sharing jobs submitted to the computer system. Similar problems may arise in satellite communication systems. This problem of optimal dynamic allocation of servers to different types of competing customers is covered by the above semi-Markov formulation with $\mu_k(i_1, i_2) = \mu_k i_k$ as the service completion rate function. In this allocation problem we have implicitly assumed that the source populations of users are infinite. However, by making a similar adjustment for the input rates as was done for the service completion rates, we can also handle the situation in which there are only finitely many users of each type where each individual user of type j generates requests for service according to a Poisson process with rate η_j whenever not in service.

To conclude, we present some numerical results for the problem of the shared use of a common waiting area. We assume the numerical data

$$\mu_1 = \mu_2 = 1 \quad \text{and} \quad \gamma_1 = \gamma_2 = 1.$$

The arrival rates λ_1 and λ_2 are varied as $(\lambda_1, \lambda_2) = (1.2, 1.0), (1.0, 1.0)$ and $(0.9, 0.7)$, and the number M of waiting places has the two values 10 and 15. We have applied both the conventional value-iteration algorithm and the modified value-iteration algorithm with the dynamic relaxation factor (3.39). In all cases the algorithms were initialized with the values $V_0(i) = 0$ for all i. In step 3(a) of the modified value-iteration algorithm the states (i_1, i_2, k) were ordered by letting first the state variable k run from 0 to 2 for i_1 and i_2 fixed, letting next the state variable i_2 run from 0 to $M - i_1$ for i_1 fixed and letting then the state variable i_1 run from 0 to M. We used both tolerance numbers $\varepsilon = 10^{-2}$ and $\varepsilon = 10^{-3}$ in the stopping criterion in order to see the effect on the number of required iterations. We found for the specific optimization problem of sharing common memory that using the modified value-iteration algorithm rather than the conventional value-iteration algorithm reduces the number of iterations on the average by a factor of 4. Using the tolerance number $\varepsilon = 10^{-3}$ rather than $\varepsilon = 10^{-2}$ increases the number of iterations only moderately. In each numerical example the same nearly optimal sharing rule was found for both $\varepsilon = 10^{-2}$ and $\varepsilon = 10^{-3}$ in either of the algorithms.

Table 3.10   Numerical results for the sharing problem

	$M = 15$					$M = 10$					
(λ_1, λ_2)	(1.2, 1.0)		(1.0, 1.0)		(0.9, 0.7)		(1.2, 1.0)		(1.0, 1.0)		(0.9, 0.7)	
l	$a_1^{(l)}$	$a_2^{(l)}$	$a_1^{(l)}$	$a_2^{(l)}$	$a_1^{(l)}$	$a_2^{(l)}$	$a_1^{(l)}$	$a_2^{(l)}$	$a_1^{(l)}$	$a_2^{(l)}$	$a_1^{(l)}$	$a_2^{(l)}$
0	11	12	12	12	13	14	7	8	8	8	8	9
1	11	12	12	12	13	13	7	7	7	7	8	8
2	10	11	11	11	12	12	6	7	7	7	7	8
3	10	11	10	10	11	12	6	7	6	6	7	7
4	9	10	10	10	10	11	5	6	6	6	6	6
5	8	10	9	9	10	10	5	5	5	5	5	5
6	8	9	9	9	9	9	4	4	4	4	4	4
7	7	8	8	8	8	8	3	3	3	3	3	3
$\geqslant 8$	7	7	7	7	7	7	2	2	2	2	2	2
g^*	0.348		0.204		0.0500		0.436		0.294		0.108	
$N(10^{-2})$	91 (271)		85 (253)		87 (431)		44 (133)		49 (124)		50 (198)	
$N(10^{-3})$	94 (407)		97 (342)		146 (578)		58 (200)		60 (166)		64 (271)	

The numerical investigations indicate that the optimal sharing rule has the intuitively reasonable property that the acceptance of a message of type 1[2] in state (i_1, i_2) implies the acceptance of that same message in state $(i_1 - 1, i_2)[(i_1, i_2 - 1)]$. A control rule with this property is conveniently represented by two non-increasing sequences $\{a_k^{(l)}, 0 \leqslant l \leqslant M\}$ of integers for $k = 1, 2$, so that a message of type k finding upon arrival l messages of the other type present is accepted only when less than $a_k^{(l)}$ messages of type k are present and the waiting area is not full. In Table 3.10 we display the numerical results. Here $N(\varepsilon)$ denotes the number of iterations required by value-iteration when the tolerance number is ε, where the first figures refer to the algorithm with relaxation and the figures between brackets refer to the algorithm without relaxation. The minimal average rejection rate g^* is estimated as $(m_N + M_N)/2$ with $N = N(10^{-3})$. $\qquad \square$

In certain applications of Markovian decision processes to inventory and queueing problems, it may happen that the number of states is unbounded. The next example shows that exploiting the structure of the specific application considered may enable us to cast the problem into a Markovian decision model with a finite state space, avoiding truncation of the set of original states.

Example 3.8 Optimal control of a stochastic service system with a variable number of operating service channels

A stochastic service system has s identical channels available for providing service, where the number of channels in operation can be controlled by turning channels on or off. For example, the service channels could be counters in a supermarket or production machines in a factory. Requests for service are sent to the service facility according to a Poisson process with rate λ. Each arriving request for service is allowed to enter the system and waits in line until an operating channel is provided. The service time of each request is exponentially distributed with mean $1/\mu$. It is assumed that the average arrival rate λ is less than the maximum service rate $s\mu$. A channel that is turned on is only able to handle one request at a time. At any time channels can be turned on or off depending on the number of requests for service in the system. A non-negative switching cost $K(a, b)$ is incurred when adjusting the number of channels on from a to b. For each channel turned on there is an operating cost at a rate of $r > 0$ per unit of time. Also, for each request a holding cost of $h > 0$ is incurred for each unit of time the message is in the system until its service is completed. The objective is to find a rule for controlling the number of channels on such that the long-run average cost per unit time is minimal.

By the lack of memory of the Poisson process and of the exponential distribution, it follows that the state of the system at any time is described by the pair (i, l), where

$i =$ the number of requests for service in the system,

$l =$ the number of channels being turned on.

The decision epochs are the epochs at which a new request for service arrives or the service of a request is completed. In this example the number of possible states is unbounded since the state variable i can assume each of the values $0, 1, \ldots$. A brute force approach would be a semi-Markov decision formulation in which the state variable i is bounded by a sufficiently large chosen integer U such that the probability of having more than U requests in the system is negligible under any reasonable control rule. This approach would lead to a very large state space when the arrival rate λ is close to the maximum service rate $s\mu$. A more efficient Markovian decision formulation is obtained by restricting the class of control rules rather than truncating the state space. It is intuitively obvious that under each reasonable control rule all of the s channels will be turned on when the number of requests in the system is sufficiently large. In other words, supposing a sufficiently large chosen integer $M \geqslant c$, it is no real restriction to assume that in the states (i, l) with $i \geqslant M$ the only feasible action is to turn on all of the s channels. However, this implies that we can restrict the control of the system only to those arrival epochs and service completion epochs at which no more than M requests remain in the system. By doing so, we obtain a semi-Markov decision formulation with as state space

$$I = \{(i, l) | 0 \leqslant i \leqslant M, 0 \leqslant l \leqslant s\},$$

and as action sets

$$A((i, l)) = \begin{cases} \{a | a = 0, \ldots, s\}, & 0 \leqslant i \leqslant M - 1, \quad 0 \leqslant l \leqslant s, \\ \{s\}, & i = M, \quad 0 \leqslant l \leqslant s. \end{cases}$$

Here action a in state (i, l) means that the number of channels on is adjusted from l to a. This semi-Markov decision formulation involves the stipulation that, when taking action $a = s$ in state (M, l), the time until the next decision epoch is defined as the time τ_M until the next service completion epoch at which no more than M requests are left behind in the system, given that all of the s channels are always on. Similarly, when taking action $a = s$ in state (M, l), the 'one-step' costs incurred until the next decision epoch are defined as the sum of the switching cost $K(l, s)$ and the holding and operating costs made during the time τ_M. Note that τ_M consists of the time until either an arrival or service completion occurs and, in case an arrival occurs first, the time needed to reduce the number of requests present from $M + 1$ to M. The semi-Markov decision formulation with an embedded state space makes sense only when it is feasible to calculate the one-step expected transition times $\tau_{(M, l)}(s)$ and the one-step expected costs $c_{(M, l)}(s)$. The calculation of these quantities is indeed not difficult when using the fact that service completions occur according to exponentially distributed interoccurrence times with the same means $1/s\mu$ as long as all of the s channels are occupied. In other words, whenever M or more requests are in the system, we can equivalently imagine that a single 'superchannel' is servicing messages one at a time at an exponential rate of $s\mu$. This analogy enables us to invoke the first result in (1.40) with $n = 1$ in section 1.5 of chapter 1, yielding

The expected time needed to reduce the number of requests present from $M + 1$ to M given that all channels are on

$$= \frac{1/(s\mu)}{1 - \lambda/(s\mu)} = \frac{1}{s\mu - \lambda}$$

and

The expected holding and operating costs incurred during the time needed to reduce the number of requests present from $M + 1$ to M given that all channels are on

$$= \frac{hM}{s\mu - \lambda} + \frac{hs\mu}{s\mu - \lambda} \left\{ \frac{1}{s\mu} + \frac{\lambda}{s\mu(s\mu - \lambda)} \right\} + \frac{rs}{s\mu - \lambda}$$

$$= \frac{h(M + 1) + rs}{s\mu - \lambda} + \frac{h\lambda}{(s\mu - \lambda)^2},$$

where the term $hM/(s\mu - \lambda)$ gives the holding costs associated with the M requests being continuously present during the time needed to reduce the number in the system from $M + 1$ to M. Thus we can conclude that

$$\tau_{(M,l)}(s) = \frac{1}{\lambda + s\mu} + \frac{\lambda}{\lambda + s\mu} \left(\frac{1}{s\mu - \lambda} \right)$$

and

$$c_{(M,l)}(s) = K(l, s) + \frac{hM + rs}{\lambda + s\mu} + \frac{\lambda}{\lambda + s\mu} \left\{ \frac{h(M + 1) + rs}{s\mu - \lambda} + \frac{h\lambda}{(s\mu - \lambda)^2} \right\}.$$

Here we used the result (1.29) in section 1.4 of chapter 1 to establish that $1/(\lambda + s\mu)$ and $\lambda/(\lambda + s\mu)$ give respectively the expected time until either an arrival or a service completion occurs and the probability that an arrival occurs earlier than a service completion, given that all of the s channels are busy. We have now done the most delicate part of the specification of the one-step transition times and the one-step costs associated with the semi-Markov decision formulation with the embedded state space I. The other $\tau_{(i,l)}(a)$ and $c_{(i,l)}(a)$ are easily verified to be given by

$$\tau_{(i,l)}(a) = \frac{1}{\lambda + \min(i, a)\mu}, \qquad 0 \leqslant i \leqslant M - 1, \quad 0 \leqslant a \leqslant s,$$

and

$$c_{(i,l)}(a) = K(l,a) + \frac{hi + ra}{\lambda + \min(i,a)\mu}, \qquad 0 \leqslant i \leqslant M - 1, \quad 0 \leqslant a \leqslant s.$$

It is left to the reader to specify the one-step transition probabilities (verify in particular that $p_{(M,l)(M-1,s)}(s) = s\mu/(\lambda + s\mu)$ and $p_{(M,l)(M,s)}(s) = \lambda/(\lambda + s\mu)$).

From now on it is straightforward to formulate the associated value-iteration algorithm. Using the data transformation with $\tau = 1/(\lambda + s\mu)$, the reader may easily verify that the recurrence relation (3.47) is in the following simple form:

$$V_n((i,l)) = \min_{0 \leqslant a \leqslant s} \left[\{\lambda + \min(i,a)\mu\}K(l,a) + hi + ra + \frac{\lambda}{\lambda + s\mu}V_{n-1}((i+1,a)) \right.$$

$$\left. + \frac{\min(i,a)\mu}{\lambda + s\mu}V_{n-1}((i-1,a)) + \frac{s\mu - \min(i,a)\mu}{\lambda + s\mu}V_{n-1}((i,l)) \right]$$

$$\text{for} \quad 0 \leqslant i \leqslant M - 1, \quad 0 \leqslant l \leqslant s$$

$$V_n((M,l)) = \frac{1}{s\mu}(\lambda + s\mu)(s\mu - \lambda)K(l,s) + \frac{h\lambda}{s\mu - \lambda} + hM + rs$$

$$+ \frac{s\mu - \lambda}{\lambda + s\mu}V_{n-1}((M-1,s)) + \frac{\lambda(s\mu - \lambda)}{s\mu(\lambda + s\mu)}V_{n-1}((M,s))$$

$$+ \frac{\lambda}{s\mu}V_{n-1}((M,l)) \qquad \qquad \text{for} \quad 0 \leqslant l \leqslant s,$$

with the convention $V_{n-1}((-1,l)) = 0$.

To conclude this example, we present some numerical results. We consider a switching cost function $K(a,b)$ of the form

$$K(a,b) = \kappa|a - b|$$

and assume the numerical data

$$s = 10, \qquad \mu = 1, \qquad r = 30 \qquad \text{and} \qquad h = 10.$$

The arrival rate λ is varied as 7 and 8, while the proportionality constant κ for the switching cost has the two values 10 and 25. In each example we take the bound $M = 20$ for the states (i,l), $i \geqslant M$, in which all of the s channels are turned on anyway. The value-iteration algorithm is started with $V_0((i,l)) = 0$ for all states (i,l) and uses the tolerance number $\varepsilon = 10^{-3}$ for its stopping criterion. Our numerical calculations indicate that for the case of linear switching costs the average cost optimal control rule is characterized by the parameters $s(i)$ and $t(i)$, $i = 0, \ldots, M$, such that in the states (i,l) with $l < s(i)$ the number of channels on is raised up to the level $s(i)$, in the states (i,l) with $s(i) \leqslant l \leqslant t(i)$ the

Table 3.11   Numerical results obtained by value-iteration

i	$\lambda = 7,\ \kappa = 10$		$\lambda = 8,\ \kappa = 10$		$\lambda = 7,\ \kappa = 25$		$\lambda = 8,\ \kappa = 25$	
	$s(i)$	$t(i)$	$s(i)$	$t(i)$	$s(i)$	$t(i)$	$s(i)$	$t(i)$
0	0	3	0	4	0	6	0	6
1	1	4	1	4	1	6	1	7
2	2	4	2	5	2	6	2	7
3	2	5	3	5	3	6	3	7
4	3	6	3	6	3	7	3	8
5	4	6	4	7	4	7	4	8
6	5	7	5	8	5	8	5	8
7	5	8	5	8	5	8	6	9
8	6	9	6	9	6	9	6	9
9	6	9	7	10	6	9	7	10
10	7	10	7	10	7	10	7	10
11	8	10	8	10	7	10	7	10
12	8	10	9	10	7	10	8	10
13	9	10	9	10	8	10	8	10
14	9	10	10	10	8	10	9	10
15	10	10	10	10	8	10	9	10
16	10	10	10	10	9	10	10	10
17	10	10	10	10	9	10	10	10
18	10	10	10	10	9	10	10	10
19	10	10	10	10	10	10	10	10
$\geqslant 20$	10	10	10	10	10	10	10	10
$N(N')$	59(174)		82(311)		87(226)		71(250)	
m_N	319.3		367.1		331.5		378.0	
M_N	319.5		367.4		331.8		378.3	

number of channels on is left unchanged and in the states (i, l) with $l > t(i)$ the number of channels on is reduced to $t(i)$.

In Table 3.11 we give for the various cases the (nearly) optimal values of $s(i)$ and $t(i)$, the number $N(N')$ of iterations required by the value-iteration algorithm with (without) the relaxation factor (3.39) and the lower and upper bounds m_N and M_N on the minimal average costs per unit time. As in the previous example, the use of a dynamic relaxation factor in the value-iteration algorithm considerably reduces the number of iterations required.   □

*3.6   SPECIAL-PURPOSE POLICY-ITERATION ALGORITHMS FOR STRUCTURED APPLICATIONS

The policy-iteration algorithm is a flexible algorithm on account of the freedom in specifying the policy-improvement procedure. In this section we demonstrate how this flexibility can be used to develop tailor-made algorithms for practical applications having a specific structure. In structured applications one typically wishes to consider only policies which have a simple form and thus are easy to implement. The structure of the application considered together with the freedom in specifying the policy-improvement procedure may be exploited to design a special-purpose policy-iteration algorithm which generates a sequence of improving policies within some class of policies having a simple form. Such an algorithm typically has the important advantage of a considerable reduction in the computational effort. In the value-determination step of the algorithm the structure of the simple policy can often be exploited to reduce the system of linear equations for the average cost and the relative values to a considerably smaller system of linear equations on an embedded set of states.

We develop below tailor-made policy-iteration algorithms for a maintained system with partial information about its state, a continuous review (s, S) inventory system with positive lead times and a queueing system with a variable service rate.

Example 3.9   Optimal inspections and replacements for a maintained system with partial information

Consider an equipment that gradually deteriorates in time and whose degree of deterioration can be observed by inspections only. Practical examples include production machines subject to stochastic breakdowns, inventory systems being depleted and maintenance of communication systems. Opportunities for inspections occur at the equidistant points in time $t = 0, 1, \dots$. An inspection reveals the exact working condition of the system. Depending on the system's degree of deterioration, an inspection may be succeeded by a replacement (or restoration). The system can be observed in one of the working conditions $1, \dots, N$ which describe increasing degrees of deterioration. The state 1 represents a new system, while the state N represents the broken state or a severe malfunction. In the absence of inspections and replacements, the working condition of the system changes according to a discrete-time Markov chain. If the system has working condition i at present, then one time unit later it will have working condition j with known probability r_{ij}. It is assumed that $r_{ij} = 0$ for $j < i$, so that

$$\sum_{j=i}^{N} r_{ij} = 1 \quad \text{for} \quad 1 \leqslant i \leqslant N,$$

that is the working condition of the system cannot improve on its own. The changes in the working condition are not observable and the exact state of the system can be revealed only by inspection. In certain applications in which the final state N represents the broken state it may be unrealistic to assume that this state is not directly observable; however, the analysis to be given below needs only slight modifications when the state N is detected immediately at its occurrence. An inspection takes a fixed, integral number of T time units and involves a fixed cost of $J > 0$. After an inspection reveals the exact working condition of the system, there is an option of either doing a replacement or leaving the system as it is. A subsequent replacement in working condition i takes a fixed, integral number of T_i time units and involves a fixed cost of $K_i > 0$. A replacement has always to be done when the final working condition N is found. After each replacement the system has working condition 1. Besides the inspection and replacement costs, the system incurs an operating cost of a_i for each unit of time it is operating in working condition i. The goal is to find a simple schedule for inspections and replacements such that the long-run average cost per unit time is minimal.

A simple rule for controlling the system is the so-called control-limit rule which is characterized by positive integers $s, s_1, \ldots, s_{N-1}$. Under this rule, replacement is done only when the working condition $i \geqslant s$ is revealed and inspection is done when s_i time units have passed since the system was lastly known to have working condition i. It is intuitively obvious that an overall average cost optimal policy will have the simple control-limit form under reasonable conditions. It is our conjecture that the following assumptions are sufficient (but surely not necessary):

(a) a_i and $R_i/(T_i \vee 1)$ are non-decreasing in i where $T_i \vee 1 = \max(1, T_i)$,
(b) $\sum_{j \geqslant i} r_{ij} a_j - a_i \geqslant \sum_{j \geqslant i} r_{ij} R_j/(T_j \vee 1) - R_i/(T_i \vee 1)$ for all i,
(c) $\sum_{j \geqslant k} r_{ij}$ is non-decreasing in i for each fixed k.

The assumption (a) is obvious and assumption (b) states that the expected increase in the occupancy costs per unit time is greater than the expected increase in the revision costs per unit time. Assumption (c) indicates that the rate at which the system deteriorates increases when the working condition becomes worse. In the sequel, we just restrict ourselves to the class of control-limit rules and our only purpose is to develop a tailor-made policy-iteration algorithm that operates on this class of simple policies.

The maintained system can be analysed in the framework of a semi-Markov decision model. The decision epochs include the epochs at which opportunities for inspections occur and the epochs at which the exact working condition of the system is revealed by an inspection. We take as state space

$$I = \{i \mid i = 1, \ldots, N\} \cup \{(i, m) \mid i = 1, \ldots, N - 1, \ m = 1, \ldots, M_i\}.$$

Here state i corresponds to the situation in which an inspection has just revealed the working condition i and state (i, m) corresponds to the situation in which m time units have passed since the system was lastly known (by inspection) to have working condition i. In order to have a finite state space, it is for practical applications no restriction to assume sufficiently large chosen integers $M_1, \ldots, M_{N-1}$ such that an inspection must always be done in the state (i, M_i), $1 \leqslant i \leqslant N - 1$. The integers M_i will have no actual impact on the algorithm. The possible actions a are denoted by

$$a = \begin{cases} 0, & \text{leave the system as it is,} \\ 1, & \text{inspect the system,} \\ 2, & \text{replace the system.} \end{cases}$$

The action $a = 0$ is the only feasible action in state 1, the actions $a = 0$ and $a = 2$ are feasible in the intermediate states i with $1 < i < N$, the action $a = 2$ is the only feasible action in state N, and the actions $a = 0$ and $a = 1$ are the feasible actions in the states (i, m) except for the states (i, M_i), where only $a = 1$ is feasible.

To give the one-step transition probabilities $p_{xy}(a)$, the one-step expected transition times $\tau_x(a)$ and the one-step expected costs $c_x(a)$, we introduce the following deterioration probabilities. For $m = 1, 2, \ldots$ and $1 \leqslant i, j \leqslant N$, define

$r_{ij}^{(m)} =$ the probability that m time units from now the system will have working condition j when the present working condition is i and no inspections and revisions are done.

Noting that a transition from a state to a lower-numbered state is not possible, it follows by conditioning that, for each $m \geqslant 1$,

$$r_{ij}^{(m)} = \sum_{k=i}^{j} r_{ik} r_{kj}^{(m-1)}, \qquad 1 \leqslant i \leqslant N \quad \text{and} \quad i \leqslant j \leqslant N,$$

where $r_{ij}^{(0)} = 1$ for $j = i$ and $r_{ij}^{(0)} = 0$ otherwise.

It is now readily verified that the one-step transition probabilities are given by

$$p_{i,(i,1)}(0) = 1, \qquad p_{i1}(2) = 1, \qquad p_{(i,m)(i,m+1)}(0) = 1, \qquad p_{(i,m),j}(1) = r_{ij}^{(m)},$$

while the other one-step transition probabilities are zero. Also,

$$\tau_i(0) = 1, \qquad \tau_i(2) = T_i, \qquad \tau_{(i,m)}(0) = 1, \qquad \tau_{(i,m)}(1) = T$$

and

$$c_i(0) = a_i, \qquad c_i(2) = K_i, \qquad c_{(i,m)}(0) = \sum_{j=i}^{N} r_{ij}^{(m)} a_j, \qquad c_{(i,m)}(1) = J.$$

We have now specified the basic elements of the semi-Markov decision model. Note that since zero inspection times and zero revision times are allowed, it may happen that two or more actions are made simultaneously. However, under each policy the number of actions made in a finite time is bounded and thus there are no obstructions to applying the policy-iteration approach.

Fix now a control-limit rule R with parameter values $s, s_1, \ldots, s_{N-1}$. We first analyse the value-determination step for this rule R. Using (3.43) and (3.44) and the above specifications, it follows that the average cost $g(R)$ and the relative values $v_x(R)$, $x \in I$, are the unique solution to the linear equations

$$v_i = \begin{cases} a_i - g + v_{(i,1)}, & \text{for } 1 \leqslant i < s, \qquad (3.48) \\ K_i - g T_i + v_1, & \text{for } s \leqslant i \leqslant N, \qquad (3.49) \end{cases}$$

$$v_{(i,m)} = \begin{cases} \displaystyle\sum_{j=i}^{N} r_{ij}^{(m)} a_j - g + v_{(i,m+1)}, & \text{for } 1 \leqslant m < s_i, \ 1 \leqslant i \leqslant N-1, \qquad (3.50) \\ J - g T + \displaystyle\sum_{j=i}^{N} r_{ij}^{(m)} v_j, & \text{for } s_i \leqslant m \leqslant M_i, \ 1 \leqslant i \leqslant N-1, \qquad (3.51) \end{cases}$$

augmented by putting one of the relative values equal to zero, say

$$v_1 = 0. \qquad (3.52)$$

By the structure of the control-limit rule, the above system of linear equations can be reduced to a considerably smaller system of linear equations in the unknowns g and v_i, $1 \leqslant i < s$. To obtain this reduced system of linear equations, observe that by a repeated application of (3.50) and by using (3.51) with $m = s_i$ we can express $v_{(i,1)}$ in terms of g and v_j, $i \leqslant j \leqslant N$. Next, by using (3.48) and (3.49), we get an embedded system of linear equations in g and v_i, $1 \leqslant i < s$. By doing these manipulations, we find

$$v_i = A_i - g B_i + \sum_{j=i}^{s-1} r_{ij}^{(s_i)} v_j, \qquad 1 \leqslant i < s, \qquad (3.53)$$

where the constants A_i and B_i are given by

$$A_i = a_i + \sum_{m=1}^{s_i-1} \sum_{j=i}^{N} r_{ij}^{(m)} a_j + J + \sum_{j=s}^{N} r_{ij}^{(s_i)} K_j, \qquad (3.54)$$

$$B_i = s_i + T + \sum_{j=s}^{N} r_{ij}^{(s_i)} T_j. \qquad (3.55)$$

The linear equations (3.53) and (3.52) together determine uniquely the average cost $g(R)$ and the relative values $v_i(R)$ for $1 \leqslant i < s$. Once we have solved these linear equations, we obtain by single-pass calculations $v_i(R)$ for $s \leqslant i \leqslant N$ from (3.49) and $v_{(i,s_i)}(R)$ for $1 \leqslant i \leqslant N-1$ from (3.51) with $m = s_i$. Next the relative values $v_{(i,m)}(R)$, $1 \leqslant m < s_i$, can be computed recursively from (3.50) when required. Also, by a single-pass calculation, each relative value $v_{(i,m)}(R)$, $s_i \leqslant m \leqslant M_i$, can be obtained from (3.51) when required. From a computational point of view it is important to note that the upper bounds M_i have no influence upon the actual computations of the average cost and the relative values.

Next we describe how to design a policy-improvement procedure which results in a new rule that is not only at least as good as the current rule R but has moreover the control-limit form. The latter requirement can be fulfilled by exploiting the freedom in specifying the policy-improvement step. Denoting the policy-improvement test quantity $T_R(x;a)$ by

$$T_R(x;a) = c_x(a) - g(R)\tau_x(a) + \sum_{y \in I} p_{xy}(a)v_y$$

and recalling that the inequality (3.41) suffices for an improved policy, we wish to determine positive integers $\bar{s}, \bar{s}_1, \ldots, \bar{s}_{N-1}$ such that

$$
\begin{aligned}
T_R(i;0) &\leqslant v_i(R) & \text{for } &1 \leqslant i < \bar{s}, \\
T_R(i;2) &\leqslant v_i(R) & \text{for } &\bar{s} \leqslant i \leqslant N,
\end{aligned}
\tag{3.56}
$$

$$T_R((i,m);0) \leqslant v_{(i,m)}(R) \quad \text{for } 1 \leqslant m < \bar{s}_i \text{ and } 1 \leqslant i \leqslant N-1, \tag{3.57}$$

$$T_R((i,m);1) \leqslant v_{(i,m)}(R) \quad \text{for } \bar{s}_i \leqslant m \leqslant M_i \text{ and } 1 \leqslant i \leqslant N-1. \tag{3.58}$$

To construct a new control-limit rule $\bar{R}$ with parameter values $\bar{s}, \bar{s}_1, \ldots, \bar{s}_{N-1}$ such that (3.56), (3.57) and (3.58) are satisfied, we use the fact that the policy-improvement test quantity $T_R(x;a)$ equals the relative value at state x when a is given by the old action R_x. Hence

$$
\begin{aligned}
T_R(i;0) &= v_i(R) & \text{for } &1 \leqslant i < s, \\
T_R(i;2) &= v_i(R) & \text{for } &s \leqslant i \leqslant N,
\end{aligned}
\tag{3.59}
$$

$$T_R((i,m);0) = v_{(i,m)}(R) \quad \text{for } 1 \leqslant m < s_i \text{ and } 1 \leqslant i \leqslant N-1, \tag{3.60}$$

$$T_R((i,m);1) = v_{(i,m)}(R) \quad \text{for } s_i \leqslant m \leqslant M_i \text{ and } 1 \leqslant i \leqslant N-1. \tag{3.61}$$

Suppose now that the integer $\bar{s}$ with $1 < \bar{s} < s$ is such that

$$T_R(i;2) < v_i(R) \qquad \text{for } \bar{s} \leqslant i < s. \tag{3.62}$$

Then, using (3.59), it follows that (3.56) holds. Similarly, when the integer $\bar{s}$ with $s < \bar{s} \leqslant N$ is such that

$$T_R(i;0) < v_i(R) \qquad \text{for } s \leqslant i < \bar{s}, \tag{3.63}$$

it follows that (3.56) holds. The same reasoning can be applied to the states (i,m), $1 \leqslant m \leqslant M_i$ with i fixed. In the case where the integer $\bar{s}_i$ with $1 \leqslant \bar{s}_i < s_i$ is such that

$$T_R((i,m);1) < v_{(i,m)}(R) \qquad \text{for } \bar{s}_i \leqslant m < s_i, \tag{3.64}$$

it follows by using (3.60) and (3.61) that (3.57) and (3.58) hold. Also, when the integer $\bar{s}_i$ with $s_i < \bar{s}_i \leqslant M_i$ is such that

$$T_R((i,m);0) < v_{(i,m)}(R) \qquad \text{for } s_i \leqslant m < \bar{s}_i, \tag{3.65}$$

it follows that (3.57) and (3.58) hold. Note that in the case where $\bar{s}_i$ has the value M_i, the chosen bound M_i may be enlarged without difficulty. The policy-improvement procedure based on the tests (3.62) to (3.65) yields the desired new control-limit rule $\bar{R}$ with parameter values $\bar{s}, \bar{s}_1, \ldots, \bar{s}_{N-1}$.

We are now in a position to formulate the special-purpose policy-iteration algorithm.

Algorithm

Step 0. Choose bounds $M_1, \ldots, M_{N-1}$ and an initial control-limit rule R with parameter values $s, s_1, \ldots, s_{N-1}$ such that $1 < s \leqslant N$ and $1 \leqslant s_i < M_i$ for $1 \leqslant i \leqslant N-1$.

Step 1. For the current control-limit rule R with parameter values $s, s_1, \ldots, s_{N-1}$, compute the unique solution $g(R)$ and $v_i(R)$, $1 \leqslant i < s$, from the linear equations (3.52) and (3.53) whose coefficients A_i and B_i are given by (3.54) and (3.55). Next compute the relative values $v_i(R)$ for $s \leqslant i \leqslant N$ from (3.49) and the relative values $v_{(i,s_i)}(R)$ for $1 \leqslant i \leqslant N-1$ from (3.51) with $m = s_i$. The other relative values are computed from (3.50) and (3.51) when required in the steps 2(b), 2(d) and 2(e) below.

Step 2(a). Test for state $i = s - 1$ to find whether

$$K_i - g(R)T_i + v_1(R) < v_i(R), \tag{3.66}$$

provided $s \neq 2$. If $s = 2$ or (3.66) does not apply for $i = s - 1$, then go to step 2(b). Otherwise, determine the smallest integer $\bar{s}$ with $1 < \bar{s} < s$ such that (3.66) holds for $\bar{s} \leqslant i < s$ and go to step 2(c).

Step 2(b). Test for state $i = s$ to find whether

$$a_i - g(R) + v_{(i,1)}(R) < v_i(R). \tag{3.67}$$

If (3.67) does not apply for $i = s$, then take $\bar{s}$ equal to s. Otherwise, determine the largest integer $\bar{s}$ with $s < \bar{s} \leqslant N$ such that (3.67) holds for $s \leqslant i < \bar{s}$.

Step 2(c). $i := 1$.

Step 2(d). Test for state (i, m) with $m = s_i - 1$ to find whether

$$J - g(R)T + \sum_{j=i}^{N} r_{ij}^{(m)} v_j(R) < v_{(i,m)}(R), \tag{3.68}$$

provided $s_i \neq 1$. If $s_i = 1$ or (3.68) does not apply for $m = s_i - 1$, then go to step 2(e). Otherwise, determine the smallest integer $\bar{s}_i$ with $1 \leqslant \bar{s}_i < s_i$ such that (3.68) holds for $\bar{s}_i \leqslant m < s_i$ and go to step 2(f).

Step 2(e). Test for state (i, m) with $m = s_i$ to find whether

$$\sum_{j=i}^{N} r_{ij}^{(m)} a_j - g(R) + v_{(i,m+1)}(R) < v_{(i,m)}(R). \tag{3.69}$$

If (3.69) does not apply for $m = s_i$, then take $\bar{s}_i$ equal to s_i. Otherwise, determine the largest integer $\bar{s}_i$ with $s_i < \bar{s}_i \leqslant M_i$ such that (3.69) holds for $s_i \leqslant m < \bar{s}_i$.

Step 2(f). $i := i + 1$. If $i < N$, go to step 2(d).

Step 3. The new control-limit rule $\bar{R}$ obtains the parameter values $\bar{s}, \bar{s}_1, \ldots, \bar{s}_{N-1}$. If $\bar{R} = R$, the algorithm is stopped. Otherwise, go to step 1 with R replaced by $\bar{R}$.

This algorithm generates a sequence of improving control-limit rules and by similar arguments as in the appendix of section 3.2 it can be shown that the algorithm converges after finitely many iterations. In accordance with the usual experience with policy-iteration algorithms, we found empirically that the number of iterations is remarkably small and is typically on the order of five to ten iterations. Although the algorithm exhibits finite convergence, we lack a theoretical proof that, under reasonable conditions on the cost structure and the deterioration probabilities, the resulting control-limit rule is average cost optimal among all possible control rules. However, we have tested many examples satisfying reasonable conditions and in each of these examples we were able to verify numerically that the algorithm converged to an overall average cost control-limit rule. This numerical verification was based on the average cost optimality equation stating that a rule R with average cost $g(R)$ and relative values $v_x(R)$, $x \in I$, is average cost optimal if $v_x(R) \leqslant T_R(x; a)$ for all states x and actions a.

As an illustration we give some numerical examples in which the number of working conditions is equal to

$$N = 7$$

and the transition matrix (r_{ij}) of deterioration probabilities is given by

$$(r_{ij}) = \begin{pmatrix} 0.9 & 0.1 & 0 & 0 & 0 & 0 & 0 \\ 0 & 0.8 & 0.2 & 0 & 0 & 0 & 0 \\ 0 & 0 & 0.8 & 0.1 & 0.1 & 0 & 0 \\ 0 & 0 & 0 & 0.7 & 0.2 & 0.1 & 0 \\ 0 & 0 & 0 & 0 & 0.6 & 0.2 & 0.2 \\ 0 & 0 & 0 & 0 & 0 & 0.5 & 0.5 \\ 0 & 0 & 0 & 0 & 0 & 0 & 1 \end{pmatrix}.$$

The numerical data for the inspection time T, the revision times T_i, the holding costs a_i and the revision costs K_i are

$$T = 0, \qquad T_i = 0 \quad \text{for} \quad 2 \leqslant i \leqslant 7, \qquad a_i = 15 + 10(i - 1)$$
$$\text{for} \quad 1 \leqslant i \leqslant 6, \qquad a_7 = 100, \qquad K_i = 300 + 10(i - 2) \quad \text{for} \quad 2 \leqslant i \leqslant 7.$$

The inspection cost J is varied as 25, 50 and 100. In Table 3.12 we give for the various values of J the control-limit rules generated by the special-purpose policy-iteration algorithm and the average costs associated with these rules. The algorithm used the bounds $M_i = 20$ for $1 \leqslant i \leqslant 6$ and was initialized by the rule R with $s = 2$ and $s_i = 1$ for $1 \leqslant i \leqslant 6$. The results in Table 3.12 confirm the usual finding that the policy-iteration algorithm converges very quickly, with the greatest improvements in costs occurring in the first few iterations. Note that intermediate rules in the algorithm may have $s_i = M_i$ for some i, even when

Table 3.12   The control-limit rules generated by the algorithm

Iteration	s	s_1	s_2	s_3	s_4	s_5	s_6	$g(R)$	
1	2	1	1	1	1	1	1	70.00	$J = 25$
2	4	20	12	7	4	1	1	49.36	
3	3	4	3	2	1	1	1	45.83	
4	3	11	6	3	1	1	1	43.14	
5	3	8	5	1	1	1	1	43.06	
6	3	9	5	2	1	1	1	42.95	
1	2	1	1	1	1	1	1	95.00	$J = 50$
2	4	20	20	17	10	6	3	54.12	
3	2	15	4	1	1	1	1	47.10	
4	3	9	7	3	1	1	1	46.05	
5	3	11	6	3	1	1	1	45.68	
1	2	1	1	1	1	1	1	145.00	$J = 100$
2	4	20	20	20	20	20	20	57.44	
3	2	19	7	1	1	1	1	52.21	
4	3	10	8	4	1	1	1	51.15	
5	3	14	8	4	1	1	1	50.19	
6	3	13	8	3	1	1	1	50.18	

the bounds M_i are chosen sufficiently large. Therefore we recommend adjusting the bounds M_i only when the final rule still has $s_i = M_i$ for some i.   □

An embedding technique

In the above maintenance problem we have seen that the computational effort in the value-determination step of the policy-iteration algorithm may be considerably reduced by exploiting the structure of the specific problem. The reduction of the linear equations (3.48) to (3.51) for the average cost and the relative values to the embedded system of linear equations (3.53) is in fact an application of a general *embedding technique*. To state this technique, generally, consider a semi-Markov decision model specified by the basic elements $(I, A(i),$ $p_{ij}(a), \tau_i(a), c_i(a))$. For a fixed rule R, let $E(= E(R))$ be an appropriately chosen embedded set of states with $E \neq I$ such that under rule R the set E of states can be reached from each initial state $i \in I$. Define now for each $i \in I$ and $j \in E$,

$p_{ij}^E(R) =$ the probability that the first entry state in the set E equals

    j when the initial state is i and rule R is used,

with the convention that for the initial state $i \in E$, the first entry state in set E should be interpreted as the state taken on at the next return to the set E. Also, define, for each $i \in I$,

$\tau_i^E(R) = $ the expected time until the first entry in the set E when the initial state is i and rule R is used,

$c_i^E(R) = $ the expected costs incurred until the first entry in the set E when the initial state is i and rule R is used (excluding the cost incurred at the epoch of the first entry in E).

Let $g(R)$ and $v_i(R)$, $i \in I$, denote the average cost and relative values associated with rule R. It will be shown below that

$$v_i(R) = c_i^E(R) - g(R)\tau_i^E(R) + \sum_{j \in E} p_{ij}^E(R)v_j(R), \qquad i \in I. \qquad (3.70)$$

Thus for rule R the average cost and relative values at the states $i \in E$ can be computed by solving the embedded system of linear equations

$$v_i = c_i^E(R) - g\tau_i^E(R) + \sum_{j \in E} p_{ij}^E(R)v_j, \qquad i \in E, \qquad (3.71)$$

together with a normalization equation $v_s = 0$ for some $s \in E$. This system of linear equations has a unique solution (why?). Next by a single-pass calculation, we can compute each relative value v_i for $i \notin E$ from (3.70). It will be clear that the above embedding approach may lead to a considerable reduction in computational effort when an appropriate embedded set E can be found which allows for an efficient computation of $p_{ij}^E(R)$, $\tau_i^E(R)$ and $c_i^E(R)$.

The proof of the relation (3.70) is not difficult. Choose a state $r \in E$ such that state r can be reached from each initial state $i \in I$ under rule R. Then, by the semi-Markov version of part (b) of Theorem 3.1, we have that the function $v_i(R)$, $i \in I$, differs only by some additive constant from the function

$$w_i(R) = K_i(R) - g(R)T_i(R), \qquad i \in I,$$

which is defined by (3.8) in section 3.1. Thus it suffices to verify that the particular relative values $w_i(R)$, $i \in I$, satisfy (3.70). By conditioning on the first entry state in the set E and using the definitions of the functions $T_i(R)$ and $K_i(R)$, it readily follows that

$$T_i(R) = \tau_i^E(R) + \sum_{\substack{j \in E \\ j \neq r}} p_{ij}^E(R)T_j(R), \qquad i \in I,$$

$$K_i(R) = c_i^E(R) + \sum_{\substack{j \in E \\ j \neq r}} p_{ij}^E(R)K_j(R), \qquad i \in I.$$

By subtracting $g(R)$ times the equation for $T_i(R)$ from the equation for $K_i(R)$ and using the definition of $w_i(R)$, we obtain

$$w_i(R) = c_i^E(R) - g(R)\tau_i^E(R) + \sum_{\substack{j \in E \\ j \neq r}} p_{ij}^E(R)w_j(R), \qquad i \in I.$$

This proves the desired result (3.70) since $w_j(R) = 0$ for $j = r$, by (3.9). $\qquad \square$

It is noted that in the above example 3.9 the set of embedding states is taken as $E = \{i | 1 \leqslant i < s\}$ for the control-limit rule R with parameter values $s, s_1, \ldots, s_{N-1}$.

The embedding principle will also be used in the next example dealing with a continuous review (s, S) inventory system. It is interesting to note that the analysis for this continuous review system applies with minor modifications equally well to the periodic review (s, S) inventory system. Heuristics for these inventory systems were already discussed in section 1.7 of chapter 1.

Example 3.10   A continuous review (s, S) inventory system with a service level constraint

Consider a single-item inventory system in which the demands for the item occur at epochs generated by a Poisson process with rate λ and the demand sizes are independent non-negative random variables having a common discrete probability distribution $\{\phi(j), j = 0, 1, \ldots\}$. The demand sizes are independent of the Poisson process. Excess demand is backlogged. The inventory position is continuously reviewed. At any time, a replenishment order of any size can be made. The replenishment lead time is a non-negative constant L. The ordering costs consist of a fixed setup cost $K \geqslant 0$ and a variable cost c per unit ordered. Also, there are holding costs at rate hk when the inventory on hand equals $k \geqslant 0$, where h is a positive constant. We wish to minimize the long-run average ordering and holding costs per unit time subject to the service level constraint that in the long-run a specified fraction of the total demand is satisfied directly from on-hand inventory. Service level constraints are typically used in practice rather than shortage costs which are usually difficult to specify.

In view of the lack of memory of the Poisson process generating the demand epochs, we consider replenishment opportunities only at the epochs at which the demands occur. Also, since excess demand is backlogged, the control of the system is based on the inventory position defined as (the inventory on hand) + (total amount on order) − (total backlog). A reasonable control rule is the (s, S) rule with $0 \leqslant s < S$. Under this rule, the inventory position is ordered up to S if at a demand epoch the inventory position drops to or below s; otherwise no ordering is done. Our goal is to develop a special-purpose policy-iteration algorithm for the computation of the best rule within the class of (s, S) rules. To handle the service level constraint, we proceed as follows. We first ignore this constraint and assume a fixed penalty cost of $\pi > 0$ for every requested unit that cannot be satisfied directly from current inventory. Then, by a special-purpose algorithm, we compute an average cost optimal (s, S) rule for the cost structure consisting of ordering, holding and penalty costs. For this particular rule, we compute the fraction of demand that is satisfied directly from on-hand inventory. If the service level constraint is not satisfied for this rule, we alter the value of the fixed penalty cost π and compute anew an average cost optimal rule for the adjusted cost structure. We continue in this way until we have found the smallest value of π for which the average cost optimal rule satisfies the service level constraint. In practical applications this Lagrangian approach will usually result in an (s, S) rule which minimizes the average ordering and holding costs subject to the service level constraint on the fraction of demand that must be satisfied directly from on-hand inventory.

The key element of the above Lagrangian approach is a special purpose policy-iteration algorithm that computes an average cost optimal (s, S) rule for the cost structure consisting of the ordering costs, the holding costs and a fixed penalty cost $\pi > 0$ for each unit shortage that occurs. For the moment we consider this cost structure and thus ignore temporarily the service level constraint. We shall now formulate a semi-Markov decision model for this particular inventory problem. In order to have a finite state space, we assume that the demand size distribution $\{\phi(j)\}$ has a finite support, that is $\sum_{j=0}^{M} \phi(j) = 1$ for some finite $M \geqslant 1$. Also, we assume upper and lower bounds U and V such that the inventory position is never raised above U, while it is always raised above V when it drops at or below V. For ease of presentation we take $V = 0$. These assumptions are not restrictive for practical applications. Finally, in our subsequent discussion we take the variable ordering cost c equal to zero. Note that, since all demand is ultimately satisfied under backlogging of excess demand, the variable ordering cost contributes to the average cost of any policy the same amount of c times the average demand per unit time. We can now specify the basic elements of the semi-Markov decision model. The decision epochs are given by the random epochs at which the demands occur. We take as state space

$$I = \{i| -M \leqslant i \leqslant U\}.$$

Here state i corresponds to the situation where a demand has just occurred leaving an inventory position of i units. We specify the action a as the inventory position just after a possible replenishment. In other words, action a in state i means the ordering of $a - i$ units. It is always required that $a \geqslant i$ and $0 < a \leqslant U$. The one-step transition probabilities $p_{ij}(a)$ and the one-step expected transition times are clearly given by

$$p_{ij}(a) = \phi(a - j)$$

with the convention that $\phi(k) = 0$ for $k < 0$, and

$$\tau_i(a) = \frac{1}{\lambda}.$$

The specification of the one-step expected costs $c_i(a)$ is rather subtle. We first observe that, since excess demand is backlogged and the lead time of each replenishment order is a constant L, the inventory on hand at any time $t + L$ is distributed as the inventory position at time t minus the total demand during a period of length L. Hence, letting $\{T_n\}$ be the sequence of decision epochs, the inventory position just after time T_n unambiguously determines the distribution of the inventory on hand at time $T_n + L$. We thus adopt the convention that when choosing action a at the decision epoch T_n, one is charged with the immediate ordering cost (if any) as well as an amount $\gamma(a)$ representing the expected holding and penalty costs incurred in $(T_n + L, T_{n+1} + L]$. Clearly, this shift in costs leaves the average cost of any policy unchanged. We now have for the one-step expected costs,

$$c_i(a) = K\delta(a - i) + \gamma(a),$$

where $\delta(j) = 1$ for $j > 0$ and $\delta(j) = 0$ otherwise. To give an expression for $\gamma(a)$, we need the probability distribution $\{r_L(j), j \geq 0\}$ of the total demand during a replenishment lead time L. The compound Poisson distribution $\{r_L(j)\}$ can be computed efficiently by the recursion scheme (1.48) with $t = L$ in section 1.6 of chapter 1. Using the lack of memory of the Poisson process generating the demand epochs, it can be shown that $\gamma(a)$ represents also the expected holding and penalty costs incurred between time L and the first decision epoch following time L under the condition that at time 0 the inventory position equals a and no replenishment orders are placed between time 0 and time L (for a proof see Federgruen and Schechner, 1983). Under the latter condition the inventory on hand minus backlogs at time L equals $a - j$ with probability $r_L(j)$ and remains constant after time L during an interval of expected length $1/\lambda$. The demand occurring at the end of this interval causes fixed penalty costs only if it exceeds the inventory on hand. Thus we obtain

$$\gamma(a) = \frac{1}{\lambda}h\sum_{j=0}^{a}(a - j)r_L(j) + \pi\left[\sum_{j=0}^{a}r_L(j)\sum_{k\geq a-j}(k - a + j)\phi(k)\right.$$
$$\left. + \left\{\sum_{j=a+1}^{\infty}r_L(j)\right\}E(D)\right], \qquad a = 0, 1, \ldots, \tag{3.72}$$

where $E(D) = \sum_j j\phi(j)$ is the average demand size. It follows easily from (3.72) that $\gamma(a)$ satisfies the recurrence relation

$$\gamma(a) = \gamma(a - 1) + \frac{1}{\lambda}\left\{h\sum_{j=0}^{a-1}r_L(j) - \pi\lambda\sum_{j=0}^{a-1}r_L(j)\sum_{k=0}^{a-1-j}\phi(k)\right\}, \qquad a = 1, 2, \ldots,$$

with $\gamma(0) = \pi E(D)$. This recurrence relation is better suited for computational purposes than the explicit expression (3.72). We have now completed the specification of the basic elements of the semi-Markov decision model.

For the inventory problem with a cost structure consisting of a fixed ordering cost and linear holding and penalty costs, it can be shown that an (s, S) rule is average cost optimal among all possible stationary rules. We omit the proof but remark only that an application of the data transformation from section 3.4 to the continuous review inventory problem results in a periodic review inventory problem for which an (s, S) rule is known to be average cost optimal (cf. also Federgruen and Zipkin, 1984).

We now develop a special-purpose policy-iteration algorithm that operates on the class of (s, S) rules. Here we restrict ourselves to (s, S) rules with $s \geqslant 0$. In practical applications the optimal value of the reorder point s will typically be non-negative. Fix now a rule R of the (s, S) type. First we discuss the computation of the average cost and the relative values associated with rule R. The average cost $g(R)$ and the relative values $v_i(R)$, $i \in I$, are the unique solution of the system of linear equations

$$
v_i = \begin{cases}
\gamma(i) - g\dfrac{1}{\lambda} + \displaystyle\sum_{j=0}^{M} v_{i-j}\phi(j) & \text{for } s < i \leqslant U, \qquad (3.73) \\[4ex]
K + \gamma(S) - g\dfrac{1}{\lambda} + \displaystyle\sum_{j=0}^{M} v_{S-j}\phi(j) & \text{for } -M \leqslant i \leqslant s, \qquad (3.74)
\end{cases}
$$

augmented by putting one of the relative values equal to zero, say

$$v_S = 0. \qquad (3.75)$$

By exploiting the structure of the (s, S) rule, these linear equations can be solved very efficiently. We first observe that (3.73) with $i = S$ and (3.74) imply

$$v_i = K + v_S, \qquad -M \leqslant i \leqslant s. \qquad (3.76)$$

This special property shows that for the embedded set of states

$$E = \{i \mid -M \leqslant i \leqslant s\}$$

the embedded system of linear equations (3.71) is trivial to solve. To be specific, define, for $i > s$,

$\tau_i(s, S) =$ the expected time until the first entry into the set E when the initial state is i and rule $R = (s, S)$ is used,

$c_i(s, S) =$ the expected costs incurred until the first entry into the set E when the initial state is i and rule $R = (s, S)$ is used (excluding the costs incurred at the epoch of the first entry into the set E).

Recalling the convention to assign 'immediate' costs $c_k(a) = K\delta(a - k) + \gamma(a)$ when action a is made in state k, it readily follows by conditioning on the size of the first demand that

$$\tau_i(s, S) = \frac{1}{\lambda} + \sum_{j=0}^{i-s-1} \tau_{i-j}(s, S)\phi(j), \qquad i = s+1, s+2, \ldots, \qquad (3.77)$$

$$c_i(s, S) = \gamma(i) + \sum_{j=0}^{i-s-1} c_{i-j}(s, S)\phi(j), \qquad i = s+1, s+2, \ldots. \qquad (3.78)$$

Hence the quantities $\tau_i(s, S)$ and $c_i(s, S)$, $i > s$, can be computed recursively starting with $\tau_{s+1}(s, S) = \lambda^{-1}/\{1 - \phi(0)\}$ and $c_{s+1}(s, S) = \gamma(s + 1)/\{1 - \phi(0)\}$. Now, by (3.70), (3.75) and (3.76),

$$v_i(R) = c_i(s, S) - g(R)\tau_i(s, S) + K, \qquad i > s. \tag{3.79}$$

In particular, by taking $i = S$ in (3.72) and using (3.75), we obtain

$$g(R) = \frac{K + c_S(s, S)}{\tau_S(s, S)}, \tag{3.80}$$

that is the average cost equals the ratio of the expected cost per cycle and the expected length of one cycle, where a cycle is the time interval between two consecutive replenishment orders.

Summarizing, the linear equations (3.73) to (3.75) allow for a simple recursive solution. First, the quantities $\tau_i(s, S)$ and $c_i(s, S)$ are computed recursively from (3.77) and (3.78) for $i = s + 1, \ldots, S$. Next we compute the average cost $g(R)$ from (3.80) and obtain the relative values $v_i(R)$ for $s + 1 \leqslant i \leqslant S$ from (3.79). Finally, each required $v_i(R)$ for $i > S$ can be obtained from (3.79) by a single-pass calculation requiring only the evaluations of $\tau_i(s, S)$ and $c_i(s, S)$ from the recurrence relations (3.77) and (3.78). Note that the upper bound U is actually not needed when computing the average cost and the relative values. The integer U is only used to bound the state space and may be adjusted during the algorithm without difficulty.

We now turn to the problem of designing a policy-improvement procedure that results in a new rule $\bar{R}$ having the desired (s, S) form. Denoting the policy-improvement test quantity $T_R(i; a)$ by

$$T_R(i; a) = c_i(a) - g(R)\tau_i(a) + \sum_{j \in I} p_{ij}(a)v_j(R),$$

we wish to determine a rule $\bar{R}$ having the (s, S) form such that

$$T_R(i; \bar{R}_i) \leqslant v_i(R) \qquad \text{for each } i \in I. \tag{3.81}$$

To do this, we first specify $T_R(i; a)$ for the relevant combinations of i and a. We always have for each state i that

$$T_R(i; a) = v_i(R) \qquad \text{when } a = R_i. \tag{3.82}$$

Using (3.75) and (3.76), we have

$$T_R(i; a) = \gamma(i) - g(R)\frac{1}{\lambda} + K, \qquad i \leqslant s \quad \text{and} \quad a = i. \tag{3.83}$$

For $a > i$ we have

$$T_R(i; a) = K + \gamma(a) - g(R)\frac{1}{\lambda} + \sum_{j=0}^{M} v_{a-j}(R)\phi(j),$$

and so, by (3.73),

$$T_R(i; a) = K + v_a(R), \qquad i \in I \qquad \text{and} \qquad a > \min(i, s). \qquad (3.84)$$

The relations (3.76), (3.81) and (3.84) suggest a search for an integer $\bar{S}$ with $s < \bar{S} \leqslant U$ such that

$$K + v_{\bar{S}}(R) < K + v_S(R) \qquad (3.85)$$

provided that such an integer exists; otherwise let $\bar{S} = S$. The integer $\bar{S}$ will be the new order-up-to level. The determination of the new reorder point $\bar{s}$ proceeds as follows. We first try to find whether the current value of the reorder point s can be decreased. The relation (3.83) and the observation $v_i(R) = K$ for $i \leqslant s$ suggest a search for an integer s' with $0 \leqslant s' < s$ such that

$$\gamma(i) - g(R)\frac{1}{\lambda} + K < K \qquad \text{for } s' < i \leqslant s.$$

If such an integer s' exists, the rule $\bar{R} = (s', \bar{S})$ can easily be shown to satisfy (3.81) (verify!). In case the reorder point s cannot be decreased, we try to increase the reorder point by searching for an integer s'' with $s < s'' < \bar{S}$, such that

$$K + v_{\bar{S}}(R) < v_i(R) \qquad \text{for } s < i \leqslant s''.$$

If such an integer s'' exists, it is readily verified by using (3.84) and (3.85) that the rule $\bar{R} = (s'', \bar{S})$ satisfies (3.81).

We are now in a position to state the algorithm for the inventory problem with penalty costs rather than a service level constraint.

Algorithm (no service level constraint)

Step 0   Choose an upper bound U and an (s, S) rule with $0 \leqslant s < S < U$. An upper bound for the optimal value of S can be shown to be the largest integer U with $\gamma(U) \leqslant K + \gamma_0$ where γ_0 is the minimum value of the unimodal function $\gamma(i)$ (cf. Veinott and Wagner, 1975).

Step 1.   Let $R = (s, S)$ be the current rule. Compute recursively the quantities $\tau_i(s, S)$ and $c_i(s, S)$ for $i = s + 1, \ldots, S$ from (3.77) and (3.78). Next compute the average cost $g(R)$ and the relative values $v_i(R)$ for $s \leqslant i \leqslant S$ from (3.80) and (3.79). The other relative values $v_i(R)$, $i > S$, are evaluated from (3.79) when required in the next step.

Step 2(a). Determine an integer $\bar{S}$ with $s < \bar{S} < U$ such that $v_{\bar{S}}(R) < 0$ provided such an integer exists; otherwise let $\bar{S} = S$.

Step 2(b). Determine s' as the smallest integer with $0 \leqslant s' < s$ such that $\gamma(i) - g(R)/\lambda < 0$ for $s' < i \leqslant s$. If such an integer s' exists, define $\bar{s} = s'$. Otherwise, let s'' be the largest integer with $s < s'' < \bar{S}$ such that $K + v_{\bar{S}}(R) < v_i(R)$ for $s < i \leqslant s''$. If such an integer s'' exists, define $\bar{s} = s''$ and otherwise let $\bar{s} = s$. This results in a new rule $\bar{R} = (\bar{s}, \bar{S})$.

Step 3.   If $\bar{R} = R$, stop; otherwise go to step 1 with R replaced by $\bar{R}$.

We state without proof that this algorithm converges after finitely many iterations to an overall average cost optimal (s, S) rule. The number of iterations required by this algorithm is remarkably small (typically less than 10) and each iteration involves only simple computations. The above algorithm applies to the inventory problem with a given cost structure including a fixed penalty cost π per unit shortage. The algorithm yields as a by-product the service level of the (s, S) rules. Therefore note that the contribution of the shortage cost π to the average cost $g(R)$ is equal to π times the average demand that goes short per unit time. Dividing this contribution by π times the average demand $\lambda E(D)$ per unit time, we find for rule R the fraction of demand that is backlogged. Hence the service level of each (s, S) rule can be computed easily. Note that in the recursive computation of the cost functions $\gamma(i)$ and $c_i(s, S)$ the different contributions of the holding and penalty costs can be dealt with separately.

Our goal is to find an (s, S) rule which minimizes the average holding and ordering costs only subject to the service level constraint that the fraction of demand to be met directly from on-hand inventory should be at least a specific value β with $0 < \beta < 1$. To search for such a rule, we apply the above algorithm repeatedly for different values of the fixed penalty cost $\pi > 0$, as in ordinary Lagrangian methods. Since the service levels of the generated (s, S) rules are non-decreasing in π, we continue until we have found the *smallest* value of π for which the associated (s, S) rule resulting from the algorithm still satisfies the service level constraint. In our experience, recovering an average cost optimal (s, S) rule when making (small) changes in the value of π requires very few iterations, provided the algorithm is restarted with the rule that was found to be optimal under the previous value of π. We now state the algorithm for the inventory problem with the service level constraint.

Algorithm (service level constraint)

Step 0. Choose a positive number π.

Step 1. For the current value of the fixed penalty cost π, compute by the preceding algorithm the (s_π, S_π) rule which minimizes the average cost per unit time (initialize this algorithm with the (s, S) rule that was found to be optimal for the previous value of π). Let β_π be the fraction of demand satisfied directly from on-hand inventory under the (s_π, S_π) rule.

Step 2. Adjust the penalty cost π until the smallest value of $\pi \geqslant 0$ is found for which $\beta_\pi \geqslant \beta$ (this procedure can be simplified by using the fact that the non-decreasing function β_π is piecewise constant since there are a finite number of intervals of π values such that within each interval the same optimal (s_π, S_π) rule applies). Go to step 1 each time π is adjusted.

This Lagrangian algorithm will usually find an (s, S) rule that minimizes the average holding and replenishment costs within the class of (s, S) rules satisfying the service level constraint. In the Lagrangian approach it may occasionally occur that the best (s, S) rule is not found, but in those rare cases the average holding and replenishment costs of the obtained (s,S) rule were always very close to the minimal average costs (cf. also the discussion on probabilistic constraints in section 3.3).

The algorithm has been developed for the specific cost structure consisting of a fixed setup cost and a linear holding cost. However, with some modifications the algorithm can also be used for other cost structures. For example when the holding costs are quadratic rather than linear we have only to modify the formula for the one-step costs $\gamma(k)$.

To conclude this example, we give some numerical results which support the approximate analysis in section 1.7 of chapter 1. In section 1.7 we derived the economic order quantity formula as an approximation for $S - s$ when the costs consist of a fixed setup cost and a linear holding cost. Also, we advocated in section 1.7 a sequential approach in which first the difference $S - s$ is calculated from the economic order quantity formula and next the reorder point s is determined by satisfying the service level constraint. This sequential approach was justified by the claim that in practical applications the simultaneous determination of the optimal values of s and S usually results in a value for $S - s$ that is fairly insensitive to the required service level and the lead time. The above algorithm can be used to verify this claim numerically. To do so we

Table 3.13   The best (s^*, S^*) rules and the approximate (s_a, S_a) rules

β	c_D^2	L	(s^*, S^*)	$\beta(s^*, S^*)$	$g(s^*, S^*)$	(s_a, S_a)	$\beta(s_a, S_a)$	$g(s_a, S_a)$	Error %
0.90	$\frac{1}{2}$	$\frac{1}{2}$	(27, 87)	0.902	54.49	(29, 79)	0.902	55.17	1.2
0.95	$\frac{1}{2}$	$\frac{1}{2}$	(35, 94)	0.950	61.92	(37, 87)	0.952	62.78	1.4
0.99	$\frac{1}{2}$	$\frac{1}{2}$	(51, 107)	0.990	77.18	(52, 102)	0.990	77.55	0.5
0.90	2	$\frac{1}{2}$	(35, 98)	0.901	63.81	(39, 89)	0.903	65.22	2.2
0.95	2	$\frac{1}{2}$	(48, 109)	0.951	75.81	(51, 101)	0.951	76.75	1.2
0.99	2	$\frac{1}{2}$	(74, 134)	0.990	101.24	(76, 126)	0.990	101.43	0.2
0.90	5	$\frac{1}{2}$	(52, 118)	0.901	82.55	(59, 109)	0.906	84.90	2.8
0.95	5	$\frac{1}{2}$	(73, 138)	0.950	102.58	(80, 130)	0.953	105.28	2.6
0.99	5	$\frac{1}{2}$	(120, 183)	0.990	148.47	(126, 176)	0.991	150.86	1.6
0.90	$\frac{1}{2}$	1	(56, 120)	0.902	59.60	(59, 109)	0.899	60.46	1.4
0.95	$\frac{1}{2}$	1	(67, 128)	0.952	69.41	(69, 119)	0.950	69.91	0.7
0.99	$\frac{1}{2}$	1	(87, 145)	0.990	88.51	(88, 138)	0.990	88.57	0.1
0.90	2	1	(67, 135)	0.902	72.53	(73, 123)	0.904	74.58	2.8
0.95	2	1	(83, 148)	0.950	86.99	(88, 138)	0.952	88.90	2.2
0.99	2	1	(115, 178)	0.990	117.99	(119, 169)	0.990	119.45	1.2
0.90	5	1	(87, 159)	0.901	95.20	(99, 149)	0.910	100.35	5.4
0.95	5	1	(113, 182)	0.950	119.17	(124, 174)	0.956	124.48	4.5
0.99	5	1	(168, 235)	0.990	172.81	(176, 226)	0.991	175.90	1.8

consider a number of examples in which the demand distribution $\phi(j)$ is given by the negative binomial distribution,

$$\phi(j) = \binom{r+j-1}{j} p^r (1-p)^j, \qquad j = 0, 1, \ldots,$$

with mean $E(D) = r(1-p)/p$ and variance $\sigma^2(D) = r(1-p)/p^2$. The parameters r and p of the negative binomial distribution are uniquely determined by $E(D)$ and $\sigma^2(D)$ provided $\sigma^2(D)/E(D) > 1$. In the examples we take

$$\lambda = 10, \qquad E(D) = 5, \qquad K = 25 \qquad \text{and} \qquad h = 1.$$

The squared coefficient of variation $c_D^2 = \sigma^2(D)/E^2(D)$ is varied as $\frac{1}{2}$, 2 and 5, while the required service level β has the values 0.90, 0.95 and 0.99 and the lead time L is varied as $\frac{1}{2}$ and 1. In Table 3.13 we give for the various examples both the best (s^*, S^*) rules obtained by the Lagrangian approach and the approximate (s_a, S_a) rules obtained by using the economic order quantity formula (1.52) for $S - s$ and the equation (1.61) for the reorder point s. Also, we give in Table 3.13 the associated actual service levels $\beta(s^*, S^*)$ and $\beta(s_a, S_a)$, the associated actual average costs $g(s^*, S^*)$ and $g(s_a, S_a)$ and the relative error percentage in costs

$$\text{Error}\% = \frac{100 \times \{g(s_a, S_a) - g(s^*, S^*)\}}{g(s^*, S^*)}\%.$$

Our numerical experiments indicate that in many practical inventory applications the approximate (s_a, S_a) rule performs quite well for the specific cost structure considered. Also, the table illustrates the important practical conclusion that the optimal value $S^* - s^*$ of the lot size depends mainly on the holding and ordering costs and only to a slight degree on the required service level, provided the grade of customer service is sufficiently high. Further, it may be interesting to note that the optimal lot size shows the tendency to increase when the lead time gets larger. □

We end this section by developing a tailor-made policy-iteration algorithm for a queueing control problem in which the queue size is controlled by varying the service rate. The next example is closely related to example 3.8 which was solved by value-iteration.

Example 3.11 A queueing control problem with a variable service rate

Messages arrive at one of the outgoing communication lines in a message-switching centre according to a Poisson process with rate λ. The message length is exponentially distributed with mean $1/\mu$. An arriving message, finding the communication line idle, is provided with service immediately; otherwise the message is stored in a buffer until access to the line can be given. The communication line is only able to transmit one message at a time, but has available two possible transmission rates σ_1 and σ_2 with $0 < \sigma_1 < \sigma_2$. It is assumed that the faster transmission rate σ_2 is larger than the offered traffic λ/μ. At any time the line may switch from one transmission rate to the other. A fixed cost of $K > 0$ is incurred when switching from rate σ_1 to the faster rate σ_2. An operating cost at rate $r_i > 0$ is incurred when the line is transmitting a message using rate σ_i, while an operating cost at rate $r_0 \geqslant 0$ is incurred when the line is idle. For each message a holding cost of $h > 0$ is incurred for each unit of time the message is in the system until its transmission is completed. The goal is to find the best rule within the class of easily implementable (i_1, i_2) rules with $0 \leqslant i_2 < i_1$. Under the two-critical numbers rule (i_1, i_2) the communication line switches from transmission rate σ_1 to σ_2 when the number of messages in the system reaches the level i_1, and switches from rate σ_2 to rate σ_1 again when the number of messages in the system reduces to the level i_2.

The problem of controlling the transmission rate as a function of the number of messages present is a semi-Markov decision problem in which the decision epochs are given by the arrival epochs and the transmission completion epochs. Since the message length is assumed to be exponentially distributed, we have by the lack of memory of the exponential distribution that the state of the system at any time is described by the pair (i, k) with

$i =$ the number of messages in the system,

$k = 1$ (2) if the line is adjusted to the transmission rate $\sigma_1 (\sigma_2)$.

The number of possible states is unbounded. However, this will offer no computational problems when using the embedding technique. To develop a tailor-made policy-iteration algorithm operating on the class of (i_1, i_2) policies, we first show how embedding enables us to calculate easily the average cost and the relative values associated with an (i_1, i_2) policy. For a given rule R of the (i_1, i_2) type, the embedded system of linear equations (3.70) becomes quite simple when choosing as embedded set

$$E = \{(i_1, 1), (i_2, 2)\}.$$

We therefore need the following two results. First, the expectation of the time needed to reduce the number of messages in the system from i to i_2 when using the transmission rate σ_2 and the expected holding and operating costs incurred during this time are easily obtained from the general result (1.40) in section 1.5 of chapter 1. Noting that under the transmission rate σ_2 the (remaining) transmission time of a message is exponentially distributed with mean $1/\sigma_2\mu$, it readily follows from (1.40) with $n = i - i_2$ that

$$\tau_{(i,1)}^E(R) = \tau_{(i,2)}^E(R), \quad i > i_1 \quad \text{and} \quad \tau_{(i,2)}^E(R) = \frac{i - i_2}{\sigma_2\mu - \lambda}, \quad i > i_2,$$

$$c_{(i,1)}^E(R) = K + c_{(i,2)}^E(R), \quad i \geqslant i_1,$$

and

$$c^E_{(i,2)}(R) = (hi_2 + r_2)\tau^E_{(i,2)}(R) + \frac{h\sigma_2\mu}{\sigma_2\mu - \lambda}$$

$$\times \left\{ \frac{1}{2\sigma_2\mu}(i - i_2)(i - i_2 - 1) + \frac{i - i_2}{\sigma_2\mu} + \frac{\lambda(i - i_2)}{\sigma_2\mu(\sigma_2\mu - \lambda)} \right\}$$

$$= \frac{(i - i_2)}{\sigma_2\mu - \lambda} \left\{ \tfrac{1}{2}h(i + i_2 + 1) + \frac{h\lambda}{\sigma_2\mu - \lambda} + r_2 \right\} \qquad i > i_2.$$

Second, for the case of Poisson arrivals and exponentially distributed message lengths, the calculation of the expected length of the first-passage time until reaching the level i_1 for the number of messages present when using the transmission rate σ_1 and the calculation of the expected holding and operating costs incurred during this time are both standard problems in probability theory. For ease of notation, put

$$\tau(i) = \tau^E_{(i,1)}(R) \qquad \text{and} \qquad c(i) = c^E_{(i,1)}(R), \qquad 0 \leqslant i < i_1.$$

Using the properties of the exponential distribution given in (1.29) in section 1.4 of chapter 1, we find by conditioning arguments the recurrence relations

$$\tau(i) = \frac{1}{\lambda + \sigma_1\mu} + \frac{\sigma_1\mu}{\lambda + \sigma_1\mu}\tau(i - 1) + \frac{\lambda}{\lambda + \sigma_1\mu}\tau(i + 1), \qquad 1 \leqslant i < i_1,$$

$$c(i) = \frac{hi + r_1}{\lambda + \sigma_1\mu} + \frac{\sigma_1\mu}{\lambda + \sigma_1\mu}c(i - 1) + \frac{\lambda}{\lambda + \sigma_1\mu}c(i + 1), \qquad 1 \leqslant i < i_1,$$

with the boundary conditions

$$\tau(0) = \frac{1}{\lambda} + \tau(1), \qquad \tau(i_1) = 0, \qquad c(0) = \frac{r_0}{\lambda} + c(1) \qquad \text{and} \quad c(i_1) = 0.$$

The solutions of these linear difference equations are standard to derive (see, for example, Miller, 1968). Omitting the tedious details of the derivation, we state only the ultimate results

$$\tau(i) = \frac{1}{(\lambda - \sigma_1\mu)^2}\left[(\lambda - \sigma_1\mu)(i_1 - i) + \sigma_1\mu\left\{ \left(\frac{\sigma_1\mu}{\lambda}\right)^{i_1} - \left(\frac{\sigma_1\mu}{\lambda}\right)^i \right\} \right], 0 \leqslant i < i_1, \quad (3.86)$$

provided $\lambda \neq \sigma_1\mu$; otherwise $\tau(i) = (2\lambda)^{-1}\{i_1(i_1 + 1) - i(i + 1)\}$. Also,

$$c(i) = \frac{1}{(\lambda - \sigma_1\mu)^3}\left[\tfrac{1}{2}h(\lambda - \sigma_1\mu)^2(i_1^2 - i^2) + \{r_1(\lambda - \sigma_1\mu)^2 \right.$$

$$- \tfrac{1}{2}h(\lambda + \sigma_1\mu)(\lambda - \sigma_1\mu)\}(i_1 - i) + \{-r_0(\lambda - \sigma_1\mu)^2 + r_1\lambda(\lambda - \sigma_1\mu)$$

$$\left. - h\lambda\sigma_1\mu\}\left\{ \left(\frac{\sigma_1\mu}{\lambda}\right)^{i_1} - \left(\frac{\sigma_1\mu}{\lambda}\right)^i \right\} \right], \qquad 0 \leqslant i < i_1, \qquad (3.87)$$

provided $\lambda \neq \sigma_1 \mu$; otherwise $c(i) = \lambda^{-1}\{h(i_1^3 - i^3)/6 + r_1(i_1^2 - i^2)/2 + (r_0 - r_1/2 - h/6)(i_1 - i)\}$.

This completes the specification of $\tau_{(i,1)}^E(R)$ and $c_{(i,1)}^E(R)$. Note that

$$\tau_{(i,2)}^E(R) = \tau_{(i,1)}^E(R) \qquad \text{and} \qquad c_{(i,2)}^E(R) = c_{(i,1)}^E(R), \qquad 0 \leqslant i \leqslant i_2.$$

The one-step transition probabilities $p_{st}^E(R)$ associated with the embedded Markov chain are trivial to specify, since the embedded Markov chain assumes alternately the states $(i_1, 1)$ and $(i_2, 2)$. Hence the embedded linear equations (3.71) have the simple form

$$v_{(i_1,1)}(R) + g(R)\frac{i_1 - i_2}{\sigma_2\mu - \lambda} - v_{(i_2,2)}(R) = K + \frac{i_1 - i_2}{\sigma_2\mu - \lambda}$$

$$\times \left\{ \tfrac{1}{2}h(i_1 + i_2 + 1) + \frac{h\lambda}{\sigma_2\mu - \lambda} + r_2 \right\},$$

$$v_{(i_2,2)}(R) + g(R)\tau(i_2) - v_{(i_1,1)}(R) = c(i_2).$$

Normalizing

$$v_{(i_2,2)}(R) = 0$$

yields

$$g(R) = \frac{K + (\sigma_2\mu - \lambda)^{-1}(i_1 - i_2)\{\tfrac{1}{2}h(i_1 + i_2 + 1) + h\lambda/(\sigma_2\mu - \lambda) + r_2\} + c(i_2)}{(i_1 - i_2)/(\sigma_2\mu - \lambda) + \tau(i_2)}$$

(3.88)

and

$$v_{(i_1,1)}(R) = -c(i_2) + g(R)\tau(i_2). \tag{3.89}$$

Next, the other relative values associated with policy R can be obtained from (3.70) by single-pass calculations. We find

$$v_{(i,1)}(R) = \begin{cases} c(i) - g(R)\tau(i) + v_{(i_1,1)}(R) & \text{for} \quad 0 \leqslant i < i_1, \\ K + v_{(i,2)}(R) & \text{for} \quad i \geqslant i_1, \end{cases}$$

(3.90)

and

$$v_{(i,2)}(R) = \begin{cases} (\sigma_2\mu - \lambda)^{-1}(i - i_2)\left\{\tfrac{1}{2}h(i + i_2 + 1) + \dfrac{h\lambda}{\sigma_2\mu - \lambda} + r_2\right\} & \text{(3.91)} \\ \quad - g(R)(i - i_2)(\sigma_2\mu - \lambda)^{-1} & \text{for } i > i_2, \\ v_{(i,1)}(R) & \text{for } 0 \leqslant i \leqslant i_2. \end{cases}$$

Next we describe how to design the policy-improvement procedure such that the resulting new policy has the same form as the current (i_1, i_2) policy. Therefore we first specify the policy-improvement quantity $T_R(x,a) = c_x(a) - g(R)\tau_x(a) + \sum_y p_{xy}(a)v_y(R)$ with action a denoting the transmission rate to be used. The reader may easily verify that

$$T_R((i,1);1) = \frac{hi + r_1}{\lambda + \sigma_1\mu} - \frac{g(R)}{\lambda + \sigma_1\mu} + \frac{\sigma_1\mu}{\lambda + \sigma_1\mu}v_{(i-1,1)}(R)$$

$$+ \frac{\lambda}{\lambda + \sigma_1\mu}v_{(i+1,1)}(R), \qquad i \geqslant i_1,$$

$$T_R((i,1);2) = K + v_{(i,2)}(R), \qquad i_2 < i < i_1,$$

$$T_R((i,2);2) = \frac{hi + r_2}{\lambda + \sigma_2\mu} - \frac{g(R)}{\lambda + \sigma_2\mu} + \frac{\sigma_2\mu}{\lambda + \sigma_2\mu}v_{(i-1,2)}(R)$$

$$+ \frac{\lambda}{\lambda + \sigma_2\mu}v_{(i+1,2)}(R), \qquad 1 \leqslant i \leqslant i_2,$$

$$T_R((i,2);1) = \begin{cases} v_{(i,1)}(R), & i_2 < i < i_1, \\ \dfrac{hi + r_1}{\lambda + \sigma_1\mu} - \dfrac{g(R)}{\lambda + \sigma_1\mu} + \dfrac{\sigma_1\mu}{\lambda + \sigma_1\mu}v_{(i-1,1)}(R) \\ \quad + \dfrac{\lambda}{\lambda + \sigma_1\mu}v_{(i+1,1)}(R), & i \geqslant i_1. \end{cases}$$

Using this specification of the policy-improvement quantity, it is now easy to design a policy-improvement procedure such that the newly constructed policy has again the (i_1, i_2) form. We first adjust the switching level i_1 such that the policy-improvement quantity is strictly less than the relative value for each of the states $(i, 1)$, with i between the current value of i_1 and the new value of i_1, and next we adjust the switching level i_2 in a similar way. This results in the following tailor-made policy-iteration algorithm.

Algorithm

Step 0. Initialize an (i_1, i_2) policy with $0 \leqslant i_2 < i_1$.

Step 1. For the current policy $R = (i_1, i_2)$, compute the average cost $g(R)$ and the relative value $v_{(i_1,1)}(R)$ from (3.88) and (3.89) with the quantities $\tau(i_2)$ and $c(i_2)$ to be calculated from (3.86) and (3.87).

Step 2. Determine i_1' as the largest integer with $i_1' > i_1$ such that

$$T_R((i,1);1) < v_{(i,1)}(R) \qquad \text{for} \quad i_1 \leqslant i < i_1'.$$

If such an integer i_1' exists, let $\bar{i}_1 = i_1'$. Otherwise, determine i_1'' as the smallest integer with $i_2 < i_1'' < i_1$ such that

$$T_R((i,1);2) < v_{(i,1)}(R) \qquad \text{for} \quad i_1'' \leqslant i < i_1,$$

and let $\bar{\imath}_1 = i_1''$ provided such an integer i_1'' exists, and otherwise let $\bar{\imath}_1 = i_1$. Next determine i_2' as the smallest integer with $0 \leqslant i_2' < i_2$ such that

$$T_R((i,2);2) < v_{(i,2)}(R) \qquad \text{for} \qquad i_2' < i \leqslant i_2.$$

If such an integer i_2' exists, define $\bar{\imath}_2 = i_2'$. Otherwise, determine i_2'' as the largest integer with $i_2 < i_2'' < \bar{\imath}_1$ such that

$$T_R((i,2);1) < v_{(i,2)}(R) \qquad \text{for} \qquad i_2 < i \leqslant i_2'',$$

and let $\bar{\imath}_2 = i_2''$ provided such an integer i_2'' exists, and otherwise let $\bar{\imath}_2 = i_2$. The relative values required in this step can be obtained from (3.90) and (3.91) by single-pass calculations.

Step 3. If the new policy $\bar{R} = (\bar{\imath}_1, \bar{\imath}_2)$ is the same as the previous policy $R = (i_1, i_2)$, then the algorithm is stopped; otherwise, go to step 1 with R replaced by $\bar{R}$.

It is our conjecture that this algorithm will always converge after finitely many iterations to an overall average cost optimal policy. In support of this conjecture are extensive numerical investigations; in all cases tested we verified numerically that the average cost optimality equation was satisfied for the obtained (i_1, i_2) policy. As an illustration we give some numerical results in Table 3.14. We assume the numerical data

$$\mu = 1, \qquad \sigma_1 = 1, \qquad \sigma_2 = 1.5 \qquad \text{and} \qquad h = 1.$$

The arrival rate λ is varied as 0.8, 0.9, 1.0, 1.1 and 1.2, and the fixed cost K has the two values 0 and 25. For each of the various cases in Table 3.14 we give the average cost optimal (i_1^*, i_2^*) policy and the minimal average costs $g(i_1^*, i_2^*)$, together with the number # of iterations required by the algorithm. In all cases the algorithm was started with the policy $(i_1, i_2) = (1, 0)$. The results in Table 3.14 confirm again the remarkably fast convergence of policy-iteration. □

Table 3.14 Numerical results obtained by the special-purpose policy-iteration method

	$K = 0$			$K = 25$		
λ	(i_1^*, i_2^*)	$g(i_1^*, i_2^*)$	#	(i_1^*, i_2^*)	$g(i_1^*, i_2^*)$	#
0.8	(10, 9)	7.656	5	(14, 7)	7.756	5
0.9	(8, 7)	1.014	5	(12, 5)	1.047	6
1.0	(7, 6)	1.322	5	(10, 3)	1.384	6
1.1	(6, 5)	1.684	5	(9, 2)	1.771	6
1.2	(5, 4)	2.108	5	(8, 2)	2.210	6

EXERCISES

3.1 Consider a periodic review production/inventory problem where the demands for a single product in the successive weeks are independent random variables with a common discrete probability distribution $\{\phi(j), j = 0,...,N\}$. Any demand in excess of on-hand inventory is lost. At the beginning of each week it has to be decided whether to start a production run or not. The lot size of each production run consists of a fixed number of Q units. The production lead time is one week so that a batch delivery of the entire lot occurs at the beginning of the next week. Due to capacity restrictions on the inventory a production run is never started when the on-hand inventory is greater than M. The following costs are involved. A fixed setup cost of $K > 0$ is incurred for a new production run started after the production facility has been idle for some time. The holding costs incurred during a week are proportional to the on-hand inventory at the end of that week, where $h > 0$ is the proportionality constant. A fixed lost-sales cost of $\pi_0 > 0$ is incurred for each unit of excess demand. Formulate the problem of finding an average cost optimal production rule as a Markov decision problem.

3.2 Consider exercise 2.3 in chapter 2. Verify by policy-iteration or linear programming that the control rule given in this exercise is average cost optimal.

3.3 A stamp machine produces six-cornered plates of the following form:

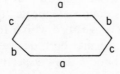

The machine has three pairs of adjustable knives. In the picture these pairs are denoted by a, b and c. Each pair of knives can fall from the correct position during the stamping of a plate. The following five situations can occur:
(1) all the three pairs have the correct position, (2) only the pairs b and c have the correct position, (3) only the pair b has the correct position, (4) only the pair c has the correct position and (5) no pair has the correct position. The probabilities q_{ij} that during a stamping a change from situation i to situation j occurs are given by

$$(q_{ij}) = \begin{pmatrix} \frac{3}{4} & \frac{1}{4} & 0 & 0 & 0 \\ 0 & \frac{1}{2} & \frac{1}{4} & \frac{1}{4} & 0 \\ 0 & 0 & \frac{3}{4} & 0 & \frac{1}{4} \\ 0 & 0 & 0 & \frac{1}{2} & \frac{1}{2} \\ 0 & 0 & 0 & 0 & 1 \end{pmatrix}.$$

After each stamping it is possible to adjust the machine such that all pairs of knives have again the correct position. The following costs are involved. The cost of bringing all pairs of knives in the correct position is 10. Each plate produced when j pairs of knives have the wrong position involves an adjustment cost of $4j$. Compute by policy-iteration or linear programming a maintenance rule which minimizes the average cost per stamping. (*Answer*: always adjust the machine when not all of the pairs of knives have the correct position; the minimal average cost per stamping is 3.5.)

3.4 An electronic equipment having two identical devices is inspected every day. Redundancy has been built into the system so that the system is still operating if only one device works. The system goes down when both devices are no longer working. The failure rate of a device depends both on its age and the condition of the other device. A device working successfully for m days will be failed the next day with respective probabilities $r_1(m)$ and $r_2(m)$ when the other device is currently being overhauled or is respectively working. It is assumed that both $r_1(m)$ and $r_2(m)$ are equal to 1 when m is sufficiently large. A device that is found in the failure state upon inspection has to be overhauled. An overhaul of a failed device takes T_0 days. Also, a preventive overhaul of a working device is possible, such an overhaul taking T_1 days. It is assumed that $1 \leqslant T_1 \leqslant T_0$. At each inspection it has to be decided to overhaul one or both the devices, or let them continue working for the next day. The goal is to minimize the long-run fraction of time the system is down. Formulate a value-iteration algorithm for the computation of an optimal maintenance rule. (*Hint*: define the states (i,j) $(i, -k)$ and $(-l, -k)$. The first states mean that both devices are working for i and j days respectively, the second state means that one device is working for i days and the other is being overhauled with a remaining overhaul time of k days, and the third state means that both devices are being overhauled with remaining overhaul times of l and k days respectively.)

3.5 Two furnaces in a steel works are used to produce pig-iron for working-up elsewhere in the factory. Each furnace needs overhauling from time to time because of failure during operation or to prevent such a failure. Assuming an appropriately chosen time unit, an overhaul of a furnace always takes a fixed number of L periods. The overhaul facility is capable of overhauling both furnaces simultaneously. A furnace just overhauled will operate successfully during i periods with probability q_i, $i = 1, \ldots, M$. A failed furnace has to be overhauled, otherwise there is the option of either a preventive overhaul or letting the furnace operate for the next period. Since other parts of the steel works are affected when not all furnaces are in action, a loss of revenue of $c(j)$ is incurred for each period during which j furnaces are out of action, $j = 1, 2$. No cost is incurred if both furnaces are working. Formulate a value-iteration algorithm for the computation of an average cost optimal overhauling policy. (This example is based on Stengos and Thomas, 1980.)

3.6 Consider a periodic review inventory system with two products that serve as substitutes for each other. The demands for product i in the successive periods are independent random variables with a common discrete probability distribution $\{\phi_i(k), k \geqslant 0\}$, $i = 1, 2$. At the end of each period unmet demand for one product is satisfied as much as possible from remaining stock (if any) of the other product. Any unmet demand that cannot be transferred is assumed to be lost. At the beginning of each period the opportunity exists to replenish the inventory of each of the products. Here there is a maximum inventory level M_i for product i, $i = 1, 2$. The lead time of any replenishment order is zero. A fixed setup cost of $K_i > 0$ is incurred for a replenishment order for product i. Also, for each unit of item i being in inventory at the beginning of a period a holding cost of $h_i > 0$ is incurred for that period, while a fixed penalty cost of $\pi_i > 0$ is incurred for each unit of lost demand for item i, $i = 1, 2$. Formulate a value-iteration algorithm for computing an average cost optimal policy. (*Note*: for ease of notation identify each decision with the stock levels just after any ordering.)

3.7 Every week a repairman server travels to customers in five towns on the successive working days of the week. The repairman visits Amsterdam (town 1) on Monday, Rotterdam (town 2) on Tuesday, Brussels (town 3) on Wednesday, Aachen (town 4) on Thursday and Arnhem (town 5) on Friday. In the different towns it may be necessary to replace a certain crucial piece of an electronic equipment rented by customers. The probability distribution of the number of replacements required at a visit to town j is given by $\{p_j(k), k \geqslant 0\}$ for $j = 1, \ldots, 5$. The numbers of required replacements on the successive days are independent of each other. The repairman is able to carry at most M spare parts. In case the number of spare parts the repairman carries is not enough to satisfy the demand in a town, another repairman has to be sent the next day to that town to complete the remaining replacements. The cost of such a special mission to town j is K_j. At the end of each day the repairman may decide to send for a replenishment of the spare parts to the town where the repairman is. The cost of sending such a replenishment to town j is a_j. Develop a value-iteration algorithm for the computation of an average cost optimal policy and indicate how to formulate converging lower and upper bounds on the minimal costs. Write a computer program and solve for the numerical data $M = 5$, $K_j = 200$ for all j, $a_1 = 60$, $a_2 = 30$, $a_3 = 50$, $a_4 = 25$, $a_5 = 100$, where the probabilities $p_j(k)$ are given as follows:

$k \backslash j$	1	2	3	4	5
0	0.5	0.25	0.375	0.3	0.5
1	0.3	0.5	0.375	0.5	0.25
2	0.2	0.25	0.25	0.2	0.25

(*Answer*: in town i the number of spare parts is raised up to the level 5 when the stock is at or below s_i and otherwise no replenishment is made, where $s_1 = s_2 = s_3 = 1$, $s_4 = 3$ and $s_5 = 0$ with average weekly costs of 39.07.)

3.8 Suppose now in example 3.1 of section 3.2 that the repair times are stochastic. A preventive repair takes either 0 or 2 days each with probability $\frac{1}{2}$, whereas a repair upon failure takes either 1, 2 or 3 days each with probability $\frac{1}{3}$. Compute by policy-iteration or linear programming an average cost optimal policy.

3.9 A cargo liner operates between the five harbours $A_1, \ldots, A_5$. A cargo shipment from harbour A_i to harbour A_j $(j \neq i)$ takes a random number τ_{ij} of days (including load and discharge) and yields a random payoff of ξ_{ij}. The shipment times τ_{ij} and the payoffs ξ_{ij} are normally distributed, with respective means $\mu(\tau_{ij})$ and $\mu(\xi_{ij})$ and respective standard deviations $\sigma(\tau_{ij})$ and $\sigma(\xi_{ij})$. We assume the following numerical data:

The values $\mu(\tau_{ij})\{\sigma(\tau_{ij})\}$

$i\backslash j$	1	2	3	4	5
1	—	$3(\frac{1}{2})$	6(1)	$3(\frac{1}{2})$	$2(\frac{1}{2})$
2	4(1)	—	$1(\frac{1}{4})$	7(1)	5(1)
3	5(1)	$1(\frac{1}{4})$	—	6(1)	8(1)
4	$3(\frac{1}{2})$	8(1)	5(1)	—	$2(\frac{1}{2})$
5	$2(\frac{1}{2})$	5(1)	9(1)	$2(\frac{1}{2})$	—

The values $\mu(\xi_{ij})\{\sigma(\xi_{ij})\}$

$i\backslash j$	1	2	3	4	5
1	—	8(1)	12(2)	6(1)	6(1)
2	20(3)	—	$2(\frac{1}{2})$	14(3)	16(2)
3	16(3)	$2(\frac{1}{2})$	—	18(3)	16(1)
4	6(1)	10(2)	20(2)	—	$6(\frac{1}{2})$
5	8(2)	16(3)	20(2)	8(1)	—

Compute by policy-iteration or linear programming a sailing route for which the long-run average reward per day is maximal. (*Answer*: $A_1 \to A_5 \to A_4 \to A_3 \to A_2 \to A_1$ with a maximal average reward of 4 per day.)

3.10 Consider exercise 1.18 of chapter 1 and assume that the project types $j = 1, \ldots, N$ are (re)numbered according to $E(\xi_j)/E(\tau_j) \geqslant E(\xi_{j+1})/E(\tau_{j+1})$ for all j. Use the policy-improvement procedure to show that the optimal acceptance set equals $A = \{1, \ldots, v\}$, where v is the smallest integer such that

$$\frac{\sum_{j=1}^{v} \lambda_j E(\xi_j)}{1 + \sum_{j=1}^{v} \lambda_j E(\tau_j)} > \frac{E(\xi_{v+1})}{E(\tau_{v+1})},$$

where $E(\xi_{N+1})/E(\tau_{N+1}) = 0$ by convention.

3.11 Consider a two-item continuous review inventory system with the possibility of coordinating replenishment orders for the items. The demand processes for items 1 and 2 are independent Poisson processes with respective demand rates λ_1 and λ_2. At each time a replenishment order for either one or both items can be placed. The lead time of any replenishment order is zero. No backlogging of excess demand is allowed so that a replenishment order for an item must be placed when the stock of that item becomes zero by a demand transaction. There is a maximum capacity C_i for the inventory of item i, $i = 1, 2$. The fixed cost of a replenishment order consisting only of item i is $K_i > 0$ ($i = 1, 2$), while the fixed cost of a replenishment order including both items is $K > 0$ with $K < K_1 + K_2$. For each unit of item i a holding cost of $h_i > 0$ is incurred per unit of time the unit is kept in stock. Formulate a value-iteration algorithm for the computation of an average cost optimal policy (be aware of possible periodicities!). Write a computer program and solve for the numerical data $\lambda_1 = 2$, $\lambda_2 = 3$, $C_1 = C_2 = 10$, $K_1 = K_2 = 10$, $K = 15$ and $h_1 = h_2 = 1$. (*Answer*: the minimal average cost is 1.733.)

3.12 Consider the control problem formulated at the end of example 2.10 in section 2.4 of chapter 2. Develop an effective value-iteration algorithm and write a computer program to verify that the optimal control rule is indeed as stated.

3.13 In example 3.6 of section 3.5, assume now $J = 3$ production lines. Using the approach with sparse transition matrices as discussed in example 3.7, develop a value-iteration algorithm for the computation of an average cost optimal policy. Write a computer program and solve for the numerical data $\mu_1 = \mu_2 = \mu_3 = 21$, $\lambda_1 = \lambda_2 = \lambda_3 = 6$, $N_1 = N_2 = N_3 = 4$, $\gamma_1 = 120$, $\gamma_2 = 370$ and $\gamma_3 = 210$. (*Answer*: the minimal average cost is 25.21.)

3.14 Consider a communication system with two processors for transmitting messages. Each of the two processors has its own waiting area, where the area for processor i consists of N_i waiting places (including one for the message (if any) being transmitted). No jockeying is allowed between the queues. Each processor is only able to transmit one message at a time, and the time needed for transmission of a message at processor i is exponentially distributed with mean $1/\mu_i$. The messages arrive at the system according to a renewal process with interarrival time density $a(x)$. A message finding upon arrival that both waiting areas are full is lost; otherwise the message is assigned to one of the processors having unoccupied waiting places. The goal is to find an assignment rule which minimizes the long-run fraction of messages that is lost per unit time. Develop a value-iteration algorithm for computing an optimal assignment rule. For the special case in which the messages arrive according to a Poisson process, formulate an alternative value-iteration algorithm with sparse transition matrices.

3.15 Suppose two independent queues that are attended by a single server. Units requesting service arrive at the ith queue according to a Poisson process with rate λ_i, and their service requirements have a general distribution with probability density $b_i(x)$. The ith queue has a finite waiting room of capacity N_i

(including any unit in service). An arrival at queue i finding that all of the waiting places there are occupied is lost. The server is only able to handle one queue at a time and provides non-preemptive service. Once the server has finished service of a unit at a particular queue, it has to be decided which queue to service next. It takes no time to switch from one queue to the other. The goal is to find a service rule minimizing the long-run fraction of arrivals that is lost. Develop a value-iteration algorithm for the computation of such a rule. For the special case in which the service requirements have an exponential distribution, formulate an alternative value-iteration algorithm with sparse transition matrices.

3.16 Suppose a service facility with two different groups of servers both able to handle requests for service that arrive according to a Poisson process with rate λ. The ith group consists of s_i identical servers and each server in the ith group has an exponentially distributed service time with mean $1/\mu_i$, $i = 1,2$. A request for service finding upon arrival that all of the servers are occupied is lost; otherwise exactly one free server has to be assigned to the request for service. A cost of γ is incurred for each request for service that is lost, and a cost of γ_i is incurred when assigning a server from the ith group to the request for service. Develop a value-iteration algorithm for the computation of an average cost optimal assignment rule.

3.17 In example 3.8 of section 3.5, suppose now that the requests for service are generated by a finite number of sources with the same average idle time $1/\eta$. That is each source generates new requests for service according to a Poisson process with rate η whenever that source has no requests waiting or being served at the service facility. Develop a value iteration algorithm for the computation of an average cost optimal control rule. Write a computer program and solve for the numerical data $N = 50$, $\eta = 0.16$, $s = 10$, $\mu = 1$, $r = 30$, $h = 10$ and $K = 10$ when $K(a,b) = K|a - b|$. (*Answer*: the minimal average cost is 299.98.)

3.18 Develop an effective value-iteration algorithm for the computation of an optimal acceptance/rejection rule for the control problem in exercise 2.26 of chapter 2 when a rejection cost of γ_i is incurred each time a customer of type i is turned away. Write a computer program and solve for the numerical data $c = 50$, $b_1 = 3$, $b_2 = 2$, $\lambda_1 = 10$, $\lambda_2 = 5$, $1/\mu_1 = 1$, $1/\mu_2 = 1$, $\gamma_1 = 3$, and $\gamma_2 = 1$. (*Answer*: the minimal average costs per unit time are 2.694.)

3.19 Consider the following modification to the control problem stated in exercise 2.32 of chapter 2. Suppose a fixed cost of $K > 0$ is incurred each time the slower communication line 2 is turned on and there is a holding cost of $h > 0$ per unit of time for each message in the system. Formulate a value-iteration algorithm for the computation of an average cost optimal policy. (*Hint*: define the states $(i,1,0)$, $(i,1,1)$, $(i,2,0)$, $(i,2,1)$ and $(i,2,2)$, where state $(i,k,0)$ corresponds to the situation that a new message arrives while i other messages are present and the lines 1 and k are on, and state (i,k,j) corresponds to the situation that a transmission at line j is completed while i messages remain in the system and the lines 1 and k are on. In the auxiliary states $(i,1,1)$, $(i,2,1)$ and $(i,2,0)$ we assume the fictitious decision of letting the system alone. Also, assume a sufficiently large integer M such that the decision of having both lines on is always made in the states $(i,1,0)$ and $(i,2,2)$ with $i \geqslant M$, and use the embedding approach followed in example 3.8 of section 3.5.)

3.20 Consider the control problem stated in exercise 2.32 of chapter 2. Suppose now that we wish to minimize the average time spent in the system by a message subject to the probabilistic constraint that the fraction of time during which the slower line 2 is used does not exceed a prespecified value α. Give a linear programming formulation for solving this problem.

3.21 Consider an equipment that gradually deteriorates in time. The system can be classified into one of the states $1, \ldots, N$, representing increasing degrees of deterioration. The holding time in each state i with $i \neq N$ is exponentially distributed with the same mean $1/\mu$, and upon completion of this holding time the system moves to the next state $i + 1$. The final state N is absorbing. At any time the system may be replaced by a spare system provided one is available. A replacement takes a negligible time and the new system has state 1. At most one spare system can be held in inventory. At any time a replenishment order for a spare system can be placed when no spare system is in inventory. The lead time of a replenishment order is constant and equals $L > 0$. The following costs are incurred. There is an operating cost of r_i per unit of time the system is in state i, $1 \leqslant i \leqslant N$. A holding cost of $h > 0$ is incurred for each unit of time a spare is kept in inventory. Also, there is a fixed replacement cost of $A_i > 0$ when replacing the system in state i. An intuitively reasonable control rule is the following (m, M) rule with $1 \leqslant m \leqslant M \leqslant N$. Under this rule a replenishment order for a spare is placed when the system reaches the state m while no spare is in inventory, and the system is replaced when the state is equal to or larger than M and a spare is available. Develop a special-purpose policy-iteration algorithm that operates on the class of (m, M) rules. (*Hint*: use exercise 1.20 of chapter 1 to specify the one-step costs.) (This problem is based on Kawai, 1983.)

BIBLIOGRAPHIC NOTES

The policy-iteration method for the discrete-time Markov decision model was developed in Howard (1960) and was extended to the semi-Markov decision model by De Cani (1964), Howard (1964) and Jewell (1963). A theoretical foundation to Howard's policy-iteration method was given in Blackwell (1962); see also Denardo and Fox (1968) and Veinott (1966). Linear programming formulations for the Markov decision model were first given by De Ghellinck (1960) and Manne (1960) and streamlined later by Denardo and Fox (1968), Derman (1970) and Hordijk and Kallenberg

(1979,1984). The computational usefulness of the value-iteration algorithm was greatly enlarged by Odoni (1969) and Hastings (1971) who introduced lower and upper bounds on the minimal average costs and on the average costs of the policies generated by the algorithm. These authors extended the original value-iteration bounds of MacQueen (1966) for the discounted cost case to the average cost case. The work of MacQueen (1966) influenced much research on value-iteration bounds; a good account of the many variants of value-iteration for the discounted cost case can be found in Hendrikx, Van Nunen and Wessels (1984) and Porteus (1980); see also Popyack (1985). Fundamental work on the convergence of value-iteration in the average cost case was done by Bather (1973), Schweitzer (1965), Schweitzer and Federgruen (1979) and White (1963), among others. The modified value-iteration algorithm with a dynamic relaxation factor is due to Popyack, Brown and White (1979); see also the closely related paper of Varaiya (1978). The data-transformation technique converting a semi-Markov decision model into an equivalent discrete-time Markov decision model was introduced in Schweitzer (1971); see also the related work of Lippman (1975) and Serfozo (1979) for continuous-time Markov decision chains. Special techniques for solving structured large-scale Markov decision problems include embedding, aggregation/disaggregation, and decomposition; the first technique is discussed in De Leve, Federgruen and Tijms (1977) and Tijms (1980), the second one in Heyman and Sobel (1984), Mendelssohn (1982), Schweitzer, Puterman and Kindle (1985) and Veugen, Van Der Wal and Wessels (1984), and the third one in Norman (1972) and Wijngaard (1979).

A survey of applications of Markov decision models is given in White (1985). The insurance application of example 3.3 is taken from Norman and Shearn (1980); see also Kolderman and Volgenant (1985). Example 3.4 dealing with a data-multiplexing problem with probabilistic constraints is inspired by Jain and Ross (1985). Another interesting application of Markovian control with probabilistic constraints can be found in Golabi, Kulkarni and Way (1982). The production control problem in example 3.6 comes from Seidman and Schweitzer (1984). The resource allocation problem in example 3.7 was motivated by Foschini and Gopinath (1983), but the solution method is new, cf. also Tijms and Eikeboom (1986). The replacement problem in example 3.9 is based on Tijms and Van Der Duyn Schouten (1985); see also Derman and Lieberman (1967), Hastings and Mello (1978), Kawai (1983) and Stengos and Thomas (1980) for other applications of Markov decision theory to replacement and maintenance. The treatment of the (s, S) inventory control model in example 3.10 is adapted from Federgruen, Groenevelt and Tijms (1984).

REFERENCES

Bather, J. (1973). 'Optimal decision procedures for finite Markov chains', Adv. Appl. Prob., 5, 521–540.

Bellman, R. (1957). *Dynamic Programming*, Princeton University Press, Princeton.

Blackwell, D. (1962). 'Discrete dynamic programming', *Ann. Math. Statist.*, **33**, 719–726.

De Cani, J. S. (1964). 'A dynamic programming algorithm for embedded Markov chains when the planning horizon is at infinity', *Management Sci.*, **10**, 716–733.

De Ghellinck, G. (1960). 'Les problèmes de decisions sequentielles', *Cahiers Centre Etudes Recherche Opér.*, **2**, 161–179.

De Leve, G., Federgruen, A., and Tijms, H. C. (1977). 'A general Markov decision method, I: model and method, II: applications', *Adv. Appl. Prob.*, **9**, 296–315, 316–335.

Denardo, E. V. (1982). *Dynamic Programming*, Prentice-Hall, Englewood Cliffs, New Jersey.

Denardo, E. V., and Fox, B. L. (1968). 'Multichain Markov renewal programs', *SIAM J. Appl. Math.*, **16**, 468–487.

Derman, C. (1970). *Finite State Markovian Decision Processes*, Academic Press, New York.

Derman, C., and Lieberman, G. J. (1967). 'A Markovian decision model for a joint replacement and stocking problem', *Management Sci.*, **13**, 609–617.

Federgruen, A., Groenevelt, H., and Tijms, H. C. (1984). 'Coordinated replenishments in a multi-item inventory system with compound Poisson demands', *Management Sci.*, **30**, 344–357.

Federgruen, A., and Schechner, Z. (1983). 'Cost formulae for continuous review inventory models with fixed delivery lags', *Operat. Res.*, **31**, 957–965.

Federgruen, A., and Zipkin, P. (1984). 'An efficient algorithm for computing optimal (s, S) policies', *Operat. Res.*, **32**, 1268–1285.

Foschini, G. J., and Gopinath, B. (1983). 'Sharing memory optimally', *IEEE Trans. Commun.*, **31**, 352–359.

Fox, B. L., and Landi, D. M. (1968). 'An algorithm for identifying the ergodic subchains and transient states of a stochastic matrix', *Commun. ACM*, **11**, 619–621.

Golabi, K., Kulkarni, R. B., and Way, C. B. (1982). 'A statewide pavement management system', *Interfaces*, **12**, no. 6, 5–21.

Hastings, N. A. J. (1971). 'Bounds on the gain of a Markov decision process', *Operat. Res.*, **19**, 240–244.

Hastings, N. A. J., and Mello, J. C. M. (1978). *Decision Networks*, Wiley, New York.

Hendrikx, M., Van Nunen, J. A. E. E., and Wessels, J. (1984). 'Some notes on iterative optimization of structured Markov decision processes with discounted rewards', *Math. Operationsforsch. u. Statist.*, **15**, 439–459.

Heyman, D. P., and Sobel, M. J. (1984). *Stochastic Models in Operations Research*, Vol. II, McGraw-Hill, New York.

Hordijk, A., and Kallenberg, L. C. M. (1979). 'Linear programming and Markov decision chains', *Management Sci.*, **25**, 352–362.

Hordijk, A., and Kallenberg, L. C. M. (1984). 'Constrained undiscounted stochastic dynamic programming', *Math. Operat. Res.*, **9**, 276–289.

Howard, R. A. (1960). *Dynamic Programming and Markov Processes*, Wiley, New York.

Howard, R. A. (1964). 'Research in semi-Markovian decision structures', *J. Oper. Res. Soc. Japan*, **6**, 163–199.

Jain, P., and Ross, K. W. (1985). 'Optimal priority assignment with hard constraint', INRIA Report No. 459, INRIA, Valbonne.

Jewell (1963), 'Markov-renewal programming: I and II', *Operat. Res.*, **11**, 938–971.

Kawai, H. (1983). 'An optimal ordering and replacement policy of a Markovian degradation system under complete observation, Part I', *J. Operat. Res. Soc. Japan*, **26**, 279–290.

Kolderman, J., and Volgenant, A. (1985). 'Optimal claiming in an automobile insurance system with bonus-malus structure', *J. Operat. Res. Soc.*, **36**, 239–247.

Lippman, S. A. (1975). 'Applying a new device in the optimization of exponential queueing systems', *Operat. Res.*, **23**, 687–710.

MacQueen, J. (1966). 'A modified dynamic programming method for Markovian decision problems', *J. Math. Appl. Math.*, **14**, 38–43.

Manne, A. (1960). 'Linear programming and sequential decisions', *Management Sci.*, **6**, 259–267.

Mendelssohn, R. (1982). 'An iterative aggregation procedure for Markov decision processes', *Operat. Res.*, **30**, 62–73.

Miller, K. S. (1968). *Linear Difference Equations*, W.A. Benjamin Inc., New York.

Norman, J. M. (1972). *Heuristic Procedures in Dynamic Programming*, Manchester University Press, Manchester.

Norman, J. M. and Shearn, D. C. S. (1980). 'Optimal claiming on vehicle insurance revisited', *J. Operat. Res. Soc.*, **31**, 181–186.

Odoni, A. (1969). 'On finding the maximal gain for Markov decision processes', *Operat. Res.*, **17**, 857–860.

Popyack, J. L. (1985). 'Accelerated convergence for successive approximations in Markov decision processes', Research Report, Drexel University, Philadelphia.

Popyack, J. L., Brown, R. L., and White, III, C. C. (1979). 'Discrete versions of an algorithm due to Varaiya', *IEEE Trans. Automat. Contr.*, **24**, 503–504.

Porteus, E. L. (1980). 'An overview of iterative methods for discounted finite Markov and semi-Markov decision chains', in *Recent developments in Markov Decision Processes* (Eds. R. Hartley, L. C. Thomas and D. J. White), pp. 1–20, Academic Press, New York.

Ross, S. M. (1970). *Applied Probability Models with Optimization Applications*, Holden-Day, San Francisco.

Ross, S. M. (1983). *Stochastic Processes*, Wiley, New York.

Schweitzer, P. J. (1965). *Perturbation Theory and Markovian Decision Processes*, Ph.D. dissertation, Massachusetts Institute of Technology.

Schweitzer, P. J. (1971). 'Iterative solution of the functional equations of undiscounted Markov renewal programming', *J. Math. Anal. Appl.*, **34**, 495–501.

Schweitzer, P. J., and Federgruen, A. (1979). 'Geometric convergence of value-iteration in multichain Markov decision problems', *Adv. Appl. Prob.*, **11**, 188–217.

Schweitzer, P. J., Puterman, M. L., and Kindle, K. W. (1985). 'Iterative aggregation-disaggregation procedures for solving discounted semi-Markovian reward processes', *Operat. Res.*, **33**, 589–605.

Seidman, A., and Schweitzer, P. J. (1984). 'Part selection policy for a flexible manufacturing cell feeding several production lines', *IIE Trans.*, **16**, 355–362.

Serfozo, R. (1979). 'An equivalence between continuous and discrete time Markov decision processes', *Operat. Res.*, **27**, 616–620.

Stengos, D., and Thomas, L. C. (1980). 'The blast furnaces problem', *European J. Operat. Res.*, **4**, 330–336.

Su. Y., and Deininger, R. (1972). 'Generalization of White's method of successive approximations to periodic Markovian decision processes', *Operat. Res.*, **20**, 318–326.

Tijms, H. C. (1980). 'An algorithm for average cost denumerable state semi-Markov decision problems with applications to controlled production and queueing systems', in *Recent Developments in Markov Decision Processes* (Eds. R. Hartley, L.C. Thomas and D.J. White), pp. 143–179, Academic Press, New York.

Tijms, H. C., and Eikeboom, A. M. (1986). 'A simple technique in Markovian control with applications to resource allocation in communication networks, *Operat. Res. Letters*, **5** (to appear).

Tijms, H. C., and Van Der Duyn Schouten, F. A. (1985). 'A Markov decision algorithm for optimal inspections and revisions in a maintenance system with partial information', *European J. Operat. Res.*, **21**, 245–253.

Varaiya, P. (1978). 'Optimal and suboptimal stationary controls for Markov chains', *IEEE Trans. Automat. Contr.*, **23**, 388–394.

Veinott, A. F., Jr. (1966). 'On finding optimal policies in discrete dynamic programming with no discounting', *Ann. Math. Statist.*, **37**, 1284–1294.

Veinott, A. F., Jr., and Wagner, H. M. (1965). 'Computing optimal (s, S) policies', *Management Sci.*, **11**, 522–555.

Veugen, L. M. M., Van Der Wal, J., and Wessels, J. (1984). 'Aggregation and dis-aggregation in Markov decision models for inventory control', *European J. Operat. Res.*, **20**, 248–254.

Wagner, H. M. (1975). *Principles of Operations Research*, 2nd ed., Prentice-Hall, Englewood Cliffs, New Jersey.

White, D. J. (1963). 'Dynamic programming, Markov chains, and the method of successive approximations', *J. Math. Anal. Appl.*, **6**, 373–376.

White, D. J. (1985). 'Real applications of Markov decision processes', *Interfaces*, **15**, no. 6 (to appear).

Wijngaard, J. (1979). 'Decomposition for dynamic programming in production and inventory control', *Engineering and Process Econom.*, **4**, 385–388.

Zijm, W. H. M. (1986). 'Asymptotic expansions for dynamic programming recursions with general nonnegative matrices', *J. Optim. Theory and Appl.*, **48** (to appear).

CHAPTER 4

Algorithms and approximations for queueing models

4.0 INTRODUCTION

Queueing models have their origin in the study of design problems of automatic telephone exchanges and were first analysed by the queueing pioneer A.K. Erlang in the early 1900s. In planning telephone systems to meet given performance criteria, questions were asked such as 'how many lines are required in order to give a certain grade of service?' or 'what is the probability that a delayed customer has to wait more than a certain time before getting a connection?' Similar questions arise in the design of many other systems: 'how many terminals are needed in a computer system in order to keep the probability of wait of a user below a prespecified value?', 'what will be the effect on the average waiting time of customers when changing the size of a maintenance staff to service leased equipment?', 'how much storage space is needed in buffers at work stations in an assembly line in order to keep the probability of blocking below a specified acceptable level?'.

These design problems and many others concern in fact facilities serving a community of users, where both the times at which the users ask for service and the lengths of the times that the requests for service will occupy facilities are stochastic, so that inevitably congestion occurs and queues may build up. In the first stage of design, the system engineer usually needs quick answers to a variety of questions as posed above. Queueing theory constitutes a basic tool for making first-approximation estimates of queue sizes and probabilities of delays. Such a simple tool should in general be preferred to simulation, especially when a large number of different configurations in the design problem is possible.

In this chapter we discuss a number of basic queueing models that have proved to be useful in analysing a wide variety of stochastic service systems. The emphasis will be on algorithms and approximations rather than on mathematical aspects. We feel that there is a need of such a treatment in view of the increased use of queueing models in modern technology. Actually, the application of queueing theory in the performance analysis of computer and communication systems has stimulated much practically oriented research on computational aspects of queueing models. It is to these aspects that the present chapter is addressed. Here considerable attention is paid to robustness results. While it was seen in chapter 2 that many loss systems (no access of arrivals finding all servers busy) are exactly or nearly insensitive to the distributional form of the service time except for its first moment, it will be demonstrated in this chapter that many delay systems (full access of arrivals) and many delay-loss systems (limited access of arrivals) allow for two-moment approximations.

In section 4.1 we introduce some basic results including the fundamental formula of Little and relationships between time-average and customer-average characteristics of queueing systems. Section 4.2 discusses algorithms for computing the state probabilities and waiting-time probabilities in single-server queueing models with full access of arrivals, where both state-dependent Markovian input and renewal input are considered and the service times are generally distributed. The algorithms discussed include a simple recursion method based on busy cycle analysis from regenerative processes and effective Markov chain methods using the powerful technique of replacing the service-time distribution by one built out of exponentially distributed components. In section 4.3 we consider several single-server queueing models with Poisson input and having a limit to either the number of customers or work in the system. For these models having applications among others in buffer design problems, we obtain tractable exact results for the special cases of deterministic and exponential service times and practically useful approximations for the case of general service times. These approximations include simple two-moment approximations based on a linear interpolation on the squared coefficient of variation of the service time. In many situations such a two-moment approximation turns out to be extremely useful. The final section 4.4 deals with multi-server queueing models. In general multi-server queues are difficult to solve exactly and therefore the emphasis is on approximations for various measures of system performance such as the delay probability and the average queue size. In particular, the multi-server queue with Poisson arrivals and general service times is discussed in detail for both cases of full access and limited access of arrivals. Throughout this chapter many numerical results are presented in order to provide insight into the performance of the solution methods.

4.1 FUNDAMENTAL RELATIONS FOR QUEUEING SYSTEMS

In this section we discuss a number of fundamental relations for queueing systems. We consider queueing systems in isolation, that is the customers require only a single server and leave the system after having received their service. However, the fundamental results below may also be useful for networks of queueing systems.

We first give Kendall's notation for a wide range of standard queueing systems. Here it is assumed that customers arrive singly and are served singly. A queueing system having waiting room for an unlimited number of customers can be described by a three-part code a/b/c. The first symbol a specifies the interarrival time distribution, the second symbol b specifies the service time distribution and the third symbol c specifies the number of servers. Some examples of Kendall's shorthand notation are:

1. M/G/1—Poisson (Markovian) input, General service time distribution, 1 server.
2. M/D/c—Poisson input, Deterministic service times, c servers.
3. GI/M/c—General, Independently distributed interarrival times, exponential (Markovian) service times, c servers.
4. GI/G/c—General, Independently distributed interarrival times, General service time distribution, c servers.

The above notation can be extended to cover other queueing systems. For example, queueing systems having waiting room only for N customers (including those in service) are often abbreviated by a four-part code a/b/c/N.

In queueing analysis it is often convenient to approximate the interarrival time (or service time) distribution by one that is built out of a finite sum or a finite mixture of exponentially distributed components, or a combination of both. This approach is usually referred to as the method of phases (stages). More generally, one might consider the so-called phase-type distributions to be interpreted as first-passage time distributions in a continuous-time Markov chain (cf. Neuts, 1981). However, for practical purposes it usually suffices to work with special phase-type distributions such as finite mixtures of Erlangian distributions with the same scale parameters and hyperexponential distributions. The hyperexponential distribution, being a mixture of exponential distributions with different means, has always a coefficient of variation greater than or equal to 1 and has a decreasing density. This distribution is particularly suited to model irregular interarrival (service) times having the feature that most outcomes tend to be small and large outcomes occur only occasionally. The set of mixtures of Erlangian distributions with the same scale parameters is much more versatile than the set of hyperexponential distributions and allows us to cover any positive value of the coefficient of variation. In particular, mixtures of E_{k-1} and E_k distributions with the same scale parameters may be used to represent regular interarrival (service) times having a coefficient of variation smaller than or equal to 1.

The theoretical basis for the use of mixtures of Erlangian distributions with the same scale parameters is provided by the following result. Let $F(t)$ be a probability distribution function with mass on $(0, \infty)$. Define for fixed $\Delta > 0$ the probability distribution function $F_\Delta(t)$ by

$$F_\Delta(t) = \sum_{k=1}^{\infty} p_{k,\Delta} \left\{ 1 - \sum_{j=0}^{k-1} e^{-t/\Delta} \frac{(t/\Delta)^j}{j!} \right\}, \qquad t \geq 0, \qquad (4.1)$$

where $p_{k,\Delta} = F(k\Delta) - F((k-1)\Delta), k = 1, 2, \ldots$. Then the following limiting result can be proved (see Schassberger, 1973):

$$\lim_{\Delta \to 0} F_\Delta(t) = F(t)$$

for every continuity point t of the distribution function F. In words, noting that the term between brackets in (4.1) is the probability distribution function of the sum of k independent exponentials with common mean Δ, the limiting result states that each positive random variable can be approximated arbitrarily closely by a random sum of independently exponentially distributed phases with the same means. This explains why finite mixtures of Erlangian distributions with the same scale parameters are widely used for queueing calculations.

The GI/G/c queue

It is convenient to consider first the GI/G/c queue when discussing some fundamental results for queueing systems. Thus we assume a multi-server queueing system with $c \geqslant 1$ identical servers where customers arrive according to positive independent interarrival times having a common probability distribution function $A(t)$. The service times of the customers are independent random variables having a common probability distribution function $B(t)$ and are also independent of the arrival process. A customer who finds upon arrival at least one of the servers free immediately enters service at a randomly chosen free server; otherwise the customer waits in queue until served. The queue discipline specifying which customer is to be served next is first-come–first-served unless stated otherwise. A server cannot be idle when customers are waiting in queue and a busy server works at unity rate. A customer leaves the system upon service completion.

We first introduce some notation. Let

$$\lambda = \text{the average arrival rate of customers,}$$
$$E(S) = \text{the average service time of a customer,}$$

where the generic variable S denotes the service time of a customer. Note that λ equals the inverse of the mean interarrival time. An important quantity is the *offered load*, which is defined as $\lambda E(S)$. This dimensionless quantity indicates the average amount of work that is offered to the system per unit time. Intuitively, in order to avoid infinitely long queues eventually building up, the offered load should be less than the maximum load the system can handle. Since each server provides service at unity rate, the maximum load the system can handle per unit time equals the number c of servers. Thus it is assumed that $\rho < 1$, where ρ is defined by

$$\rho = \frac{\lambda E(S)}{c}. \tag{4.2}$$

The quantity ρ is called the *server utilization*. It will be seen below that ρ gives the long-run fraction of time that an arbitrary server is busy. Also, we make the technical assumption that the probability that an interarrival time is larger than a service time is positive. It is stated without proof that this assumption and (4.2) guarantee that the mean cycle time is finite with a cycle being defined as the time elapsed between two consecutive arrivals finding the system empty of customers.

Next we define the following random variables:

$L(t) =$ the number of customers in the system at time t (including those in service),
$L_q(t) =$ the number of customers in the queue at time t (excluding those in service),
$D_n =$ the delay in queue of the nth customer (excluding service time),
$R_n =$ the amount of time spent by the nth customer in the system (including service time),
$V(t) =$ the work in the system (workload) at time t, that is the sum of the service times of all customers in the queue plus the sum of remaining service times of the customers in service at time t.

Under the assumptions made above it can be shown that each of the above random variables has a limiting distribution as $t \to \infty$ or $n \to \infty$, where the limiting distributions are proper probability distributions and are independent of the initial condition of the system. In particular, let

$$p_j = \lim_{t \to \infty} P\{L(t) = j\}, \qquad j = 0, 1, \ldots,$$

$$W_q(x) = \lim_{n \to \infty} P\{D_n \leqslant x\}, \qquad x \geqslant 0.$$

Roughly speaking, p_j gives the probability that an outside observer looking at the system when it has been in operation for a very long time will find j customers in the system at an arbitrary point in time. Also, p_j can be interpreted as the long-run fraction of time that j customers will be in the system. Similarly, $W_q(x)$ gives the probability that a customer entering the system when it has been in operation for a very long time will be delayed in queue not more than a time x.

It is important to note that the distribution of the number of customers in the system is invariant to the order of service when the queue discipline is *service-time independent* and *work conserving*. Here service-time independent means that the rule for selecting a next customer to be served does not depend on the service time of a customer, while work conserving means that the work or service requirement of a customer is not affected by the queue discipline. Queue disciplines having these properties include first-come–first-served, last-come–first-served and service in random order. The waiting-time distribution will obviously depend on the order of service, whereas the workload distribution is invariant among the work-conserving queue disciplines.

Let the random variables L, L_q, W_q, W and V have the limiting distributions of the respective processes $\{L(t)\}$, $\{L_q(t)\}$, $\{D_n\}$, $\{R_n\}$ and $\{V(t)\}$. By the theory of regenerative processes, it can be seen that the following averages apply with probability 1,

$$\lim_{t \to \infty} \frac{1}{t} \int_0^t L(u)du = E(L), \qquad \lim_{t \to \infty} \frac{1}{t} \int_0^t L_q(u)\,du = E(L_q),$$

$$\lim_{n \to \infty} \frac{1}{n} \sum_{k=1}^{n} D_k = E(W_q), \qquad \lim_{n \to \infty} \frac{1}{n} \sum_{k=1}^{n} R_k = E(W),$$

$$\lim_{t \to \infty} \frac{1}{t} \int_0^t V(u)\,du = E(V).$$

In words, supposing that the system is going to operate for a very long time, the long-run average number of customers in queue and the long-run average delay in queue per customer will be equal to the constants $E(L_q)$ and $E(W_q)$ with probability 1. A similar statement applies to $E(L)$, $E(W)$ and $E(V)$. In the discussion below the constants $E(L)$, $E(L_q)$, $E(W)$, $E(W_q)$ and $E(V)$ will always be interpreted as long-run averages.

Little's formula for the GI/G/c queue

One of the most useful results for queueing systems is the so-called formula of Little, which relates certain averages like the average number of customers in queue and the average delay in queue of a customer. The formula of Little is valid for almost any queueing system. In particular, for the GI/G/c queue, we have the following fundamental relations between the average number in queue (system) and the average wait in queue (system),

$$E(L_q) = \lambda E(W_q) \qquad \text{and} \qquad E(L) = \lambda E(W). \tag{4.3}$$

Following Ross (1983), we give a simple heuristic argument to explain these relations. However, the heuristic derivations to be given below can be made rigorously (see Stidham, 1974, and Franken et al., 1983). The heuristic argument is quite flexible and is applicable to a wide variety of queueing situations. The basic idea is to superimpose a reward structure on the system by imagining that each customer entering the system pays money to the system according to an appropriately chosen rule. We can then apply the principle

The average reward earned by the system per unit time
= (the average number of customers entering the system per unit time)
× (the average amount paid per entering customer). (4.4)

To argue the first relation in (4.3), imagine that each customer pays at a rate of 1 per unit time while in queue. Then the system earns a reward at rate j whenever j customers are in queue and so the average reward earned by the system per unit time is equal to the average number of customers in queue. On the other hand, the average amount paid per entering customer equals the average delay in queue of a customer, while the average number of customers entering the system per unit time equals the arrival rate λ. In this way we obtain $E(L_q) = \lambda E(W_q)$. The other relation in (4.3) follows similarly. It is important to point out that the relations in (4.3) apply for any queue discipline.

As another example of Little's formula, we have the relation

$$\text{The average number of busy servers} = \lambda E(S). \tag{4.5}$$

Thus, since each of the c servers carries the same load, the long-run fraction of time that an arbitrary server is busy equals $\lambda E(S)/c$. To see (4.5) imagine that each customer pays at a rate of 1 per unit time while in service. Then, by applying the principle (4.4), we find that the average number of customers in service equals $\lambda E(S)$. Each customer is served singly and so the average number of busy servers is the same as the average number of customers in service yielding (4.5). Note that the relation (4.5) holds for any queue discipline that is work conserving.

The average number of busy servers is also given by $\sum_{j=1}^{c-1} jp_j + c\sum_{j=c}^{\infty} p_j$, since p_j gives the fraction of time of having j customers in the system. Thus we obtain the useful identity

$$\sum_{j=1}^{c-1} jp_j + c\left(1 - \sum_{j=0}^{c-1} p_j\right) = \lambda E(S). \tag{4.6}$$

In particular, we have for the GI/G/1 queue the general result that $p_0 = 1 - \lambda E(S)$. Another application of the principle (4.4) is the following relation between the average work in the system and the average delay in queue of a customer:

$$E(V) = \lambda\{E(S)E(W_q) + \tfrac{1}{2}E(S^2)\}. \tag{4.7}$$

To argue this relation imagine that each customer in the system pays at a rate of y per unit time when its remaining service time is y. Under this reward structure the average reward earned by the system per unit time is equal to the average amount of work in the system. A customer whose delay in queue is W_q and service time is S pays at a constant rate of W_q during a time S and pays at the rate $S - x$ when having received an amount x of service. Thus

$$\text{The average amount paid per customer} = E\left(SW_q + \int_0^S (S - x)\,dx\right)$$

$$= E(S)E(W_q) + \tfrac{1}{2}E(S^2),$$

where the latter equality uses the fact that the queueing time and service time of a customer are mutually independent because of service in order of arrival. The relation (4.7) now follows. Note that this relation is valid for any queue discipline that is service-time independent and work conserving.

Finally, for the M/G/c queue the Little relation $E(L_q) = \lambda E(W_q)$ generalizes to

$$E(L_q(L_q - 1) \cdots (L_q - k + 1)) = \lambda^k E(W_q^k), \qquad k = 1, 2, \ldots, \qquad (4.8)$$

when service is in order of arrival. A proof of this useful relation (cf. also Haji and Newell, 1971) proceeds as follows. We first establish the following relationship between the waiting-time distribution and the queue-size distribution in the M/G/c queue:

$$P\{L_q = j\} = \begin{cases} P\{W_q = 0\} + \displaystyle\int_0^\infty e^{-\lambda x} w_q(x)\, dx, & j = 0, \\[3mm] \displaystyle\int_0^\infty e^{-\lambda x} \frac{(\lambda x)^j}{j!} w_q(x)\, dx, & j = 1, 2, \ldots \end{cases} \qquad (4.9)$$

Here $w_q(x)$ denotes the density of $P\{W_q \leq x\}$ for $x > 0$. Next, it is a matter of simple algebra to obtain (4.8) from (4.9). To derive the relationship (4.9), consider a marked customer arriving when the system has been in operation for a very long time and thus finding the system in statistical equilibrium. Then the delay in queue of the marked customer is distributed as W_q. The following two observations are now made. First, since service is in order of arrival, the customers left behind in queue when the marked customer enters service are exactly those having arrived during the delay in queue of the marked customer. Second, some reflections show that between any two service beginnings at which j customers remain in queue there must be an arrival finding j customers in queue, and conversely. Thus the long-run fraction of customers finding upon arrival j other customers in queue must be equal to the long-run fraction of customers leaving j other customers behind in queue when entering service. In other words, the number in queue just prior to the arrival of the marked customer is distributed as the number in queue just after the marked customer enters service. Combining the above observations, we find that the number of customers arriving during the queueing time W_q of the marked customer is distributed as the number of customers in queue seen by the marked customer upon arrival. Next, using the property that Poisson arrivals see time averages, observe that the latter number in queue is distributed as L_q, being the number in queue at a random point in time. Thus we conclude that L_q is distributed as the number of arrivals during W_q and so, by conditioning on W_q, we get (4.9).

Relationships between time-average and customer-average probabilities

From the viewpoint of a customer it may be more important to know the distribution of the number of customers in the system just prior to arrival rather than the distribution of the number of customers present at an arbitrary point of time. Thus, denoting by L_n the number of customers in the system just prior to the nth arrival, we define

$$\pi_j = \lim_{n \to \infty} P\{L_n = j\}, \qquad j = 0, 1, \ldots.$$

Roughly speaking, π_j gives the probability that an arbitrary customer arriving when the system has been in operation for a very long time will find j other customers in the system. Also, we can interpret π_j as the long-run fraction of customers finding upon arrival j other customers present. In particular, the delay probability Π_W which is defined by

$$\Pi_W = 1 - \sum_{j=0}^{c-1} \pi_j,$$

gives the long-run fraction of arrivals that have to wait in queue.

In general the customer-average probabilities π_j are different from the time-average probabilities p_j. One important exception is the M/G/c queue for which

$$\pi_j = p_j \qquad \text{for} \quad j = 0, 1, \ldots. \tag{4.10}$$

This identity is generally valid for queueing systems with Poisson input and is usually referred to as 'Poisson arrivals see time averages' (see section 1.8 of chapter 1).

For special cases of the GI/G/c queue a relationship between the time-average and customer-average probabilities can be established by using the following up- and down-crossings result for the stochastic process $\{L(t)\}$. To state this result, we say that an upcrossing from $j-1$ to j occurs when the number of customers in the system goes up from $j-1$ to j by an arrival. A downcrossing from j to $j-1$ occurs when the number of customers in the system goes down from j to $j-1$ by a service completion. For each GI/G/c queueing system, we have, for $j = 1, 2, \ldots,$

The long-run average number of upcrossings from $j-1$ to j per unit time
= the long-run average number of downcrossings from j to $j-1$
per unit time. (4.11)

This result follows directly by observing that between two consecutive up-crossings from $j-1$ to j a downcrossing from j to $j-1$ must occur, and conversely. Hence in any time interval the number of upcrossings from $j-1$ to j can differ at most 1 from the number of downcrossings from j to $j-1$. We note that the above up- and down-crossing result is in fact valid for any queueing system in which customers arrive one at a time and are served one at a time.

As an application of (4.11), we prove that for the GI/M/c queue with exponential service rate μ the following relation holds:

$$\min(j, c)\mu p_j = \lambda \pi_{j-1}, \qquad j = 1, 2, \ldots \tag{4.12}$$

To do so, note that by the lack of memory of the exponential distribution, service completions occur according to a Poisson process with rate $\min(j, c)\mu$ whenever j customers are in the system. Thus, using the relation (1.63) in section 1.8 of chapter 1, the average number of downcrossings from j to $j - 1$ per unit time equals $\min(j, c)\mu p_j$. Next we establish that the average number of upcrossings from $j - 1$ to j per unit time equals $\lambda \pi_{j-1}$, yielding (4.12). Therefore we use the following basic relation:

the fraction of arrivals finding the system in some state
= the average number of arrivals per unit time finding the system in that state divided by the average number of arrivals per unit time.

Hence, since the arrival rate is λ and π_{j-1} can be interpreted as the fraction of arrivals finding $j - 1$ other customers present, the average number of upcrossings from $j - 1$ to j per unit time equals $\lambda \pi_{j-1}$.

In the case where the interarrival time and service time distributions are of the phase type, a somewhat other approach can be used to relate the time-average and the customer-average (micro) state probabilities. To illustrate this, consider the $H_2/E_2/c$ queue. This queueing system can be modelled as a continuous-time Markov chain by interpreting the interarrival time as a mixture of two exponentials of the types 1 and 2 with respective rates λ_1 and λ_2 and the service time as a sum of two independent exponential phases with a common mean. Let p_{jkm} denote the time-average probability that j customers are present, the number of services being in progress and having a remaining service time of one phase is k and the interarrival time in progress is of type m. Also, let π_{jk} be the customer-average probability of finding the system in state (j, k). Then, using the above basic relation, we find $\pi_{jk} = (\lambda_1 p_{jk1} + \lambda_2 p_{jk2})/\lambda$. The probabilities π_{jk} enable us to obtain the waiting time distribution as a weighted sum of conditional waiting time distributions. The latter distributions can be calculated as first-passage time distributions in a modified continuous-time Markov chain in which new arrivals are disregarded when service is in order of arrival. This approach is generally applicable (cf. also Kühn, 1972).

Asymptotic expansion of the waiting-time probabilities in the GI/G/c queue

The GI/G/c queueing system has the important property that under weak conditions the waiting-time distribution function has an asymptotically exponential tail. This property may be quite useful for practical purposes since in most applications of practical interest the asymptotic expansion for the waiting-time distribution function $W_q(x)$ gives accurate estimates already for moderate values of x. To state the asymptotic expansion, we assume for ease that the interarrival time distribution function $A(x)$ and the service-time distribution function $B(x)$ have respective densities $a(x)$ and $b(x)$. Under the condition that the service-time density $b(x)$ decreases at least exponentially fast to 0 as $x \to \infty$, define δ as the unique positive solution to the characteristic equation

$$\int_0^\infty e^{-\delta x} a(x)\,dx \int_0^\infty e^{(\delta/c)y} b(y)\,dy = 1. \tag{4.13}$$

Note that $0 < \delta < c\mu$ for a probability density $b(x)$ being of the order $e^{-\mu x}$ as $x \to \infty$. Also, define the constant τ by

$$\tau = \left\{ \int_0^\infty e^{-\delta x} a(x)\,dx \right\}^{-1} \tag{4.14}$$

It was shown in Takahashi (1981) that

$$\frac{p_{j-1}}{p_j} \approx \tau \qquad \text{for } j \text{ large}, \tag{4.15}$$

and, assuming service in order of arrival,

$$P\{W_q > x\} \approx \frac{\delta\eta}{\lambda(\tau-1)^2 \tau^{c-1}} e^{-\delta x} \qquad \text{for } x \text{ large}, \tag{4.16}$$

where the constant η is defined by

$$\eta = \lim_{j \to \infty} \tau^j p_j. \tag{4.17}$$

It is noted that for the particular case of the GI/M/c queue a stronger result than (4.16) holds, namely W_q is exponentially distributed.

For the M/G/c queue the characteristic equation (4.13) can easily be rewritten as

$$\lambda \int_0^\infty e^{\delta y} \{1 - B(cy)\}\,dy = 1 \tag{4.18}$$

in agreement with the defining equation (1.75) in section 1.9 of chapter 1 for the special case of $c = 1$. Also, it follows from (4.14) that for the M/G/c queue

$$\tau = 1 + \frac{\delta}{\lambda}. \tag{4.19}$$

The decay parameter δ of the asymptotic expansion for $W_q(x)$ can be determined quite easily by solving the equation (4.14) by some standard numerical method. Note that for generalized Erlangian distributions the equation (4.14) reduces to a polynomial equation in δ. However, the difficulty is to find the constant η. For the case of phase-type distributions the constant η may be computed by solving the balance equations for an appropriately defined continuous-time Markov chain model for the queueing system.

The finite capacity GI/G/c/N queueing system

The GI/G/c/N queueing system differs from the GI/G/c queueing system only in the stipulation that a customer finding upon arrival N other customers in the system is rejected and has no further influence on the system. Here it is assumed that $N \geqslant c$.

The stochastic processes $\{L(t), \ t \geqslant 0\}$, $\{L_q(t), \ t \geqslant 0\}$, $\{V(t), \ t \geqslant 0\}$, $\{D_n, \ n \geqslant 1\}$ and $\{R_n, \ n \geqslant 1\}$ are defined as before except that D_n and R_n represent now the delay in queue and the time in system of the nth *entering* customer. The time-average probabilities $p_j (0 \leqslant j \leqslant N)$ and the customer-average probabilities $\pi_j (0 \leqslant j \leqslant N)$ are also defined as before. Note that π_j refers to the probability that an *arbitrary* customer finds upon arrival j other customers in the system. In particular,

$$\pi_N = \text{the long-run fraction of arrivals that is rejected,} \qquad (4.20)$$

and thus

$$\lambda(1 - \pi_N) = \text{the arrival rate of } \textit{entering} \text{ customers.} \qquad (4.21)$$

The Little relations (4.3), (4.5) and (4.7) can be seen to apply to the GI/G/c/N queue provided we replace λ by $\lambda(1 - \pi_N)$ (see Franken *et al.*, 1983). Thus

$$E(L_q) = \lambda(1 - \pi_N)E(W_q),$$
$$E(L) = \lambda(1 - \pi_N)E(W),$$

The average number of busy servers $= \lambda(1 - \pi_N)E(S)$,

$$E(V) = \lambda(1 - \pi_N)\{E(S)E(W_q) + \tfrac{1}{2}E(S^2)\}.$$

As a consequence of the third relation, the fraction of time that an arbitrary server is busy equals $\lambda(1 - \pi_N)E(S)/c$.

For the particular cases of the M/G/c/N queue and the GI/M/c/N queue we also have simple relations between the time-average and customer-average probabilities. By the property that 'Poisson arrivals see time averages', we have for the M/G/c/N queue that

$$\pi_j = p_j \qquad \text{for } j = 0, \dots, N, \qquad (4.22)$$

while for the GI/M/c/N queueing system with exponential service rate μ the reader may verify analogously to (4.12) that

$$\min(j,c)\mu p_j = \lambda \pi_{j-1} \quad \text{for } j = 1,\ldots,N. \tag{4.23}$$

4.2 SINGLE-SERVER QUEUEING SYSTEMS WITH FULL ACCESS OF ARRIVALS

This section discusses algorithms to compute the queue size and waiting-time distributions in a variety of single-server queueing systems with full access of arrivals. In what follows we shall treat the cases of Poisson input, state-dependent Markovian input and general input, where the service times have a general distribution. The queue discipline is first-come–first-served unless stated otherwise.

4.2.1 The M/G/1 queue

Consider the M/G/1 queueing system in which customers arrive according to a Poisson process with rate λ and the service time S of a customer has a general probability distribution function $B(x) = P\{S \leqslant x\}$ with $B(0) = 0$. It is assumed that the offered load $\rho = \lambda E(S)$ is smaller than 1.

We first establish a computational scheme for the time-average probabilities p_j yielding as well the customer-average probabilities π_j (recall that $\pi_j = p_j$ for all j in the M/G/1 queue). To do so, we rely upon the theory of regenerative processes and the property that 'Poisson arrivals see time averages'. This regenerative approach is a versatile method that leads to numerically stable recursion algorithms.

The regenerative approach for the queue-size distribution

The stochastic process $\{L(t), t \geqslant 0\}$ describing the number of customers in the system is regenerative. This process regenerates itself each time an arrival occurs that finds the system empty of other customers. Letting a cycle be the time elapsed between two consecutive arrivals finding the system empty, we define

$T =$ the length of a cycle

$T_k =$ the amount of time in a cycle during which k customers are in the system, $k = 0, 1, \ldots$.

Note that $T = \sum_{k=0}^{\infty} T_k$. Then, by the theory of regenerative processes,

$$p_k = \frac{E(T_k)}{E(T)}, \quad k = 0, 1, \ldots, \tag{4.24}$$

that is the long-run fraction of time the process is in state k equals the expected amount of time in a cycle that the process is in state k divided by the expected length of a cycle. In particular, using the fact that $E(T_0) = 1/\lambda$ by the lack of memory of the Poisson arrival process,

$$p_0 = \frac{1}{\lambda E(T)}. \tag{4.25}$$

The following simple idea is crucial for the derivation of a recurrence relation for the probabilities p_k. Divide a cycle into a random number of disjoint intervals separated by the service completion epochs and calculate $E(T_k)$ as the sum of the contributions to $E(T_k)$ of the various intervals. Thus, define the random variable

$N_j =$ the number of service completion epochs in a cycle at which j
  customers are left behind, $j = 0, 1, \ldots,$

and define the quantity

$A_{jk} =$ the expected amount of time that k customers are present during
  a service time that is started with j customers present.

Noting that the first service in a cycle starts with one customer present and noting that during a service it is only possible to have k customers present when that service starts with no more than k customers present, it follows that

$$E(T_k) = A_{1k} + \sum_{j=1}^{k} E(N_j)A_{jk}, \qquad k = 1, 2, \ldots. \tag{4.26}$$

To be precise we should point out that we have used Wald's equation in appendix A to establish that $E(N_j)A_{jk}$ is the contribution to $E(T_k)$ of those service intervals starting with j customers present (verify!).

To find another relation between $E(T_k)$ and $E(N_j)$, the following observation is important. Since the customers enter the system one at a time and are served one at a time, it follows from physical considerations that the number of arrivals in a cycle finding upon entrance j other customers present must be equal to the number of departures in a cycle leaving upon service completion j other customers behind. Hence

The expected number of arrivals in a cycle finding upon entrance j
  other customers present $= E(N_j), \qquad j = 0, 1, \ldots.$

On the other hand, since the arrival process is a Poisson process, we have by the basic relation (1.63) in section 1.8 of chapter 1 that

The expected number of arrivals in a cycle finding upon entrance j
  other customers present $= \lambda E(T_j).$

Thus we find the useful relation

$$E(N_j) = \lambda E(T_j), \qquad j = 0, 1, \ldots. \tag{4.27}$$

The relations (4.26) and (4.27) together yield

$$E(T_k) = A_{1k} + \sum_{j=1}^{k} \lambda E(T_j) A_{jk}, \qquad k = 1, 2, \ldots.$$

Dividing both sides of this equation by $E(T)$ and using (4.24) and (4.25), we find the recurrence relation

$$p_k = \lambda p_0 A_{1k} + \sum_{j=1}^{k} \lambda p_j A_{jk}, \qquad k = 1, 2, \ldots. \tag{4.28}$$

This relation allows the recursive computation of the probabilities $p_0, p_1, p_2, \ldots$ once we have evaluated the constants A_{jk}. The starting probability p_0 is explicitly given by

$$p_0 = 1 - \rho, \tag{4.29}$$

as follows from (4.8) with $c = 1$. The recursion scheme (4.28) is numerically stable since the calculations involve only additions and multiplications with positive numbers and thus can cause no loss of significant digits. Also, from a computational point of view it is important to note that the recursive calculations in (4.28) may be halted earlier by using the asymptotic result (4.15) which usually applies already for relatively small values of j. The constant τ in (4.15) can be computed beforehand from (4.18) and (4.19) with $c = 1$.

It remains to specify the constants A_{jk}. An explicit expression for A_{jk} can be found by the following technique. Assuming that at epoch 0 a new service starts while j customers are present, define $\chi_t = 1$ if this service is still in progress at time t and k customers are present at time t; otherwise define $\chi_t = 0$. Then we represent A_{jk} as

$$A_{jk} = E\left(\int_0^\infty \chi_t \, dt \right).$$

Since $E(\chi_t) = P\{\chi_t = 1\} = \{1 - B(t)\} e^{-\lambda t} (\lambda t)^{k-j} / (k-j)!$ using the fact that the number of arrivals in a time t is Poisson distributed with mean λt, we find

$$A_{jk} = \int_0^\infty \{1 - B(t)\} e^{-\lambda t} \frac{(\lambda t)^{k-j}}{(k-j)!} dt, \qquad k \geq j \quad \text{and} \quad j \geq 1. \tag{4.30}$$

It is important to note that A_{jk} depends on j and k only through the difference $k - j$. The integral in (4.30) is easy to evaluate for service-time distributions of practical interest. For example, when the service time is deterministic and equals the constant D, we find by using the identity (1.23) in section 1.4 of chapter 1 that

$$A_{jk} = \frac{1}{\lambda} \left\{ 1 - \sum_{h=0}^{k-j} e^{-\lambda D} \frac{(\lambda D)^h}{h!} \right\}.$$

Also, an explicit expression for A_{jk} can be given when the service time has a generalized Erlangian distribution. For the important case of a service-time distribution function

$$B(t) = \sum_{i=1}^{r} q_i \left\{ 1 - \sum_{h=0}^{i-1} e^{-\mu t} \frac{(\mu t)^h}{h!} \right\}, \qquad t \geq 0, \tag{4.31}$$

we obtain by using again the identity (1.23) that

$$A_{jk} = \sum_{i=1}^{r} q_i \sum_{h=0}^{i-1} \binom{k-j+h}{h} \frac{\lambda^{k-j}\mu^h}{(\lambda+\mu)^{k-j+h+1}}.$$

For the special case of pure Erlangian service (that is $q_r = 1$) a simpler recursion scheme for the state probabilities is provided by (2.51) and (2.52) in section 2.3 of chapter 2. Also, we note that for the case of phase-type service an alternative computational approach is given in Neuts (1982).

Using the generating function approach discussed in section 1.6 of chapter 1 we can derive from (4.28) and (4.30) the moments of the queue size L_q. Omitting the algebraic derivations, we find for the first two moments,

$$E(L_q) = \frac{\lambda^2 E(S^2)}{2(1-\rho)}, \qquad E(L_q(L_q-1)) = \frac{\lambda^3 E(S^3)}{3(1-\rho)} + 2\{E(L_q)\}^2. \tag{4.32}$$

The waiting-time distribution

Computational methods for the probability distribution of W_q, the delay in queue of a customer, will be given under the assumption of the first-come–first-served queue discipline. First we note that the delay probability $P\{W_q > 0\}$ equals $1 - \pi_0$ and so, by (4.10) and (4.29),

$$P\{W_q > 0\} = \rho. \tag{4.33}$$

Further, by (4.8) and (4.32) we find

$$E(W_q) = \frac{\lambda E(S^2)}{2(1-\rho)}, \qquad E(W_q^2) = \frac{\lambda E(S^3)}{3(1-\rho)} + 2\{E(W_q)\}^2. \tag{4.34}$$

Denoting by c_S^2 the squared coefficient of variation of the service time, it is important to point out that the Pollaczek–Khintchine formula for the average delay $E(W_q)$ allows the interesting representation

$$E(W_q) = (1 - c_S^2)E_{\text{det}}(W_q) + c_S^2 E_{\text{exp}}(W_q). \tag{4.35}$$

Here $E_{\text{det}}(W_q)$ and $E_{\text{exp}}(W_q)$ denote the average delay $E(W_q)$ for the special cases of deterministic service times and exponentially distributed service times with the same means $E(S)$. This representation is easily verified by writing $c_S^2 = E(S^2)/E^2(S) - 1$ and noting that $E(S^2)$ has the values $E^2(S)$ and $2E^2(S)$ for the respective cases of deterministic and exponential services. It will be seen later that relations of the type (4.35) using a linear interpolation on the squared coefficient of variation of the service time are extremely useful for developing two-moment approximations in more complex queueing systems; the usefulness of such two-moment approximations was already noticed in Cox (1955).

Tractable expressions for the waiting-time probabilities $P\{W_q > x\}$ can be given for many service-time distributions of practical interest. To do so, we recall that the function $P\{W_q > x\}$ is the same as the function $q(x)$ being extensively studied in section 1.9 of chapter 1; see relation (1.67) in which the servicing rate σ should be taken equal to 1. Therefore it follows that explicit expressions for $P\{W_q > x\}$ are given by the equations (A.32) and (A.33) with $\sigma = 1$ in appendix C when the service time has a K_2 distribution. In particular, we have for exponential service

$$P\{W_q > x\} = \rho e^{-(1 - \rho)x/E(S)}, \qquad x \geqslant 0. \tag{4.36}$$

For deterministic service with constant service time D, we have by (1.73) in section 1.9 of chapter 1 the explicit result

$$P\{W_q > x\} = 1 - (1 - \rho) \sum_{j=0}^{k} (-1)^j \frac{(\lambda x - \rho j)^j}{j!} e^{\lambda x - \rho j},$$

$$kD \leqslant x < (k + 1)D, \qquad k = 0, 1, \ldots. \tag{4.37}$$

It should be pointed out that the evaluation of $P\{W_q > x\}$ from (4.37) may cause numerical difficulties for large values of x, in particular when ρ increases. The reason is that roundoff errors may occur when taking differences of large numbers being close to each other. In the more general context of the M/D/c queue we give in subsection 4.3.2 an alternative to (4.37) that is much better suited for numerical purposes. Also, for larger values of x the asymptotic expansion (4.40) below provides an excellent alternative to (4.37), in particular when ρ is close to 1.

A simple computational method for the waiting-time probabilities can be given for the important case in which the service-time distribution is represented by (4.31). For this particular case a minor modification of the analysis in example 2.5 in section 2.3 of chapter 2 yields

$$P\{W_q > x\} = \sum_{j=1}^{\infty} f_j \sum_{k=0}^{j-1} e^{-\mu x} \frac{(\mu x)^k}{k!}, \qquad x \geqslant 0, \tag{4.38}$$

where the probabilities f_j are recursively computed from

$$\mu f_j = \sum_{k=(j-r)^+}^{j-1} f_k \left(\lambda \sum_{i=j-k}^{r} q_i \right), \qquad j = 1, 2, \ldots, \tag{4.39}$$

starting with $f_0 = 1 - \rho$, where $(j - r)^+ = \max(0, j - r)$. Noting that the service time can be represented with probability q_i as the sum of i independent exponential phases with common mean $1/\mu$, the recursion relation (4.39) follows by equating the rate at which the system leaves the macrostate of having at least j uncompleted phases in the system to the rate at which the system enters that macrostate.

For larger values of x the waiting-time probabilities can be computed from an asymptotic expansion for $P\{W_q > x\}$, assuming that the tail $1 - B(y)$ of the service-time distribution is of order e^{-By} for some $B > 0$ as $y \to \infty$. Under this assumption it was shown in section 1.9 of chapter 1 that

$$P\{W_q > x\} \approx \gamma e^{-\delta x} \qquad \text{for large} \quad x, \tag{4.40}$$

where δ and γ are given by (1.75) and (1.79) in which $\sigma = 1$ is put. Also, the following approximation $W_q^{\text{app}}(x)$ for $W_q(x) = P\{W_q \leqslant x\}$ was suggested in section 1.9:

$$W_q^{\text{app}}(x) = 1 - (\alpha e^{-\beta x} + \gamma e^{-\delta x}), \qquad x \geqslant 0, \tag{4.41}$$

where α and β are given by (1.82) in which $\sigma = 1$ is put. The approximation (4.41) can be used only when $\beta > \delta$, a condition which has been verified experimentally for a wide class of service-time distributions of practical interest. The approximation (4.41) is in accordance with (4.40) when $\beta > \delta$ and reflects the exact formulae for $P\{W_q > 0\}$ and $E(W_q)$. Also, the approximation (4.41) is exact when the service time has a K_2 distribution, as can be seen from the formulae (A.32) and (A.33) in appendix C.

The numerical investigations done in section 1.9 show that in practical applications the asymptotic expansion (4.40) applies for x already in the range of $E(S)$ to $2E(S)$, while the approximation (4.41) gives good results for all $x \geqslant 0$ provided ρ is not too small (say, $\rho \geqslant 0.2$). The performances of (4.40) and (4.41) improve when ρ increases. In addition to the numerical results in section 1.9, we give in Table 4.1 the exact values of $P\{W_q > x\}$ to see how sensitive the

Table 4.1   Exact values for the waiting-time probabilities $P\{W_q > x\}$

x	0.5	1	2	4	7	10	15	20
$\rho = 0.2$ E_2	0.1246	0.0715	0.0214	0.0018				
$E_{1,3}$	0.1270	0.0729	0.0206	0.0014				
H_2^b	0.1384	0.1013	0.0610	0.0269	0.0087	0.0028	0.0004	
H_2^g	0.1438	0.1105	0.0677	0.0258	0.0061	0.0014	0.0001	
$\rho = 0.5$ E_2	0.3728	0.2662	0.1311	0.0311	0.0036	0.0004		
$E_{1,3}$	0.3771	0.2700	0.1309	0.0300	0.0033	0.0004		
H_2^b	0.3969	0.3251	0.2314	0.1294	0.0570	0.0252	0.0065	0.0017
H_2^g	0.4063	0.3430	0.2497	0.1332	0.0520	0.0203	0.0042	0.0009
$\rho = 0.8$ E_2	0.7120	0.6243	0.4758	0.2755	0.1213	0.0534	0.0136	0.0035
$E_{1,3}$	0.7152	0.6282	0.4780	0.2755	0.1206	0.0528	0.0133	0.0034
H_2^b	0.7292	0.6721	0.5819	0.4493	0.3092	0.2132	0.1147	0.0618
H_2^g	0.7359	0.6865	0.6019	0.4638	0.3137	0.2122	0.1106	0.0577

waiting-time probabilities are to more than the first two moments of the service-time distribution. The squared coefficient of variation c_S^2 of the service time is varied as 0.5 and 2, while the offered load ρ has the values 0.2, 0.5 and 0.8. For $c_S^2 = 0.5$ we consider both the E_2 distribution and the $E_{1,3}$ distribution, while for $c_S^2 = 2$ we consider both the H_2 distribution with the normalization of balanced means (H_2^b) and the H_2 distribution with the gamma normalization (H_2^g); see appendix B for a specification of the parameters of these distributions. In all examples we take $E(S) = 1$. Our numerical investigations indicate that the waiting-time probabilities are fairly insensitive to more than the first two moments of the service time, provided c_S^2 is not too large and the service-time density satisfies a reasonable shape constraint. The sensitivity is mainly manifested in the ultimate tail of the waiting-time distribution, as is to be expected. The waiting-time probabilities become increasingly sensitive to the underlying service-time distribution when c_S^2 becomes larger and ρ becomes smaller.

Two-moment approximations for the waiting times

In applications it often happens that only the first two moments of the service time are available. In these situations two-moment approximations may be very helpful provided c_S^2 is not too large. Roughly stated, two-moment approximations may be used when $0 \leqslant c_S^2 \leqslant 2$. Such approximations should not be used blindly. It depends to some extent on the performance measure of interest whether the use of a two-moment approximation is justifiable. For example, more care should be exercised in using a two-moment approximation for the 99.9 per cent percentile of the waiting-time distribution than one for the standard deviation of the waiting time.

We now discuss two approaches to approximating the waiting-time probabilities when only the first two moments of the service time are available and c_S^2 is not too large. The first approach is based on the fact that the waiting-time probabilities are easy to compute for generalized Erlangian distributions. An $E_{k-1,k}$ distribution according to (A.11) and (A.12) in appendix B may be fitted to the service time when $0 < c_S^2 \leqslant 0.5$, in which case the waiting-time probabilities may be approximated by using either the computational scheme (4.38) or the approximate formula (4.41). A K_2 distribution according to (A.15) and (A.17) may be fitted to the service time when $0.5 < c_S^2 \leqslant 2$, in which case the waiting-time probabilities may be approximated by using the formula (A.33) in appendix C.

The second approach is based on a two-moment approximation of the form (4.35). A natural question that arises is whether such an approximation applies as well to the waiting-time probabilities. Numerical experiments show that the answer is in the negative. However, although the waiting-time probabilities themselves do not allow for a two-moment approximation of the form (4.35), we found that the waiting-time *percentiles* do allow for such a two-moment approximation. This approximation for the waiting-time percentiles yields indirectly a two-moment approximation for the waiting-time probabilities. The pth percentile $\xi(p)$ of a continuous probability distribution function $F(x)$ of a non-negative random variable is defined by

$$F(\xi(p)) = p, \quad F(0) < p < 1.$$

In analogy to (4.35) we suggest to approximate the pth percentile $\xi(p)$ of the waiting-time distribution function $P\{W_q \leqslant x\}$ by

$$\xi_{app}(p) = (1 - c_S^2)\xi_{det}(p) + c_S^2\xi_{exp}(p), \quad 1 - \rho < p < 1. \tag{4.42}$$

Here $\xi_{det}(p)$ and $\xi_{exp}(p)$ denote the corresponding percentiles for the respective cases of deterministic and exponential services. These percentiles are easy to compute exactly. Our numerical investigations show that (4.42) is an excellent approximation to $\xi(p)$ for *all* values of p, provided c_S^2 is not too large. Noting that

$$P\{W_q > x \,|\, W_q > 0\} = \frac{1}{\rho}P\{W_q > x\}, \quad x > 0,$$

it follows that a two-moment approximation of the form (4.42) applies as well to the percentiles of the conditional waiting-time distribution function $1 - P\{W_q > x \,|\, W_q > 0\}$; the latter percentiles are defined for all $0 < p < 1$.

Next we show that the two-moment approximation (4.42) and the asymptotic expansion (4.40) together yield the following two-moment approximations δ_{app} and γ_{app} for the coefficients δ and γ in (4.40),

$$\delta_{app} = \frac{\delta_{det}\delta_{exp}}{(1 - c_S^2)\delta_{exp} + c_S^2\delta_{det}} \tag{4.43}$$

and

$$\gamma_{app} = \gamma_{exp}^q \gamma_{det}^{1-q} \quad \text{with} \quad q = c_S^2\delta_{app}/\delta_{exp}. \tag{4.44}$$

Here $\delta_{det}(\delta_{exp})$ and $\gamma_{det}(\gamma_{exp})$ are the values of δ and γ for the particular case of deterministic (exponential) service times. To argue these approximations, we first note that the pth percentile $\xi(p)$ of $P\{W_q \leqslant x\}$ tends to infinity as $p \to 1$ and so, by the asymptotic expansion (4.40),

$$\xi(p) = \frac{1}{\delta}\ln\left(\frac{\gamma}{1-p}\right) \quad \text{as} \quad p \to 1.$$

In particular, this asymptotic expansion applies to both $\xi_{det}(p)$ and $\xi_{exp}(p)$. Suppose that the relation $\xi(p) = (1 - c_S^2)\xi_{det}(p) + c_S^2\xi_{exp}(p)$ was to hold exactly for all p. Then, by substituting the asymptotic expansions of the respective percentiles,

$$\frac{1}{\delta}\ln\left(\frac{\gamma}{1-p}\right) = (1-c_S^2)\frac{1}{\delta_{\text{det}}}\ln\left(\frac{\gamma_{\text{det}}}{1-p}\right) + c_S^2\frac{1}{\delta_{\text{exp}}}\ln\left(\frac{\gamma_{\text{exp}}}{1-p}\right) \qquad \text{as} \quad p \to 1,$$

or, equivalently,

$$\ln\left\{\left(\frac{\gamma}{1-p}\right)^{1/\delta}\right\} = \ln\left\{\gamma_{\text{det}}^a \gamma_{\text{exp}}^b \left(\frac{1}{1-p}\right)^{a+b}\right\}$$

$$\text{with} \quad a = \frac{1-c_S^2}{\delta_{\text{det}}}, \quad b = \frac{c_S^2}{\delta_{\text{exp}}},$$

as $p \to 1$. This relation suggests the approximations (4.43) and (4.44). The quantities δ_{det}, γ_{det}, δ_{exp} and γ_{exp} are easy to compute. By (4.36),

$$\delta_{\text{exp}} = \frac{1-\rho}{E(S)} \quad \text{and} \quad \gamma_{\text{exp}} = \rho, \tag{4.45}$$

while it is readily derived from (1.75) and (1.79) in section 1.9 of chapter 1 that δ_{det} is the unique positive solution of the equation

$$e^{\delta_{\text{det}}E(S)} = 1 + \frac{\sigma\delta_{\text{det}}}{\lambda}, \tag{4.46}$$

with $\sigma = 1$, and γ_{det} is given by

$$\gamma_{\text{det}} = (1-\rho)\{\delta_{\text{det}}E(S) - (1-\rho)\}^{-1}. \tag{4.47}$$

The approximations (4.43) and (4.44) in conjunction with the asymptotic expansion (4.40) provide the following asymptotic two-moment estimate for the pth percentile $\xi(p)$ of $P\{W_q \leqslant x\}$,

$$\xi_{\text{asy}}(p) = \frac{1}{\delta_{\text{app}}}\ln\left(\frac{\gamma_{\text{app}}}{1-p}\right) \qquad \text{for } p \text{ sufficiently close to 1.} \tag{4.48}$$

This asymptotic approximation is easier to compute than (4.42) and is obtained from (4.42) by replacing $\xi_{\text{det}}(p)$ by $(1/\delta_{\text{det}})\ln\{\gamma_{\text{det}}/(1-p)\}$ when p is close enough to 1 (note from (4.36) that $\xi_{\text{exp}}(p) = (1/\delta_{\text{exp}})\ln\{\gamma_{\text{exp}}/(1-p)\}$ for all $p > 1 - \rho$). How large p should be in order to use (4.48) rather than (4.42) depends to some extent on the offered load ρ. Roughly stated, the asymptotic percentile (4.48) may be used for practical purposes when

$$1 - p \leqslant \tfrac{1}{2}\rho,$$

provided ρ is not very small. In other words,

$$\eta_{\text{asy}}^c(p) = \frac{1}{\delta_{\text{app}}}\ln\left\{\frac{\gamma_{\text{app}}}{\rho(1-p)}\right\} \tag{4.49}$$

Table 4.2   Exact and approximate percentiles for $P\{W_q \leqslant x | W_q > 0\}$

		$c_S^2 = 0.5$					$c_S^2 = 2$				
p		0.2	0.5	0.9	0.99	0.999	0.2	0.5	0.9	0.99	0.999
$\rho = 0.2$	exa	0.25	0.54	2.06	3.90	5.73	0.32	1.20	4.53	9.30	14.08
	app	0.26	0.56	1.98	3.87	5.76	0.31	1.14	4.67	9.52	14.39
	app*	0.33	0.56	2.02	3.89	5.76	0.19	1.18	4.60	9.49	14.39
$\rho = 0.5$	exa	0.39	1.09	3.34	6.54	9.75	0.54	2.00	7.12	14.46	21.80
	app	0.41	1.10	3.33	6.55	9.77	0.53	1.96	7.16	14.53	21.91
	app*	0.42	1.08	3.33	6.55	9.77	0.49	2.00	7.16	14.53	21.91
$\rho = 0.8$	exa	0.91	2.64	8.52	16.95	25.37	1.53	5.14	17.49	35.16	52.83
	app	0.93	2.63	8.52	16.95	25.38	1.50	5.14	17.50	35.18	52.86
	app*	0.91	2.63	8.52	16.95	25.38	1.53	5.14	17.50	35.18	52.86

may be used to approximate the pth percentile of the conditional waiting-time distribution when $p \geqslant \frac{1}{2}$. In Table 4.2 we give the exact values, the approximate values (4.42) and the approximately asymptotic values (4.49) of the percentiles of the conditional waiting-time distribution for two different service-time densities when ρ and p are varied. These values are respectively denoted by exa, app and app*. We consider the E_2 density ($c_S^2 = 0.5$) and the H_2 density with the gamma normalization and $c_S^2 = 2$. It should be pointed out that for the H_2 distribution the performance of the two-moment approximation (4.42) depends to some extent on the normalization used; the performance is surprisingly good for the 'natural' gamma normalization (cf. also Table 4.3).

To conclude the discussion on two-moment approximations, we give a more systematic treatment of the relative errors introduced by using δ_{app} and γ_{app} rather than δ and γ. Replacing δ and γ in (4.40) by δ_{app} and γ_{app} introduces a relative error percentage of

$$100 \left| \frac{\gamma_{app} e^{-\delta_{app}x} - \gamma e^{-\delta x}}{\gamma e^{-\delta x}} \right| = 100 \left| \frac{\gamma_{app}}{\gamma} e^{(\delta - \delta_{app})x} - 1 \right|$$

in the asymptotic estimate for the waiting-time probability. Numerical calculations show that the deviations of δ_{app} and γ_{app} from δ and γ are quite acceptable in practical applications for the relevant range of x values provided ρ is not too small. The performance of δ_{app} and γ_{app} improves when ρ increases. In Table 4.3 we give $\Delta_\delta = \delta - \delta_{app}$ and $r_\gamma = \gamma_{app}/\gamma$ for several values of ρ and c_S^2 and for various service-time distributions. We consider the Weibull distribution (Weib), the gamma distribution (gamma), a mixture of E_1 and E_3 distributions with the same scale parameters ($E_{1,3}$), a shifted exponential distribution (shif), an H_2 distribution with the gamma normalization (H_2^g) and an H_2 distribution with balanced means (H_2^b) (cf. appendix B). Observe from Table 4.3 that the factors γ_{app}/γ and $\exp\{(\delta - \delta_{app})x\}$ determining the

Table 4.3   The deviations Δ_δ and r_γ for various distributions

		$\rho = 0.2$		$\rho = 0.5$		$\rho = 0.8$	
		Δ_δ	r_γ	Δ_δ	r_γ	Δ_δ	r_γ
$c_S^2 = 0.5$	Weib	0.1172	0.762	0.0214	0.944	0.0020	0.990
	gamma	0.0296	0.878	0.0039	0.979	0.0002	0.998
	$E_{1,3}$	0.1072	0.791	0.0220	0.946	0.0023	0.990
	shif	-0.1139	1.135	-0.0305	1.052	-0.0038	1.015
$c_S^2 = 2$	gamma	-0.0177	1.302	-0.0031	1.043	-0.0002	1.004
	H_2^g	0.0115	0.984	0.0018	0.998	0.0001	1.000
	H_2^b	-0.0968	1.468	-0.0408	1.226	-0.0063	1.061

relative error tend to correct each other for the values of x of practical interest. For example, for $\rho = 0.5$, $x = 3.34$ and an Erlang-2 (Weibull) service-time distribution with $E(S) = 1$ and $c_S^2 = 0.5$, the two-moment approximations (4.43) and (4.44) introduce a relative error of 0.9 (13) per cent in the value 0.050 (0.0490) of the asymptotic estimate (4.40). The degree of acceptability of some relative error in a small probability depends strongly on the absolute magnitude of that probability; e.g. an estimate of 0.0012 for a probability of 0.0010 has a relative error of 20 per cent but is yet practically useful. Finally, Table 4.3 demonstrates that the approximations δ_{app} and γ_{app} perform better for the H_2 distribution with the gamma normalization than for the H_2 distribution with the normalization of balanced means.

The distribution of the workload

In the $M/G/1$ queueing system the waiting-time distribution function $W_a(x)$ also represents the limiting distribution of $V(t)$ which is defined as the amount of work in the system at time t, that is $V(t)$ is the sum of the remaining service requirements of the customers in the system at time t. To see this, note that $V(t)$ can also be interpreted as the amount of time that a customer would have to wait in queue if that customer arrived at time t and service is in order of arrival. Thus, using the property 'Poisson arrivals see time averages', the limiting distribution of $V(t)$ is equal to that of the waiting time of a customer.

The fact that the distribution of the work in the system is invariant among work-conserving queue disciplines can be used to obtain the average waiting time for other queue disciplines than the first-come–first-served. To the end of this subsection we consider a priority queue discipline and the processor-sharing queue discipline and their effect on the average waiting time of customers. These queue disciplines are quite useful in the scheduling of jobs in computer operating systems.

The $M/G/1$ queue with the non-preemptive priority queue discipline

Consider the $M/G/1$ queueing system in which the customers can be divided into p groups where customers of group j have priority of service over customers in groups above $j(j = 1, \ldots, p)$. The priority queue discipline is assumed to be *non-preemptive*, that is the service of a customer is not preempted when a customer of higher priority arrives but is continued until completion. Within each priority group the queue discipline is first-come–first-served. The probability that an arriving customer belongs to group j is $r_j(j = 1, \ldots, p)$. Thus, letting λ be the overall arrival rate of customers, the customers belonging to priority group j arrive according to a Poisson process with rate

$$\lambda_j = \lambda r_j, \qquad j = 1, \ldots, p.$$

Each priority group has its own service-time distribution. The service time of a customer in group j is distributed as the random variable S_j having finite first two moments. It is assumed that the offered load ρ defined by

$$\rho = \sum_{j=1}^{p} \lambda_j E(S_j) \qquad (4.50)$$

is smaller than 1. Denote by $E(W_{qj})$ the average delay in queue of a customer in priority group j. Using the fact that the work in the system is the same as it would be if the customers were served in order of arrival, it can be shown that (see for a proof Kleinrock, 1976)

$$E(W_{qj}) = \frac{\sum_{j=1}^{p} \lambda_j E(S_j^2)}{2(1 - \rho_j)(1 - \rho_{j-1})}, \qquad j = 1, \ldots, p, \qquad (4.51)$$

where $\rho_0 = 0$ and $\rho_j = \sum_{k=1}^{j} \lambda_k E(S_k), j = 1, \ldots, p$. Consequently, the overall average delay in queue of a customer is given by

$$\sum_{j=1}^{p} r_j E(W_{qj}), \qquad (4.52)$$

as opposed to $\lambda \sum_{j=1}^{p} r_j E(S_j^2) / \{2(1 - \rho)\}$ for the first-come–first-served queue discipline. Using a priority queue discipline rather than the first-come–first-served discipline may lead to a considerable reduction in the average waiting time of customers; a separation into two groups discriminating between large and small service times often suffices to achieve this effect. An extension of the above results to the multi-server non-preemptive priority queue with exponential services can be found in Kella and Yechiali (1985).

The $M/G/1$ processor-sharing queueing system

The processor-sharing queueing system provides simultaneous service to all customers in the system where the processing rate of the server is equally divided between all customers present. This system can be used to approximate some time-shared computer systems. The processor-sharing queue discipline allows us to discriminate among jobs on the basis of their required service times in favour of the shorter jobs. To state this precisely, consider the following generalized processor-sharing rule for the $M/G/1$ queue with arrival rate λ and service time S of a customer. If there are i customers in the system, each is provided with service at a rate of $f(i)$ per unit time, that is the attained service time of each customer grows with an amount of $f(i)\Delta x$ in a time Δx with Δx small. Here $f(i)$ is some positive function. Denoting by p_i the limiting probability of having i customers in the system and letting $E(W|s)$ be the average time spent in the system by a customer whose required service time is s, it can be shown that

$$p_i = \frac{(\rho^i/i!)\phi(i)}{\sum_{k=0}^{\infty}(\rho^k/k!)\phi(k)}, \qquad i = 0, 1, \ldots, \tag{4.53}$$

and

$$E(W|s) = \frac{s\sum_{k=0}^{\infty}(\rho^k/k!)\phi(k+1)}{\sum_{k=0}^{\infty}(\rho^k/k!)\phi(k)}, \qquad s > 0, \tag{4.54}$$

where ρ denotes the offered load $\lambda E(S)$ and $\phi(i)$ is defined by

$$\phi(i) = \left\{\prod_{j=1}^{i} f(j)\right\}^{-1}, \qquad i = 0, 1, \ldots,$$

with $\phi(0) = 1$ by convention. Here it is assumed that $\sum_{k=0}^{\infty}\rho^k\phi(k)/k!$ is finite. It is remarkable that the results (4.53) and (4.54) depend on the service-time distribution only through the first moment. A proof of these results can be found in Cohen (1979); cf. also Kleinrock (1976). The first reference gives many other useful results for processor-sharing queueing systems including closed queueing networks. The standard $M/G/1$ processor-sharing queueing system corresponds to the special case of

$$f(i) = \frac{1}{i}, \qquad i = 1, 2, \ldots.$$

In this case we have $\phi(i) = i!$ for $i = 0, 1, \ldots$, and the formulae (4.53) and (4.54) reduce to

$$p_i = (1 - \rho)\rho^i, \qquad i = 0, 1, \ldots, \tag{4.55}$$

and

$$E(W|s) = \frac{s}{1 - \rho}, \qquad s > 0. \tag{4.56}$$

The result (4.55) states that in the standard $M/G/1$ processor-sharing queueing system with general service times the number of customers in the system is distributed as in the M/M/1 first-come–first-served queueing system with exponential service times. The equations (4.54) and (4.56) show that processor sharing has the nice property that the average conditional response time is linear in the required service time, which implies that a customer requiring twice as much service time as some other will spend on the average twice as much time in the system. For results on the response-time distribution function, see Ott (1984) and references therein.

4.2.2 A single-server queueing system with state-dependent Markovian input

Consider a single-server queueing system at which customers arrive singly according to a state-dependent Markovian input process with rate λ_j whenever j customers are in the system, that is with probability $\lambda_j \Delta t + o(\Delta t)$ a customer arrives in the next time Δt when j customers are in the system at the present time. As usual $o(\Delta t)$ represents any quantity that tends to zero faster than Δt as $\Delta t \rightarrow 0$. Each arriving customer is assumed to join the system and waits until served. The service times of the customers are independent positive random variables with a common probability distribution function B. The customers are served in order of arrival. Denoting by the generic variable S the service time of a customer, it is assumed that $\lambda_j E(S) < 1$ for all j sufficiently large. This condition guarantees that no infinitely long queue can build up.

The single-server queueing system with state-dependent Markovian input covers a number of important queueing models. We give the following two models as they are useful in computer science applications.

Example 4.1 *A closed two-stage cyclic queueing model of multi-programming*

Consider a closed queueing system in which N customers (jobs) always circulate between two single-server stations as schematically shown in Figure 4.1. This queueing model could represent a multi-programming computer system under heavy traffic where there is always a job waiting outside the system ready to enter when there is a departure from the system. Here server 1 is a central processing unit (CPU) and server 2 is an input–output device (I/O device). The service time of a job at server 1 has a general probability distribution function

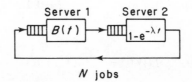

N jobs

Figure 4.1 A closed two-stage cyclic queueing system

$B(x)$, while the service time of a job at server 2 is exponentially distributed with mean $1/\lambda$. It is assumed that either of the two servers can handle only one job at a time and uses the first-come–first-served discipline.

The distribution of the jobs present at the various stations is completely specified by giving the number of jobs at server 1 (say). Since server 2 provides exponential service with rate λ whenever jobs are there, it follows that the arrival process of jobs at server 1 is a Poisson process with rate λ as long as the number of jobs at server 1 is less than N. Thus, isolating the subsystem consisting of server 1, this queueing system has a state-dependent Markovian input with rate

$$\lambda_j = \begin{cases} \lambda, & 0 \leqslant j < N, \\ 0, & j \geqslant N. \end{cases} \tag{4.57}$$

$\square$

Example 4.2   *The machine repairman model of an interactive computer system*

Consider a queueing system in which requests for service are generated by a finite number N of identical sources and the requests are handled by a single server as schematically shown in Figure 4.2. The sources could constitute a group of machines maintained by a single repairman or a group of terminals generating jobs for a computer system. The service times of the requests generated by the sources have a general probability distribution function B. It is assumed that the server can handle only one request at a time and uses a first-come–first-served service discipline. New requests for service can be generated only by idle sources, which are sources having no previous request waiting or being served at the server. A source idle at the present time will generate a request in the next Δt time units with probability $\eta \Delta t + o(\Delta t)$, independently of the states of the other sources. In other words, the time until an idle source generates a request (the think time) is exponentially distributed with mean $1/\eta$.

It is easily seen that in the above closed queueing system new requests arrive

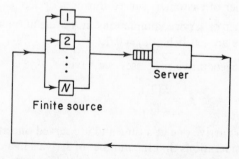

Figure 4.2 The machine repairman model

at the server according to a state-dependent Markovian input process with rate

$$\lambda_j = \begin{cases} (N-j)\eta, & 0 \leqslant j < N, \\ 0, & j \geqslant N. \end{cases} \tag{4.58}$$

whenever j jobs are waiting or being served at the server.   □

The time-average and customer-average state probabilities

We now give a unifying algorithmic analysis for the state probabilities for the general class of single-server queueing systems with state-dependent Markovian input. The analysis will proceed along the same lines as for the M/G/1 queue. As in section 4.1 we define the time-average probability p_j as the probability of having j customers in the system at a random point in time and the customer-average probability π_j as the probability of an arrival finding j other customers in the system. We first derive the following relationships for the distributions $\{p_j\}$ and $\{\pi_j\}$:

$$\pi_j = \frac{\lambda_j p_j}{\sum_{k=0}^{\infty} \lambda_k p_k}, \qquad j = 0, 1, \ldots, \tag{4.59}$$

$$\sum_{k=0}^{\infty} \lambda_k p_k = \frac{1 - p_0}{E(S)}. \tag{4.60}$$

To prove these relations note that the stochastic process describing the number of customers in the system regenerates itself each time an arriving customer finds the system empty of other customers. Letting a cycle be the time elapsed between two consecutive arrivals finding the system empty, we define the random variables

T = the length of a cycle,

T_k = the amount of time in a cycle during which k customers are in the system, $k = 0, 1, \ldots,$

N = the number of customers served during a cycle,

N_j = the number of service completions in a cycle at which j customers are left behind, $j = 0, 1, \ldots$.

By the theory of regenerative processes, we have

$$p_k = \frac{E(T_k)}{E(T)}, \qquad k = 0, 1, \ldots. \tag{4.61}$$

Since the customers arrive one at a time and are served one at a time, it follows from physical considerations that the number of arrivals in a cycle finding upon entrance j other customers in the system must be equal to the number of departures in a cycle leaving upon service completion j other customers behind. Hence

The expected number of arrivals in a cycle finding upon entrance j other customers in the system $= E(N_j),$ $j = 0, 1, \ldots .$ (4.62)

On the other hand, noting that the arrival process of customers is a Poisson process with rate λ_j whenever j customers are in the system and using the fundamental relation (1.63) in section 1.8 of chapter 1, it can be seen that the expected number of arrivals in a cycle finding upon entrance j other customers in the system equals $\lambda_j E(T_j)$. Thus we find

$$E(N_j) = \lambda_j E(T_j), \qquad j = 0, 1, \ldots . \tag{4.63}$$

As an immediate consequence of (4.62) and (4.63),

The expected number of arrivals in a cycle $= \displaystyle\sum_{j=0}^{\infty} \lambda_j E(T_j).$ (4.64)

The long-run fraction of customers finding upon arrival j other customers in the system equals the expected number of arrivals in a cycle finding upon entrance j other customers present divided by the expected number of arrivals in a cycle. Hence, by (4.62) to (4.64),

$$\pi_j = \frac{\lambda_j E(T_j)}{\displaystyle\sum_{k=0}^{\infty} \lambda_k E(T_k)}, \qquad j = 0, 1, \ldots .$$

Dividing the numerator and denominator of the right side of this relation by $E(T)$ and applying (4.61), we get the desired result (4.59). To verify the relation (4.60), note that $1 - p_0$ represents the fraction of time the server is busy and thus equals the ratio of the expected amount of time in a cycle during which the server is busy and the expected length of a cycle. Hence

$$1 - p_0 = \frac{E(T - T_0)}{E(T)} .$$

On the other hand, using Wald's equation in appendix A, it can be seen that the expected time in a cycle during which the server is busy equals the expected number of customers served in a cycle times the expected service time of a customer. That is $E(T - T_0) = E(N)E(S)$. Hence, using the representation $N = \sum_{j=0}^{\infty} N_j$ and the relation (4.63), we find

$$E(T - T_0) = E(S) \sum_{j=0}^{\infty} \lambda_j E(T_j). \tag{4.65}$$

Dividing both sides of this identity by $E(T)$ yields the relation (4.60). It may be instructive to give the following heuristic derivation of (4.60). This alternative derivation is based on the observation that the average arrival rate of entering customers must be equal to the average throughput, which is defined as the average number of customers served per unit time. Dividing both sides of (4.64) by $E(T)$ and using (4.61), we have

$$\text{The average arrival rate of entering customers} = \sum_{j=0}^{\infty} \lambda_j p_j. \qquad (4.66)$$

The average rate at which the server processes customers whenever the server is busy equals $1/E(S)$, while the fraction of time the server is busy is given by $1 - p_0$. Thus

$$\text{The average throughput} = \frac{1 - p_0}{E(S)},$$

in agreement with (4.60) and (4.66). It is important to note that this relation for the average throughput is generally valid for each service discipline that is work conserving. Also, using (4.60) and (4.66), we find

$$E(L_q) = \frac{1 - p_0}{E(S)} E(W_q) \qquad (4.67)$$

where $E(L_q)$ is computed from $E(L_q) = \sum_{j=1}^{\infty} (j - 1)p_j$.

Next we derive a recursion scheme to compute the time-average probabilities p_j which in turn yield the customer-average probabilities π_j by applying (4.59) and (4.60). To do so, we define as before the quantity A_{jk} as the expected amount of time that k customers are present during a service time that is started with j customers present. The computation of the quantities A_{jk} will be discussed below. Proceeding as in the derivation of (4.26), we divide the cycle into disjunct intervals according to the service completion epochs. By doing so, we can compute $E(T_k)$ as

$$E(T_k) = A_{1k} + \sum_{j=1}^{k} E(N_j) A_{jk}, \qquad k = 1, 2, \ldots, \qquad (4.68)$$

while by the lack of memory of the Markovian input process

$$E(T_0) = \frac{1}{\lambda_0}. \qquad (4.69)$$

Invoke now (4.63) in (4.68) and divide both sides of the resulting relation by $E(T)$. Using (4.61) and noting that $\lambda_0 p_0 = 1/E(T)$ by (4.69), we obtain the recurrence relation

$$p_k = \lambda_0 p_0 A_{1k} + \sum_{j=1}^{k} \lambda_j p_j A_{jk}, \qquad k = 1, 2, \ldots. \qquad (4.70)$$

This numerically stable scheme allows the recursive computation of p_1/p_0,

$p_2/p_0, \ldots$. The unknown p_0 next follows by using the normalization equation $\sum_{k=0}^{\infty} p_k/p_0 = 1/p_0$.

It remains to specify the quantities A_{jk}. In general we cannot give a closed-form expression for A_{jk} that is computationally tractable. Notable exceptions are the particular models considered in examples 4.1 and 4.2. By making a slight modification to the derivation of formula (4.30), we find for the particular specification (4.57) of the arrival rates λ_j that

$$
A_{jk} = \begin{cases} \int_0^{\infty} \{1 - B(t)\} e^{-\lambda t} \dfrac{(\lambda t)^{k-j}}{(k-j)!}\, dt, & 1 \leqslant j < N, j \leqslant k < N, \\[2mm] \int_0^{\infty} \{1 - B(t)\}\left\{1 - \sum_{i=0}^{N-j-1} e^{-\lambda t}\dfrac{(\lambda t)^i}{i!}\right\} dt, & 1 \leqslant j \leqslant N,\ k = N. \end{cases} \tag{4.71}
$$

Also, for the specification (4.58) of the arrival rates λ_j, we have

$$
A_{jk} = \int_0^{\infty} \{1 - B(t)\}\binom{N-j}{k-j}(1 - e^{-\eta t})^{k-j}(e^{-\eta t})^{N-k}\, dt, \qquad 1 \leqslant j \leqslant k \leqslant N, \tag{4.72}
$$

as follows by noting that the binomial probability in the latter integral represents the probability that $k-j$ requests are generated by $N-j$ idle sources within a time t.

To give a generally applicable computational method for the quantities A_{jk} we assume that the service-time distribution is given by

$$
B(t) = \sum_{i=1}^{r} q_i \left\{1 - \sum_{j=0}^{i-1} e^{-\mu t}\frac{(\mu t)^j}{j!}\right\}, \tag{4.73}
$$

where $q_i \geqslant 0$ for $1 \leqslant i \leqslant r$ and $\sum_{i=1}^{r} q_i = 1$. Recall from section 4.1 that the distribution function of a positive random variable can be approximated arbitrarily closely by a mixture of Erlangian distributions with the same scale parameters. The assumption (4.73) allows the service time with probability q_i to be represented as the sum of i independent exponential phases which are processed consecutively. Using this representation, we can exploit the lack of memory of the exponential distribution to derive a recursion scheme for the quantities A_{jk}. For that purpose we define

$A_{jki} = $ the expected amount of time that k customers are present during a service time that consists of i exponential phases and is started with j customers present.

Obviously,

$$
A_{jk} = \sum_{i=1}^{r} q_i A_{jki}.
$$

We now derive a recurrence relation for the quantities A_{jki} with k fixed. To do so, consider the situation in which k customers are in the system and the residual service time of the service in progress consists of i phases still to be completed. Using a well-known property of the exponential distribution (cf. (1.29) in section 1.4 of chapter 1), we have that with probability $\lambda_j/(\lambda_j + \mu)$ a new arrival occurs before the current phase in service is completed, in which case we get the situation of $j + 1$ customers in the system and a residual service time consisting of i phases to be completed. In the case where the current phase in service is completed before the next arrival, we get the situation of j customers in the system and a residual service time consisting of $i - 1$ phases to be completed. Thus, using the lack of memory of the Markovian input process and of the exponential processing time of a phase, it can be seen that, for fixed k,

$$A_{jki} = \frac{\lambda_j}{\lambda_j + \mu} A_{j+1,k,i} + \frac{\mu}{\lambda_j + \mu} A_{j,k,i-1}, \qquad 1 \leqslant j < k \quad \text{and} \quad 1 \leqslant i \leqslant r. \quad (4.74)$$

The quantity A_{kki} is to be computed from

$$A_{kki} = \frac{1}{\lambda_k + \mu} + \frac{\mu}{\lambda_k + \mu} A_{k,k,i-1}, \qquad 1 \leqslant i \leqslant r, \quad (4.75)$$

where the contribution $1/(\lambda_k + \mu)$ to A_{kki} represents the expected time until either an arrival occurs or a phase is completed. We have the boundary conditions

$$A_{jk0} = 0, \qquad 1 \leqslant j \leqslant k.$$

The recursion relations (4.74) and (4.75) provide a numerically stable and efficient scheme to compute the quantities A_{jki} for any fixed $k \geqslant 1$. First we recursively compute A_{kki} for $i = 1, \ldots, r$ from (4.75) starting with $A_{kk0} = 0$. Then, successively for $j = k - 1, \ldots, 1$, we recursively compute A_{jki} for $i = 1, \ldots, r$ from (4.74). It is noted that the above algorithm needs only an obvious modification when the service time has an H_2 distribution.

A simpler algorithm for the state probabilities for pure Erlangian service

Consider the particular case of (4.73) with $q_r = 1$, that is the service time has an Erlang-r distribution with mean r/μ. Then we can formulate an alternative algorithm that is much simpler than (4.70) in conjunction with (4.74) and (4.75). For pure Erlangian service the number of uncompleted phases in the system determines unambiguously the number of customers in the system. Moreover, the process describing the number of uncompleted phases in the system is a continuous-time Markov chain. Denoting by $\{f_j, j \geqslant 0\}$ the equilibrium distribution of this continuous-time Markov chain, we have

$$p_0 = f_0, \qquad p_j = \sum_{k=(j-1)r+1}^{jr} f_k \qquad \text{for } j \geqslant 1. \quad (4.76)$$

Letting

$$\gamma_0 = \lambda_0, \qquad \gamma_k = \lambda_m \qquad \text{for } (m-1)r < k \leqslant mr, \tag{4.77}$$

we have that customers arrive according to a Poisson process with rate γ_k whenever k uncompleted phases are in the system. Equating the rate at which the process leaves the macrostate of having at least j uncompleted phases in the system to the rate at which the system enters that macrostate yields

$$\mu f_j = \sum_{k=(j-r)^+}^{j-1} \gamma_k f_k, \qquad j = 1, 2, \ldots, \tag{4.78}$$

where $(j-r)^+ = \max(j-r, 0)$. This relation provides a simple recursion method to compute the probabilities f_j.

Example 4.1 (continued) A closed two-stage cyclic queueing model of multi-programming

The state probabilities p_j, $j = 0, \ldots, N$, of the number of jobs present at server 1 can be computed by applying an algorithm as discussed above (cf. also subsection 4.3.1 for the closely related model of the finite capacity $M/G/1$ queue). In particular, the probability p_0 giving the fraction of time server 1 is idle is a useful quantity. The average number of jobs processed per unit time at server 1 is $(1 - p_0)/E(S)$, implying that

$$\text{The average cycle time of a job} = \frac{NE(S)}{1 - p_0}.$$

Here the cycle time C of a job is defined as the time elapsed between two consecutive completions of that job at server 1. The above equation for $E(C)$ follows by noting that with a fixed number N of jobs in the system the average throughput at server 1 must be equal to N times the reciprocal of the average cycle time of a job. The mean cycle time $E(C)$ is invariant among the work-conserving queue disciplines. The higher moments of the cycle time depend of course on the queue discipline. A rather good approximation to the cycle time distribution is often obtained as follows. Supposing a sufficiently large number N of jobs, the cycle time is approximately distributed as $\tau_1 + \cdots + \tau_N$, where $\tau_1, \ldots, \tau_N$ are independent random variables each being distributed as the service time at the station with the slowest average service rate (cf. Boxma, 1985).

The state probabilities p_j, $0 \leqslant j \leqslant N$, can explicitly be given when server 1 also provides exponential service. For this particular case, the reader may easily verify from (4.78) that

$$p_j = \frac{\{1 - \lambda E(S)\}\{\lambda E(S)\}^j}{1 - \{\lambda E(S)\}^{N+1}}, \qquad j = 0, \ldots, N, \tag{4.79}$$

provided $\lambda E(S) \neq 1$; otherwise $p_j = 1/(N + 1)$ for $j = 0, \ldots, N$. This result assumes that server 1 provides exponential service on a first-come–first-served basis. In applications to computer systems it may be more realistic to assume the processor-sharing queue discipline at server 1. Under this discipline all jobs present at server 1 are served simultaneously where the unit work rate of the service is equally divided over all of the jobs present. It can be shown that the result (4.79) also holds when the service times of jobs at server 1 have a general distribution and server 1 uses the processor-sharing discipline (see section 2.5, chapter 2). ◻

Example 4.2 (continued) The machine repairman model of an interactive computer system

In the machine repairman model having general service times for the requests we can compute the distribution $\{p_j, \ 0 \leqslant j \leqslant N\}$ of the number of requests present at the server by applying an algorithm as discussed above. We consider here in more detail the important case of exponential service times for the requests. It is convenient to write $\{p_j(N)\}$ and $\{\pi_j(N)\}$ for the distributions $\{p_j\}$ and $\{\pi_j\}$ in order to make clear the dependence of these distributions on the number N of sources.

A specialization of (4.78) to the machine repairman model having exponential service with rate $\mu = 1/E(S)$ yields

$$p_j(N) = \frac{N!}{(N - j)!} \{\eta E(S)\}^j p_0(N), \qquad j = 0, \ldots, N.$$

Then, using the normalization equation $\sum_{j=0}^{N} p_j(N) = 1$,

$$p_j(N) = \frac{\{\eta E(S)\}^j/(N - j)!}{\sum_{k=0}^{N} \{\eta E(S)\}^k/(N - k)!}, \qquad j = 0, \ldots, N. \tag{4.80}$$

Using (4.59), (4.60) and (4.80), it follows after some straightforward manipulations that the customer-average probabilities $\pi_j(N)$ are given by

$$\pi_j(N) = \frac{\{\eta E(S)\}^j/(N - 1 - j)!}{\sum_{k=0}^{N-1} \{\eta E(S)\}^k/(N - 1 - k)!}, \qquad j = 0, \ldots, N - 1. \tag{4.81}$$

Hence we find for the case of exponential service the remarkable result

$$\pi_j(N) = p_j(N - 1), \qquad j = 0, \ldots, N - 1.$$

This result has the following important interpretation. The state distribution seen by a newly arriving request at the server is the same as the state distribution would be at a randomly chosen epoch in a system having one source less and thus having $N - 1$ circulating customers. It appears that a result of this type is generally valid for a large class of closed queueing networks (see Sevcik and Mitrani, 1981).

In view of applications to computer systems it is important to point out that the results (4.80) and (4.81) hold for think and service times both having general distributions with respective means $1/\eta$ and $E(S)$ when the server uses the processor-sharing queue discipline (see section 2.5, chapter 2). In this context we also point out that for the machine repairman model with the first-come–first-served queue discipline the state probabilities are also insensitive to the distributional form of the think times when the repair times are exponentially distributed.

Further, we point out here that the result (4.81) and Little's law provide the basis for an extremely useful recursive solution method for the computation of the average throughput and average queue lengths in closed queueing networks with exponential servers or processor sharing. This recursive solution method is known as the *mean value analysis*. We shall use the machine repairman model as a vehicle to introduce the basic ideas of the mean value analysis. To do so we assume that the think and service times are both exponentially distributed and that service is in order of arrival. For the machine repairman model with N sources, define

$\lambda(N) = $ the average throughput at the server,

$EW(N) = $ the average time a request spends at the server,

$EL(N) = $ the average number of requests at the server.

Using the lack of memory of the exponential service-time distributions and noting that service is in order of arrival, we find

The average time a request spends at the server
$= E(S) + E(S) \times$ (the average number of other requests seen upon arrival).

Thus, using (4.81),

$$EW(N) = E(S) + E(S)EL(N-1). \tag{4.82}$$

Noting that $\lambda(N)$ also represents the average arrival rate of requests at the server, we have by Little's law that

$$EL(N) = \lambda(N)EW(N). \tag{4.83}$$

An extra relation between $EL(N)$ and $EW(N)$ is obtained by using the fact that the number of sources generating requests is fixed. Thus the average throughput $\lambda(N)$ must be equal to N times the reciprocal of the average cycle time between two requests generated by a same source. This yields

$$\lambda(N) = \frac{N}{EW(N) + 1/\eta}. \tag{4.84}$$

The equations (4.82) to (4.84) provide a simple recursion scheme for the computation of the averages $EW(N)$, $\lambda(N)$ and $EL(N)$ for $N = 1, 2, \ldots$. Supposing the average queue length $EL(N-1)$ for a system with $N-1$ sources is known, we calculate the averages $EW(N)$, $\lambda(N)$ and $EL(N)$ for a system with N sources by applying subsequently the equations (4.82), (4.84) and (4.83). The recursion is obviously initialized with $EL(0) = 0$. The recursion scheme (4.82) to (4.84) can be extended to a large class of closed queueing networks, cf. De Soua e Silva, Lavenberg and Muntz (1984), Reiser and Lavenberg (1980) and Sauer and Chandy (1981).   □

The waiting-time distribution

We next turn to the computation of the waiting-time distribution in the single-server queueing system with state-dependent Markovian input. Here we assume the first-come–first-served queue discipline. To compute the probability distribution function $W_q(x)$ of the delay in queue of a customer, it is in general not sufficient to know the customer-average probability distribution $\{\pi_j\}$. We also need information about the residual service time of the service in progress (if any). This required information can be represented efficiently when the service time has a phase-type distribution. In the discussion below we assume that the service-time distribution function is given by (4.73). Then the stochastic process $\{X(t)\}$ describing jointly the number of customers in the system and the number of uncompleted phases of the service in progress (if any) is a continuous-time Markov chain. For this stochastic process we define the time-average and the customer-average probabilities p_{ji} and π_{ji} by

$p_{ji} = \lim_{t \to \infty} P\{j$ customers are in the system at time t and the residual
$\qquad$ service time of the service in progress at time t consists of i phases$\}$

and

$\pi_{ji} = \lim_{n \to \infty} P\{$the nth customer sees upon arrival j other customers
$\qquad$ in the system and a service in progress whose residual
$\qquad$ service time consists of i phases$\}$,

with the convention $p_{00} = p_0$ and $\pi_{00} = \pi_0$. The customer-average probabilities π_{ji} determine the waiting-time distribution, as will be shown later. We first discuss the computation of the probabilities p_{ji} and π_{ji}. By making an obvious modification to the proof of (4.59) (define N_{ji} as the number of arrivals in a cycle finding the system in state (j, i)),

$$\pi_{ji} = \frac{\lambda_j p_{ji}}{\sum_{k=0}^{\infty} \lambda_k p_k}. \tag{4.85}$$

Equating the rate at which the process $\{X(t)\}$ leaves any state to the rate at which the process enters that state yields the balance equations

$$\lambda_0 p_0 = \mu p_{11},$$
$$(\mu + \lambda_1)p_{1i} = \mu p_{1,i+1} + \mu q_i p_{21} + \lambda_0 q_i p_0, \qquad 1 \leqslant i \leqslant r,$$
$$(\mu + \lambda_j)p_{ji} = \mu p_{j,i+1} + \mu q_i p_{j+1,1} + \lambda_{j-1} p_{j-1,i}, \qquad j \geqslant 2, 1 \leqslant i \leqslant r, \tag{4.86}$$

with $p_{j,r+1} = 0$ by convention. In principle the microstate probabilities could be computed by solving this system of linear equations using truncation when this system is infinite. However, we can develop a much more efficient recursion scheme to compute the probabilities p_{ji} by combining the above balance equations and the recursion scheme for the macrostate probabilities p_j. To do so we need the macrostate balance equation

$$\mu p_{j+1,1} = \lambda_j p_j, \qquad j \geqslant 0, \tag{4.87}$$

which follows by equating the rate at which the process $\{X(t)\}$ leaves the macrostate of having at least $j + 1$ customers in the system to the rate at which the process enters that state.

The following recursion scheme can now be given to compute the probabilities $p_{ji}(1 \leqslant i \leqslant r)$ successively for $j = 1, 2, \ldots$. Once we have computed p_j from an appropriate recursion scheme discussed earlier, we compute $p_{j+1,1}$ from (4.87) and next compute p_{ji} from (4.86) using the previously computed values $p_{j+1,1}$ and $p_{j,i+1}$. Note that this computational scheme is numerically stable since no differences are evaluated during the computations. Finally, the desired probability π_{ji} follows from p_{ji} by applying (4.85) and (4.60).

We now turn to the specification of the waiting-time distribution function $W_q(x)$. To do so define z_k as the steady-state probability that a customer finds upon arrival k phases yet to be completed in the system. The waiting time in queue of a customer finding upon arrival k uncompleted phases in the system is distributed as the sum of k independent exponentials with common mean $1/\mu$, and thus has an Erlang-k distribution. Here we assume service in order of arrival. Thus we find

$$P\{W_q > x\} = \sum_{k=1}^{\infty} z_k \sum_{j=0}^{k-1} e^{-\mu x} \frac{(\mu x)^j}{j!}, \qquad x \geqslant 0. \tag{4.88}$$

The probabilities z_k can be computed from the probabilities π_{ji}. Therefore, denote by $q_t^{(n)}$ the probability that n customers waiting in queue represent a total of t phases to be completed. That is $q_0^{(0)} = 1$ and $q_t^{(0)} = 0$ for $t \neq 0$, while for $n \geqslant 1$ the probabilities $q_t^{(n)}$ can be obtained from the convolution formula

$$q_t^{(n)} = \sum_{i=1}^{r} \hat{q}_i q_{t-i}^{(n-1)}, \qquad t = n, \ldots, nr.$$

Note that $q_t^{(n)} = 0$ for $t < n$. It now follows that

$$z_k = \sum_{j=1}^{k} \sum_{i=1}^{r} \pi_{ji} q_{k-i}^{(j-1)}, \qquad k = 1, 2, \ldots.$$

The first moment of W_q is computed from (4.67). It is left to the reader to verify that

$$E(W_q^2) = \sum_{k=1}^{\infty} z_k \frac{k(k+1)}{\mu^2}.$$

The computation of the probabilities z_k can be simplified considerably when $q_r = 1$, that is when the service time has an Erlang-r distribution. In this particular case the time-average probability f_j of having j uncompleted phases in the system is computed from the simple recursion scheme (4.78), while the customer-average probability z_j is related to f_j by (verify!)

$$z_j = \frac{\gamma_j f_j}{\sum_{k=0}^{\infty} \lambda_k p_k} = \frac{\gamma_j f_j E(S)}{1 - f_0}, \qquad j = 0, 1, \ldots,$$

with γ_j given by (4.77).

To conclude this subsection, we give for both models of the examples 4.1 and 4.2 the waiting-time probabilities for two different service-time densities in order to see how sensitive these probabilities are to the higher moments of the service time. We consider the E_2 and $E_{1,3}$ densities each having $E(S) = 1$ and $c_S^2 = 0.5$. In Table 4.4 we give $P\{W_q > x\}$ for the two-stage cyclic queueing model of multi-programming with $K = 10$ jobs where the load factor $\rho = \lambda E(S)$ is varied

Table 4.4   The probabilities $P\{W_q > x\}$ for the cyclic queueing model with $K = 10$ jobs

	x	0	1	2	3	5	7	10	$E(W_q)$	$E(W_q^2)$
$\rho = 0.5$	E_2	0.4999	0.2660	0.1309	0.0638	0.0150	0.0034	0.0003	0.748	2.105
	$E_{1,3}$	0.4999	0.2699	0.1307	0.0625	0.0142	0.0031	0.0003	0.748	2.069
$\rho = 0.8$	E_2	0.7876	0.6010	0.4433	0.3225	0.1607	0.0690	0.0121	2.401	12.325
	$E_{1,3}$	0.7877	0.6054	0.4459	0.3238	0.1608	0.0689	0.0118	2.409	12.315
$\rho = 1.5$	E_2	0.9976	0.9917	0.9801	0.9580	0.8432	0.5761	0.1550	7.478	62.448
	$E_{1,3}$	0.9974	0.9912	0.9788	0.9557	0.8393	0.5771	0.1552	7.462	62.243

Table 4.5   The probabilities $P\{W_q > x\}$ for the machine repairman model with $K = 10$ sources

	x	0	1	2	3	5	7	10	$E(W_q)$	$E(W_q^2)$
$\eta = 0.05$	E_2	0.4249	0.1929	0.0760	0.0284	0.0035	0.0004	—	0.503	1.071
	$E_{1,3}$	0.4248	0.1966	0.0755	0.0271	0.0030	0.0003	—	0.504	1.051
$\eta = 0.10$	E_2	0.7516	0.5024	0.3005	0.1662	0.0412	0.0079	0.0005	1.484	4.849
	$E_{1,3}$	0.7510	0.5081	0.3036	0.1669	0.0402	0.0073	0.0004	1.490	4.828
$\eta = 0.20$	E_2	0.9811	0.9207	0.8093	0.6554	0.3246	0.1107	0.0120	4.094	22.046
	$E_{1,3}$	0.9800	0.9200	0.8099	0.6579	0.3277	0.1106	0.0110	4.100	22.049

as 0.5, 0.8 and 1.5. In Table 4.5 we give $P\{W_q > x\}$ for the machine repairman model of an interactive computer system with $K = 10$ sources where the average think time $1/\eta$ is varied as 20, 10 and 5. We indeed find that the waiting-time probabilities are rather insensitive to more than the first two moments of the service time, provided c_S^2 is not too large. For the machine repairman model we also find that in many cases the delay probability and the average throughput are fairly insensitive to the distributional form of the repair time.

4.2.3   The GI/G/1 queueing system

This subsection considers the GI/G/1 queueing system in which the interarrival and service times both have general probability distributions. Here it is assumed that the offered load ρ is less than 1. In general the GI/G/1 queue is very difficult to analyse and simple analytical results can be obtained only for special cases. However, many interarrival-time and service-time distributions of practical interest allow for an algorithmic analysis resulting in tractable computational methods. Besides, useful approximations can be given. The development of computational methods will be the subject of this subsection; cf. also subsection 4.4.7 for additional discussions.

An embedded Markov chain approach for phase-type service

In many applications the service-time distribution may be represented by a mixture of Erlang distributions with the same scale parameters. Under the assumption that the service-time distribution function $B(t)$ is given by

$$B(t) = \sum_{i=1}^{r} q_i \left\{ 1 - \sum_{j=0}^{i-1} e^{-\mu t} \frac{(\mu t)^j}{j!} \right\}, \qquad t \geq 0,$$

we can develop an effective computational method for the waiting-time probabilities when service is in order of arrival. As usual we interpret the service time as a random sum of independent exponential phases with common mean $1/\mu$. Then, by the memoryless property of the exponential distribution, the embedded process describing the number of uncompleted phases in the system just prior to the arrival epochs is a discrete-time Markov chain. Also, this Markov chain is aperiodic. Letting, for $n = 1, 2, \ldots,$

$X_n =$ the number of uncompleted phases in the system just prior
to the arrival of the nth customer,

we have that, for any $i = 0, 1, \ldots,$

$$z_i = \lim_{n \to \infty} P\{X_n = i\}$$

exists and is independent of the initial state of the system. Since the delay in queue of a customer finding upon arrival i uncompleted phases in the system is distributed as the sum of i independent exponentials with common mean $1/\mu$, it follows with (1.22) in section 1.4 of chapter 1 that

$$P\{W_q > x\} = \sum_{i=1}^{\infty} z_i \sum_{k=0}^{i-1} e^{-\mu x} \frac{(\mu x)^k}{k!}, \quad x \geq 0. \tag{4.89}$$

In particular,

$$P\{W_q > 0\} = 1 - z_0 \quad \text{and} \quad E(W_q) = \sum_{i=1}^{\infty} \frac{i}{\mu} z_i.$$

To find the limiting probabilities z_i of the Markov chain $\{X_n\}$, we first have to determine the one-step transition probabilities $p_{ij} = P\{X_{n+1} = j | X_n = i\}$ of this Markov chain. To do so note that the service completions of phases occur according to a Poisson process with rate μ as long as the server is busy. Letting $a(t)$ denote the probability density of the interarrival time A and noting that an arrival increases the number of phases in the system by k with probability q_k, it follows by the usual conditioning arguments that for $j \neq 0$,

$$p_{ij} = \sum_{k=1}^{r} q_k \int_0^{\infty} P\left\{ \begin{matrix} \text{there occur } i+k-j \text{ phase completions during an} \\ \text{interarrival time } A | A = t \end{matrix} \right\} a(t)\,dt$$

$$= \sum_{k=1}^{r} q_k \int_0^{\infty} e^{-\mu t} \frac{(\mu t)^{i+k-j}}{(i+k-j)!} a(t)\,dt, \quad i \geq 0, \quad 1 \leq j \leq i+r. \tag{4.90}$$

Some care should be exercised in evaluating p_{i0}. The easiest way to find p_{i0} is to use

$$p_{i0} = 1 - \sum_{j=1}^{i+r} p_{ij}, \quad i \geq 0.$$

Clearly, $p_{ij} = 0$ for $j > i+r$. The integrals in (4.90) are easily reduced to finite sums for deterministic, Erlangian and hyperexponential interarrival times by using the identity (1.23) in section 1.4 of chapter 1.

Next, by (2.11) in section 2.1 of chapter 2, the limiting probabilities z_j can be computed as the unique solution to the linear equations

$$z_j = \sum_{i=0}^{\infty} z_i p_{ij}, \quad j = 0, 1, \ldots,$$

together with the normalization equation

$$\sum_{j=0}^{\infty} z_j = 1.$$

To solve this infinite system of linear equations, one should truncate this system by choosing a sufficiently large integer N such that $\sum_{i>N} z_i < \varepsilon$ for some small number ε (for example $\varepsilon = 10^{-7}$). The resulting finite system of linear equations can be solved by the successive overrelaxation method discussed in appendix

D. In general the truncation integer N has to be found by trial and error. For 'reasonable' arrival processes having less variability than the Poisson process, one might expect that the tail probabilities $\sum_{i>k} z_i$, $k \geqslant 1$, are smaller than the corresponding tail probabilities in the M/G/1 queue with the same arrival rate and the same service-time distribution as the GI/G/1 queue. Then an upper bound for N follows by applying first the recursion scheme (4.39) until the sum of the probabilities f_j exceeds $1 - \varepsilon$.

It is important to point out that for the special case of Erlang-r service (that is $q_r = 1$) the number of phases in the system determines unambiguously the number of customers present so that the probability π_i of finding i other customers in the system is given by

$$\pi_0 = z_0, \qquad \pi_i = \sum_{j=(i-1)r+1}^{ir} z_j \qquad \text{for } i \geqslant 1. \tag{4.91}$$

A similar remark applies to the finite capacity GI/E$_r$/1 queue.

Using the algorithm (4.89) we computed in Table 4.6 the performance measures $\Pi_W = P\{W_q > 0\}$ and $E(W_q)$ for various interarrival and service-time densities in order to see how sensitive these measures are to the higher moments of the interarrival and service times. The densities considered include the E_4 density, the E_2 density, the $E_{1,3}$ density with squared coefficient of variation $\frac{1}{2}$ and the H_2 densities with balanced means (H_2^b) and the gamma normalization (H_2^g), each having a squared coefficient of variation of 2. The offered load $\rho = E(S)/E(A)$ is varied as 0.1, 0.3, 0.5 and 0.7, where in each example $E(S) = 1$ is taken. It appears that the shape of the interarrival time density has a considerably larger effect on the performance measures than the shape of the service-time density, in particular for light traffic being characterized by a small value of the delay probability Π_W. Numerical investigations indicate that Π_W and $E(W_q)$ are fairly insensitive to more than the first two moments of the interarrival and service times, provided c_A^2 and c_S^2 are not too large and the

Table 4.6   Sensitivity results for the delay probability and the average delay

	$\rho = 0.1$		$\rho = 0.3$		$\rho = 0.5$		$\rho = 0.7$	
	Π_W	$E(W_q)$	Π_W	$E(W_q)$	Π_W	$E(W_q)$	Π_W	$E(W_q)$
$E_2/E_2/1$	0.0240	0.0166	0.1600	0.1310	0.3596	0.3904	0.5972	1.0391
$E_{1,3}/E_2/1$	0.0544	0.0433	0.1882	0.1700	0.3727	0.4293	0.5997	1.0757
$E_2/E_{1,3}/1$	0.0242	0.0162	0.1615	0.1292	0.3625	0.3875	0.6003	1.0353
$E_{1,3}/E_{1,3}/1$	0.0543	0.0432	0.1882	0.1687	0.3740	0.4263	0.6020	1.0712
$H_2^g/E_4/1$	0.1864	0.1451	0.4695	0.5786	0.6603	1.3045	0.8078	2.8684
$H_2^b/E_4/1$	0.1318	0.0951	0.3828	0.3909	0.6071	0.9827	0.7927	2.4626
$H_2^g/E_2/1$	0.1842	0.1709	0.4597	0.6554	0.6501	1.4498	0.8015	3.1751
$H_2^b/E_2/1$	0.1315	0.1137	0.3802	0.4626	0.6012	1.1427	0.7860	2.8022

traffic is not too light (cf. also Klincewicz and Whitt, 1984). Also, the numerical results reveal the surprising result that for irregular interarrival times with $c_A^2 > 1$ the delay probability Π_W decreases when c_S^2 increases (cf. also Table 4.7). This behaviour is just the way it works and a satisfactorily intuitive explanation seems difficult to give. A theoretical proof for the case of H_2 input can be found in Whitt (1984a).

The regenerative approach for phase-type arrivals

The regenerative approach introduced in subsection 4.2.1 for the M/G/1 queue can in principle be extended to the GI/G/1 queue with phase-type arrivals.

Table 4.7   Exact and approximate values for Π_W and $E(W_q)$

		$\rho = 0.2$		$\rho = 0.5$		$\rho = 0.8$	
		Π_W	$E(W_q)$	Π_W	$E(W_q)$	Π_W	$E(W_q)$
D/E$_4$/1	exa	3.2×10^{-6}	9.3×10^{-7}	0.0473	0.0173	0.4457	0.3187
	Fre	3.2×10^{-6}	9.3×10^{-7}	0.0462	0.0170	0.4308	0.3119
	KLB	4.7×10^{-3}	7.3×10^{-7}	0.0909	0.0087	0.4571	0.2567
D/E$_2$/1	exa	5.0×10^{-4}	2.7×10^{-4}	0.1164	0.0784	0.5476	0.7568
	Fre	5.0×10^{-4}	2.7×10^{-4}	0.1139	0.0771	0.5378	0.7481
	KLB	8.7×10^{-3}	3.0×10^{-4}	0.1429	0.0659	0.5565	0.7165
D/M/1	exa	0.00698	0.00703	0.2032	0.2550	0.6286	1.6927
	Fre	0.00698	0.00703	0.2032	0.2550	0.6286	1.6927
	KLB	0.01538	0.00869	0.2000	0.2567	0.6244	1.6930
E$_4$/D/1	exa	0.0092	0.0021	0.1633	0.0497	0.5783	0.3861
	KLB	0.0214	0.0001	0.1875	0.0279	0.6214	0.3436
E$_2$/D/1	exa	0.0645	0.0242	0.3233	0.1767	0.7019	0.9033
	KLB	0.0636	0.0165	0.3125	0.1791	0.7189	0.9200
H$_2$/D/1	exa	0.3500	0.2811	0.6724	1.1552	0.8785	4.2546
	KLB	0.2552	0.1676	0.6250	0.7788	0.8719	3.6193
E$_2$/E$_2$/1	exa	0.0813	0.0604	0.3596	0.3904	0.7262	1.8653
	KLB	0.0720	0.0642	0.3409	0.4232	0.7309	1.9184
E$_2$/M/1	exa	0.0938	0.1035	0.3820	0.6180	0.7399	2.8440
	KLB	0.0786	0.1202	0.3571	0.6711	0.7370	2.9178
E$_2$/H$_2$/1	exa	0.1103	0.2026	0.4052	1.0948	0.7522	4.8247
	KLB	0.0882	0.2394	0.3750	1.1694	0.7431	4.9174
H$_2$/E$_2$/1	exa	0.3362	0.3868	0.6501	1.4498	0.8701	5.2809
	KLB	0.2547	0.2559	0.6212	1.1031	0.8688	4.7561
H$_2$/M/1	exa	0.3255	0.4827	0.6340	1.7321	0.8631	6.3029
	KLB	0.2542	0.3282	0.6176	1.3801	0.8660	5.8033

Although the resulting recursion scheme for the state probabilities turns out not to be computationally tractable, useful results for the delay probability and the average queue length can be deduced from this recursion scheme by using the technique of generating functions. The extension of the regenerative approach to the GI/G/1 queue with phase-type arrivals is based on the observation that at the service completion epochs the status of the system is completely described by the number of customers left behind and the current phase of the interarrival time in progress. The extended regenerative approach will be outlined for the $H_2/G/1$ queue in which the interarrival time density $a(t)$ is given by

$$a(t) = p_1 \lambda_1 e^{-\lambda_1 t} + p_2 \lambda_2 e^{-\lambda_2 t}, \qquad t \geq 0, \qquad (4.92)$$

with $0 \leq p_1, p_2 \leq 1$. It is helpful to imagine the following physical realization of the interarrival times. After the occurrence of an arrival, the time until the next arrival is determined to be exponentially distributed with either mean $1/\mu_1$ or mean $1/\mu_2$ according to a lottery with respective probabilities p_1 and p_2. Let us say that the arrival process is in phase $i (= 1, 2)$ if the time until the next arrival is exponentially distributed with mean $1/\mu_i$. The system is said to be in state (j, i) whenever j customers are in the system and the arrival process is in phase i. The following probabilities are defined:

$$p_{ji} = \lim_{t \to \infty} P\{\text{the system is in state } (j, i) \text{ at time } t\},$$

$$\pi_{ji} = \lim_{n \to \infty} P\{\text{the system is in state } (j, i) \text{ just prior to the arrival of}$$
$$\text{the } n\text{th customer}\},$$

$$q_{ji} = \lim_{n \to \infty} P\{\text{the system is in state } (j, i) \text{ just behind the departure of}$$
$$\text{the } n\text{th customer}\}.$$

The time-average probabilities p_{ji} and the customer-average probabilities π_{ji} are related by

$$\pi_{ji} = \frac{\lambda_i p_{ji}}{\lambda}, \qquad j \geq 0, \quad i = 1, 2, \qquad (4.93)$$

where λ denotes the average arrival rate and is given by

$$\lambda = \left(\frac{p_1}{\lambda_1} + \frac{p_2}{\lambda_2} \right)^{-1}.$$

This relationship can be seen as follows. The probability π_{ji} can be interpreted as the long-run fraction of arrivals finding the system in state (j, i). This fraction is equal to the average number of arrivals per unit time finding the system in state (j, i) divided by the average arrival rate λ. Since customers arrive according to a Poisson process with rate λ_i whenever the system is in state (j, i) and Poisson arrivals see time averages, the average number of arrivals per unit time finding the system in state (j, i) equals $\lambda_i p_{ji}$. In addition to (4.93), we have that

$$q_{j1} + q_{j2} = \pi_{j1} + \pi_{j2}, \qquad j \geq 0, \tag{4.94}$$

as follows by noting that the long-run average number of arrivals finding j other customers in the system must be equal to the long-run average number of departures leaving j other customers behind in the system.

The stochastic process describing the state of the system regenerates itself each time an arrival finds the system empty. Assuming that epoch 0 is such a regeneration epoch, define

T = the next epoch at which an arrival finds the system empty,
T_{ki} = the amount of time the system is in state (k, i) during the cycle $(0, T]$,
N = the number of service completions in $(0, T]$,
N_{ji} = the number of service completions in $(0, T]$ at which the system is left behind in state (j, i).

By the theory of regenerative processes,

$$p_{ji} = \frac{E(T_{ji})}{E(T)} \quad \text{and} \quad q_{ji} = \frac{E(N_{ji})}{E(N)}, \qquad j \geq 0, i = 1, 2. \tag{4.95}$$

Also, define for $1 \leq j \leq k$ and $i, l = 1, 2$,

A_{jk}^{li} = the expected amount of time the system spends in state (k, i) during a service started when the system is in state (j, l).

Analogously to (4.26) it can be seen that

$$E(T_{ki}) = p_1 A_{1k}^{1i} + p_2 A_{1k}^{2i} + \sum_{j=1}^{k} \{E(N_{j1})A_{jk}^{1i} + E(N_{j2})A_{jk}^{2i}\}, \quad k \geq 1, i = 1, 2. \tag{4.96}$$

Next we derive a second set of relations between the quantities $E(T_{ji})$ and $E(N_{ji})$. To do so we use the balance equation stating that, for any subset S of states,

The expected number of transitions out of the set S during a cycle
= the expected number of transitions into the set S during a cycle.

By choosing $S = \{(0, 1), \ldots, (k, 1)\}$ and noting that customers arrive according to a Poisson process with rate λ_i whenever the system is in some state (j, i), it follows that

$$p_2 \lambda_1 \sum_{j=0}^{k-1} E(T_{j1}) + \lambda_1 E(T_{k1}) = p_1 \lambda_2 \sum_{j=0}^{k-1} E(T_{j2}) + E(N_{k1}), \qquad k \geq 0. \tag{4.97}$$

Similarly, by choosing $S = \{(0, 2), \ldots, (k, 2)\}$,

$$p_1 \lambda_2 \sum_{j=0}^{k-1} E(T_{j2}) + \lambda_2 E(T_{k2}) = p_2 \lambda_1 \sum_{j=0}^{k-1} E(T_{j1}) + E(N_{k2}), \qquad k \geq 0. \tag{4.98}$$

Using (4.95) and the identity $E(N)/E(T) = \lambda$ (why?), the relations (4.96) to (4.98) provide four sets of recurrence equations for the probabilities p_{j1}, p_{j2}, q_{j1} and q_{j2}, $j \geqslant 0$. These equations are not to be recommended for the computation of the state probabilities (see subsection 4.4.7 for a general discussion of better suited methods). Nevertheless, they enable us to derive explicit expressions for the delay probability and the average queue size by using the technique of generating functions. This derivation can be found in Van Hoorn and Seelen (1983). Here we state only the ultimate results. Letting ξ_0 be the unique solution to the characteristic equation

$$\lambda_1 \lambda_2 + \xi^2 - (\lambda_1 + \lambda_2)\xi - \left(\int_0^\infty e^{-\xi t} b(t) \, dt \right) \{\lambda_1 \lambda_2 - (p_1 \lambda_1 + p_2 \lambda_2)\xi\} = 0,$$

in $\xi \geqslant \min(\lambda_1, \lambda_2)$ with $b(t)$ denoting the service time density, we have

$$\Pi_W = 1 - \frac{(1 - \rho)\lambda_1 \lambda_2}{\xi_0} \tag{4.99}$$

and

$$E(L_q) = \frac{\lambda^2}{2(1 - \rho)} \left\{ E(S^2) + \frac{2p_1}{\lambda_1^2} \right.$$
$$\left. + \frac{2p_2}{\lambda_2^2} + 2E(S)\frac{p_1 \lambda_1 + p_2 \lambda_2}{\lambda_1 \lambda_2} - \frac{2(\lambda_1 + \lambda_2)}{\lambda_1 \lambda_2} \right\} + \frac{\lambda}{\xi_0}. \tag{4.100}$$

These results are in agreement with the more general result (5.205) in chapter 5 of Cohen (1982). Actually, it follows from the latter result that the formulae (4.99) and (4.100) also apply when the interarrival density is a K_2 density of the form (4.92); such densities allow for a squared coefficient of variation c_A^2 larger than $\frac{1}{2}$ (cf. also appendix B). Incidentally, a K_2 density with c_A^2 very close to $\frac{1}{2}$ (say, $c_A^2 = 0.50001$) is for practical purposes a sufficiently close approximation to the E_2 density. The result (5.205) in Cohen (1982) also covers the case of interarrival times having a mixture of Erlangian distributions in which case, however, the calculation of complex (conjugate) roots of a characteristic equation is required when $c_A^2 < \frac{1}{2}$.

The D/G/1 queue

The D/G/1 queueing model is particularly useful in the analysis of appointment systems or clocked schedules for computer systems. In the case where the service-time density is given by the K_2 density

$$b(t) = p_1 \mu_1 e^{-\mu_1 t} + p_2 \mu_2 e^{-\mu_2 t}, \qquad t \geqslant 0,$$

the following exact results for the $D/K_2/1$ queue are obtained from the general expression (5.191) in Cohen (1982):

$$P\{W_q > 0\} = 1 - \frac{\xi_1 \xi_2}{\mu_1 \mu_2}, \qquad E(W_q) = -\frac{\mu_1 + \mu_2}{\mu_1 \mu_2} + \frac{1}{\xi_1} + \frac{1}{\xi_2}, \tag{4.101}$$

where ξ_1 and ξ_2 with $0 < \xi_1 < \min(\mu_1, \mu_2) \leqslant \xi_2$ are the two positive roots of the characteristic equation

$$\mu_1 \mu_2 + \xi^2 - (\mu_1 + \mu_2)\xi - e^{-\xi D}\{\mu_1 \mu_2 - (p_1 \mu_1 + p_2 \mu_2)\xi\} = 0.$$

The expression (5.191) in Cohen (1982) also covers the case of generalized Erlangian service with $c_S^2 < \frac{1}{2}$. However, for the particular case of the D/G/1 queue with $c_S^2 < \frac{1}{2}$ it is easier to compute the following excellent approximations due to Fredericks (1982):

$$\Pi_W^{Fre} = \frac{1 - B(D)}{\int_D^\infty e^{\delta(t-D)} b(t)\, dt},$$

$$E^{Fre}(W_q) = \frac{\int_D^\infty (t-D) b(t)\, dt}{B(D) - 1 + \int_D^\infty e^{\delta(t-D)} b(t)\, dt}, \tag{4.102}$$

where δ is the unique positive solution to the equation

$$e^{-\delta D} \int_0^\infty e^{\delta y} b(y)\, dy = 1.$$

It is remarkable how extremely accurate the approximations in (4.102) are for the case of light traffic (see Table 4.7). In this case the induced approximation for $E(W_q \mid W_q > 0)$ is also very accurate, although the average conditional waiting time is quite sensitive to errors in the delay probability when this probability is very small.

Two-moment approximations for Π_W and $E(W_q)$

Using a hybrid combination of basic queueing results and experimental analysis, the following two-moment approximations for Π_W and $E(W_q)$ were obtained by Krämer and Langenbach-Belz (1976):

$$\Pi_W^{KLB} = \rho + (c_A^2 - 1)\rho(1-\rho) \times \begin{cases} \dfrac{1 + c_A^2 + \rho c_S^2}{1 + \rho(c_S^2 - 1) + \rho^2(4c_A^2 + c_S^2)} & \text{if } c_A^2 \leqslant 1, \\[4mm] \dfrac{4\rho}{c_A^2 + \rho^2(4c_A^2 + c_S^2)} & \text{if } c_A^2 > 1, \end{cases} \tag{4.103}$$

$$E^{KLB}(W_q) = \frac{\rho E(S)}{2(1-\rho)}(c_A^2 + c_S^2) \times \begin{cases} \exp\left\{\dfrac{-2(1-\rho)(1-c_A^2)^2}{3\rho(c_A^2 + c_S^2)}\right\} & \text{if } c_A^2 \leqslant 1, \\[4mm] \exp\left\{\dfrac{-(1-\rho)(c_A^2 - 1)}{c_A^2 + 4c_S^2}\right\} & \text{if } c_A^2 > 1. \end{cases} \tag{4.104}$$

These approximations are useful for quick estimates in practical engineering provided c_A^2 is not too large and Π_W is not too small (cf. also Table 4.7). It is our numerical experience that for H_2 input ($c_A^2 > 1$) the approximations perform rather well for the normalization of balanced means but may perform less satisfactorily for other normalizations. In the case where a two-moment approximation for the average conditional waiting time $E(W_q | W_q > 0)$ is required, we recommend the use of the special-purpose approximation (4.230) with $c = 1$ in subsection 4.4.7 rather than the ratio of the above two approximations for $E(W_q)$ and Π_W.

In Table 4.7 we give the exact and approximate values of Π_W and $E(W_q)$ for a number of GI/G/1 queueing systems. The interarrival-time and service-time distributions include the deterministic distribution, Erlang-k distributions and the H_2 distribution with the gamma normalization and a squared coefficient of variation of 2. The offered load ρ is varied as 0.2, 0.5 and 0.8. In all examples we take $E(S) = 1$. In Table 4.7 the exact results are denoted by exa, the approximations in (4.102) by Fre and the two-moment approximations (4.103) and (4.104) by KLB.

An approximation for the tail probabilities of the conditional waiting times

Numerical experiments indicate that in many practical applications of the GI/G/1 queueing model the asymptotically exponential expansion (4.16) of the waiting-time distribution function already applies for x in the range of $E(S)$ to $2E(S)$ provided the delay probability is not too small (say, $\Pi_W \geqslant 0.2$). Therefore we would like to have a simple method for the computation of the coefficients of the asymptotic expansion. The decay parameter of the asymptotically exponential tail is easily computed, but the computation of the amplitude coefficient is not so simple and requires in general the calculation of the state probabilities. However, when considering the exponential tail of the *conditional* waiting distribution, a rather easily computable approximation to its amplitude coefficient may be given by using analytical results for the special cases of the M/G/1 queue and the GI/M/1 queue.

Letting the decay parameter δ be the unique solution to the readily solved equation (4.13) with $c = 1$, we have by (4.16) that

$$P\{W_q > x \,|\, W_q > 0\} \approx \phi e^{-\delta x} \qquad \text{for } x \text{ large,} \qquad (4.105)$$

where the amplitude coefficient ϕ can be characterised as

$$\phi = \frac{\delta \eta}{\lambda (\tau - 1)^2 \Pi_W}. \qquad (4.106)$$

Here λ is the average arrival rate and the constants τ and η are given by (4.14) and

(4.17). For the particular case of the M/G/1 queue we have by (4.33) and (4.40) that

$$\phi = \frac{1-\rho}{\rho\lambda\delta}\left[\int_0^\infty ye^{\delta y}\{1 - B(y)\}\,dy\right]^{-1}. \tag{4.107}$$

For the particular case of the GI/M/1 queue it follows from (2.19) in section 2.2 of chapter 2 that, for some $0 < \omega < 1$,

$$P\{W_q > 0\} = \omega \quad \text{and} \quad P\{W_q > x\,|\,W_q > 0\} = e^{-x(1-\omega)/E(S)} \quad \text{for} \quad x > 0,$$

implying that

$$\phi = 1. \tag{4.108}$$

On the other hand, for the GI/M/1 queue we can write (4.106) as

$$\phi = \frac{\delta\tau E(S)}{\tau - 1}. \tag{4.109}$$

To see this, note that, by (4.12) and (2.18), $p_j = \lambda E(S)(1-\omega)\omega^{j-1}$ for $j \geqslant 1$. Hence, by (4.15) and (4.17), $\tau = 1/\omega$ and $\eta = \lambda E(S)(1-\omega)/\omega$. This gives $\Pi_W = 1/\tau$ and $\eta = \lambda E(S)(\tau - 1)$. Substituting these identities into (4.107) yields the representation (4.109) for the GI/M/1 queue.

The idea is now to approximate the amplitude coefficient ϕ for the GI/G/1 queue by $\delta\tau E(S)/(\tau - 1)$ times a correction factor, being chosen such that the resulting approximation is exact both for the M/G/1 queue and for the GI/M/1 queue. To do so, define $\delta(M, G)$ and $\tau(M, G)$ as the particular values of δ and τ for the M/G/1 queue with arrival rate λ and service time S. Note that, by (4.19), $\tau(M, G) = 1 + \lambda^{-1}\delta(M, G)$. We propose the following approximation to the amplitude coefficient ϕ of the asymptotic expansion (4.105) for the GI/G/1 queue:

$$\phi_{\text{app}} = \frac{\delta\tau}{\tau - 1}\frac{\tau(M, G) - 1}{\delta(M, G)\tau(M, G)}\frac{1-\rho}{\rho\lambda\delta(M, G)}\left[\int_0^\infty ye^{\delta(M,G)y}\{1 - B(y)\}\,dy\right]^{-1}. \tag{4.110}$$

We remark that the integrals in (4.13), (4.14) and (4.110) are easily evaluated for phase-type distributions of practical interest. It is obvious from (4.107) and (4.110) that the approximation ϕ_{app} is exact for the M/G/1 queue. To verify that ϕ_{app} is also exact for the GI/M/1 queue, substitute the result $\delta(M, M) = (1 - \rho)/E(S)$ into (4.110) and use the identities (4.108) and (4.109).

The approximation ϕ_{app} was tested extensively in Seelen and Tijms (1984). It was found for practical queueing systems that the relative error in ϕ_{app} is usually below 5 per cent provided the delay probability is not too small (say, $\Pi_W \geqslant 0.2$) and c_A^2 is not too close to zero (say, $c_A^2 \geqslant \frac{1}{2}$). In Table 4.8 we give for a number of GI/G/1 queueing systems the exact and approximate values of the percentiles of the conditional waiting-time distribution function. The approximate value of the

Table 4.8   Exact and approximate values of the conditional waiting-time
percentiles

	p	0.5	0.8	0.9	0.95	0.99	Π_W
$D/E_4/1$	exa	0.26	0.59	0.83	1.07	1.61	0.0473
	app	0.53	0.81	1.03	1.25	1.75	
$D/E_2/1$	exa	0.48	1.09	1.54	1.99	3.01	0.1164
	app	0.69	1.27	1.70	2.14	3.15	
$D/H_2/1$	exa	1.66	3.86	5.52	7.19	11.05	0.2892
	app	1.26	3.46	5.13	6.79	10.65	
$E_2/E_4/1$	exa	0.62	1.29	1.77	2.26	3.37	0.3440
	app	0.74	1.38	1.86	2.34	3.46	
$E_2/E_2/1$	exa	0.79	1.73	2.43	3.13	4.74	0.3596
	app	0.88	1.80	2.49	3.19	4.80	
$E_2/H_2/1$	exa	1.83	4.39	6.33	8.26	12.76	0.4052
	app	1.64	4.20	6.13	8.07	12.57	
$H_2/E_4/1$	exa	1.44	3.09	4.34	5.59	8.48	0.6603
	app	1.31	2.96	4.21	5.46	8.36	
$H_2/E_2/1$	exa	1.60	3.54	5.00	6.46	9.86	0.6501
	app	1.52	3.46	4.92	6.39	9.78	
$H_2/H_2/1$	exa	2.46	6.07	8.80	11.54	17.88	0.6142
	app	2.66	6.28	9.01	11.74	18.09	

pth percentile is computed from

$$\zeta_{app}(p) = \frac{1}{\delta}\ln\left(\frac{\phi_{app}}{1-p}\right). \tag{4.111}$$

The interarrival-time and service-time distributions include the deterministic distribution, Erlang distributions and the H_2 distribution. For the latter distribution we have taken a squared coefficient of variation of 2 and the gamma normalization. Also, in all examples we have taken $E(S) = 1$ and $\rho = 0.5$. The exact values of the delay probability are also included in Table 4.8.

4.3   SINGLE-SERVER QUEUEING SYSTEMS WITH LIMITED ACCESS OF ARRIVALS

Many practical queueing systems have a limit to either the number of customers or the work in the system. The limits will usually be such that the results for the infinite capacity queueing systems studied in the previous section cannot be applied. However, in many cases the analysis of queueing systems with limited access of arrivals closely parallels that of similar queueing systems without capacity restrictions. In this section we discuss several basic queueing models with restricted accessibility that have proved to be useful in a variety of practical problems.

4.3.1 The M/G/1 queueing system with finite waiting room

In this system a limit of N customers is allowed and customers arriving to find N other customers present are turned away. The arrival pattern of customers is a Poisson process with rate λ. Each customer who is turned away has no further influence on the system. The service time S of an entering customer has a general probability distribution function $B(x)$ with finite first two moments. In this model the finite waiting room acts as a regulator on the queue size and so no *a priori* assumption about the load factor $\rho = \lambda E(S)$ is needed.

We denote by $p_j(N)$ and $\pi_j(N)$ the time-average probability of having j customers in the system and the customer-average probability of finding upon arrival j other customers in the system $(j = 0, \ldots, N)$. Since Poisson arrivals see time averages, we have

$$\pi_j(N) = p_j(N), \qquad j = 0, \ldots, N.$$

In particular,

$$p_N(N) = \text{the long-run fraction of arrivals that are rejected}$$

and

$$\lambda(1 - p_N(N)) = \text{the average arrival rate of entering customers.}$$

The state probabilities $p_j(N)$ can be found by invoking results of subsection 4.2.2 as the present model is in fact a special case of the general queueing model studied in subsection 4.2.2. This follows by noting that the arrival process of *entering* customers can be described by a Poisson process with state-dependent arrival rate λ_j when j customers are in the system, where

$$\lambda_j = \begin{cases} \lambda, & 0 \leqslant j \leqslant N - 1, \\ 0, & j = N. \end{cases} \tag{4.112}$$

In other words, the *entering* customers in the M/G/1/N queueing model can be identified with the arriving customers in the particular queueing model of subsection 4.2.2 with the state-dependent arrival rate (4.112). Obviously, the time-average probabilities are the same in both models, as opposed to the customer-average probabilities which differ a factor, being the probability that a customer is not turned away. Thus the state probabilities $p_j(N)$, $0 \leqslant j \leqslant N$, can be computed from the recursion scheme (4.70) using the specification (4.112). Recall that for the special case of Erlangian service a simpler algorithm is provided by (4.76) to (4.78).

The algorithm (4.70) applies for any value of the offered load $\rho = \lambda E(S)$. For the case of $\rho < 1$, a useful relation exists between the time-average probabilities of the finite capacity M/G/1 queue and those of the infinite capacity M/G/1 queue. Denoting by $p_j(\infty)$ the time-average probabilities for the latter queueing system, we have

$$p_j(N) = \gamma p_j(\infty), \qquad 0 \leqslant j \leqslant N - 1, \tag{4.113}$$

with

$$\gamma = \left\{ 1 - \rho + \rho \sum_{k=0}^{N-1} p_k(\infty) \right\}^{-1}.$$

We briefly sketch a proof of this result. Denote by T and T^∞ the lengths of a cycle in the finite and infinite capacity $M/G/1$ queueing systems. Also, let the corresponding random variables T_j and T_j^∞ denote the amount of time that j customers are in the system during one cycle. By the theory of regenerative processes,

$$p_j(N) = \frac{E(T_j)}{E(T)} \quad \text{and} \quad p_j(\infty) = \frac{E(T_j^\infty)}{E(T^\infty)}. \tag{4.114}$$

Some reflections show that the distribution of T_j for $j \leqslant N - 1$ does not depend on the limit on the queue size because of the lack of memory of the Poisson arrival process. Hence,

$$E(T_j) = E(T_j^\infty), \qquad 0 \leqslant j \leqslant N - 1, \tag{4.115}$$

implying $p_j(N) = \gamma p_j(\infty)$, $0 \leqslant j \leqslant N - 1$ with $\gamma = E(T^\infty)/E(T)$. Now, invoke the identity (4.65) with the specification (4.112), divide both sides of this identity by $E(T^\infty)$ and use (4.114) and (4.115) in order to obtain the constant γ.

As a consequence of (4.113) and the normalization equation $\sum_{j=0}^{N} p_j(N) = 1$, the rejection probability $p_N(N)$ can be computed from

$$p_N(N) = \left\{ 1 - \rho + \rho \sum_{j=0}^{N-1} p_j(\infty) \right\}^{-1} \left\{ 1 - \rho + (\rho - 1) \sum_{j=0}^{N-1} p_j(\infty) \right\}.$$

This relation is particularly useful when calculating $p_N(N)$ for a range of values of N in order to find a value of N for which $p_N(N)$ is acceptably low. To the end of this subsection we consider this interesting problem of buffer design in more detail. For any rejection level v with $0 < v < 1$, denote by $N(v)$ the smallest integer N for which

$$p_N(N) \leqslant v.$$

In practical applications v will typically be close to zero. The assumption of $\rho < 1$ guarantees that $N(v)$ exists for all v sufficiently small. Our numerical investigations reveal that $N(v)$ is remarkably insensitive to more than the first two moments of the service time S provided c_S^2 is not too large (say $0 \leqslant c_S^2 \leqslant 1$) and the service-time density satisfies a reasonable shape constraint and has an at least exponentially fast decreasing tail. The latter requirement is particularly needed for very small v and excludes distributions with extremely long tails such as the lognormal distribution. Also, in analogy with (4.42), we found that under the above regularity conditions $N(v)$ can be approximated quite well by

$$N_{app}(v) = (1 - c_S^2) N_{det}(v) + c_S^2 N_{exp}(v), \tag{4.116}$$

Table 4.9   The buffer capacity $N(v)$ for various service-time densities with $c_S^2 = \frac{1}{2}$

	$\rho = 0.5$					$\rho = 0.8$				
v	10^{-2}	10^{-3}	10^{-4}	10^{-5}	10^{-6}	10^{-2}	10^{-3}	10^{-4}	10^{-5}	10^{-6}
E_2	5	8	10	13	16	11	19	27	35	42
$E_{1,3}$	5	8	10	13	15	11	19	27	34	42
Weib	5	8	10	13	15	11	19	27	34	42
logn	6	9	13	17	21	13	22	31	41	50
app	5	8	11	13	16	11	19	27	35	42

Table 4.10   The buffer capacities $N_{det}(v)$ and $N_{exp}(v)$

	ρ	0.1	0.2	0.3	0.4	0.5	0.6	0.7	0.8	0.9
$v = 10^{-2}$	det	2	3	3	4	4	5	6	8	13
	exp	2	3	4	5	6	8	10	14	23
$v = 10^{-3}$	det	3	4	4	5	6	7	10	13	23
	exp	3	5	6	7	9	12	16	24	44
$v = 10^{-4}$	det	4	5	5	7	8	10	13	19	35
	exp	4	6	8	10	13	17	23	35	66
$v = 10^{-5}$	det	4	5	7	8	10	12	16	24	46
	exp	5	8	10	13	16	21	29	45	88
$v = 10^{-6}$	det	5	6	8	9	12	15	20	29	57
	exp	6	9	12	15	19	26	36	55	110
$v = 10^{-7}$	det	6	7	9	11	13	17	23	35	68
	exp	7	10	14	18	23	30	42	66	132

where $N_{det}(v)$ and $N_{exp}(v)$ denote the value of $N(v)$ for the respective cases of deterministic and exponential services with the same mean service time $E(S)$. It is remarkable that this two-moment approximation still performs very well for extremely small rejection levels v. To illustrate this, we give in Table 4.9 the exact value of $N(v)$ for various service-time densities together with the approximate value $N_{app}(v)$ to be denoted by app in the table. Taking $E(S) = 1$ and $c_S^2 = \frac{1}{2}$, we consider the E_2 density (E_2), the mixture of E_1 and E_3 densities with the same scale parameters ($E_{1,3}$), the Weibull density (Weib) and the lognormal density (logn). Also, in Table 4.10 we give $N_{det}(v)$ and $N_{exp}(v)$ for a range of values of ρ and v.

The next example demonstrates the relationship between queueing systems and production/inventory systems.

Example 4.3   A discrete production/inventory problem

Consider a production system which operates only intermittently to produce a single product. Production is stopped if the inventory is sufficiently high and production is restarted if the inventory has dropped sufficiently low. Demands for the product occur at epochs generated by a Poisson process with rate λ and each demand is for one unit of product. Any demand that cannot be satisfied directly from on-hand inventory is lost. The units of the product are manufactured one at a time. The production time to complete one unit is distributed as the random variable T_p. The following control rule involving a single switching level m is used. Production is stopped when inventory reaches the level m, while production is restarted as soon as the inventory falls below m.

We wish to find the smallest value of the switching level m for which the fraction of demand that is lost does not exceed a prespecified value v. The required level m can be found by modelling the production/inventory problem with switching level m as a finite capacity M/G/1 queueing system. To do so identify the production facility with a single server, the production times with the service times of customers, and m minus the on-hand inventory with the number of customers being served or awaiting service. In the queueing terminology, the fraction of demand that is lost is represented by the fraction of customers that is turned away.

As a numerical example, assume a demand rate $\lambda = 0.9$, a production time T_p with $E(T_p) = 1$ and var $(T_p) = \frac{1}{2}$ and a rejection level $v = 0.01$. Then, using (4.116) and Table 4.10, the switching level m is taken as 18.   □

4.3.2   A finite capacity M/G/1 queueing system with buffer overflow

Consider a single-server queueing system where customers arrive according to a Poisson process with rate λ. Each customer brings a positive amount S of work having a general probability distribution function $B(x)$ with finite first two moments. The work is put into a buffer that can store only some fixed amount K. A customer bringing along more work than the remaining capacity of the buffer causes an overflow where the excess of work is lost. The buffer is emptied at a constant rate of $\sigma > 0$ whenever there is work in the buffer.

This queueing model has a variety of applications such as dam and production/inventory systems with a finite storage space and communication systems with a finite buffer storage for incoming messages. We are interested in getting tractable results for the following operating characteristics:

$\pi(K) =$ the long-run fraction of arrivals that cause an overflow,

$EV(K) =$ the long-run average amount of work in the buffer.

To obtain these measures, let $V_K(x)$, $0 \leqslant x \leqslant K$, be the limiting probability distribution function of the amount of work in the buffer. Note that $V_K(x)$ has a positive mass at $x = 0$ and has density $v_K(x)$ for $x > 0$. Using the property

'Poisson arrivals see time averages', it follows that the steady-state probability of an arrival causing an overflow is given by

$$\pi(K) = \{1 - B(K)\}V_K(0) + \int_0^K \{1 - B(K - x)\}v_K(x)\,dx. \qquad (4.117)$$

Also, we have

$$EV(K) = \int_0^K xv_K(x)\,dx. \qquad (4.118)$$

Under the assumption that the offered load ρ defined by

$$\rho = \frac{\lambda E(S)}{\sigma}$$

is smaller than unity, we can express $V_K(x)$ in terms of $V_\infty(x)$ which is defined as the limiting distribution function of the workload in the system with infinity capacity $K = \infty$. To guarantee the existence of the latter distribution function it is required that $\rho < 1$. The function $V_\infty(x)$, being the distribution function of the workload in the M/G/1 queue with the processing rate σ, is exactly the function $1 - q(x)$ studied in section 1.9 of chapter 1; see relation (1.68).

We next establish that

$$V_K(x) = \frac{V_\infty(x)}{V_\infty(K)}, \qquad 0 \leqslant x \leqslant K. \qquad (4.119)$$

The proof, being similar to that of (4.113), proceeds as follows. For the finite capacity queueing system underlying $V_K(x)$, define T as the length of a cycle and $T(x)$ as the amount of time that the workload is not greater than x during one cycle. Here a cycle is the time elapsed between two consecutive moments at which the system becomes empty. Similarly, define T_∞ and $T_\infty(x)$ for the infinite capacity queueing system underlying $V_\infty(x)$. Noting that both queueing systems behave identically whenever the workload is not more than K and using the lack of memory of the Poisson input process, it can be seen that $T(x)$ has the same distribution as $T_\infty(x)$ for $0 \leqslant x \leqslant K$ and so

$$E(T(x)) = E(T_\infty(x)), \qquad 0 \leqslant x \leqslant K.$$

Also, by the theory of regenerative processes,

$$V_K(x) = \frac{E(T(x))}{E(T)} \qquad \text{and} \qquad V_\infty(x) = \frac{E(T_\infty(x))}{E(T_\infty)}.$$

The above relations together imply that for the constant $c = E(T_\infty)/E(T)$,

$$V_K(x) = cV_\infty(x), \qquad 0 \leqslant x \leqslant K.$$

Using the fact that $V_K(K) = 1$, we get the desired result (4.119).

Invoking (4.119) in (4.117) and assuming that the input distribution function $B(x)$ has everywhere a density $b(x)$, we obtain with partial integration

$$\pi(K) = \frac{1}{V_\infty(K)}\left[\{1 - B(K)\}V_\infty(0) + \{1 - B(K - x)\}V_\infty(x)\bigg|_0^K - \int_0^K V_\infty(x)b(K - x)\,dx \right]$$

$$= \frac{1}{V_\infty(K)}\{V_\infty(K) - \int_0^K V_\infty(K - x)b(x)\,dx\}. \tag{4.120}$$

Using the integro-differential equation (1.70) for $q(x) = 1 - V_\infty(x)$ in section 1.9 of chapter 1 it is readily verified that the term between brackets in (4.120) equals $(\sigma/\lambda)V_\infty'(x)$ and so

$$\pi(K) = \frac{(\sigma/\lambda)V_\infty'(K)}{V_\infty(K)}. \tag{4.121}$$

In the case where the tail of $b(x)$ tends at least exponentially fast to zero as $x \to \infty$, we obtain as a consequence of (4.121) and the asymptotic expansion (1.78) that

$$\pi(K) \approx \frac{\sigma\delta}{\lambda}\gamma e^{-\delta K} \quad \text{for} \quad K \text{ large,} \tag{4.122}$$

where δ and γ are defined by (1.75) and (1.79). Also, using (4.118) and (4.119), we obtain after partial integration

$$EV(K) = \frac{1}{V_\infty(K)}\{KV_\infty(K) - \int_0^K V_\infty(x)\,dx\}. \tag{4.123}$$

The exact formulae (4.120) to (4.122) can further be elaborated for the special cases of deterministic and exponential inputs and enable us to derive tractable approximations for the case of general input. We write $\pi_{\text{det}}(K)$ and $EV_{\text{det}}(K)$ for $\pi(K)$ and $EV(K)$ when the input is deterministic. Similarly, we write $\pi_{\text{exp}}(K)$ and $EV_{\text{exp}}(K)$ for exponential input.

Deterministic input

For the case in which the input is a constant D, we have by (4.120) that $\pi_{\text{det}}(K) = \{V_\infty(K) - V_\infty(K - D)\}/V_\infty(K)$. Hence, using (1.72) in section 1.9 of chapter 1, we find

$$\pi_{\text{det}}(K) = 1 - \left\{\sum_{j=0}^r (-1)^j \frac{(\lambda K/\sigma - \rho j)^j}{j!}e^{-\rho j}\right\}^{-1}$$

$$\times \left[\sum_{j=0}^{r-1} (-1)^j \frac{\{\lambda K/\sigma - \rho(j+1)\}^j}{j!}e^{-\rho(j+1)}\right], \tag{4.124}$$

where r is the integer satisfying $rD \leqslant K < (r+1)D$. Also, using (4.122) and (1.72), we find after considerable algebra that

$$EV_{\text{det}}(K) = K - \left\{ \sum_{j=0}^{r} (-1)^j \frac{(\lambda K/\sigma - \rho j)^j}{j!} e^{\lambda K/\sigma - \rho j} \right\}^{-1} \sigma(1-\rho)\lambda^{-1}$$

$$\times \left\{ -(r+1) + \sum_{j=0}^{r} e^{\lambda K/\sigma - \rho j} \sum_{i=0}^{j} (-1)^i \frac{(\lambda K/\sigma - \rho j)^i}{i!} \right\}. \qquad (4.125)$$

We state without proof that the expressions (4.124) and (4.125) hold not only for $\rho < 1$ but also apply when $\rho \geqslant 1$.

Exponential input

For the case of exponential input, we obtain from (4.121), (4.122) and (1.71) that

$$\pi_{\text{exp}}(K) = \frac{(1-\rho)e^{-(1-\rho)K/E(S)}}{1 - \rho e^{-(1-\rho)K/E(S)}} \qquad (4.126)$$

and

$$EV_{\text{exp}}(K) = \frac{\rho E(S)/(1-\rho) - \{E(S)/(1-\rho) + K\}\rho e^{-(1-\rho)K/E(S)}}{1 - \rho e^{-(1-\rho)K/E(S)}}. \qquad (4.127)$$

We state without proof that the expressions (4.126) and (4.127) also apply for $\rho \geqslant 1$ where, for $\rho = 1$, L'Hospital's rule should be used.

General input

The exact expressions (4.120) and (4.121) are in general not computationally tractable. Invoking the approximation (1.80) for $q(x) = 1 - V_\infty(x)$ yields the following simple approximations for $\pi(K)$ and $EV(K)$:

$$\pi_{\text{app}}(K) = \frac{(\sigma/\lambda)(\alpha\beta e^{-\beta K} + \gamma\delta e^{-\delta K})}{1 - \alpha e^{-\beta K} - \gamma e^{-\delta K}} \qquad (4.128)$$

and

$$EV_{\text{app}}(K) = K - \left\{ \frac{K - (\alpha/\beta)(1 - e^{-\beta K}) - (\gamma/\delta)(1 - e^{-\delta K})}{1 - \alpha e^{-\beta K} - \gamma e^{-\delta K}} \right\}, \qquad (4.129)$$

where α, β, γ and δ are given by (1.82), (1.79) and (1.75). Here it is assumed that the tail of the probability density $b(x)$ of the input tends at least exponentially fast to zero as $x \to \infty$. The approximations (4.128) and (4.129) are exact when the input has a K_2 distribution.

Our numerical investigations indicate that (4.128) and (4.129) are excellent approximations for practical purposes when the input has a general distribution. For both Erlang-10 and Erlang-4 input we give in Table 4.11 the exact and

Table 4.11   Exact and approximate values of $\pi(K)$ and $EV(K)$

| | | K = 1.5 | | | | K = 3 | | | |
| | | Erlang-10 | | Erlang-4 | | Erlang-10 | | Erlang-4 | |
		$\pi(K)$	$EV(K)$	$\pi(K)$	$EV(K)$	$\pi(K)$	$EV(K)$	$\pi(K)$	$EV(K)$
$\rho = 0.2$	exa	0.1596	0.111	0.2182	0.107	0.0058	0.136	0.0161	0.150
	app	0.1556	0.112	0.2101	0.108	0.0058	0.136	0.0160	0.150
$\rho = 0.5$	exa	0.3032	0.291	0.3258	0.276	0.0529	0.469	0.0725	0.488
	app	0.3038	0.291	0.3224	0.277	0.0530	0.469	0.0725	0.488
$\rho = 0.8$	exa	0.4417	0.468	0.4331	0.433	0.1769	0.960	0.1860	0.949
	app	0.4428	0.468	0.4314	0.443	0.1769	0.960	0.1860	0.949

approximate values of $\pi(K)$ and $EV(K)$ for several values of ρ and K. In all examples we take $\sigma = 1$ and $E(S) = 1$. In the case where the input distribution is a finite mixture of Erlangian distributions with the same scale parameters, the exact values of $\pi(K)$ and $EV(K)$ can be computed by using the equations (4.121) and (4.123) together with the algorithmic method (4.38).

Two-moment approximations

In accordance with earlier experience with related queueing systems, our numerical experiments indicate that $\pi(K)$ and $EV(K)$ are both rather insensitive to more than the first two moments of the input distribution provided c_S^2 is not too large and the input density satisfies a reasonable shape constraint. As is to be expected, the overflow probability is more sensitive to the distributional form of the input than the average amount of work in the buffer, in particular when the overflow probability is very small. In general a two-moment approximation for the overflow probability $\pi(K)$ should be used only when $0 \leqslant c_S^2 \leqslant 1$ and ρ is not very small. As an illustration, we give in Table 4.12 the exact values of $\pi(K)$ for several values of ρ and K and for various input densities

Table 4.12   Sensitivity results for $\pi(K)$ for several input densities

| | $\rho = 0.2$ | | | $\rho = 0.5$ | | | $\rho = 0.8$ | | |
	K = 1.5	K = 3	K = 4.5	K = 2	K = 4	K = 7	K = 3	K = 7	K = 15
E_2	0.248	0.039	0.0059	0.216	0.046	0.0052	0.194	0.047	0.0047
Weib	0.259	0.037	0.0049	0.222	0.046	0.0049	0.196	0.047	0.0047
logn	0.220	0.039	0.0085	0.200	0.046	0.0064	0.187	0.047	0.0050
shif	0.229	0.042	0.0077	0.206	0.047	0.0059	0.190	0.047	0.0049
$E_{1,3}$	0.260	0.037	0.0049	0.221	0.046	0.0049	0.196	0.047	0.0046

having the same first two moments but differing in tail behaviour. Taking $E(S) = 1$ and $c_S^2 = 0.5$, we consider the E_2 density (E_2), the Weibull density (Weib), the lognormal density (logn), the shifted exponential density (shif) and the mixture of E_1 and E_3 densities with the same scale parameters $(E_{1,3})$. In all examples we assume $\sigma = 1$. The exact values of $\pi(K)$ for the various input distributions have been computed by a laborious discretization algorithm to be discussed in the appendix to this subsection.

In view of the results in Table 4.12, we may suggest the following two-moment approximations to $\pi(K)$ and $EV(K)$ when only the first two moments of the input are available with $0 \leqslant c_S^2 \leqslant 1$. In the case $1/k < c_S^2 \leqslant 1/(k-1)$ for some $k \geqslant 3$, the distribution specified by (A.11) and (A.12) in appendix B is fitted to the input and the corresponding equations (4.128) and (4.129) are used. In the case $\frac{1}{2} < c_S^2 \leqslant 1$, a K_2 distribution specified by (A.15) and (A.17) is fitted to the input and the corresponding results (4.128) and (4.129) are applied.

An alternative two-moment approximation to $EV(K)$ is provided by

$$EV_{\text{app2}}(K) = (1 - c_S^2)EV_{\text{det}}(K) + c_S^2 EV_{\text{exp}}(K) \qquad (4.130)$$

Such a two-moment approximation is in general not to be recommended for $\pi(K)$, in particular when $\pi(K)$ is small. However, in analogy to the M/G/1 queue in which a two-moment approximation of the form (4.130) should be used for the waiting-time percentiles rather than for the waiting-time probabilities, the overflow probability $\pi(K)$ can indirectly be approximated as follows. For any $0 < v < 1$, define $K(v)$ as the solution to

$$\pi(K) = v;$$

that is $K(v)$ is the smallest value of the buffer capacity such that the overflow probability achieves a prespecified value v. Then an excellent two-moment approximation to $K(v)$ is provided by $K_{\text{app}}(v) = (1 - c_S^2)K_{\text{det}}(v) + c_S^2 K_{\text{exp}}(v)$ where $K_{\text{det}}(v)$ and $K_{\text{exp}}(v)$ are the corresponding values of $K(v)$ for the special cases of deterministic input and exponentially distributed input. For the practically important case in which a very small overflow probability v is required (assuming $\rho < 1$), this approximation $K_{\text{app}}(v)$ is nearly the same as

$$K_{\text{asy}}(v) = \frac{1}{\delta_{\text{app}}} \ln\left(\frac{\sigma \delta_{\text{app}} \gamma_{\text{app}}}{\lambda v}\right), \qquad (4.131)$$

in agreement with the asymptotic expansion (4.122) in which δ and γ are replaced by δ_{app} and γ_{app}, where δ_{app} and γ_{app} are defined by (4.43) and (4.44). How small v should be in order to apply (4.131) depends on the offered load ρ to some extent. Roughly stated, we may apply (4.131) for $v \leqslant 10^{-2}$ when $\rho \leqslant 0.8$ and for $v \leqslant 10^{-3}$ otherwise. For several values of ρ and v, we give in Table 4.13 the exact and approximate values $K = K(v)$ and $K_{\text{asy}} = K_{\text{asy}}(v)$ when the input has an Erlang-2 distribution.

So far the discussion has mainly been restricted to the case of an offered load

Table 4.13   Exact and approximate values of $K(v)$ for Erlang-2 input

| v | $\rho = 0.2$ | | $\rho = 0.5$ | | $\rho = 0.8$ | | $\rho = 0.9$ | |
	K	K_{app}	K	K_{app}	K	K_{app}	K	K_{app}
0.05	2.80	2.75	3.89	3.83	6.81	6.31	9.58	7.45
0.01	4.08	4.06	6.09	6.08	12.31	12.20	19.86	19.38
0.005	4.63	4.62	7.05	7.05	14.79	14.74	24.76	24.52
10^{-3}	5.91	5.93	9.29	9.30	20.64	20.63	36.50	36.45
10^{-4}	7.74	7.80	12.49	12.52	29.05	29.06	53.52	53.52
10^{-5}	9.57	9.67	15.69	15.74	37.47	37.49	70.58	70.59
10^{-6}	11.39	11.54	18.89	18.96	45.90	45.91	87.65	87.66

of $\rho < 1$. The exact results (4.124) and (4.125) and (4.126) and (4.127) for the special cases of deterministic input and exponential input hold also for $\rho \geqslant 1$. Thus the two-moment approximation (4.130) to $EV(K)$ and the similar two-moment approximation $K_{app}(v)$ to $K(v)$ apply as well for values for $\rho \geqslant 1$, provided c_S^2 is not too large. The two-moment approximation $K_{app}(v)$ to $K(v)$ yields indirectly a two-moment approximation to $\pi(K)$. Also, for the case of $\rho > 1$, an alternative approximation to $\pi(K)$ can be given. This alternative approximation due to De Kok and Van Ommeren (1986) uses the whole input density and is given by

$$\pi(K) \approx \frac{\sigma s^*}{\lambda}\left[1 + \frac{\{1 + (\lambda/\sigma)\mathscr{L}'(s^*)\}e^{-s^*K}}{\rho - 1 - \{1 + (\lambda/\sigma)\mathscr{L}'(s^*)\}e^{-s^*K}}\right], \qquad (4.132)$$

where s^* is the unique positive solution to

$$s + \frac{\lambda}{\sigma_1}\int_0^\infty e^{-sx}b(x)\,dx - \frac{\lambda}{\sigma_1} = 0$$

and $\mathscr{L}'(s)$ is the derivative of the Laplace transform $\mathscr{L}(s) = \int_0^\infty e^{-sx}b(x)\,dx$. The result (4.132) is asymptotically exact as $K \to \infty$, but can be used for practical purposes already for relatively small values of K.

As an illustration, we give in Table 4.14 the exact and approximate values of $\pi(K)$ and $EV(K)$ for several values of $\rho(>1)$ and K when the input has an

Table 4.14   Exact and approximate values of $\pi(K)$ and $EV(K)$ when $\rho > 1$

| | $\rho = 1.25$, $K = 4$ | | $\rho = 1.25$, $K = 8$ | | $\rho = 2$, $K = 4$ | | $\rho = 2$, $K = 8$ | |
	$\pi(K)$	$EV(K)$	$\pi(K)$	$EV(K)$	$\pi(K)$	$EV(K)$	$\pi(K)$	$EV(K)$
exa	0.334	2.26	0.277	5.48	0.620	3.21	0.617	7.19
app	0.335	2.30	0.277	5.56	0.620	3.21	0.618	7.19

E_2 distribution, where the approximate values of $\pi(K)$ and $EV(K)$ are based on (4.132) and (4.130). In the examples we take $E(S) = 1$ and $\sigma = 1$.

Appendix. A discretization algorithm

This appendix discusses a discretization algorithm for the computation of the performance measures $\pi(K)$ and $EV(K)$ when the input has a general distribution. This discretization algorithm is based on the fact that the input variable can be approximated arbitrarily closely by a random sum of independent exponentially distributed random variables. Using such a representation the work in the system can be approximated by a discrete number of 'packets' of work, each having an exponential length with mean Δ for some small number Δ. The discretized model allows us to describe the number of packets in the system as a continuous-time Markov chain for which powerful computational methods are available.

The representation (4.1) for the input distribution function $B(x)$ underlies the discretization algorithm. To improve the accuracy of the computations we slightly modify (4.1) such that the first moment $E(S)$ of the input is exactly matched. Letting, for $k = 0, 1, \ldots,$

$$b'_k(\Delta) = B(k\Delta) - B((k-1)\Delta) \qquad \text{and} \qquad b''_k(\Delta) = B((k+1)\Delta) - B(k\Delta),$$

with $B(-\Delta) = 0$ by convention, it is readily verified that

$$\sum_{k=1}^{\infty} k\,\Delta b''_k(\Delta) \leqslant E(S) \leqslant \sum_{k=1}^{\infty} k\,\Delta b'_k(\Delta) \qquad \text{and} \qquad \sum_{k=1}^{\infty} k\,\Delta b'_k(\Delta) = \Delta + \sum_{k=1}^{\infty} k\,\Delta b''_k(\Delta).$$

Define now, for $k = 0, 1, \ldots,$

$$b_k(\Delta) = \omega\{B(k\Delta) - B((k-1)\Delta)\} + (1 - \omega)\{B((k+1)\Delta) - B(k\Delta)\},$$

with

$$\omega = \frac{E(S)}{\Delta} - \sum_{k=1}^{\infty} kb''_k(\Delta).$$

It is a matter of simple algebra to verify that

$$\sum_{k=0}^{\infty} b_k(\Delta) = 1 \qquad \text{and} \qquad \sum_{k=1}^{\infty} k\,\Delta b_k(\Delta) = E(S).$$

Next the work input S of any arrival is represented with probability $b_k(\Delta)$ by a batch of k independent packets each having an exponentially distributed length with mean Δ. The buffer has only space for $K(\Delta)$ packets with

$$K(\Delta) = \left[\frac{K}{\Delta}\right],$$

where $[x]$ is the largest integer contained in x. From an arriving batch of packets only those packets are accepted for which there is still a place in the buffer. An overflow occurs when not all packets of an arrival can be put into the buffer. The packets in the buffer are processed one at a time and the processing time of each packet has an exponential distribution whose mean $1/\mu_\Delta$ is given by

$$\frac{1}{\mu(\Delta)} = \frac{\Delta}{\sigma}.$$

The process describing the number of packets in the buffer is a continuous-time Markov chain. Denoting by $\{p_i(\Delta), i \geqslant 0\}$ the equilibrium distribution of this Markov chain, we find by the standard technique of equating the rate at which the process leaves the set of states $\{i, i+1, \ldots\}$ to the rate at which the process enters that set,

$$\mu(\Delta)p_i(\Delta) = \sum_{j=0}^{i-1}\left\{\lambda \sum_{k \geqslant i-j} b_k(\Delta)\right\}p_j(\Delta), \qquad i = 1, \ldots, K(\Delta). \qquad (4.133)$$

By the property 'Poisson arrivals see time averages' the time-average probability $p_i(\Delta)$ also gives the steady-state probability that a batch arrival finds i packets in the buffer. Thus the sought overflow probability $\pi(K)$ and the average workload $EV(K)$ may be approximated in any desired accuracy by

$$\pi(K) \approx \sum_{j=0}^{K(\Delta)} p_j(\Delta) \sum_{k > K(\Delta)-j} b_k(\Delta), \qquad EV(K) \approx \sum_{j=1}^{K(\Delta)} j\Delta p_j(\Delta)$$

provided Δ is chosen sufficiently small. The probabilities can be computed recursively from (4.133) together with the normalization equation $\sum_{i=0}^{K(\Delta)} p_i(\Delta) = 1$. Although this algorithm is very simple the computational effort increases quadratically when the grid size Δ is halved. The question remains how small to choose Δ in order to obtain sufficiently accurate answers. In general one should search for a grid size Δ such that the results for the grid sizes Δ and $\Delta/2$ are sufficiently close to each other. The results in Tables 4.12 and 4.14 were computed with a grid size $\Delta/2 = 0.0025E(S)$.

It will be clear that the above discretization technique is generally applicable. For example, in subsection 4.3.3 we consider a related queueing system in which an arriving customer enters the system only if the work in the system is not more than K. For this queueing system with bounded waiting time the discretization algorithm (4.133) needs only a slight modification. The probabilities $p_i(\Delta)$ with $1 \leqslant i \leqslant K(\Delta)$ satisfy the same relation (4.133), while the probabilities $p_i(\Delta)$ with $i > K(\Delta)$ satisfy

$$\mu(\Delta)p_i(\Delta) = \sum_{j=0}^{K(\Delta)}\left\{\lambda \sum_{k \geqslant i-j} b_k(\Delta)\right\}p_j(\Delta), \qquad i > K(\Delta).$$

Actually, this latter relation and the normalization equation $\sum_{j=0}^{\infty} p_j(\Delta) = 1$

are not needed for the computation of the rejection probability $1 - \sum_{i=0}^{K(\Delta)} p_i(\Delta)$. To compute $p_i(\Delta)$ for $0 \leqslant i \leqslant K(\Delta)$ it suffices to use (4.133) in conjunction with Little's result

$$1 - p_0(\Delta) = \frac{\{\lambda \sum_{i=0}^{K(\Delta)} p_i(\Delta)\} E(S)}{\sigma},$$

stating that the fraction of time the server is busy equals the average arrival rate of entering customers times the average processing time of an entering customer.

As another application of the discretization algorithm we mention a queueing system in which the service time of a customer depends on the waiting time (cf. exercise 4.19). Such queueing systems occur, for example, in a steel plant in which hot ingots are rolled one at a time into sheets of steel and an ingot must be reheatened to the required rolling temperature when it has cooled down during its waiting before rolling.

4.3.3   An M/G/1 queueing system with impatient customers

A queueing system often encountered in practice is that in which customers wait for service for a limited time only and leave the system if not served during that time. Such queueing systems not only occur in telephone systems but are also related to inventory systems with perishable goods (cf. also exercises 4.20 and 4.21). In this subsection we consider a single-server queueing system in which customers arrive according to a Poisson process with rate λ. The service or work requirement of each customer is distributed as the positive random variable S having a general probability distribution function $B(x)$ with finite first two moments. Each arriving customer enters the system but is willing to wait in queue only for a fixed time τ. A customer who waited for a time τ without service beginning leaves the system and becomes a lost customer. The single server can handle only one customer at a time and provides service at a constant rate of $\sigma > 0$ whenever the system is not empty. Thus the service requirement S of a customer has a processing time of S/σ.

We are interested in the following performance measures:

$\pi(\tau) =$ the long-run fraction of customers that are lost,

$p_0(\tau) =$ the long-run fraction of time the server is idle,

$W_q(\tau) =$ the long-run average delay in queue of a customer
            (with respect to both served and lost customers),

$W_{qs}(\tau) =$ the long-run average delay in queue of a served customer.

The measures $p_0(\tau)$ and $W_{qs}(\tau)$ are obtained easily from $\pi(\tau)$ and $W_q(\tau)$. Noting that each lost customer spends a time τ in the queue, we find

$$W_q(\tau) = \tau \pi(\tau) + W_{qs}(\tau)\{1 - \pi(\tau)\}.$$

By essentially the same arguments as used to obtain (4.5), we have

$$1 - p_0(\tau) = \lambda\{1 - \pi(\tau)\}\frac{E(S)}{\sigma}.$$

This formula states that the fraction of time the server is busy equals the average arrival rate of served customers times the average processing time of a customer.

In order to obtain $\pi(\tau)$ and $W_q(\tau)$, it is helpful to consider a closely related queueing system. In the queueing system under consideration an arriving customer always enters the system but is lost after a time τ if upon arrival the amount of work in the system is greater than $\sigma\tau$, the latter being equivalent to a waiting time greater than τ. Consider as a modified system the same queueing system as above except that a customer is now turned away immediately if upon arrival the amount of work in the system is greater than $\sigma\tau$. Some reflections show that the admitted customers in the modified system are exactly the served customers in the original system. The performance measures $\pi(\tau)$ and $W_q(\tau)$ for the original system will be expressed in terms of the limiting distribution function $V_\tau(x)$ of the amount of work in the modified system. The probability distribution function $V_\tau(x)$ has a positive mass at $x = 0$ and a density $v_\tau(x)$ for $x > 0$. Using the property 'Poisson arrivals see time averages', it is easily seen that

$$\pi(\tau) = 1 - V_\tau(\sigma\tau), \qquad W_q(\tau) = \tau\{1 - V_\tau(\sigma\tau)\} + \int_0^{\sigma\tau}\frac{x}{\sigma}v_\tau(x)\,dx. \qquad (4.134)$$

These relations require $V_\tau(x)$ only for $0 \leqslant x \leqslant \sigma\tau$. It is possible to express $V_\tau(x)$ in terms of $V_\infty(x)$, which is defined as the limiting distribution function of the amount of work in the modified system with $\tau = \infty$. Here it is assumed that the offered load ρ defined by

$$\rho = \frac{\lambda E(S)}{\sigma}$$

is smaller than 1. Obviously, $V_\infty(x)$ is the well-studied limiting distribution of the workload in the standard $M/G/1$ queueing system with processing rate σ. We next prove that

$$V_\tau(x) = \frac{V_\infty(x)}{1 - \rho + \rho V_\infty(\sigma\tau)}, \qquad 0 \leqslant x \leqslant \sigma\tau. \qquad (4.135)$$

To do so, we first note that by exactly the same arguments as used to derive (4.119), it can be verified that for some constant $c > 0$,

$$V_\tau(x) = cV_\infty(x), \qquad 0 \leqslant x \leqslant \sigma\tau. \qquad (4.136)$$

To find the constant c we establish the following integral equation for $v_\tau(x)$:

$$\sigma v_\tau(x) = \lambda\{1 - B(x)\}V_\tau(0) + \lambda \int_0^{\min(x, \sigma\tau)} \{1 - B(x - y)\}v_\tau(y)\mathrm{d}y, \qquad x > 0. \qquad (4.137)$$

A simple way to obtain this integral equation is to apply a useful up- and downcrossings technique studied in Brill and Posner (1977) and Cohen (1977). This technique has many interesting applications to applied probability problems. We only give a rough sketch of the derivation of (4.137) by the up- and downcrossings technique. The key observation is that, for any $x > 0$,

The long-run average number of downcrossings of level x of the workload process per unit time = the long-run average number of upcrossings of level x of the workload process per unit time.

Next we argue that

$$\sigma v_\tau(x) = \text{the long-run average number of downcrossings of level } x \text{ of}$$
$$\text{the workload process per unit time.} \qquad (4.138)$$

A heuristic derivation of this basic relation proceeds as follows (see Cohen, 1977, for a rigorous proof). Let $D_t(x)$ be the expected number of downcrossings of level x of the workload process in $(0, t]$ and let $d_\infty(x) = \lim_{t \to \infty} D_t(x)/t$ be the long-run average number of downcrossings of level x per unit time. Noting that it takes a time $\Delta x/\sigma$ to reduce the workload with an amount of Δx, it can be seen that, for small Δx,

The expected amount of time the workload is in $(x, x + \Delta x]$
$$\text{during } (0, t] \approx D_t(x)\frac{\Delta x}{\sigma}.$$

Dividing both sides of this relation by t and letting $t \to \infty$, we obtain

The long-run fraction of time the workload is in $(x, x + \Delta x] \approx d_\infty(x)\frac{\Delta x}{\sigma}$.

On the other hand, the long-run fraction of time the workload is in the interval $(x, x + \Delta x]$ equals $V_\tau(x + \Delta x) - V_\tau(x)$. Hence, we find

$$d_\infty(x) \approx \frac{\sigma\{V_\tau(x + \Delta x) - V_\tau(x)\}}{\Delta x}$$

for small Δx. Next, by letting $\Delta x \to 0$, we get the desired result (4.138). To find the rate of upcrossings, note that upcrossings occur only at those arrival epochs at which the workload process is at a level not above $\sigma\tau$. A positive jump occurring at some level y exceeds the level $x \ (> y)$ with probability $1 - B(x - y)$. Also, since arrivals occur according to a Poisson process with rate λ and the fraction of time the workload is between y and $y + \Delta y$ equals $v_\tau(y)\Delta y$, the average number of upcrossings per unit time starting in the interval $(y, y + \Delta y]$ is given by $\lambda v_\tau(y)\Delta y$ for small Δy when $0 \leqslant y \leqslant \sigma\tau$. Also, the rate of upcrossings starting from level 0 is $\lambda V_\tau(0)$. Thus the long-run average number of upcrossings of level x per unit time is given by the right side of (4.137), which completes the verification of the integral equation for $v_\tau(x)$.

In particular, using partial integration, we obtain from (4.137) with $\tau = \infty$ after some manipulations that

$$V_\infty(x) = V_\infty(0) + \frac{\lambda}{\sigma} \int_0^x V_\infty(x - y)\{1 - B(y)\}\,\mathrm{d}y, \qquad x > 0, \qquad (4.139)$$

in agreement with equation (1.73) in section 1.9 of chapter 1. It is now a matter of some algebra to obtain the constant c in (4.136). Using the normalization equation $V_\tau(0) + \int_0^\infty v_\tau(x)\,\mathrm{d}x = 1$ and the relations (4.136) and (4.137) with $x \leqslant \sigma\tau$, we find after some straightforward manipulations

$$1 = cV_\infty(\sigma\tau) + c\frac{\lambda}{\sigma} V_\infty(0) \int_{\sigma\tau}^\infty \{1 - B(x)\}\,\mathrm{d}x$$

$$+ c\frac{\lambda}{\sigma} \int_0^{\sigma\tau} v_\infty(y)\,\mathrm{d}y \int_{\sigma\tau - y}^\infty \{1 - B(u)\}\,\mathrm{d}u,$$

where $v_\infty(y)$ denotes the density of $V_\infty(y)$ for $y > 0$. The latter term in the right side of this equation can be elaborated further as

$$c\frac{\lambda}{\sigma} \int_0^{\sigma\tau} \{1 - B(u)\}\,\mathrm{d}u \int_{\sigma\tau - u}^{\sigma\tau} v_\infty(y)\,\mathrm{d}y + c\frac{\lambda}{\sigma} \int_{\sigma\tau}^\infty \{1 - B(u)\}\,\mathrm{d}u \int_0^{\sigma\tau} v_\infty(y)\,\mathrm{d}y$$

$$= c\frac{\lambda}{\sigma} \int_0^{\sigma\tau} \{1 - B(u)\}\{V_\infty(\sigma\tau) - V_\infty(\sigma\tau - u)\}\,\mathrm{d}u$$

$$+ c\frac{\lambda}{\sigma} \int_{\sigma\tau}^\infty \{1 - B(u)\}\{V_\infty(\sigma\tau) - V_\infty(0)\}\,\mathrm{d}u$$

$$= c\rho V_\infty(\sigma\tau) - c\frac{\lambda}{\sigma} \int_0^{\sigma\tau} V_\infty(\sigma\tau - u)\{1 - B(u)\}\,\mathrm{d}u$$

$$- c\frac{\lambda}{\sigma} V_\infty(0) \int_{\sigma\tau}^\infty \{1 - B(u)\}\,\mathrm{d}u,$$

where the latter equality uses $\int_0^\infty \{1 - B(u)\}\,\mathrm{d}u = E(S)$. Thus, collecting terms and using (4.139), we find

$$1 = cV_\infty(\sigma\tau) + c\rho V_\infty(\sigma\tau) - c\frac{\lambda}{\sigma} \int_0^{\sigma\tau} V_\infty(\sigma\tau - u)\{1 - B(u)\}\,\mathrm{d}u = c\rho V_\infty(\sigma\tau) + cV_\infty(0),$$

implying with $V_\infty(0) = 1 - \rho$ that $c = 1/\{1 - \rho + \rho V_\infty(\sigma\tau)\}$. This completes the verification of the relation (4.135).

The relations (4.134) and (4.135) together provide exact expressions for $\pi(\tau)$ and $W_q(\tau)$, using the fact that $V_\infty(x)$ is given by the function $1 - q(x)$ studied in section 1.9 of chapter 1. We next show that these expressions lead to computationally tractable results for the special cases of deterministic and exponential service requirements and to useful approximations for the case of general service requirements.

Deterministic service

Writing $\pi^{\mathrm{det}}(\tau)$ and $W_q^{\mathrm{det}}(\tau)$ for $\pi(\tau)$ and $W_q(\tau)$ when the service requirement is deterministic, we obtain after considerable algebra from (4.134), (4.135) and (1.72) that

$$\pi^{\mathrm{det}}(\tau) = \frac{1}{\rho}\left[\rho - 1 + \left\{ 1 + \rho e^{\lambda\tau} \sum_{j=0}^{r} (-1)^j \frac{(\lambda\tau - \rho j)^j}{j!} e^{-\rho j} \right\}^{-1} \right] \qquad (4.140)$$

and

$$W_q^{\mathrm{det}}(\tau) = \tau - \frac{1}{\lambda}\left\{ 1 + \rho e^{\lambda\tau} \sum_{j=0}^{r} (-1)^j \frac{(\lambda\tau - \rho j)^j}{j!} e^{-\rho j} \right\}^{-1}$$

$$\times \left\{ -(r+1) + e^{\lambda\tau} \sum_{j=0}^{r} e^{-\rho j} \sum_{i=0}^{j} (-1)^i \frac{(\lambda\tau - \rho j)^i}{i!} \right\}, \qquad (4.141)$$

where r is the integer satisfying

$$rE(S) \leqslant \sigma\tau < (r+1)E(S).$$

These formulae were obtained under the assumption of $\rho < 1$. However, using results in Hokstad (1979), it can be shown that the formulae (4.140) and (4.141) apply as well for $\rho \geqslant 1$.

Exponential service

Writing $\pi^{\mathrm{exp}}(\tau)$ and $W_q^{\mathrm{exp}}(\tau)$ for $\pi(\tau)$ and $W_q(\tau)$ when the service requirement is exponentially distributed, we obtain from (4.134), (4.135) and (1.72) that

$$\pi^{\mathrm{exp}}(\tau) = \frac{(1-\rho)\rho e^{-\sigma(1-\rho)\tau/E(S)}}{1 - \rho^2 e^{-\sigma(1-\rho)\tau/E(S)}} \qquad (4.142)$$

and

$$W_q^{\mathrm{exp}}(\tau) = \frac{\rho E(S)/\{\sigma(1-\rho)\} - [E(S)/\{\sigma(1-\rho)\} + \rho\tau]\rho e^{-\sigma(1-\rho)\tau/E(S)}}{1 - \rho^2 e^{-\sigma(1-\rho)\tau/E(S)}}. \qquad (4.143)$$

Also, these formulae hold not only for $\rho < 1$ but apply as well for $\rho \geqslant 1$ where for $\rho = 1$ L'Hospital's rule should be used (cf. also Gnedenko and Kovalenko, 1968).

General service

The exact results (4.134) and (4.135) are in general not computationally tractable. Invoking the approximation (1.80) for $q(x) = 1 - V_\infty(x)$ yields the following simple approximations for $\pi(\tau)$ and $W_q(\tau)$,

$$\pi^{\text{app}}(\tau) = \frac{(1 - \rho)(\alpha e^{-\beta\sigma\tau} + \gamma e^{-\delta\sigma\tau})}{1 - \rho(\alpha e^{-\beta\sigma\tau} + \gamma e^{-\delta\sigma\tau})} \tag{4.144}$$

and

$$W_q^{\text{app}}(\tau) = \frac{\alpha/(\sigma\beta) - \alpha\{1/(\sigma\beta) + \rho\tau\}e^{-\beta\sigma\tau} + \gamma/(\sigma\delta) - \gamma\{1/(\sigma\delta) + \rho\tau\}e^{-\delta\sigma\tau}}{1 - \rho(\alpha e^{-\beta\sigma\tau} + \gamma e^{-\delta\sigma\tau})}, \tag{4.145}$$

where the constants α, β, γ and δ are given by (1.82), (1.79) and (1.75). Here it is assumed that the tail of the probability density $b(x)$ of the service requirement tends at least exponentially fast to zero as $x \to \infty$. The approximations (4.144) and (4.145) are exact when the service requirement has a K_2 distribution.

Also, using the asymptotic expansion (1.78), we readily obtain from (4.134) and (4.135) that

$$\pi(\tau) \approx (1 - \rho)\gamma e^{-\sigma\tau\delta} \quad \text{for large } \sigma\tau. \tag{4.146}$$

This asymptotic expansion may be used for practical purposes when $\pi(\tau)$ is on the order of 10^{-2} or less.

In Table 4.15 we give the exact and approximate values of $\pi(\tau)$ and $W_q(\tau)$ for various distributions with the same first two moments. Taking $E(S) = 1$ and $c_S^2 = 0.5$, we consider the Erlang-2 density, the Weibull density, the lognormal density and the shifted exponential density. In all examples we assume $\tau = 2$ and $\sigma = 1$, while the load factor ρ is varied as 0.2, 0.5 and 0.8. The exact values have been computed by a discretization algorithm as discussed in the appendix to subsection 4.3.2, while the approximate values correspond to the approxi-

Table 4.15   Exact and approximate results for various service-time densities

ρ		Erlang-2		Weibull		Lognormal		Shifted exponential	
		$\pi(\tau)$	$W_q(\tau)$	$\pi(\tau)$	$W_q(\tau)$	$\pi(\tau)$	$W_q(\tau)$	$\pi(\tau)$	$W_q(\tau)$
0.2	exa	0.0172	0.162	0.0169	0.164	0.0175	0.158	0.0181	0.159
	app	0.0172	0.162	0.0164	0.165	—	—	0.0181	0.159
0.5	exa	0.0701	0.467	0.0703	0.471	0.0683	0.457	0.0700	0.458
	app	0.0701	0.467	0.0701	0.472	—	—	0.0700	0.458
0.8	exa	0.1537	0.804	0.1547	0.808	0.1496	0.795	0.1515	0.794
	app	0.1537	0.804	0.1546	0.808	—	—	0.1515	0.794

mations (4.144) and (4.145). Note that these approximations do not apply to the lognormal density because of its long tail. The numerical investigations indicate that the measures of system performance are quite insensitive to more than the first two moments of the service-time distribution, provided c_S^2 is not too large and ρ is not very small (cf. also De Kok and Tijms, 1985a, for additional numerical results).

An alternative approximation to $W_q(\tau)$ is given by

$$W_q^{\text{app}}(\tau) = (1 - c_S^2)W_q^{\text{det}}(\tau) + c_S^2 W_q^{\text{exp}}(\tau),$$

provided c_S^2 is not large. Such a two-moment approximation is in general not to be recommended for small values of $\pi(\tau)$, but may be used for the 'quantile' $\tau(v)$ which is defined as the solution to $\pi(\tau) = v$, $0 < v < 1$. Since the exact results (4.140) and (4.141) and (4.142) and (4.143) for the special cases of deterministic service and exponential service also hold for values of $\rho \geqslant 1$, the above two-moment approximations apply as well when $\rho \geqslant 1$. Also, for the case of $\rho > 1$, we have the following excellent approximation to $\pi(\tau)$:

$$\pi(\tau) \approx \left(1 - \frac{1}{\rho}\right)\left[1 + \frac{\{1 + (\lambda/\sigma)\mathscr{L}'(s^*)\}e^{-s^*\tau}}{(\rho - 1)\rho - \{1 + (\lambda/\sigma)\mathscr{L}'(s^*)\}e^{-s^*\tau}}\right], \qquad (4.147)$$

where s^* and $\mathscr{L}'(s^*)$ are as in (4.132). This result, being due to De Kok and Van Ommeren (1986), is asymptotically exact as $\tau \to \infty$.

4.3.4   An M/G/1 queueing system with bounded sojourn time

Consider a single-server queueing system where customers arrive according to a Poisson process with rate λ. The customers bring along amounts of work that are sampled independently from a general distribution with given mean $E(S)$ and squared coefficient of variation c_S^2. The amount of work of an arriving customer is accepted only if the sum of the amount of work in the system just prior to arrival and the work requirement of that customer does not exceed some fixed amount K. A rejected customer has no further influence on the system. The amount of work brought in by an accepted arrival is put into a buffer with capacity K. The buffer is emptied at a constant rate of $\sigma > 0$ per unit time whenever there is work in the buffer.

This queueing model with bounded sojourn time is a useful model in communication systems to analyse storage buffers emptying at a constant rate and receiving messages from a high-speed data channel with the stipulation that each arriving message is rejected whose length plus the length of all messages in storage would exceed the buffer capacity. Another application of the above model is a finite capacity production/inventory system in which production occurs at a constant rate whenever the inventory is below its maximum level, the demand process is a compound Poisson process and each demand unable to be met in its whole from on-hand inventory is lost.

A general problem of practical interest is to find the buffer capacity K such that the probability of an arrival being rejected is not more than a specific value. In applications in communication systems the required rejection probability is typically very small (say, on the order of 10^{-5} or less). A tractable exact solution of this problem seems possible only for the special cases of deterministic and exponential work requirements. Also, in practice it often happens that only the first two moments of the work requirements are available. Therefore we focus on the determination of simple two-moment approximations for the buffer capacity in order to achieve a given rejection probability. Since in practical applications the required rejection level is typically small, it is reasonable to assume that the offered load ρ defined by

$$\rho = \frac{\lambda E(S)}{\sigma}$$

is smaller than 1.

Letting

$$\pi(K) = \text{the long-run fraction of arrivals that are rejected}$$
$$\text{when the buffer capacity is } K,$$

our primary goal is to find a solution $K(v)$ to the equation

$$\pi(K) = v, \tag{4.148}$$

with v a prescribed number between 0 and 1. That is $K(v)$ is the buffer capacity required to achieve a rejection level v. In practical applications involving a small rejection level v, the equation (4.148) will have a solution $K(v)$. However, the equation (4.148) need not be solvable in some (unrealistic) cases. To see this, consider the case in which the work requirement of each customer is a same constant D. Then $\pi(K) = 1$ for $K < D$, while $\pi(D)$ equals the fraction of time the buffer is not empty and so $\pi(D) = (D/\sigma)/(D/\sigma + 1/\lambda) = \rho/(1 + \rho)$. Hence for deterministic work requirements the equation (4.148) has a unique solution only for $0 < v \leqslant \rho/(1 + \rho)$.

The two-moment approximations for $K(v)$ will as usual be based on exact results for the special cases of deterministic and exponential work requirements. We first derive exact results for these particular cases, including results for the average workload defined by

$$EV(K) = \text{the long-run average amount of work in the buffer when}$$
$$\text{the buffer has capacity } K.$$

We write $\pi_{\text{det}}(K)$ and $EV_{\text{det}}(K)$ for $\pi(K)$ and $EV(K)$ when the work requirement is deterministic. Similarly, we write $\pi_{\text{exp}}(K)$ and $EV_{\text{exp}}(K)$ when the work requirement is exponentially distributed.

Deterministic work requirements

For the case of deterministic work requirements with $D = E(S)$, the present queueing system is in fact identical to the modified queueing system with impatient customers dealt with in the previous subsection. In this latter queueing system a customer is admitted only if upon arrival the amount of work in the system is not more than $\sigma\tau$ for some fixed τ. Because the work requirement of each customer is the same constant D, the arrivals entering the present queueing system with buffer capacity K are exactly those arrivals entering the modified queueing system with $\sigma\tau = K - D$. Consequently, by substituting $\tau = (K - D)/\sigma$ in formula (4.140), we obtain

$$\pi_{\det}(K) = \frac{1}{\rho}\left(\rho - 1 + \left[1 + \rho e^{\lambda(K-D)/\sigma}\sum_{j=0}^{s-1}(-1)^j\frac{\{\lambda K/\sigma - \rho(j+1)\}^j}{j!}e^{-\rho j}\right]^{-1}\right),$$

(4.149)

where s is the integer satisfying

$$sD \leqslant K < (s+1)D.$$

Further, using the fact that for the deterministic case the limiting distribution function of the amount of work in the buffer is given by the distribution function $V_\tau(x)$ studied in the previous subsection, it can be shown after considerable algebra that

$$EV_{\det}(K) = \lambda\{1 - \pi_{\det}(K)\}\left[\frac{D\{a(K) - (K-D)\pi_{\det}(K)/\sigma\}}{1 - \pi_{\det}(K)} + \frac{D^2}{2\sigma}\right], \quad (4.150)$$

where $a(K)$ is given by the right side of (4.141) in which τ is replaced by $(K - D)/\sigma$ and r is replaced by $s - 1$. The formulae (4.149) and (4.150) hold for any value of ρ (cf. Hokstad, 1979).

A useful asymptotic expansion for $\pi_{\det}(K)$ can be given. By (4.146) with $\tau = (K - D)/\sigma$, we find

$$\pi_{\det}(K) \approx \eta_{\det}e^{-\delta_{\det}K} \qquad \text{for } K \text{ large} \qquad (4.151)$$

with

$$\eta_{\det} = (1 - \rho)\gamma_{\det}e^{\delta_{\det}D},$$

where $\delta_{\det}$ and $\gamma_{\det}$ are defined by (4.46) and (4.47). For the computation of small values of $\pi_{\det}(K)$ (say, $\pi(K) \leqslant 10^{-3}$) it is recommended that the asymptotic expansion (4.151) is used rather than the explicit expression (4.149). Numerical difficulties may occur when using (4.149) to compute very small values of $\pi(K)$.

Exponential work requirement

For the case of exponential work requirement the limiting distribution function of the amount of work in the buffer can explicitly be determined. Denote by $W_K(x), 0 \leqslant x \leqslant K$, this limiting distribution function and let $w_K(x)$ be the density of $W_K(x)$ for $x > 0$. An integral equation for $W_K(x)$ is obtained by applying the up- and downcrossings technique discussed in the previous subsection. As before, the rate of downcrossings of level $x > 0$ of the workload process is $\sigma w_K(x)$, using the fact that $w_K(x)\Delta x$ is the fraction of time the workload is between x and $x + \Delta x$ for small Δx. The rate of upcrossings of level x follows from the following two observations. First, an arrival occurring when the workload is at level $y < x$ causes an upcrossing of level x only if the work requirement of that arrival is between $x - y$ and $K - y$, which latter event has probability $B(K - y) - B(x - y)$. Here B denotes the probability distribution function of the work requirement of an arrival. Second, since arrivals occur according to a Poisson process with rate λ, the rate of upcrossings starting from some interval $(y, y + \Delta y]$ is $\lambda w_K(y)\Delta y$ for $y > 0$ and small Δy, while the rate of upcrossings starting from level 0 equals $\lambda W_K(0)$. Thus, by equating the rate at which the workload process downcrosses the level x to the rate at which the workload process upcrosses the level x, we find

$$\sigma w_K(x) = \lambda\{B(K) - B(x)\}W_K(0) + \lambda \int_0^x \{B(K - y) - B(x - y)\}w_K(y)\mathrm{d}y,$$

$$0 < x < K.$$

For the exponential case with

$$B(x) = 1 - \mathrm{e}^{-x/E(S)}, \qquad x \geqslant 0,$$

this integral equation allows for an explicit solution. We only sketch the derivation of the solution. Letting

$$h(x) = \frac{w_K(x)}{W_K(0)}, \qquad 0 < x < K, \tag{4.152}$$

we find by differentiation of the above integral equation that

$$h'(x) = -\frac{\lambda}{\sigma}b(x) + \frac{\lambda}{\sigma}B(K - x)h(x) - \frac{\lambda}{\sigma}\int_0^x b(x - y)h(y)\mathrm{d}y, \qquad 0 < x < K,$$

where $b(x)$ denotes the density of $B(x)$. For the special case of exponential work requirements this integro-differential equation can be converted to a second-order linear differential equation by the standard trick of the change of dependent variable $h(x) = \exp\{-x/E(S)\}g'(x)$ with $g(0) = 0$. The resulting differential equation in $g(x)$ is easily solved. We refer the reader for details to Gavish and Schweitzer (1977). Here we state only the final result,

$$h(x) = \frac{\rho}{E(S)}\beta_0^{1-\rho}(\beta_x^{\rho-1} - \beta_x^\rho)\mathrm{e}^{\rho(\beta_0 - \beta_x)}, \qquad 0 \leqslant x \leqslant K, \tag{4.153}$$

where

$$\beta_x = \mathrm{e}^{-(K-x)/E(S)}, \qquad 0 \leqslant x \leqslant K. \tag{4.154}$$

Next we readily obtain the constant $W_K(0)$, being the probability of an empty buffer. Using (4.152), (4.153) and the equation $W_K(0) + \int_0^K w_K(y)\,dy = 1$, it is a matter of simple algebra to verify that

$$W_K(0) = (1 - \rho)[1 - \rho\beta_0^{1-\rho}e^{\rho(\beta_0 - 1)} - (\rho\beta_0)^{1-\rho}e^{\rho\beta_0}\{\gamma(\rho, \rho) - \gamma(\rho, \rho\beta_0)\}]^{-1},$$

(4.155)

provided $\rho \neq 1$. Here

$$\gamma(\alpha, x) = \int_0^x u^{\alpha - 1}e^{-u}\,du, \qquad \alpha, x > 0,$$

denotes the well-known incomplete gamma function (cf. also appendix B). The rejection probability $\pi_{exp}(K)$ and the average workload $EV_{exp}(K)$ are calculated from

$$\pi_{exp}(K) = \{1 - B(K)\}W_K(0) + \int_0^K \{1 - B(K - y)\}w_K(y)\,dy$$

and

$$EV_{exp}(K) = \int_0^K yw_K(y)\,dy,$$

resulting after some tedious calculations in

$$\pi_{exp}(K) = W_K(0)\beta_0^{1-\rho}e^{\rho(\beta_0 - 1)}$$

(4.156)

and

$$EV_{exp}(K) = W_K(0)\rho E(S)\beta_0^{1-\rho}e^{\rho\beta_0}\left[\frac{K}{(\rho - 1)E(S)}\{e^{-\rho} + \gamma(\rho, \rho)\rho^{-\rho}\}\right.$$
$$\left. + \frac{\beta_0^{\rho - 1}e^{-\rho\beta_0} - e^{-\rho}}{(\rho - 1)^2} - \frac{\gamma(\rho, \rho) - \gamma(\rho, \rho\beta_0)}{(\rho - 1)^2\rho^{\rho - 1}} - \frac{\int_{\rho\beta_0}^{\rho} x^{-1}\gamma(\rho, x)\,dx}{(\rho - 1)\rho^{\rho}}\right].$$

(4.157)

Efficient numerical procedures are available to evaluate the incomplete gamma function $\gamma(\rho, x)$, while the integral appearing in $EV_{exp}(K)$ can be evaluated numerically by standard Gaussian integration.

The above formulae for $\pi_{exp}(K)$ and $EV_{exp}(K)$ apply for any value of ρ. For the case of $\rho < 1$, a useful asymptotic expansion for $\pi_{exp}(K)$ is readily obtained from (4.156). The probability $W_K(0)$ of having an empty buffer tends to $1 - \rho$ for $K \to \infty$, as can be seen by noting that for very large K the queueing system behaves like an M/G/1 queueing system with infinite capacity. Thus, by (4.154) to (4.156),

$$\pi_{exp}(K) \approx \eta_{exp}e^{-\delta_{exp}K} \qquad \text{for} \quad \text{large } K,$$

(4.158)

where

$$\delta_{\exp} = \frac{1 - \rho}{E(S)} \quad \text{and} \quad \eta_{\exp} = (1 - \rho)e^{-\rho}.$$

General work requirement

For the case of a generally distributed work requirement we are not able to obtain explicit results. The numerical values of $\pi(K)$ and $EV(K)$ could be computed in principle by a discretization algorithm similar to the one discussed in the appendix to subsection 4.3.2. However, such a discretization algorithm will be computationally infeasible for problems involving a very small rejection level and a large buffer capacity. Therefore we resort to approximations. As stated before, our primary goal is to find approximations for problems involving a small rejection level. Numerical experiments show that the usual two-moment approximation based on a linear interpolation on c_S^2 does not work for $\pi(K)$. However, we found that the buffer capacity $K(v)$ needed to achieve a rejection level v lends itself quite well to such a two-moment approximation. In analogy to (4.116), the buffer capacity $K(v)$ may be approximated by

$$K_{\text{app}}(v) = (1 - c_S^2)K_{\det}(v) + c_S^2 K_{\exp}(v), \tag{4.159}$$

provided c_S^2 is not too large (say, $0 \leqslant c_S^2 \leqslant 1$) and the work requirement density satisfies a reasonable shape constraint. Here $K_{\det}(v)$ and $K_{\exp}(v)$ denote the exact values of $K(v)$ for the respective cases of deterministic and exponential work requirements. The approximation (4.159) may be applied for all values of v including extremely small values. Our numerical experiments indicate that the two-moment approximation $K_{\text{app}}(v)$ is quite useful for practical purposes, in particular when the load factor ρ is not too small. In Table 4.16 we give the buffer capacity $K_{\text{app}}(v)$ for several values of v and ρ and for various work requirement densities having the same first two moments. In all examples we take $\sigma = 1$, $E(S) = 1$ and $c_S^2 = \frac{1}{2}$. We consider the Erlang-2 density, the Weibull density, the lognormal density, the shifted exponential density and a mixture of Erlang-1 and Erlang-3 densities with the same scale parameters. Also in Table 4.16 we give the actual rejection probabilities π_{E_2}, π_{Weib}, π_{logn}, π_{shif} and $\pi_{E_{1,3}}$ associated with the various work requirement densities when the buffer capacity is $K_{\text{app}}(v)$. A remarkable finding from the numerical investigations is that, for a rejection level v up to 10^{-3} and for a load factor $\rho \geqslant 0.5$, the two-moment approximation performs rather well for the lognormal density, although this density has no exponentially fast decreasing tail unlike the other densities.

The analytical results (4.151) and (4.158) together with our numerical investigations lead to the conjecture that for generally distributed work

Table 4.16   The actual rejection probabilities for various densities with $c_S^2 = \frac{1}{2}$

ρ	v	$K_{app}(v)$	π_{E_2}	π_{Weib}	π_{logn}	π_{shif}	$\pi_{E_{1,3}}$
0.2	0.01	3.79	0.012	0.011	0.014	0.014	0.011
0.5	0.01	5.34	0.011	0.011	0.012	0.011	0.011
0.8	0.01	9.63	0.011	0.011	0.010	0.010	0.011
0.2	0.005	4.36	0.0059	0.0050	0.0080	0.0075	0.0050
0.5	0.005	6.29	0.0055	0.0054	0.0061	0.0060	0.0053
0.8	0.005	12.01	0.0053	0.0054	0.0052	0.0052	0.0054
0.2	0.001	5.65	0.0012	0.0009	0.0024	0.0018	0.0009
0.5	0.001	8.53	0.0011	0.0010	0.0015	0.0013	0.0010
0.8	0.001	17.76	0.0011	0.0011	0.0011	0.0011	0.0011

requirements the rejection probability $\pi(K)$ has the asymptotic expansion

$$\pi(K) \approx \eta e^{-\delta K} \qquad \text{for} \quad \text{large } K, \tag{4.160}$$

provided the tail of the density $b(x)$ of the work requirement tends at least exponentially fast to zero as $x \to \infty$. Here the decay parameter δ is the unique positive solution to the equation (1.75) in section 1.9 of chapter 1, while the amplitude factor η is some unknown positive constant. In the same way as in the derivation of (4.43) and (4.44), it can be argued that a consistency of the two-moment approximation (4.159) with the asymptotic expansion (4.160) suggests the following two-moment approximations for δ and η:

$$\delta_{app} = \frac{\delta_{det}\delta_{exp}}{(1 - c_S^2)\delta_{exp} + c_S^2\delta_{det}},$$

$$\eta_{app} = \eta_{exp}^q \eta_{det}^{1-q} \qquad \text{with} \quad q = c_S^2 \frac{\delta_{app}}{\delta_{exp}}.$$

Thus, for v sufficiently small, the buffer capacity $K(v)$ may alternatively be approximated by

$$K_{asy}(v) = \frac{1}{\delta_{app}} \ln\left(\frac{\eta_{app}}{v}\right). \tag{4.161}$$

The approximation $K_{asy}(v)$ gives practically the same results as $K_{app}(v)$ when v is sufficiently small (say, $v \leqslant 10^{-3}$). The approximation $K_{asy}(v)$ is much easier to compute than $K_{app}(v)$ and avoids the numerically unstable expression (4.149). In Table 4.17 we give for several values of v the values of $K(v)$ and $K_{asy}(v)$ for the cases of deterministic and exponential work input, where in all examples we take $\sigma = 1$ and $E(S) = 1$.

To conclude this subsection, we note that the average workload $EV(K)$ for a given buffer capacity K may be approximated as usual by

$$EV_{app}(K) = (1 - c_S^2)EV_{det}(K) + c_S^2 EV_{exp}(K).$$

Table 4.17 The values of $K(v)$ and $K_{asy}(v)$ for deterministic and exponential inputs

		$\rho = 0.2$		$\rho = 0.5$		$\rho = 0.8$		$\rho = 0.9$	
		det	exp	det	exp	det	exp	det	exp
$v = 0.01$	exa	2.34	5.24	3.79	6.90	7.71	11.55	12.20	16.31
	asy	2.33	5.23	3.78	6.82	7.62	10.98	11.78	14.03
$v = 0.005$	exa	2.62	6.10	4.34	8.25	9.28	14.74	15.34	22.16
	asy	2.59	6.09	4.34	8.21	9.23	14.44	15.13	20.96
$v = 10^{-3}$	exa	3.20	8.11	5.62	11.44	12.97	22.55	22.94	37.30
	asy	3.20	8.11	5.63	11.43	12.96	22.49	22.90	37.05
$v = 10^{-4}$	exa	4.06	10.98	7.45	16.04	18.31	34.01	34.02	60.10
	asy	4.06	10.98	7.45	16.03	18.31	34.00	34.01	60.08
$v = 10^{-5}$	exa	4.93	13.86	9.28	20.64	23.65	45.52	45.13	83.11
	asy	4.93	13.86	9.28	20.64	23.65	45.52	45.13	83.10

It was found that this approximation performs very well, including for values of $\rho \geqslant 1$, (cf. De Kok and Tijms, 1985b, for numerical results).

4.4 MULTI-SERVER QUEUEING SYSTEMS

Multi-server queues are notoriously difficult and a simple algorithmic analysis is possible only for special cases. In principle any practical queueing process could be modelled as a Markov process by incorporating sufficient information in the state description, but the dimensionality of the state space would grow quickly beyond any practical bound. Nevertheless, using special numerical techniques to solve large systems of equilibrium equations for Markov chains, the continuous-time Markov chain approach has proved to be extremely useful for multi-server queueing systems with special phase-type interarrival-time and service-time distributions, provided the number of servers is not very large. In many situations, however, one resorts to approximation methods for calculating measures of system performance. Useful approximations for complex queueing systems are often obtained by an appropriate application of exact results for simpler related queueing systems.

In this section we discuss both exact and approximate solution methods for the state probabilities and the waiting-time probabilities in multi-server queues. We first present exact methods for the tractable models of the M/M/c queue and the M/D/c queue. Next we consider the M/G/c queue with general service times and give several approximations including two-moment approximations based on exact results for the M/M/c queue and the M/D/c queue. Special attention is paid to the M/G/∞ queue, being a very useful model for practical applications with a large number of servers. Also, some approximate results are presented for the finite capacity $M/G/c$ queue and the machine repair model with multiple repairmen. Finally, we discuss computational methods for the $GI/G/c$ queueing model with general interarrival and service times.

4.4.1 The M/M/c queue

This queueing system assumes a Poisson arrival process with rate λ, exponentially distributed service times with mean $E(S) = 1/\mu$ and c identical servers are available. It is supposed that the server utilization

$$\rho = \frac{\lambda E(S)}{c}$$

is smaller than 1. By the memoryless property of the Poisson process and of the exponential distribution, the process describing the number of customers in the system is a continuous-time Markov chain. Using the familiar technique of equating the rate at which the process enters the set of states of having at least j customers present to the rate at which the system leaves that set, we obtain for the time-average probabilities p_j the recursion relation

$$\lambda p_{j-1} = \min(j, c)\mu p_j, \qquad j = 1, 2, \ldots.$$

It is easy to show that this recursion relation allows for the explicit solution,

$$p_j = \begin{cases} \dfrac{(c\rho)^j}{j!} p_0, & j = 0, \ldots, c-1, \\[2ex] \dfrac{(c\rho)^j}{c! c^{j-c}} p_0, & j \geqslant c, \end{cases} \qquad (4.162)$$

with

$$p_0 = \left\{ \sum_{k=0}^{c-1} \frac{(c\rho)^k}{k!} + \frac{(c\rho)^c}{c!(1-\rho)} \right\}^{-1}.$$

Recall from (4.10) that for this queueing system the time-average probabilities p_j give also the customer-average probabilities π_j. It is a matter of simple algebra to verify that the delay probability $\Pi_W = \sum_{j=c}^{\infty} \pi_j$ and the average queue size $E(L_q) = \sum_{j=c}^{\infty} (j-c)p_j$ are given by

$$\Pi_W(\exp) = \frac{(c\rho)^c}{c!(1-\rho)} \left\{ \sum_{k=0}^{c-1} \frac{(c\rho)^k}{k!} + \frac{(c\rho)^c}{c!(1-\rho)} \right\}^{-1} \qquad (4.163)$$

and

$$E(L_q) = \frac{(c\rho)^c \rho}{c!(1-\rho)^2} \left\{ \sum_{k=0}^{c-1} \frac{(c\rho)^k}{k!} + \frac{(c\rho)^c}{c!(1-\rho)} \right\}^{-1}. \qquad (4.164)$$

The formula (4.163) is called *Erlang's delay formula*. It has long been known that Erlang's delay formula provides an excellent approximation to the delay probability in the M/G/c queue with non-exponential service times (cf. also subsection 4.4.3).

The waiting-time distribution function can also be given explicitly when service is in order of arrival. Therefore note that the conditional waiting time of a customer finding upon arrival j other customers present is distributed as the sum of $j - c + 1$ independent exponentials with the same means $1/c\mu$ and thus has an E_{j-c+1} distribution. Since the steady-state probability π_j that an arbitrary customer finds upon arrival j other customers present equals p_j, it follows by the law of total probability that

$$P\{W_q > x\} = \sum_{j=c}^{\infty} p_j \sum_{k=0}^{j-c} e^{-c\mu x} \frac{(c\mu x)^k}{k!}, \qquad x \geqslant 0.$$

Hence, after some algebra,

$$P\{W_q > x\} = \Pi_W(\exp) e^{-c\mu(1-\rho)x}, \qquad x \geqslant 0. \tag{4.165}$$

In particular, the average delay in queue of a customer equals

$$E(W_q) = \frac{(c\rho)^c}{c! \, c\mu(1-\rho)^2} \left\{ \sum_{k=0}^{c-1} \frac{(c\rho)^k}{k!} + \frac{(c\rho)^c}{c!(1-\rho)} \right\}^{-1}, \tag{4.166}$$

in agreement with (4.164) and Little's formula $E(L_q) = \lambda E(W_q)$.

4.4.2   The M/D/c queue

In this model the arrival process of customers is a Poisson process with rate λ, the service time of each customer is the same constant D and c identical servers are available. It is assumed that the server utilization $\rho = \lambda D/c$ is smaller than 1.

An exact algorithmic analysis of the M/D/c queueing system is based on the following observation. Since the service times are constant, any customer in service at some time t will have left the system at time $t + D$, while the customers present at time $t + D$ are exactly those customers either waiting in queue at time t or arriving in $(t, t + D]$. Hence, letting $p_j(u)$ be the probability of having j customers in the system at time u and noting that the number of arrivals in a time D is Poisson distributed with mean λD, it follows by conditioning on the state at time t that

$$p_0(t + D) = \sum_{k=0}^{c} p_k(t) e^{-\lambda D},$$

$$p_j(t + D) = \sum_{k=0}^{c} p_k(t) e^{-\lambda D} \frac{(\lambda D)^j}{j!}$$

$$+ \sum_{k=c+1}^{c+j} p_k(t) e^{-\lambda D} \frac{(\lambda D)^{j-k+c}}{(j-k+c)!}, \qquad j \geqslant 1.$$

Next, by letting $t \to \infty$ in these equations and noting that $p_j(u) \to p_j$ as $u \to \infty$, we find that the time-average probabilities p_j satisfy the linear equations

$$p_j = e^{-\lambda D} \frac{(\lambda D)^j}{j!} \sum_{k=0}^{c} p_k + \sum_{k=c+1}^{c+j} p_k e^{-\lambda D} \frac{(\lambda D)^{j-k+c}}{(j-k+c)!}, \qquad j \geqslant 0, \qquad (4.167)$$

where $\sum_{k=c+1}^{c} = 0$ by convention. Also, we have the normalizing equation

$$\sum_{j=0}^{\infty} p_j = 1. \qquad (4.168)$$

Before discussing how to solve numerically the equations (4.167) and (4.168), we derive the first two moments of the queue size L_q from these equations by using the technique of generating functions. Therefore we define

$$P(z) = \sum_{j=0}^{\infty} p_j z^j \quad \text{and} \quad P_q(z) = \sum_{j=c}^{\infty} p_j z^{j-c}, \qquad |z| \leqslant 1.$$

Obviously,

$$E(L_q) = P'_q(1) \quad \text{and} \quad E(L_q(L_q - 1)) = P''_q(1). \qquad (4.169)$$

Also, we have the relation

$$P(z) = \sum_{j=0}^{c-1} p_j z^j + z^c P_q(z). \qquad (4.170)$$

Put for abbreviation

$$Q_i = \sum_{k=0}^{i} p_k, \qquad i = 0, \ldots, c. \qquad (4.171)$$

Multiplying both sides of (4.167) by z^j, summing over j and using an interchange of the order of summation yields

$$P(z) = Q_c e^{\lambda D(z-1)} + \sum_{j=1}^{\infty} z^j \sum_{i=1}^{i} p_{c+i} e^{-\lambda D} \frac{(\lambda D)^{j-i}}{(j-i)!}$$

$$= Q_c e^{\lambda D(z-1)} + \sum_{i=1}^{\infty} p_{c+i} z^i \sum_{j=i}^{\infty} e^{-\lambda D} \frac{(\lambda D)^{j-i}}{(j-i)!} z^{j-i}$$

$$= Q_c e^{\lambda D(z-1)} + e^{\lambda D(z-1)} \sum_{k=c+1}^{\infty} p_k z^{k-c},$$

implying

$$P(z) = \frac{\sum_{k=0}^{c} p_k z^k - Q_c z^c}{1 - z^c e^{\lambda D(1-z)}}. \qquad (4.172)$$

Next, by differentiation and using (4.169) and (4.170) together with L'Hospital's rule, we find after considerable algebra

$$E(L_q) = \frac{(c\rho)^2 - c(c-1) + \sum_{j=0}^{c-1} \{c(c-1) - j(j-1)\} p_j}{2c(1-\rho)} \qquad (4.173)$$

and

$$E(L_q(L_q - 1))$$
$$= \frac{(c\rho)^3 - c(c-1)(c-2) + \sum_{j=0}^{c-1}\{c(c-1)(c-2) - j(j-1)(j-2)\}p_j}{3c(1-\rho)}$$
$$- \frac{(c-1-c\rho^2)}{1-\rho}E(L_q). \tag{4.174}$$

It is noteworthy that these expressions for the moments of the queue size involve the first c state probabilities only. This fact is computationally useful, since in iterative methods for solving the equations (4.167) and (4.168) the first c state probabilities will usually converge earlier than the other state probabilities.

We next discuss computational matters involved with the numerical solution of the linear equations (4.167) and (4.168). These linear equations can be solved by successive overrelaxation when this infinite system is truncated in an appropriate way to a finite system. The truncation may be based on the inequality,

$$\sum_{j=k}^{\infty} p_j^{\text{det}} \leqslant \sum_{j=k}^{\infty} p_j^{\text{exp}}, \qquad k \geqslant 1, \tag{4.175}$$

where p_j^{det} and p_j^{exp} denote the time-average probabilities p_j in the respective cases of the M/D/c queue and the M/M/c queue. This inequality is intuitively reasonable by noting that the M/D/c queue involves less variability than the M/M/c queue. The probabilities p_j^{exp} are easy to compute from the recursion scheme (4.162). The truncation of the infinite system (4.167) and (4.168) may be based on the smallest integer L for which $\sum_{j=L}^{\infty} p_j^{\text{exp}}$ is less than some small number ε (for example $\varepsilon = 10^{-8}$). The resulting finite system of linear equations can be solved effectively by the iterative method of successive overrelaxation described in appendix D, where the probabilities p_j^{exp} provide a good starting point for the algorithm. For the M/D/c queue an even more efficient successive overrelaxation method can be formulated by using the asymptotic result (see the relations 4.15, 4.18 and 4.19).

$$\frac{p_{j-1}}{p_j} \to \tau \qquad \text{as} \quad j \to \infty, \tag{4.176}$$

where τ is given by

$$\tau = 1 + \frac{\delta}{\lambda} \tag{4.177}$$

with δ being the unique positive solution to the equation

$$\lambda(e^{\delta D/c} - 1) - \delta = 0. \tag{4.178}$$

The convergence in (4.176) is usually quite fast when the traffic is not very light. Here light traffic corresponds to a small value of the delay probability. In this context is should be noted that in multi-server queueing systems the server utilization is in general not an appropriate measure for the traffic load on the system, as the reader may see by looking up in Table 4.18 below the values of the delay probability for various values of c with ρ kept fixed. The convergence result (4.176) allows for a more efficient truncation of the linear equations (4.167) and (4.168) by using an integer $N(>c)$ such that the approximation

$$p_j \approx \tau^{N-j} p_N \quad \text{for} \quad j \geqslant N \tag{4.179}$$

is reasonably believed to be accurate. For non-light traffic, such a truncation integer N is typically much smaller than the one based on (4.175). The problem is of course that a right value of N satisfying (4.179) is not known beforehand and that an initial guess with N very large would be inefficient. The problem of the numerical determination of N is solved by the use of an adaptive procedure starting with a 'low' estimate of N and increasing this estimate when necessary. Thus the iterative approach using (4.179) solves a sequence of finite systems of linear equations wih increasing sizes. The solution obtained for the system corresponding to some value of N provides a starting point for the system corresponding to the next value of N. Each finite system of linear equations is solved by overrelaxation, being somewhat modified in order to avoid divergence problems when the current estimate of N is too small. The efficiency of the algorithm is further improved by choosing adaptively the tolerance number in the stopping criterion of the overrelaxation method for some specific value of N; this tolerance number should be chosen smaller as the estimate for N increases. This special-purpose overrelaxation method for the M/D/c queue is extremely efficient and is described in detail in Seelen and Tijms (1985). It is important to have very efficient methods for the calculation of the state probabilities and the waiting-time probabilities in the M/D/c queue, since the results for the M/D/c queue will enable us to give useful approximations for measures of system performance in the M/G/c queue with general service times.

The computation of the waiting-time probabilities

We next show how to use the state probabilities p_j for the computation of the waiting-time probabilities when service is in order of arrival. Using an ingenious probabilistic argument to be given below, it was shown by Crommelin (1932) that, for any $m \geqslant 0$,

$$P\{W_q \leqslant x\} = \sum_{i=0}^{c-1} Q_i \sum_{j=0}^{m} e^{\lambda(x-mD)} \frac{\{-\lambda(x-mD)\}^{jc+c-1-i}}{(jc+c-1-i)!},$$

$$mD \leqslant x < (m+1)D, \tag{4.180}$$

where the probabilities $Q_i, 0 \leqslant i \leqslant c - 1$, are given by (4.171). The terms in (4.180) alternate in sign and in general are much larger than their sum. Thus the numerical evaluation of the sum may be hampered by roundoff errors due to the loss of significant digits, in particular when the traffic is not light (that is Π_W not being small). The numerical difficulty in using (4.180) was already realized by Crommelin (1932) and as a way out he suggested the following equivalent representation of (4.180):

$$P\{W_q \leqslant x\} = \sum_{j=1}^{\infty} \sum_{i=0}^{c-1} Q_i e^{-\lambda\{(m+j)D - x\}} \frac{[\lambda\{(m+j)D - x\}]^{(m+j)c+c-1-i}}{\{(m+j)c + c - 1 - i\}!}.$$

This sum contains infinitely many terms but each term is positive. Unfortunately the representation with the infinite sum is not of much help for non-light traffic. The numerical problem is now the slow convergence of the infinite sum; before convergence is achieved the calculations may be halted by the occurrences of exponent underflow and overflow. Nevertheless, we can provide an alternative that suffices for practical purposes. A computationally more useful representation of the waiting-time probabilities is actually contained in Crommelin's original derivation of the waiting-time distribution, but was apparently overlooked in the aim at a closed-form solution.

The derivation of a computational scheme for the waiting-time probabilities proceeds as follows. Consider a marked customer arriving when the system has been in operation for a very long time and thus finding the system in statistical equilibrium. Fix $x \geqslant 0$ with $mD \leqslant x < (m + 1)D$ for some integer $m \geqslant 0$ and write $x = mD + u$ for some $0 \leqslant u < D$. Noting that exactly mc customers will be served during a time mD when all c servers are continuously busy, some reflections show that the delay in queue of the marked customer is not more than $x = mD + u$ if and only if at most $mc + c - 1$ customers among those present just prior to the arrival of the marked customer will remain in the system a time u later. Under the assumption that the system is in statistical equilibrium, define

$b_v(u) = P\{$at most v customers among those present at an arbitrary

epoch remain in the system a time u later$\}$.

Then, by the above argument and the property that 'Poisson arrivals see time averages', it follows that

$$P\{W_q \leqslant x\} = b_{mc+c-1}(u) \qquad \text{for} \quad x = mD + u \quad \text{with} \quad 0 \leqslant u < D. \qquad (4.181)$$

It remains to find the probabilities $b_v(u)$. Therefore note that for any $t > u$ and $j \geqslant 0$,

$P\{$at most j customers are in the system at time $t\}$

$$= \sum_{k=0}^{j} P\{k \text{ arrivals occur in } (t - u, t] \text{ and at most } j - k$$

customers among those present at time $t - u$ remain at time $t\}$

$$= \sum_{k=0}^{j} P\{\text{at most } j-k \text{ customers among those present at time } t-u \text{ remain}$$

$$\text{at time } t\} \, e^{-\lambda u} \frac{(\lambda u)^k}{k!},$$

where the latter equality uses the lack of memory of the Poisson arrival process. Hence, by letting $t \to \infty$,

$$\sum_{i=0}^{j} p_i = \sum_{k=0}^{j} b_{j-k}(u) e^{-\lambda u} \frac{(\lambda u)^k}{k!}, \qquad j = 0, 1, \dots. \tag{4.182}$$

Next, using the technique of generating functions, Crommelin (1932) obtained from (4.172), (4.181) and (4.182) the closed-form solution (4.180). However, from a computational point of view, the intermediate results (4.181) and (4.182) are much more useful than the closed-form solution (4.180). The probabilities $b_j(u)$, $j \geq 0$, can be computed recursively from

$$b_0(u) = e^{\lambda u} p_0$$

$$b_j(u) = e^{\lambda u} \sum_{i=0}^{j} p_i - \sum_{k=0}^{j-1} b_k(u) \frac{(\lambda u)^{j-k}}{(j-k)!}, \qquad j = 1, 2, \dots$$

Although this recursion scheme also requires the taking of differences, it offers much less numerical difficulties than the closed-form expression (4.180). It turns out that a computational improvement of the recursion scheme is obtained by rewriting it as

$$b_j(u) = e^{\lambda u} p_j - \sum_{k=0}^{j-1} b_k(u) \frac{(\lambda u)^{j-1-k}}{(j-1-k)!} \left(\frac{\lambda u}{j-k} - 1 \right), \qquad j = 1, 2, \dots \tag{4.183}$$

The latter equation is obtained by subtracting the equations in (4.182) corresponding to j and $j-1$.

Summarizing, the waiting-time probability $P\{W_q \leq x\}$ may be calculated as follows once the state probabilities $p_j, j \geq 0$, have been computed. Representing x as $x = mD + u$ for some integer $m \geq 0$ and $0 \leq u < D$, the recursion scheme (4.183) is applied until we obtain $b_{mc+c-1}(u)$ yielding the desired probability $P\{W_q \leq x\}$. The following comments on this procedure should be made. The recursion scheme (4.183) will also ultimately be hampered by roundoff errors for large values of λu, in particular when c and ρ increase (note that $\lambda = c\rho/D$). In practice, however, this is no problem at all, since the easily computed asymptotic expansion for $P\{W_q > x\}$ usually works long before the recursion scheme (4.183) offers numerical difficulties. Thus the recursion scheme (4.183) should be used in an appropriate combination with the asymptotic expansion for $P\{W_q > x\}$ to be discussed next.

We now discuss in more detail the asymptotic expansion for the waiting-time probabilities in the M/D/c queue. By the general relation (4.16) we have that

$$P\{W_q > x\} \approx \frac{\delta \eta}{\lambda(\tau - 1)^2 \tau^{c-1}} e^{-\delta x} \qquad \text{for } x \text{ sufficiently large,} \qquad (4.184)$$

where τ and δ are computed from (4.177) and (4.178) and η may be computed from

$$\eta = \lim_{j \to \infty} \tau^j p_j. \qquad (4.185)$$

A better way to compute η for the M/D/c queue is to use the equation

$$\eta = \frac{\sum_{i=0}^{c-1} p_i(\tau^{i-1} - \tau^{c-1})}{c/\tau - \lambda D}, \qquad (4.186)$$

involving only the first c state probabilities. To prove (4.186), note that the representation (4.172) of the generating function $P(z)$ has the same form as (A.34) in appendix C. Considering the equation $1 - z^c e^{\lambda D(1-z)} = 0$, it is easily shown that this equations has a unique solution z_1 in the interval $(1, \infty)$ on the real axis. By (4.177) and (4.178), we must have $z_1 = \tau$. Further, it is not difficult to verify that $|z^c e^{\lambda D(1-z)}| > 1$ for all complex $z \neq z_1$ with $1 < |z| \leq z_1$. Thus an application of the asymptotic expansion (A.35) yields

$$p_j \approx \frac{\sum_{k=0}^{c} p_k \tau^k - \sum_{k=0}^{c} p_k \tau^c}{c\tau^{c-1} e^{\lambda D(1-\tau)} - \lambda D \tau^c e^{\lambda D(1-\tau)}} \tau^{-j-1} \qquad \text{for } j \text{ large.} \qquad (4.187)$$

Substituting the identity $\tau^c e^{\lambda D(1-\tau)} = 1$ in (4.187) and using (4.185) yields the result (4.186).

It turns out from extensive numerical experiments that the asymptotic expansion (4.184) applies already for x as large as $D/\sqrt{c}$ provided the delay probability Π_W is not too small (say, $\Pi_W \geq 0.2$). Note that for the case of very light traffic the asymptotic expansion (4.184) has little practical meaning, since in that case the waiting-time probabilities almost fade away before the asymptotic expansion applies. Also, we found that for the practically important range of $p \geq 0.9$ the conditional waiting-time percentile can accurately be calculated by using the asymptotic expansion for $P\{W_q > x | W_q > 0\}$ when the traffic is not very light. These empirical findings are confirmed by the numerical results in Table 4.18 in which the exact and asymptotic percentiles of the conditional waiting-time distribution function $P\{W_q \leq x | W_q > 0\}$ are given for several values of ρ and c. In all examples we take the deterministic service time $D = 1$.

Remark 4.1. The M/D/c queue can also be used to compute the mean waiting time and the waiting time probabilities in the multi-server queue with Erlangian input and deterministic services. It can easily be shown that the $E_r/D/c$ queue and the M/D/cr queue with the same server utilizations have identical waiting

Table 4.18   Exact and asymptotic percentiles for the conditional waiting time

			p	0.2	0.5	0.7	0.9	0.95	0.99	Π_W
$c = 2$	$\rho = 0.5$	exa		0.182	0.466	0.684	1.089	1.378	2.014	0.3233
		asy		0.269	0.456	0.659	1.096	1.372	2.013	
$c = 2$	$\rho = 0.8$	exa		0.392	0.931	1.533	2.808	3.613	5.480	0.7019
		asy		0.395	0.940	1.533	2.808	3.613	5.480	
$c = 5$	$\rho = 0.5$	exa		0.074	0.208	0.328	0.528	0.632	0.868	0.1213
		asy		0.180	0.255	0.336	0.511	0.622	0.878	
$c = 5$	$\rho = 0.8$	exa		0.162	0.419	0.651	1.161	1.482	2.229	0.5336
		asy		0.195	0.413	0.650	1.160	1.482	2.229	
$c = 10$	$\rho = 0.5$	exa		0.039	0.113	0.184	0.312	0.380	0.512	0.0331
		asy		0.150	0.187	0.228	0.315	0.370	0.499	
$c = 10$	$\rho = 0.8$	exa		0.083	0.225	0.353	0.603	0.765	1.139	0.3847
		asy		0.122	0.231	0.349	0.604	0.765	1.139	
$c = 15$	$\rho = 0.7$	exa		0.040	0.114	0.184	0.311	0.380	0.532	0.1283
		asy		0.100	0.147	0.198	0.306	0.374	0.533	
$c = 15$	$\rho = 0.8$	exa		0.056	0.156	0.248	0.419	0.524	0.774	0.2955
		asy		0.096	0.169	0.248	0.418	0.525	0.774	
$c = 25$	$\rho = 0.7$	exa		0.025	0.073	0.119	0.205	0.253	0.351	0.0572
		asy		0.086	0.114	0.144	0.209	0.250	0.346	
$c = 25$	$\rho = 0.8$	exa		0.035	0.099	0.159	0.270	0.334	0.480	0.1900
		asy		0.074	0.118	0.165	0.267	0.331	0.481	
$c = 35$	$\rho = 0.8$	exa		0.025	0.073	0.119	0.204	0.251	0.356	0.1297
		asy		0.064	0.096	0.130	0.202	0.248	0.355	
$c = 35$	$\rho = 0.9$	exa		0.045	0.126	0.201	0.352	0.448	0.770	0.4104
		asy		0.065	0.130	0.200	0.352	0.448	0.770	
$c = 50$	$\rho = 0.8$	exa		0.018	0.053	0.087	0.152	0.187	0.263	0.0776
		asy		0.057	0.079	0.103	0.154	0.186	0.261	
$c = 50$	$\rho = 0.9$	exa		0.032	0.091	0.146	0.253	0.320	0.475	0.3355
		asy		0.052	0.097	0.147	0.253	0.320	0.475	

time distributions when service is in order of arrival, see Iversen (1983). This result is also useful for obtaining approximations to the mean waiting time and the waiting time percentiles in the $GI/D/c$ queue when the interarrival time has a squared coefficient of variation c_A^2 that is not greater than 1. In case the delay probability is not too small, these performance measures can be approximated accurately by a linear interpolation of the corresponding performance measures in the $E_{k-1}/D/c$ queue and the $E_k/D/c$ queue, where k is the integer with $1/k < c_A^2 \leqslant 1/(k-1)$ and the linear interpolation is based on the squared coefficient of variation of the interarrival times, compare also the similar relations (4.198) and (4.201) in the next subsection.

4.4.3   The M/G/c queue

For this queueing model with c identical servers the arrival process of customers is a Poisson process with rate λ and the service time S of a customer has a general probability distribution function $B(x)$ with finite first two moments. It is assumed that the server utilization $\rho = \lambda E(S)/c$ is less than 1.

The M/G/c queue with general service times permits no simple analytical solution, not even for the average waiting time. Useful approximations can be obtained by the regenerative approach discussed in subsection 4.2.1. In applying this approach to the multi-server queue we encounter the difficulty that the number of customers left behind at a service completion epoch does not provide sufficient information to describe the system completely. In fact we need the additional information of the elapsed service times of the other services (if any) still in progress. A full inclusion of this information in the state description would lead to an intractable analysis. However, as an approximation, we will aggregate the information of the elapsed service times in such a way that the resulting approximate model enables us to carry through the regenerative analysis. A closer look to the regenerative approach reveals that we need only a suitable approximation to the time until the next service completion epoch. We now make the following approximation assumption with regard to the behaviour of the process at the service completion epochs.

Approximation assumption

(a)   If at a service completion epoch k customers are left behind in the system with $1 \leqslant k < c$, then the time until the next service completion epoch is distributed as $\min(S_1^e, \dots, S_k^e)$, where $S_1^e, \dots, S_k^e$ are independent random variables having each the residual life distribution function

$$B_e(t) = \frac{1}{E(S)} \int_0^t \{1 - B(x)\}\,dx, \qquad t \geqslant 0, \tag{4.188}$$

as the probability distribution function.

(b)   If at a service completion epoch k customers are left behind in the system with $k \geqslant c$, then the time until the next service completion is distributed as S/c, where S denotes the original service time of a customer.

This approximation assumption may be motivated as follows. First, if not all c servers are busy, the M/G/c queueing system may be treated as an M/G/∞ queueing system in which a server is immediately provided to each arriving customer. For this latter queueing system it was shown in Takács (1962) that at an arbitrary epoch in the steady state the remaining service time of any busy server is distributed as the residual life in a renewal process with the service times as the interoccurrence times, and thus has $B_e(t)$ as the probability distribution function. Second, if all of the c servers are busy, then the M/G/c queueing system may be approximated by an $M/G/1$ queueing system in which the single server works c times as fast as each of the c servers in the original system. It should be pointed out that the approximation assumption is easily shown to hold exactly when the service times are exponentially distributed.

Approximations to the state probabilities and the moments of the queue size

Under the approximation assumption we can derive analogously as in subsection 4.2.1 a recurrence relation of the type (4.28). Here the quantity A_{jk} should now be defined as the expected amount of time that k customers are present until the next service completion epoch, given that a service has just been completed with j customers left behind. For the multi-server queue the quantities A_{jk} are rather complicated at first sight, since they involve multi-dimensional integrals when $j < c$. However, it is possible to rework the basic relation of the type (4.28) with the tricky A_{jk}'s to a simple recursion scheme. The details of the tedious manipulations are left out here but can be found in Tijms, Van Hoorn and Federgruen (1981) (see also Van Hoorn, 1984). Thus the following recursion scheme yielding approximations p_j^{app} to the time-average probabilities p_j ($=$ the customer-average probabilities π_j) can be established:

$$p_j^{app} = \frac{(c\rho)^j}{j!} p_0^{app}, \qquad\qquad j = 0,\ldots,c-1, \qquad (4.189)$$

$$p_j^{app} = \lambda\alpha_{j-c}p_{c-1}^{app} + \lambda\sum_{k=c}^{j}\beta_{j-k}p_k^{app}, \qquad j = c, c+1,\ldots, \qquad (4.190)$$

where

$$\alpha_n = \int_0^\infty \{1 - B_e(t)\}^{c-1}\{1 - B(t)\}e^{-\lambda t}\frac{(\lambda t)^n}{n!}\,dt, \qquad n = 0,1,\ldots,$$

$$\beta_n = \int_0^\infty \{1 - B(ct)\}e^{-\lambda t}\frac{(\lambda t)^n}{n!}\,dt, \qquad n = 0,1,\ldots.$$

Using the normalization equation $\sum_{j=0}^{\infty}p_j = 1$, we easily derive from (4.189) and (4.190) the explicit expression

$$p_0^{app} = \left\{\sum_{k=0}^{c-1}\frac{(c\rho)^k}{k!} + \frac{(c\rho)^c}{c!(1-\rho)}\right\}^{-1}. \qquad (4.191)$$

The above recursion scheme for the state probabilities is easy to apply in practice. In general the constants α_n and β_n have to be evaluated by numerical integration. However, this is no serious problem for service-time distributions of practical interest. For the important case that the service-time distribution is a finite mixture of Erlangian distributions the Gauss–Laguerre quadrature is highly recommended (see appendix C in Van Hoorn, 1984, for details of this easily implementable procedure). Also, for the special case of H_2 service, the integrals for α_n and β_n can be reduced to finite sums.

It follows from (4.189) to (4.191) and (4.162) that

$$p_j^{\text{app}} = p_j^{\text{exp}} \qquad \text{for } j = 0, \ldots, c-1, \tag{4.192}$$

where p_j^{app} denotes the time-average probability p_j in the corresponding M/M/c queue. That is the approximate queueing system behaves like an M/M/c queue when not all of the c servers are busy. It is noteworthy that Miyazawa (1986) shows that a weakened version of the first part of the approximation assumption is in fact sufficient for obtaining the M/M/c approximation to the M/G/c queue when not all of the servers are busy. As a consequence of (4.192), we find the following approximation to the delay probability $\Pi_W = 1 - \sum_{j=0}^{c-1} p_j$.

$$\Pi_W^{\text{app}} = \frac{(c\rho)^c}{c!(1-\rho)} \left\{ \sum_{k=0}^{c-1} \frac{(c\rho)^k}{k!} + \frac{(c\rho)^c}{c!(1-\rho)} \right\}^{-1}. \tag{4.193}$$

That is we find the widely used Erlang delay probability $\Pi_W(\exp)$ which is known to be a generally good approximation to the delay probability in the M/G/c queue (cf. also Table 4.19 below). Further support for the quality of the approximations to the state probabilities is provided by the result that $p_{j-1}^{\text{app}}/p_j^{\text{app}}$ is asymptotically exact as $j \to \infty$. That is

$$\frac{p_{j-1}^{\text{app}}}{p_j^{\text{app}}} \to \tau \qquad \text{as } j \to \infty, \tag{4.194}$$

with τ as in (4.15). The proof of this result will be given later in this subsection. Since the constant τ is easily computed and the convergence in (4.194) is usually quite fast, the asymptotic result $p_{j-1}^{\text{app}}/p_j^{\text{app}} \approx \tau$ for j large enough may be computationally useful when applying the recursion scheme (4.190). Finally, we note that the approximations are consistent with low-traffic and heavy-traffic results in Burman and Smith (1983) and Köllerström (1974).

We next derive from the recursion scheme approximations to the moments of the queue size by using the powerful technique of generating functions. Defining

$$P_q(z) = \sum_{j=c}^{\infty} p_j^{\text{app}} z^{j-c}, \qquad |z| \leq 1,$$

we readily obtain from (4.190) that

$$P_q(z) = \lambda p_{c-1}^{\text{app}} \frac{\alpha(z)}{1 - \lambda \beta(z)}, \tag{4.195}$$

where

$$\alpha(z) = \int_0^{\infty} \{1 - B_e(t)\}^{c-1} \{1 - B(t)\} e^{-\lambda t(1-z)} \, dt,$$

$$\beta(z) = \int_0^{\infty} \{1 - B(ct)\} e^{-\lambda t(1-z)} \, dt.$$

Table 4.19 Exact and approximate results for the M/G/c queue

c		$c_S^2 = 0$			$c_S^2 = 0.5$			$c_S^2 = 2$			
		Π_W	$E(L_q)$	$cv(L_q)$	Π_W	$E(L_q)$	$cv(L_q)$	Π_W	$E(L_q)$	$cv(L_q)$	
2	exa	0.3233	0.177	3.110	0.3308	0.256	2.925	0.3363	0.487	2.736	$\rho = 0.5$
	app	0.3333	0.194	2.986	0.3333	0.260	2.900	0.3333	0.479	2.760	
5	exa	0.1213	0.077	5.050	0.1279	0.104	4.811	0.1335	0.181	4.600	
	app	0.1304	0.087	4.778	0.1304	0.107	4.736	0.1304	0.176	4.688	
10	exa	0.0331	0.024	9.611	0.0352	0.030	9.249	0.0373	0.048	8.940	
	app	0.0361	0.025	9.152	0.0361	0.030	9.088	0.0361	0.047	9.141	
2	exa	0.7019	1.445	1.515	0.7087	2.148	1.485	0.7141	4.231	1.460	$\rho = 0.8$
	app	0.7111	1.517	1.471	0.7111	2.169	1.475	0.7111	4.196	1.469	
5	exa	0.5336	1.156	1.787	0.5484	1.693	1.758	0.5611	3.250	1.747	
	app	0.5541	1.256	1.706	0.5541	1.723	1.737	0.5541	3.191	1.770	
10	exa	0.3847	0.879	2.163	0.4021	1.265	2.127	0.4179	2.365	2.127	
	app	0.4092	0.952	2.057	0.4092	1.286	2.098	0.4092	2.326	2.158	
25	exa	0.1900	0.477	3.196	0.2033	0.661	3.132	0.2164	1.173	3.142	
	app	0.2091	0.495	3.029	0.2091	0.663	3.084	0.2091	1.178	3.185	
50	exa	0.0776	0.214	5.123	0.0840	0.282	5.005	0.0908	0.471	5.020	
	app	0.0870	0.207	4.840	0.0870	0.277	4.921	0.0870	0.488	5.086	

In particular, the derivatives $P'_q(1)$ and $P''_q(1)$ yield approximations for $E(L_q)$ and $E(L_q(L_q - 1))$. By differentiation of (4.195), we find after lengthy algebra the approximations

$$E^{app}(L_q) = \left\{ (1 - \rho)\gamma_1 \frac{c}{E(S)} + \rho \frac{E(S^2)}{2E^2(S)} \right\} E_{exp}(L_q) \qquad (4.196)$$

and

$$E^{app}(L_q(L_q - 1)) = \left\{ (1 - \rho)\gamma_2 \frac{c^2}{E^2(S)} + \rho \frac{E(S^3)}{3E^3(S)} \right\} \frac{\rho^2}{1 - \rho} \Pi_W^{app}$$

$$+ \frac{\rho^2}{1 - \rho} \frac{E(S^2)}{E^2(S)} E^{app}(L_q), \qquad (4.197)$$

where $E^k(S)$ is the shorthand notation for $\{E(S)\}^k$ and the constants γ_1 and γ_2 are defined by

$$\gamma_1 = \int_0^\infty \{1 - B_e(t)\}^c \, dt \qquad \text{and} \qquad \gamma_2 = 2 \int_0^\infty t \{1 - B_e(t)\}^c \, dt.$$

Here $E_{exp}(L_q)$ denotes the average queue size in the M/M/c queue and is given by the explicit formula (4.164). The results (4.196) and (4.197) imply approximations for the first two moments of the waiting time W_q when service is in order of arrival. By (4.8) we have

$$E(L_q) = \lambda E(W_q) \qquad \text{and} \qquad E(L_q(L_q - 1)) = \lambda^2 E(W_q^2).$$

The above approximations to the state probabilities and the moments of the queue size are well suited for practical purposes. The computational effort of the approximate algorithm depends only to a slight degree on the number c of servers, as opposed to exact methods for which the computing times quickly increase when c becomes large. Extensive numerical experiments in Van Hoorn (1984) indicate that the approximations are sufficiently accurate for use in practice. As an illustration we give in Table 4.19 for several examples the exact and approximate values of the delay probability Π_W, the average queue size $E(L_q)$ and the coefficient of variation $cv(L_q)$ of the queue size. Here we consider deterministic service ($c_S^2 = 0$), E_2 service ($c_S^2 = 0.5$) and H_2 service with the gamma normalization and $c_S^2 = 2$. The number c of servers is varied as 2, 5, 10, 25 and 50 and the server utilization ρ has the two values 0.5 and 0.8. The approximate values (app) have been calculated from (4.193), (4.196) and (4.197), while the exact values (exa) have been computed by using an algorithm in Seelen (1986). Note from the results in Table 4.19 that the Erlang delay probability provides indeed a good approximation of Π_W.

Several other approximations are known for the mean queue size besides the approximation (4.196). In particular, we mention the approximations

$$E^{app2}(L_q) = (1 - c_S^2)E_{det}(L_q) + c_S^2 E_{exp}(L_q) \qquad (4.198)$$

and

$$E^{\text{app3}}(L_q) = \tfrac{1}{2}(1 + c_S^2)\frac{2E_{\text{det}}(L_q)E_{\text{exp}}(L_q)}{2aE_{\text{det}}(L_q) + (1 - a)E_{\text{exp}}(L_q)},\qquad(4.199)$$

where $a = 1$ when $c = 1$ and otherwise

$$a = \frac{1}{c - 1}\left\{\frac{E(S^2)}{\gamma_1 E(S)} - c - 1\right\}.$$

Here $E_{\text{det}}(L_q)$ denotes the average queue size in the corresponding M/D/c queue. Although extensive tables with exact values of $E_{\text{det}}(L_q)$ are available (see the tablebooks of Kühn, 1976, and Seelen, Tijms and Van Hoorn, 1985), it may be helpful to have the following excellent approximation to $E_{\text{det}}(L_q)$:

$$E^{\text{app}}_{\text{det}}(L_q) = \tfrac{1}{2}\left\{1 + (1 - \rho)(c - 1)\frac{\sqrt{4 + 5c} - 2}{16\rho c}\right\}E_{\text{exp}}(L_q).$$

This special-purpose approximation was developed by Cosmetatos (1975). The appealing two-moment approximation (4.198) advocated by Cosmetatos (1976) and Page (1972) should be applied only when c_S^2 is not too large, say $0 \leqslant c_S^2 \leqslant 2$ (cf. also Hokstad, 1986, for a sensitivity analysis for $E(L_q)$ when c_S^2 becomes large). The approximation (4.199) was developed by Boxma, Cohen and Huffels (1979). This approximation, as well as the approximation (4.196), involves the service-time distribution by the constant γ_1. The practical utility of the latter two approximations is considerably enhanced by using the following two-moment approximation to γ_1:

$$\gamma_1^{\text{app}} = (1 - c_S^2)\frac{E(S)}{c + 1} + c_S^2\frac{E(S)}{c},\qquad(4.200)$$

provided c_S^2 is not too large, say $0 \leqslant c_S^2 \leqslant 2$. This approximation is easily verified to be exact for the special cases of deterministic service ($c_S^2 = 0$) and exponential service ($c_S^2 = 1$). Further, in support of (4.200), it can be shown that $E(S)/(c + 1) \leqslant \gamma_1 \leqslant E(S)/c$ when the service time has an increasing failure rate. For H_2 service ($c_S^2 \geqslant 1$) with density $p\mu_1 e^{-\mu_1 t} + (1 - p)\mu_2 e^{-\mu_2 t}$, the constant γ_1 allows for the explicit expression

$$\gamma_1 = \sum_{i=0}^{c}\frac{\binom{c}{i}r^i(1 - r)^{c-i}}{i\mu_1 + (c - i)\mu_2}\quad\text{where}\quad r = \frac{p/\mu_1}{p/\mu_1 + (1 - p)/\mu_2},$$

as follows by using $B_e(t) = \{(1/E(S))\}\{p(1 - e^{-\mu_1 t}) + (1 - p)(1 - e^{-\mu_2 t})\}$. The various approximations to $E(L_q)$ were extensively studied in Van Hoorn (1984). It was empirically verified that in practical situations the mean queue size is quite insensitive to more than the first two moments of the service time provided c_S^2 is not too large. Here it was found that each of the above approximations to $E(L_q)$ performs very well and has relative error percentages being typically in the range of 0 to 2 per cent when $0 \leqslant c_S^2 \leqslant 1$. Also, it turns out that for the case of $c_S^2 > 1$ the approximation (4.199) usually yields the best results where the relative errors are somewhat larger than in the case of $c_S^2 \leqslant 1$.

The approximations for the average queue size $E(L_q)$ imply approximations for the average waiting time $E(W_q)$ by Little's formula $E(L_q) = \lambda E(W_q)$. In certain applications it may be required to approximate the average conditional waiting time $E(W_q | W_q > 0)$. This performance measure may be approximated as in (4.198), but a simpler alternative is provided by the results in Seelen and Tijms (1984) dealing with the GI/G/c queue. For the particular case of the M/G/c queue, the two-moment approximation to the average conditional waiting time $TW = E(W_q | W_q > 0)$ is

$$
TW^{\mathrm{app}} = \begin{cases}
\left(1 - \tfrac{1}{2}\rho - \tfrac{1}{2}\rho^2\right)\left(\dfrac{1 - c_S^2}{c+1} + \dfrac{c_S^2}{c}\right)E(S) \\[3mm]
\quad + \dfrac{(3\rho - \rho^3)(1 + c_S^2)}{4(1-\rho)c}E(S) & \text{if } c_S^2 \leqslant 1, \\[5mm]
(1 + \rho)\left(\dfrac{1 - c_S^2}{c+1} + \dfrac{c_S^2}{c}\right)E(S) \\[3mm]
\quad + \dfrac{\rho^2(1 + c_S^2)}{2(1-\rho)c}E(S) & \text{if } c_S^2 > 1.
\end{cases}
$$

We next give an example in which two-moment approximations are applied. The example, being adapted from Martin (1972), demonstrates the usefulness of the principle of pooling for queueing systems, that is, in general, it is better to have a centralized group of servers than a decentralized group of servers.

Example 4.4   Pooling of agents in an airline reservation system

Consider an airline company at which calls arrive according to a Poisson process with an average of 25 calls per hour. The calls are to enquire about flights or to make reservations. The airline has both foreign and domestic flights. The probability that a call is for a domestic flight is 2/3. The handling time of a call for a domestic flight has a mean of 3 minutes and a standard deviation of 2 minutes, while the handling time of a call for a foreign flight has a mean of 5 minutes and a standard deviation of 5 minutes. Incoming calls finding all handling agents busy are asked to wait. The service requirement is that the probability a call has to wait is not more than 0.10 and the average wait of a delayed call of either type is at most 100 seconds. We are interested to know how many agents are required when foreign and domestic calls are handled by separate agents and how many when the agents can handle either foreign or domestic calls.

We first analyse the case of separate agents. Taking the minute as the time unit, the total call rate is $\lambda = 25/60$ calls per minute. Splitting the Poisson input process into foreign and domestic calls yields Poisson processes with respective rates of $\lambda_F = \frac{1}{3}\lambda$ and $\lambda_D = \frac{2}{3}\lambda$ (see section 1.4 of chapter 1). If we consider the foreign calls that have an average service time $E(S_F) = 5$, the offered load is $\lambda_F E(S_F) = 25/36$, implying a server utilization of $\lambda_F = 25/(36c_F)$ when c_F agents are available. To find approximately the minimum number c_F^* of agents required for foreign calls, we first seek the minimum c_F such that the corresponding $\Pi_W^{app} \leqslant 0.1$ and next determine the smallest $c_F^* \geqslant c_F$ such that the corresponding $TW^{app} \leqslant 1\frac{2}{3}$. Simple calculations based on (4.193) yield $c_F = 3$ with $\Pi_W^{app} = 0.036$. For $c_F = 3$, we find $TW^{app} = 2.17$ and so we try four agents. With four agents, we indeed find that both design criteria are satisfied. Thus we find for the foreign calls,

$$c_F^* = 4 \quad \text{with} \quad \Pi_W^{app} = 0.006, \qquad TW^{app} = 1.51.$$

In the same way, we find for the domestic calls that

$$c_D^* = 3 \quad \text{with} \quad \Pi_W^{app} = 0.058, \qquad TW^{app} = 1.11.$$

Next consider the case of pooled agents. Denoting by S the overall service time of a call, it follows from $E(S^k) = (1/3)E(S_F^k) + (2/3)E(S_D^k)$ for $k = 1$, 2 that $E(S) = 3\frac{2}{3}$ and $c_S^2 = 107/121$. The total offered load is $\lambda E(S) = 55/36$. By the same reasoning as above, we find for the case of pooled agents

$$c^* = 4 \quad \text{with} \quad \Pi_W^{app} = 0.079, \qquad TW^{app} = 1.42.$$

Here it suffices to have four agents when pooling the agents rather than to have seven agents when dealing separately with the foreign and domestic calls. □

Approximations for the waiting-time probabilities

Similarly as for the M/G/1 queue, the waiting-time probabilities do not allow a direct approximation but may be approximated via the waiting-time percentiles. Recall that the pth waiting-time percentile $\xi(p)$ of a continuous probability distribution function $F(x)$ is defined by $F(\xi(p)) = p$ for $0 < p < 1$. Both for the unconditional waiting-time distribution function $P\{W_q \leqslant t\}$ and the conditional waiting-time distribution function $P\{W_q \leqslant t \mid W_q > 0\}$ the pth percentile may be approximated by

$$\xi_{app}(p) = (1 - c_S^2)\xi_{det}(p) + c_S^2\xi_{exp}(p), \tag{4.201}$$

provided c_S^2 is not too large. Here $\xi_{det}(p)$ and $\xi_{exp}(p)$ denote the corresponding waiting-time percentiles for the respective cases of the M/D/c queue and the M/M/c queue. The waiting-time percentile $\xi_{exp}(p)$ is trivially computed from (4.165), while in subsection 4.3.2 an efficient algorithm is given for the computation of $\xi_{det}(p)$. The approximation (4.201) applies for any

Table 4.20 Numerical results for the conditional waiting-time percentiles

		$c_S^2 = 0.5$					$c_S^2 = 2$				
$\rho = 0.5$	p	0.2	0.5	0.9	0.95	0.99	0.2	0.5	0.9	0.95	0.99
$c = 2$	exa	0.200	0.569	1.713	2.196	3.316	0.256	0.930	3.482	4.586	7.150
	app	0.203	0.580	1.696	2.187	3.309	0.264	0.920	3.517	4.613	7.197
$c = 5$	exa	0.082	0.240	0.722	0.919	1.370	0.099	0.339	1.318	1.758	2.783
	app	0.082	0.243	0.725	0.915	1.355	0.104	0.346	1.314	1.765	2.816
$c = 10$	exa	0.042	0.124	0.380	0.482	0.714	0.049	0.161	0.616	0.829	1.336
	app	0.042	0.126	0.386	0.489	0.717	0.051	0.164	0.609	0.818	1.330
$\rho = 0.8$	p	0.2	0.5	0.9	0.95	0.99	0.2	0.5	0.9	0.95	0.99
$c = 2$	exa	0.467	1.341	4.286	5.554	8.498	0.724	2.512	8.688	11.347	17.523
	app	0.475	1.332	4.282	5.551	8.497	0.724	2.534	8.705	11.366	17.546
$c = 5$	exa	0.193	0.554	1.735	2.242	3.420	0.274	0.962	3.430	4.493	6.964
	app	0.192	0.556	1.732	2.239	3.417	0.284	0.967	3.445	4.510	6.981
$c = 10$	exa	0.098	0.285	0.881	1.135	1.724	0.133	0.461	1.687	2.219	3.454
	app	0.097	0.286	0.877	1.131	1.721	0.140	0.468	1.699	2.231	3.466
$c = 25$	exa	0.040	0.118	0.364	0.467	0.703	0.052	0.174	0.649	0.861	1.354
	app	0.040	0.119	0.365	0.466	0.701	0.055	0.179	0.651	0.865	1.362
$c = 50$	exa	0.020	0.061	0.189	0.241	0.361	0.025	0.083	0.309	0.413	0.658
	app	0.020	0.061	0.191	0.243	0.362	0.027	0.085	0.309	0.412	0.658

feasible p. Note that the (conditional) probability of waiting not more than a given time x_0 may be approximated by the value p_0 resulting from an iterative application of (4.201) until $\xi_{app}(p)$ is sufficiently close to x_0.

Using the extensive material in the tablebook of Seelen, Tijms and Van Hoorn (1985), we verified that (4.201) provides a useful approximation for practical purposes. As an illustration in Table 4.20 we give for several examples the exact and approximate values for the conditional waiting-time percentiles for a number of values of p. We consider Erlang-2 service ($c_S^2 = 0.5$) and H_2 service with the gamma normalization and $c_S^2 = 2$. The number c of servers is varied as 2, 5, 10, 25 and 50 and the server utilization ρ has the two values 0.5 and 0.8. The (nearly) exact values of the conditional waiting-time percentiles have been computed from a very accurate approximation to the waiting-time distribution function by a sum of a number of exponential functions whose coefficients were obtained from the exactly computed state probabilities (see Seelen, 1986).

Also, our numerical investigations indicate that the asymptotic expansion (4.16) for the waiting-time distribution may be applied for practical purposes already for x as large as $E(S)/\sqrt{c}$, provided the delay probability is not too small (say $\Pi_W \geqslant 0.2$). The asymptotic expansion (4.16) applies sooner as Π_W increases. As pointed out earlier, for very light traffic the asymptotic expansion (4.16) has little practical meaning since then the waiting-time probabilities almost fade away before the asymptotic expansion applies.

The decay parameter δ of the asymptotic expansion

$$P\{W_q > x\} \approx Ae^{-\delta x} \qquad \text{for large } x$$

is easily computed from (4.18) and therefore it would be helpful to have a tractable approximation to the amplitude factor A. Such an approximation is given by

$$A_{app} = \frac{(1-\rho)\Pi_W(\exp)\int_0^\infty e^{\delta t}\{1 - B_e(t)\}^{c-1}\{1 - B(t)\}\,dt}{\rho\delta} \cdot \frac{1}{\int_0^\infty te^{\delta t}\{1 - B(ct)\}\,dt}. \qquad (4.202)$$

Here δ is the unique positive solution to (see equation 4.18)

$$\lambda \int_0^\infty e^{\delta y}\{1 - B(cy)\}\,dy = 1.$$

Note that for phase-type services the integrals in (4.202) are rather easily evaluated by using the Gauss–Laguerre procedure discussed in Van Hoorn (1984); an alternative numerical integration procedure is given in Jagerman (1985), although some care should be exercised in using this procedure when c_S^2 becomes large. To derive the approximation (4.202), note that by (4.16), (4.17) and (4.19),

$$A = \frac{\eta\delta}{\lambda(\tau - 1)^2\tau^{c-1}} \qquad \text{with} \quad \tau = 1 + \frac{\delta}{\lambda} \qquad \text{and} \qquad \eta = \lim_{j \to \infty} \tau^j p_j.$$

Table 4.21   The ratio of the approximate and exact values of A

c	$\rho = 0.5$			$\rho = 0.8$			$\rho = 0.9$		
	2	5	10	10	15	25	25	35	50
$c_S^2 = 0$	1.21	1.41	1.17	1.09	1.03	0.90	1.01	0.97	0.92
$c_S^2 = \frac{1}{3}$	1.05	1.03	0.85	1.01	0.98	0.92	0.99	0.97	0.95
$c_S^2 = \frac{1}{2}$	1.03	1.00	0.85	1.00	0.98	0.94	0.99	0.98	0.96
$c_S^2 = 2$	0.98	0.99	1.09	1.00	1.01	1.06	1.01	1.02	1.04
$c_S^2 = 10$	0.97	0.96	1.15	0.99	1.04	1.16	1.03	1.07	1.12

An approximation to η follows from the asymptotic expansion of the approximate state probabilities p_j^{app}. This asymptotic expansion is obtained from the representation (4.195) of the generating function $P_q(z) = \sum_{j=0}^{\infty} p_{c+j}^{\text{app}} z^j$. The equation (4.195) has the same form as (A.34) in appendix C. Similarly to the derivation of (4.187), we find by applying (A.35) that

$$p_{c+j}^{\text{app}} \approx \frac{p_{c-1}^{\text{app}} \int_0^{\infty} e^{-\lambda(1-\tau)t} \{1 - B_e(t)\}^{c-1} \{1 - B(t)\} \, dt}{\lambda \int_0^{\infty} t e^{-\lambda(1-\tau)t} \{1 - B(ct)\} \, dt} \tau^{-j-1} \qquad \text{for large } j.$$

Incidentally, this asymptotic expansion verifies (4.194). By (4.189), (4.191) and (4.192), we have

$$p_{c-1}^{\text{app}} = \frac{1-\rho}{\rho} \Pi_W(\exp).$$

Using the above relations, we now readily obtain the desired result (4.202).

The approximation (4.202) is well-suited for use in practice provided the delay probability is not too small. To illustrate this, we give in Table 4.21 the ratio of the approximate and exact values of the amplitude factor A for several service-time distributions and for several values of ρ and c. We consider the deterministic distribution $(c_S^2 = 0)$, the E_3 distribution $(c_S^2 = \frac{1}{3})$, the E_2 distribution $(c_S^2 = \frac{1}{2})$ and the H_2 distribution with the gamma normalization $(c_S^2 = 2, 10)$.

4.4.4   The M/G/∞ queue

In this queueing model with Poisson arrivals and general service times a server is immediately available for each arriving customer. The queueing model with infinitely many servers may be a useful approximation to practical queueing systems with a large number of servers. Also this model has interesting applications to inventory systems.

The $M/G/\infty$ queueing system has the pleasant robustness property whereby measures of system performance depend on the service time only through the first moment. Denoting by λ the arrival rate of customers and by S the service time of a customer, we obtain informally from (4.189) and (4.191) with $c = \infty$ that the state probabilities p_j are given by

$$p_j = e^{-\lambda E(S)} \frac{\{\lambda E(S)\}^j}{j!}, \qquad j = 0, 1, \ldots . \tag{4.203}$$

That is the number of busy servers is Poisson distributed with mean $\lambda E(S)$. We shall give a direct proof that this result is indeed true for any value of the offered load. The exactness of (4.203) is in further support of the approximations (4.189) to (4.191). To prove (4.203), we define, for $j = 0, 1, \ldots$,

$$p_{0j}(t) = P\{j \text{ servers are busy at time } t | \text{the system is empty at time } 0\},$$

and establish a backward differential equation for $p_{0j}(t)$. To find $p_{0j}(t + \Delta t)$ for Δt small, we condition upon the state of the system at time Δt. No arrival occurs in $(0, \Delta t]$ with probability $1 - \lambda \Delta t + o(\Delta t)$, in which case the conditional probability of having j busy servers at time $t + \Delta t$ equals $p_{0j}(t)$. If an arrival occurs in $(0, \Delta t]$, we distinguish between the two mutually exclusive events: (a) the service of that arrival is completed at time t and (b) the service of that arrival is not completed at time t. These two events have respective probabilities $B(t)$ and $1 - B(t)$, with $B(t)$ denoting the probability distribution function of the service time S. In the case of the former event, the conditional probability of having j busy servers at time t is $p_{0j}(t)$, while in the case of the latter event this probability equals $p_{0,j-1}(t)$. Here we essentially use the fact that the number of new services initiated in $(\Delta t, t + \Delta t]$ is independent of the state of the system at time Δt since there are infinitely many servers. Thus, for any $t > 0$ and $j = 0, 1, \ldots$,

$$p_{0j}(t + \Delta t) = (1 - \lambda \Delta t)p_{0j}(t) + \lambda \Delta t B(t)p_{0j}(t) + \lambda \Delta t \{1 - B(t)\} p_{0,j-1}(t) + o(\Delta t),$$

implying

$$p'_{0j}(t) = \lambda \{1 - B(t)\} \{p_{0,j-1}(t) - p_{0,j}(t)\}, \qquad t > 0, \qquad j = 0, 1, \ldots,$$

with $p_{0,-1}(t) = 0$ by convention. It is left to the reader to verify by induction that this system of differential equations allows for the explicit solution

$$p_{0j}(t) = \exp\left[-\lambda \int_0^t \{1 - B(x)\} \, dx\right] \frac{[\lambda \int_0^t \{1 - B(x)\} \, dx]^j}{j!}, \qquad t > 0, \quad j = 0, 1, \ldots .$$

Next, by letting $t \to \infty$ and using $\int_0^\infty \{1 - B(x)\} \, dx = E(S)$, we get (4.203).

Remark 4.2

An alternative proof of (4.203) can be based on the property of the Poisson arrival process that, given n arrivals in $(0, t)$, the n arrival epochs are distributed as the order statistics of n independent uniform random variables on $(0, t)$ (see Takács, 1962). This alternative proof allows us to extend the above result to the case of batch arrivals occurring according to a Poisson process with rate λ, where the service times of the customers in the batches are independent of each other and have the common probability distribution function $B(x)$. In particular, denoting by N the number of busy servers when the system has reached statistical equilibrium, we have that the mean and the variance of N are given by

$$E(N) = \lambda \alpha E(S),$$

$$\text{var}(N) = \lambda \alpha \int_0^\infty \{1 - B(x)\} B(x) \, dx + \lambda(\alpha^2 + \beta^2) \int_0^\infty \{1 - B(x)\}^2 \, dx, \quad (4.204)$$

where α and β denote the mean and the standard deviation of the batch size.

The following example based on Parikh (1977) illustrates how the $M/G/\infty$ queueing system may be used to estimate measures of system performance in the $M/G/c$ queue with a large number of servers.

Example 4.5 A stochastic allocation problem and fleet sizing

Suppose a nationwide company has a large number C of transport vehicle units that are used to ship customer orders. The vehicles must be allocated to a certain number F of fleets. The fleets operate independently of each other and each fleet services its own group of customers. A customer order is shipped singly by one transport vehicle unit. If a customer order finds upon arrival no transport vehicle available, the shipment of the order is delayed until a transport vehicle of the concerning fleet returns. For each fleet the service criterion is the fraction of orders that are delayed. The goal of the company is to assign the total number of transport vehicles over the various fleets in order to achieve the result that all fleets provide as nearly as possible a uniform level of service. This design problem assumes that customer orders for fleet i arrive according to a Poisson process with rate λ_i and involve an estimated load of $E_i = \lambda_i E(S_i)$, $i = 1, \ldots, F$. Here the service time S_i of a customer order for fleet i should be interpreted as the time from shipment of that order until the transport vehicle used is available for another shipment. The coefficients of variation of the service times S_i are assumed to be not more than 1.

To find the allocation $(c_1, \ldots, c_F)$ with $c_1 + \cdots + c_F = C$ of the C transport vehicles to the F fleets, we first discuss an approximation of the $M/G/\infty$ queue to the $M/G/c$ queue with c large. Consider the $M/G/c$ queue with an offered load of $E = \lambda E(S)$ and suppose we wish to determine the minimum number of servers in order to achieve a delay probability of not more than a prespecified value v. As pointed out in subsection 4.3.3, the Erlang delay probability $\Pi_W(\exp)$ from the $M/M/c$ queue is an excellent approximation to the delay probability

Π_W in the M/G/c queue. Also, when the service time has a coefficient of variation of not more than 1, there is much numerical evidence in support of the conjecture $\Pi_W \leqslant \Pi_W(\exp)$. Another way to approximate Π_W in the M/G/c queue is to use the inequality

$$\sum_{j=c}^{\infty} e^{-E}\frac{E^j}{j!} \leqslant \Pi_W.$$

This inequality will be intuitively obvious by noting that its left side represents the probability of having at least c customers in the M/G/∞ queue while the right side represents the corresponding probability in the M/G/c queue. A rigorous proof of the inequality can be found in Stoyan (1983). Thus in many practical M/G/c queueing systems with an offered load of E we have for the delay probability Π_W the bounds

$$\sum_{j=c}^{\infty} e^{-E}\frac{E^j}{j!} \leqslant \Pi_W \leqslant \Pi_W(\exp).$$

Define now $c^{\exp}(v)$ and $c^{\infty}(v)$ as the smallest values of c for which $\Pi_W(\exp) \leqslant v$ in the M/M/c queue and $\sum_{j=c}^{\infty} e^{-E}E^j/j! \leqslant v$ in the M/G/∞ queue. In Table 4.22 we give $c^{\exp}(v)$ and $c^{\infty}(v)$ for a range of values of v and E. It appears that for v small enough and E sufficiently large the critical levels $c^{\exp}(v)$ and $c^{\infty}(v)$ are very close to each other. The conclusion is that for practical purposes $c^{\infty}(v)$ may be used in the design of the fleet sizes. In its turn $c^{\infty}(v)$ may be approximated reasonably well by $c^{N}(v)$ to be calculated from a normal approximation to the cumulative

Table 4.22 The critical levels $c^{\exp}(v)$, $c^{\infty}(v)$ and $c^{N}(v)$

	E	2	5	10	15	25	50	100	150	200
$v = 0.50$	$c^{\exp}(v)$	3	7	12	18	28	54	106	157	208
	$c^{\infty}(v)$	3	6	11	16	26	51	101	151	201
	$c^{N}(v)$	2.5	5.5	10.5	15.5	25.5	50.5	100.5	150.5	200.5
$v = 0.20$	$c^{\exp}(v)$	4	8	14	20	31	58	111	164	216
	$c^{\infty}(v)$	4	8	14	19	30	57	109	161	213
	$c^{N}(v)$	3.7	7.4	13.2	18.8	29.7	56.5	108.9	160.8	212.4
$v = 0.10$	$c^{\exp}(v)$	5	9	16	22	33	61	115	168	221
	$c^{\infty}(v)$	5	9	15	21	33	60	114	167	219
	$c^{N}(v)$	4.3	8.4	14.6	20.5	31.9	59.6	113.3	166.2	218.6
$v = 0.05$	$c^{\exp}(v)$	6	10	17	23	35	64	119	173	226
	$c^{\infty}(v)$	6	10	16	23	34	63	118	171	225
	$c^{N}(v)$	4.8	9.2	15.7	21.9	33.7	62.1	116.9	170.6	223.8
$v = 0.01$	$c^{\exp}(v)$	7	12	19	26	39	68	125	181	235
	$c^{\infty}(v)$	7	12	19	26	38	68	125	180	235
	$c^{N}(v)$	5.8	10.7	17.9	24.5	37.1	66.9	123.8	179.0	233.4

Poisson probabilities (see the relation below). The numbers $c^N(v)$ are also included in Table 4.22.

The problem of the allocation of the total number C of vehicles to the F fleets in order to achieve nearly the same service level over the fleets can be solved as follows. Try several values of v until the desired value of v is found for which the sum of the fleet sizes $c_i(v)$ $(i = 1, \ldots, F)$ obtained from the appropriate values $c^\infty(v)$ is equal to C. A good approximation to the desired value of v may be computed beforehand by using the well-known normal approximation to the Poisson distribution. Denoting by $\Phi(x)$ the standard normal probability distribution function, we have

$$\sum_{j=c}^{\infty} e^{-E} \frac{E^j}{j!} \approx 1 - \Phi\left(\frac{c - E - 0.5}{\sqrt{E}}\right),$$

provided E is not too small. Thus, letting k_v with $1 - \Phi(k_v) = v$ denote the $(1 - v)$th percentile of the standard normal distribution, we may approximate $c^\infty(v)$ by

$$c^N(v) = E + k_v\sqrt{E} + 0.5.$$

Approximating the required fleet sizes $c_i(v)$ by $c_i^N(v) = E_i + k_v\sqrt{E_i} + 0.5$ and summing over i yields the approximation

$$k_v \approx \frac{C - \sum_{i=1}^{F} E_i - 0.5F}{\sum_{i=1}^{F} \sqrt{E_i}}.$$

Thus, by the definition of k_v, the value of v for which the service level is nearly the same over all fleets can be approximated by

$$v \approx 1 - \Phi\left(\frac{C - \sum_{i=1}^{F} E_i - 0.5F}{\sum_{i=1}^{F} \sqrt{E_i}}\right). \tag{4.205}$$

As an illustration, consider the numerical data

$$C = 250, \qquad F = 5, \qquad E_1 = 10, \qquad E_2 = 25, \qquad E_3 = 35,$$
$$E_4 = 50 \qquad \text{and} \qquad E_5 = 75.$$

An application of (4.205) yields the estimate of 0.039 for the 'uniform' service level. By trying several values of v in the neighbourhood of this estimate, we ultimately find the following assignment:

$$c_1 = 16(0.057), \qquad c_2 = 34(0.060), \qquad c_3 = 46(0.051), \qquad c_4 = 63(0.051),$$
$$c_5 = 91(0.048).$$

The figures between brackets denote the corresponding values of $\Pi_W(\exp)$, being a good approximation to the delay probability. $\qquad\qquad\qquad\qquad\qquad\qquad\square$

The next example that is based on Sherbrooke (1968) is another illustration of the relationships between queueing and inventory systems.

Example 4.6   A repairable item inventory system

Many repairable item inventory systems can be modelled as a two-echelon inventory system consisting of a central depot and a number N of regional bases that operate independently of each other. Failed items arrive at the base level and are either repaired at the base or at the central depot, depending on the complexity of the repair. More specifically, failed items arrive at the bases $1, \ldots, N$ according to independent Poisson processes with respective rates $\lambda_1, \ldots, \lambda_N$. A failed item at base j can be repaired at the base with probability q_j; otherwise the item must be repaired at the depot. The average repair time of an item is R_j at base j and R_0 at the depot. It takes an average time of U_j to ship an item from base j to the depot and back. The base immediately replaces a failed item from base stock if available; otherwise the replacement of the failed item is backordered until an item becomes available at the base. If a failed item from base j arrives at the depot for a repair, the depot sends immediately a replacement item to the base j from depot stock if available; otherwise the replacement is backordered until a repaired item becomes available. In the two-echelon system a total of J spare parts are available. The goal is to spread these parts over the bases and the depot in order to minimize the total average number of backorders outstanding at the bases.

Suppose that S_j spare parts have been assigned to base $j(j = 1, \ldots, N)$ and S_0 spare parts to the depot such that $S_0 + S_1 + \cdots + S_N = J$. The key element in analysing the inventory system is the analogy between this system and the infinite server queueing system. For each base j the failed items arriving at the base j can be thought of as customers entering service immediately in a queueing system with infinitely many servers. Here the service time should be defined as the repair time in case of a repair at the base and as the time until receipt of a replacement from the depot otherwise. Some reflections show that backorders exist only at base j when the number of customers in service is greater than S_j. Thus, by an application of (4.203), we find, for each j,

The average number of backorders at base $j = \displaystyle\sum_{k=S_j}^{\infty} (k - S_j) e^{-\lambda_j \tau_j} \frac{(\lambda_j \tau_j)^k}{k!}$,   (4.206)

where

$$\tau_j = \text{the average service time of a customer at base } j.$$

To find the unknowns $\tau_j (1 \leq j \leq N)$, we first note that by the very definition of service time,

$$\tau_j = q_j R_j + (1 - q_j)\{U_j + W_0\}, \qquad j = 1, \ldots, N,   (4.207)$$

where W_0 is defined as

$W_0 = $ the average time elapsed between the arrival of a failed item at the depot and the shipment of a replacement for that item.

It remains to determine W_0. To do so we note that, by well-known properties of the Poisson process (see section 1.4 in chapter 1), failed items arrive at the depot according to a Poisson process with rate

$$\lambda_0 = \sum_{j=1}^{N} \lambda_j (1 - q_j).$$

Next, letting

L_0 = the average number of backorders outstanding at the depot,

we establish the following application of Little's formula:

$$L_0 = \lambda_0 W_0. \tag{4.208}$$

An easy way to see this relationship is to interpret L_0 as the long-run average cost per unit time when a cost at rate 1 per unit of time is incurred for each backorder outstanding at the depot. On the other hand, we have under this cost structure that the long-run average cost per unit time equals the average arrival rate λ_0 of failed items at the depot multiplied by the average time W_0 until a replacement order is shipped for a failed item. This proves (4.208). The quantity L_0 is easily found. Similarly as above, any failed item arriving at the depot can be thought of as a customer entering service immediately in an infinite-server queueing system. Here the service time is now defined as the repair time of the failed item at the depot. Hence, by another application of (4.203), the number of items in repair at the depot is Poisson distributed with mean $\lambda_0 R_0$. Since backorders exist at the depot only if more than S_0 items are in repair, it follows that

$$L_0 = \sum_{i=S_0}^{\infty} (i - S_0) e^{-\lambda_0 R_0} \frac{(\lambda_0 R_0)^i}{i!}. \tag{4.209}$$

The relations (4.206) to (4.209) enable us to calculate the total average number of backorders outstanding at the bases for the given assignment $(S_0, S_1, \ldots, S_N)$ of the J spare parts. Next, by some search procedure, the optimal values of S_0, $S_1, \ldots, S_N$ may be found (cf. also Muckstadt, 1978). □

4.4.5 The finite capacity M/G/c queue

In subsection 4.2.4 some of the results for the M/G/1 queue were extended to the case of a single-server queue with state-dependent Markovian input. Under the approximation assumption stated in subsection 4.4.3, the approximate results (4.189) and (4.190) can be extended analogously to the case of a multi-server queue with state-dependent Markovian input. The extended results are in general not computationally tractable with the notable exceptions of the finite capacity M/G/c model and the machine repair model with multiple repairmen. The approximate results for these two useful queueing models are discussed in this subsection and the next one.

We first consider the M/G/c/N queueing model including as a special case the important M/G/c loss system. This queueing model has a Poisson input with rate λ, a general service time S, c identical servers and $N(\geqslant c)$ waiting places for the customers (including those in service). An arriving customer who finds N other customers in the system is turned away and has no further influence on the system. Since the number of customers in the system is bounded by N, we need no *a priori* assumption on the load factor $\rho = \lambda E(S)/c$. We already established in section 4.1 some relationships for several measures of system performance. In particular, letting $\{p_j, j = 0, \ldots, N\}$ denote the time-average probabilities, we have by the property 'Poisson arrivals see time averages' that

$$\text{The long-run fraction of arrivals that is rejected} = p_N. \quad (4.210)$$

Under the approximation assumption stated in subsection 4.3.3, the following approximation to the state probabilities p_k was obtained in Tijms and Van Hoorn (1982):

$$p_k^{\text{app}} = \begin{cases} \dfrac{(c\rho)^k}{k!} p_0^{\text{app}}, & 0 \leqslant k \leqslant c - 1, \\[2mm] \lambda p_{c-1}^{\text{app}} \alpha_{k-c} + \lambda \displaystyle\sum_{j=c}^{k} p_j^{\text{app}} \beta_{k-j}, & c \leqslant k \leqslant N - 1, \end{cases} \quad (4.211)$$

$$p_N^{\text{app}} = \rho p_{c-1}^{\text{app}} - (1 - \rho) \sum_{j=c}^{N-1} p_j^{\text{app}},$$

where α_n and β_n are defined below (4.190). From these relations we compute recursively the numbers $p_k^{\text{app}}/p_0^{\text{app}}$ for $k = 1, \ldots, N$ and next obtain p_0^{app} by applying the normalization equation $\sum_{k=0}^{N} p_k = 1$. In particular, approximations to the rejection probability and the average queue size are obtained from the calculated numbers p_k^{app} (see also Miyazawa, 1986, for somewhat simpler, but less accurate, approximations to these performance measures).

For the case of $\rho < 1$ we can establish for the approximations a similar relation as (4.113). Writing $p_k^{\text{app}}(N)$ for the approximations p_k^{app} in the M/G/c/N queue and $p_k^{\text{app}}(\infty)$ for the corresponding approximations (4.189) to (4.191) in the M/G/c queue, it can be shown that

$$p_k^{\text{app}}(N) = \left\{ 1 - \rho + \rho \sum_{j=0}^{N-1} p_j^{\text{app}}(\infty) \right\}^{-1} p_k^{\text{app}}(\infty), \qquad 0 \leqslant k \leqslant N - 1. \quad (4.212)$$

Numerical calculations indicate that for the case of $c > 1$ the equation (4.212) holds only approximately when we consider the exact values of $p_j(N)$ and $p_j(\infty)$ (cf. also (4.113) for the case of $c = 1$). However, in agreement with (4.15) and (4.212), it appears that the rejection probability satisfies $p_{N+1}(N+1) \approx (1/\tau)p_N(N)$

for N large enough, where τ is given by (4.19). This asymptotic result requires of course that $\rho < 1$.

The above approximations are exact for the special case of exponential service since then the approximation assumption holds exactly. For the case of exponential service we can give a simple alternative to the recursion (4.211). By the familiar argument of equating the rate at which the system leaves the set $\{j,\ldots,N\}$ of states to the rate at which the system enters that set, we obtain the recursion relation

$$\lambda p_{j-1} = \min(j,c)\mu p_j, \qquad 1 \leqslant j \leqslant N, \tag{4.213}$$

where $\mu = 1/E(S)$ denotes the exponential service rate. Also, for exponential service we have a simple expression for the distribution function of the delay in queue of an *entering* customer. Noting that an entering customer finds upon arrival j other customers in the system with probability $p_j/(1 - p_N)$ (verify!), we find by a minor modification to the derivation of (4.165) that

$$P\{W_q > t\} = \frac{1}{1 - p_N} \sum_{j=c}^{N-1} p_j \sum_{k=0}^{j-c} e^{-c\mu t} \frac{(c\mu t)^k}{k!}, \qquad t \geqslant 0. \tag{4.214}$$

Our numerical investigations indicate that the approximations (4.211) are accurate enough for practical purposes. To illustrate this, we give in Table 4.23 the exact and approximate values of the rejection probability $P_{\text{rej}}(= p_N)$ for various service-time distributions and several values of ρ and N. The service-time distributions consist of the deterministic distribution $(c_S^2 = 0)$, the E_2 distribution $(c_S^2 = \frac{1}{2})$, the exponential distribution $(c_S^2 = 1)$ and the H_2 distribution with the gamma normalization and $c_S^2 = 2$. In the examples we take $c = 5$ and we vary ρ as 0.5, 0.8 and 1.5 and N as $c + 1$, $c + 2$ and $c + 5$. For the cases of E_2 service and H_2 service the exact values of P_{rej} were computed by an algorithm discussed in Seelen (1986), whereas for the case of deterministic service, computer simulation was used to find the exact values of P_{rej}. In each example we simulated 500,000 customer arrivals, except for the example with $\rho = 0.5$ and $N = c + 5$, in which example we simulated 2,000,000 customer arrivals. In Table 4.23 the notation $0.0287(\pm .0006)$ means that the simulated value is 0.0287 with $[0.0281, 0.0293]$ as the 95 per cent confidence interval. The numerical investigations indicate that for the case of $\rho \geqslant 1$ the $M/M/c/N$ queue provides a reasonably good approximation to the $M/G/c/N$ queue. In this context we also state the conjecture that for the $M/G/c/N$ queue (and more generally for the $GI/G/c/N$ queue) with $\rho > 1$ the rejection probability tends to $1 - 1/\rho$ as $N \to \infty$ when the other parameters are kept fixed (see also Sobel, 1980). This asymptotic estimate being independent of the distributional forms of the interarrival and service time is already accurate for relatively small values of $N - c$ provided ρ is not too close to 1. The following elegant argument due to Heyman (1980) can be given in support of the above conjecture. The average arrival rate (λ) is equal to the sum of the average arrival rate of rejected

Table 4.23 Exact and approximate values of P_{rej} for the M/G/c/N queue with $c = 5$

	N	$\rho = 0.5$			$\rho = 0.8$			$\rho = 1.5$	
		$c+1$	$c+2$	$c+5$	$c+1$	$c+2$	$c+5$	$c+1$	$c+5$
$c_S^2 = 0$	app	0.0286	0.0105	0.00036	0.1221	0.0742	0.0179	0.3858	0.3348
	exa	0.0287	0.0108	0.00035	0.1224	0.0746	0.0175	0.3870	0.3352
		(± 0.0006)	(± 0.0004)	(± 0.00003)	(± 0.0012)	(± 0.0011)	(± 0.0007)	(± 0.0016)	(± 0.0020)
$c_S^2 = \frac{1}{2}$	app	0.0311	0.0135	0.0010	0.1306	0.0882	0.0308	0.3975	0.3395
	exa	0.0314	0.0137	0.0010	0.1318	0.0895	0.0314	0.4000	0.3400
$c_S^2 = 1$	exa	0.0337	0.0166	0.0020	0.1374	0.0990	0.0425	0.4046	0.3453
$c_S^2 = 2$	app	0.0370	0.0211	0.0046	0.1450	0.1123	0.0603	0.4114	0.3555
	exa	0.0366	0.0206	0.0044	0.1435	0.1103	0.0587	0.4092	0.3537

customers (λP_{rej}) and the average arrival rate of entering customers, the latter term being equal to the average number of customers served per unit time. Denoting by v the average number of busy servers, the average number of customers served per unit time can be seen to be equal to $v/E(S)$ and thus we obtain the conservation equation

$$\lambda = \lambda P_{rej} + \frac{v}{E(S)}.$$

Noting that always $v \leqslant c$, this equation implies the lower bound $P_{rej} \geqslant 1 - 1/\rho$. Intuitively, for the case of $\rho > 1$, this inequality will become an equality as $N \to \infty$ since then $v \to c$.

The M/G/c loss system

In this model no queueing is allowed and a customer finding upon arrival all of the c servers busy is lost. The M/G/c loss model is one of the most useful queueing models. Several applications of this model have already been discussed in example 2.8 of section 2.4 of chapter 2.

The following famous result holds for the M/G/c loss model:

$$p_j = \frac{\{\lambda E(S)\}^j/j!}{\sum_{k=0}^{c} \{\lambda E(S)\}^k/k!}, \qquad j = 0,\ldots,c, \tag{4.215}$$

so that the state probabilities depend on the service-time distribution only through its mean. Also, note from equations (4.203) and (4.205) that the probability of having j busy servers in the M/G/c loss system equals the conditional probability of having j busy servers in the M/G/∞ system, given that no more than c servers are busy. A proof of the result (4.215) can be found in, for example, Cohen (1976). Many stochastic service systems in which arrivals are lost when not getting immediate access to service have the property that the state probabilities are insensitive to the distributional form of the service times when there is finite-source input or Poisson input. This insensitivity property is very useful in practice.

Using the property 'Poisson arrivals see time averages', it follows from (4.215) that

$$\text{The fraction of arrivals that is lost} = \frac{\{\lambda E(S)\}^c/c!}{\sum_{k=0}^{c} \{\lambda E(S)\}^k/k!}. \tag{4.216}$$

This formula is known as *Erlang's loss formula*. Also, the average arrival rate of entering customers equals $\lambda(1 - p_c)$ and so, by making an obvious modification to the derivation of (4.5),

$$\text{The average number of busy servers} = \lambda(1 - p_c)E(S). \tag{4.217}$$

It is noted that for the case of $\rho > 1$ the loss probability tends to $1 - 1/\rho$ when $\lambda E(S) \to \infty$ and $c \to \infty$ such that $\rho = \lambda E(S)/c$ remains constant. This result is also true for the GI/G/c loss system (see Whitt, 1984b).

An important application of the M/G/c loss system is discussed in the following example.

Example 4.7   A base-stock policy for a continuous-review inventory system

The so-called base-stock policy is often used in inventory systems of expensive, slow-moving items for which unit demands occur. Under this control rule a replenishment order for exactly one unit is placed each time the on-hand inventory decreases with one unit by the occurrence of a demand. Suppose such a control rule is applied to a continuous-review inventory system in which the demand process for a certain item is a Poisson process with demand rate λ. Each demand that cannot be satisfied directly from on-hand inventory is lost. At any time a replenishment order may be placed. It is assumed that the lead times of the replenishment orders are independent and identically distributed random variables with a mean of τ time units. The inventory control of the item is exercised by a base-stock policy with a given base-stock level R. That is the on-hand inventory plus the amount on order is always kept equal to R so that a replenishment order for one unit is placed each time the on-hand inventory decreases by one unit. The following costs are involved. There is a holding cost of $h > 0$ per unit of time for each unit kept in stock, while a fixed penalty cost of $\pi > 0$ is incurred for each demand that is lost.

We are interested in how to choose the base-stock level R as a function of the parameters λ, τ, h and π in order to minimize the long-run average cost per unit time. It will be seen that for this particular inventory control problem the concept of *indifference curves* is very helpful. These curves show under what conditions we are indifferent between neighbouring values of the control variable R.

First we obtain the average cost per unit time under a given base-stock policy with level R. To do so we establish a correspondence between the inventory process and the process describing the number of busy servers in an M/G/c loss system. By identifying outstanding orders with busy servers and lead times with service times, it is readily seen that the number of outstanding orders is distributed as the number of busy servers in an M/G/c loss system. Hence, using that

The on-hand inventory $= R -$ (the number of outstanding orders),

it follows from (4.215) that

The fraction of time the on-hand inventory is $j = \dfrac{(\lambda\tau)^{R-j}/(R-j)!}{\sum_{k=0}^{R}(\lambda\tau)^k/k!}, \qquad 0 \leqslant j \leqslant R.$

In particular, since Poisson arrivals see time averages,

The fraction of demand that is lost $= p(R;\lambda\tau)$,

where we put for abbreviation

$$p(R;\lambda\tau) = \frac{(\lambda\tau)^R/R!}{\sum_{k=0}^{R}(\lambda\tau)^k/k!}.$$

Hence the average demand that is lost per unit time equals $\lambda p(R;\lambda\tau)$. The average on-hand inventory is easily verified to be given by

$$R - \lambda\{1 - p(R;\lambda\tau)\}\tau.$$

A direct way to see this result is to apply the relation (4.217).

We can now conclude that the average cost per unit time under a base-stock policy with level R is given by

$$g(R) = \pi\lambda p(R;\lambda\tau) + h[R - \lambda\{1 - p(R;\lambda\tau)\}\tau].$$

Next we develop the indifference curves. We are indifferent between the use of R and $R + 1$ when

$$g(R) = g(R + 1).$$

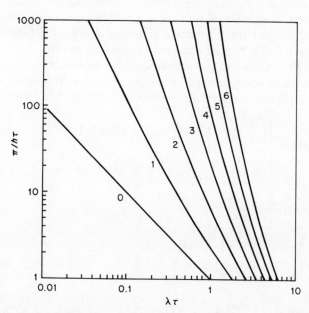

Figure 4.3 The indifference curves

It is a matter of simple algebra to verify that this equation is equivalent to

$$\lambda\tau\{p(R;\lambda\tau) - p(R+1;\lambda\tau)\} = \left(1 + \frac{\pi}{h\tau}\right)^{-1}. \qquad (4.218)$$

The latter equation is in terms of the dimensionless quantities $\lambda\tau$ and $\pi/(h\tau)$ and shows that the optimal choice of R depends only upon these two parameters. For a given value of R the equation (4.218) determines a curve in a two-dimensional graph of $\pi/(h\tau)$ versus $\lambda\tau$. Thus the indifference curve between R and $R+1$ is obtained by fixing R in (4.218) and finding $(1 + \pi/h\tau)^{-1}$ from the left side of (4.218) for various values of $\lambda\tau$. In Figure 4.3 we show the indifference curves for $R = 0, \ldots, 6$.

The indifference curves provide much insight into the sensitivity of the optimal choice of R to the various parameters. For example, when $\lambda\tau$ and h are fixed there is a wide range of values of the penalty cost π for which the same optimal choice of R results, in particular when the offered load $\lambda\tau$ is not too large. This is a useful observation since in practice the precise value of the penalty cost π is often difficult to specify. For this inventory control problem the use of indifference curves for quick engineering purposes was suggested in Silver and Smith (1977). $\qquad\qquad\square$

4.4.6 The machine repair model with multiple repairmen

This model extends the model in example 4.2 in subsection 4.2.2 to the case of multiple servers. We assume N sources each having an exponentially distributed think time with mean $1/\eta$, a general probability distribution function $B(x)$ for the service time S of a request and $c(\leqslant N)$ identical servers.

Under the approximation assumption stated in subsection 4.3.3, the following approximation can be established for the time-average probability p_k of having k requests waiting or being served at the servers (see Tijms and Van Hoorn, 1982):

$$p_k^{\text{app}} = \begin{cases} \binom{N}{k}\{\eta E(S)\}^k p_0^{\text{app}}, & 0 \leqslant k \leqslant c-1, \\[2mm] (N-c+1)\eta\alpha_{ck}p_{c-1}^{\text{app}} + \sum_{j=c}^{k}(N-j)\eta\beta_{jk}p_j^{\text{app}}, & c \leqslant k \leqslant N, \end{cases}$$

$$(4.219)$$

where

$$\alpha_{ck} = \int_0^\infty \{1 - B_e(t)\}^{c-1}\{1 - B(t)\}\phi_{ck}(t)\,dt, \qquad \beta_{jk} = \int_0^\infty \{1 - B(ct)\}\phi_{jk}(t)\,dt$$

with

$$\phi_{jk}(t) = \binom{N-j}{k-j}(1 - e^{-\eta t})^{k-j}e^{-\eta t(N-k)}, \qquad t > 0, \quad k \geqslant j \geqslant c.$$

Here the equilibrium distribution function $B_e(t)$ is given by (4.188). The approximation to the probabilities p_j is exact for the special case of exponential service. In this case a simple recursion scheme is provided by

$$(N - j + 1)\eta p_{j-1} = \min(j, c)\mu p_j, \qquad 1 \leqslant j \leqslant N,$$

where $\mu = 1/E(S)$ is the average service rate.

Denoting by L_q the number of requests queueing at the servers, the moments of L_q are directly computed from the state probabilities p_j. The customer-average probability π_j that an arriving request finds j other requests at the servers is easily obtained from the time-average probabilities p_k. Noting that new requests arrive at the servers according to a Poisson process with rate $(N - j)\eta$ whenever j requests are there, it can be shown in the same way as in the proof of (4.59) that

$$\pi_j = \frac{(N - j)\eta p_j}{\sum_{k=0}^{N}(N - k)\eta p_k}, \qquad 0 \leqslant j \leqslant N - 1. \tag{4.220}$$

In particular, the long-run fraction of requests that is delayed equals

$$\Pi_W = \sum_{j=c}^{N-1} \pi_j. \tag{4.221}$$

Also, by making a minor modification to the derivation of (4.5), we have

$$\text{The average number of busy servers} = \left\{ \sum_{k=0}^{N}(N - k)\eta p_k \right\} E(S). \tag{4.222}$$

The utilization of each server equals $(1/c)$ times the above average.

Our numerical investigations indicate that the approximation (4.219) is accurate enough for practical purposes. In Table 4.24 we give the exact and approximate values of Π_W, $E(L_q)$ and the coefficient of variation $cv(L_q)$ of L_q for various service-time distributions and for several values of N. We consider as service-time distributions the deterministic distribution $(c_S^2 = 0)$, the E_2 distribution $(c_S^2 = \frac{1}{2})$, the exponential distribution $(c_S^2 = 1)$ and the H_2 distribution with the gamma normalization and $c_S^2 = 2$. In the examples we take $c = 5$ and $\eta = 0.4$ and we vary N as $2c$, $3c$ and $4c$. For the cases with non-deterministic service the exact values in Table 4.24 were computed by an algorithm discussed in Seelen (1986), whereas for the case of deterministic service, computer simulation was used to find the exact values. In each example 500,000 service requests were simulated. In Table 4.24 the notation 0.0857 ($\pm$ .0010) means that the simulated value is 0.0857 with [0.0847, 0.0867] as the 95 percent confidence interval.

4.4.7 The GI/G/c queue

It seems obvious that the general GI/G/c queue offers enormous difficulties in getting practical useful results. However, the capabilities of current computers

Table 4.24 Exact and approximate results for the machine repairmen model with $c = 5$

		$N = 2c$			$N = 3c$			$N = 4c$		
		P_{rej}	$E(L_q)$	$cv(L_q)$	P_{rej}	$E(L_q)$	$cv(L_q)$	P_{rej}	$E(L_q)$	$cv(L_q)$
$c_S^2 = 0$	app	0.0909	0.052	5.14	0.4729	0.613	1.67	0.8965	2.945	0.717
	exa	0.0857	0.049	5.36	0.4581	0.577	1.75	0.8937	2.920	0.735
		±.0010	±.001		±.0028	±.007		±.0022	±.025	
$c_S^2 = 0.5$	app	0.0894	0.059	4.99	0.4529	0.709	1.68	0.8535	3.124	0.781
	exa	0.0880	0.059	5.04	0.4478	0.703	1.71	0.8507	3.119	0.790
$c_S^2 = 1$	exa	0.0875	0.068	4.84	0.4351	0.794	1.69	0.8183	3.267	0.825
$c_S^2 = 2$	app	0.0846	0.082	4.69	0.4086	0.918	1.71	0.7666	3.474	0.890
	exa	0.0863	0.082	4.63	0.4153	0.923	1.68	0.7728	3.477	0.874

and special iterative solution methods for large-scale systems of linear equations arising in Markov chains enable us nowadays to obtain the numerical solution of many GI/G/c queueing systems of practical interest. As already pointed out in the introduction to section 4.4, the queueing system may be analysed by a continuous-time Markov chain model when the interarrival-time and service-time distributions are of the phase-type (Ph/Ph/c queue). By a detailed state description involving sufficient information about the number of customers present and the status of the service processes and the arrival process, it is possible to set up the equilibrium equations for the microstate probabilities. This system of linear equations possesses a structure enabling the application of special iterative methods to solve the equations numerically. An aggregation/disaggregation algorithm for the numerical solution of the equilibrium equations arising in the Ph/Ph/c queue was first presented by Takahashi and Takami (1976). The approach of this algorithm may also be applied successfully to other problems with an identifiable structure so that aggregation of states within blocks reflects this structure (see Schweitzer and Kindle, 1985). The algorithm alternately solves an aggregated version and a disaggregated version of the problem. Thus, by solving in each step a sequence of small sets of linear equations, the algorithm avoids the diseconomies of scale of solving one large set of linear equations. The specialization of the algorithm to the Ph/Ph/c queue, having no bound on the queue size, involves a clever truncation of the infinite system of linear equations by using the ultimately geometric behaviour of the state probabilities. An improved algorithm for the iterative solution of the equilibrium equations in Ph/Ph/c queueing systems was proposed by Seelen (1986). This algorithm resembles some aspects of the conventional aggregation/disaggregation method, but the main differences are a simplified structure due to an adaptation of the disaggregation step and the use of a dynamically adjusted relaxation factor. The algorithm is a highly effective interweave of simple numerical tools exploiting structure with qualitative results from probabilistic analysis; the algorithm exploits the structure of the Ph/Ph/c queue only to such an extent that the algorithm needs only slight modifications for multi-server queueing systems with finite waiting room and state-dependent input. Like the conventional aggregation/disaggregation method, the algorithm is remarkably robust in the sense that the number of iterations required is fairly insensitive to the number of states and to the starting point. The number of iterations is, of course, strongly dependent on the relative precision required in the final answers; in fact, in case there is no need for a high precision in the answers the iterative method may be used as an approximation method. Also, the computing time of the algorithm depends very much on the number of servers and the number of phases needed to describe the service time, since these factors mainly determine the size of the system of linear equations to be solved. Note that the number of phases of a phase-type distribution becomes large when the coefficient of variation comes close to zero.

The algorithm of Seelen (1986) was instrumental in compiling the tablebook of Seelen, Tijms and Van Hoorn (1985). For a wide spectrum of practical multi-server queueing models with infinite and finite waiting room, the tablebook gives the exact values of measures of system performance such as the delay probability, the blocking probability, the probability of having all servers busy and the first two moments of the queue size. Also, the tablebook gives the (nearly) exact values of the waiting-time percentiles for the multi-server queue with infinite waiting room and service in order of arrival. The specific interarrival-time and service-time distributions include mixtures of E_k and E_{k-1} distributions with the same scale parameters and H_2 distributions for a practical range of the coefficients of variation and for many values of the number of servers (up to $c = 50$ servers). Also, for the particular cases of the Ph/D/c queue and the Ph/M/c queue tables are given up to $c = 200$ servers. These tables may be useful for interpolation purposes. The results for the Ph/D/c queue were computed by an efficient algorithm given in Van Hoorn (1985) that is an extension of the algorithm for the M/D/c queue discussed in subsection 4.4.2. For the Ph/M/c queue two very efficient algorithms are available, namely the well-known embedded Markov chain algorithm (see, for examples, Cooper, 1981) and the matrix-geometric solution method presented in Neuts (1982). The embedded Markov chain algorithm for the GI/M/c queue is a rather straightforward extension of the one for the GI/M/1 queuen discussed in example 2.2 in section 2.2 of chapter 2. The algorithm for the Ph/D/c queue and the Ph/M/c queue are computationally feasible for very large values of c. To the end of the discussion on exact algorithms for the GI/G/c queue, we mention the rather involved algorithms of De Smit (1983), Ishikawa (1979, 1984) and Ramaswami and Lucantoni (1985) for the respective cases of the GI/H₂/c queue, the GI/E_k/c queue and the GI/Ph/c queue.

We next concentrate on simple approximate methods for the GI/G/c queue.

Approximations

Numerical investigations reveal that in many practical GI/G/c queueing systems both the average queue length $E(L_q)$ and the average conditional waiting time $TW = E(W_q | W_q > 0)$ may reasonably well be approximated by using the familiar interpolation formula

$$P^{\text{app}}_{\text{GI/G/c}} = (1 - c_S^2)P_{\text{GI/D/c}} + c_S^2 P_{\text{GI/M/c}} \qquad (4.223)$$

provided c_S^2 is not too large. Here $P_{\text{GI/D/c}}$ and $P_{\text{GI/M/c}}$ denote the exact values of the appropriate performance measure P for the special cases of the GI/D/c queue and the GI/M/c queue. Recall that the latter two cases allow for a relatively simple algorithmic analysis, compare also remark 4.1 at the end of subsection 4.4.2. The linear interpolation formula (4.223) is in general not to be recommended for the delay probability Π_W, particularly not when c_A^2 is close to zero. For example for the $E_{10}/G/5$ queue with $\rho = 0.8$ the delay probability has the values 0.0776, 0.2934, 0.3285 and 0.3896 when c_S^2 has the respective values 0, $\frac{1}{3}$, $\frac{1}{2}$ and 1 corresponding to deterministic and Erlangian services. An interpolation formula like (4.223) should always be issued together with a caveat against its blind application. In particular, it may be hazardous to use the interpolation formula when the traffic load on the system is very light. For a more elaborate discussion on interpolation formulae for the multi-server queue, the reader is referred to Seelen, Tijms, and Van Hoorn (1985).

The approximation (4.223) reflects the empirical finding that measures of system performance are in general more sensitive to the interarrival-time distribution than to the service-time distribution, in particular when the traffic is light (cf. also Neuts, 1985, and Wolff, 1977, for an interesting discussion on the effect of variability on queues). To illustrate this finding, we give in Table 4.25 the exact values of the delay probability Π_W, the probability P_B of having all servers busy, the average queue size $E(L_q)$ and the average conditional waiting time TW. The interarrival-time and service-time distributions include the deterministic distribution, the E_2 distribution, the mixture of E_1 and E_3 distributions with the same scale parameters ($E_{1,3}$) with a squared coefficient of variation of $\frac{1}{2}$, the exponential distribution and the H_2 distribution with a

Table 4.25   Exact results for the GI/G/5 queue

	$\rho = 0.5$				$\rho = 0.8$			
	Π_W	P_B	$E(L_q)$	TW	Π_W	P_B	$E(L_q)$	TW
$E_2/D/5$	0.0331	0.0517	0.012	0.143	0.3847	0.4445	0.439	0.285
$E_2/E_2/5$	0.0591	0.0811	0.035	0.239	0.4532	0.4985	0.967	0.534
$E_2/M/5$	0.0708	0.0927	0.057	0.324	0.4778	0.5167	1.469	0.769
$E_2/H_2^a/5$	0.0843	0.1053	0.103	0.489	0.5021	0.5347	2.480	1.234
$E_2/H_2^b/5$	0.0809	0.1023	0.092	0.455	0.4977	0.5316	2.394	1.203
$E_{1,3}/D/5$	0.0408	0.0577	0.017	0.171	0.3843	0.4416	0.466	0.303
$E_{1,3}/E_2/5$	0.0634	0.0835	0.040	0.251	0.4536	0.4978	0.984	0.542
$E_{1,3}/M/5$	0.0744	0.0944	0.061	0.330	0.4789	0.5165	1.483	0.774
$E_{1,3}H_2^a/5$	0.0874	0.1065	0.107	0.488	0.5038	0.5349	2.490	1.236
$E_{1,3}/H_2^b/5$	0.0841	0.1037	0.096	0.455	0.4991	0.5318	2.406	1.205
$H_2^a/D/5$	0.2725	0.2123	0.290	0.425	0.6675	0.6159	2.770	1.037
$H_2^a/E_2/5$	0.2510	0.1959	0.300	0.478	0.6583	0.6079	3.272	1.243
$H_2^a/M/5$	0.2373	0.1871	0.324	0.546	0.6504	0.6029	3.800	1.461
$H_2^a/H_2^a/5$	0.2226	0.1777	0.373	0.670	0.6425	0.5985	4.845	1.885
$H_2^a/H_2^b/5$	0.2295	0.1819	0.360	0.628	0.6480	0.6027	4.778	1.844
$H_2^b/D/5$	0.2252	0.1859	0.183	0.324	0.6729	0.6191	2.478	0.920
$H_2^b/E_2/5$	0.2145	0.1793	0.218	0.407	0.6561	0.6087	3.039	1.158
$H_2^b/M/5$	0.2065	0.1745	0.253	0.490	0.6453	0.6027	3.598	1.394
$H_2^b/H_2^a/5$	0.1966	0.1685	0.314	0.639	0.6340	0.5971	4.679	1.845
$H_2^b/H_2^b/5$	0.2004	0.1707	0.297	0.593	0.6400	0.6011	4.599	1.797

squared coefficient of variation of 2 for both the normalization of balanced means (H_2^b) and the gamma normalization (H_2^g). In the examples we take $c = 5$ and vary the server utilization ρ as 0.5 and 0.8. Note that the results in Table 4.25 support the claim that in many practical situations the use of two-moment approximations is justified provided the coefficients of variation are not too large and the traffic is not very light. Also, the numerical results in Table 4.25 might suggest that the $H_2/G/c$ queue has the property that the delay probability always decreases when c_S^2 becomes larger with $E(S)$ kept fixed. It was noted in subsection 4.2.3 that this property always holds when $c = 1$, but numerical investigations indicate that for $c > 1$ this property seems to hold only when c_A^2 has passed some crossover point larger than 1.

Next we turn to approximations involving both the interarrival-time and service-time distributions. The approximations obtained for the $M/G/c$ queue by the regenerative approach were extended by Van Hoorn and Seelen (1986) to the $GI/G/c$ queue with phase-type arrivals. They derived several computationally tractable approximation methods for the cases of Erlangian input and hyperexponential input, where the approximation methods involve an exact modelling of the arrival process, a fact which is important because of the large impact of the arrival process on measures of system performance. In particular, approximations are obtained for performance measures such as the delay probability, the probability of having all servers busy and the average queue size. The computational procedure for the case of Erlangian input is more involved than the one for the case of hyperexponential input. The algorithm for Erlangian input with $c_A^2 < \frac{1}{2}$ requires the computation of (conjugate) complex roots of some characteristic equation. For phase-type services this characteristic equation reduces to a polynomial equation for which effective numerical procedures are available for the calculation of the complex roots. We refer to Van Hoorn and Seelen (1986) for a description of the approximate algorithm for the case of generalized Erlangian input when the interarrival-time distribution is a mixture of Erlangian distributions with the same scale parameters. Here we present only one of their algorithms (namely algorithm II) for the case of H_2 input with interarrival-time density

$$a(t) = r_1 \lambda_1 e^{-\lambda_1 t} + r_2 \lambda_2 e^{-\lambda_2 t}, \qquad t \geqslant 0, \tag{4.224}$$

where $0 \leqslant r_1, r_2 \leqslant 1$ and $r_1 + r_2 = 1$. The H_2 density has always a squared coefficient of variation greater than or equal to 1. However, by weakening the requirement that $0 \leqslant r_1, r_2 \leqslant 1$, a density of the form (4.224) may also be used to represent a probability density (K_2 density) with a squared coefficient of variation between $\frac{1}{2}$ and 1 (see appendix B). Surprisingly, it turns out from numerical experiments that the algorithm derived in Van Hoorn and Seelen (1986) for H_2 input applies as well to the case of K_2 input. We state only the steps of this algorithm and refer to Van Hoorn and Seelen (1986) for the probabilistic arguments used to obtain this algorithm.

Approximate algorithm for the $K_2/G/c$ queue

Step 1 (to be done only when $c_A^2 \neq 1$). Compute ξ_0 as the unique solution to the equation

$$\lambda_1 \lambda_2 + \xi^2 - (\lambda_1 + \lambda_2)\xi - \left\{ \int_0^\infty e^{-(\xi/c)t} b(t)\,dt \right\} \{ \lambda_1 \lambda_2 - (r_1 \lambda_1 + r_2 \lambda_2)\xi \} = 0$$

in $\xi \geq \min (\lambda_1, \lambda_2)$, where $b(t)$ denotes the service-time density. Next calculate the constants

$$v_0 = -(r_1 \lambda_1 + r_2 \lambda_2)\left(\frac{r_1 \lambda_1}{\lambda_1 - \xi_0} + \frac{r_2 \lambda_2}{\lambda_2 - \xi_0} \right)^{-1} + \lambda_1 + \lambda_2 - \xi_0,$$

$$\alpha_1 = (\lambda_2 - \lambda_1)\frac{r_2}{\lambda_2 - v_0},$$

$$\alpha_2 = (\lambda_1 - \lambda_2)\frac{r_1}{\lambda_1 - v_0},$$

$$x_0 = (r_1 \alpha_1 + r_2 \alpha_2)\left(\frac{r_1 \lambda_1}{\lambda_1 - \xi_0} + \frac{r_2 \lambda_2}{\lambda_2 - \xi_0} \right)^{-1}.$$

Step 2. Let $n := c - 1$. In the case $c_A^2 \neq 1$, let

$$q_{n1} := \frac{\alpha_2 - x_0}{\alpha_2 - \alpha_1} \quad \text{and} \quad q_{n2} := \frac{\alpha_1 - x_0}{\alpha_1 - \alpha_2};$$

otherwise let $q_{ni} := r_i$ for $i = 1, 2$.

Step 3. Denoting by $\lambda = (r_1/\lambda_1 + r_2/\lambda_2)^{-1}$ and $\mu = 1/E(S)$ the average arrival rate and the average service rate, calculate

$$p_{n1} := \frac{\lambda\{q_{n1}(\lambda_2 + r_1 n\mu) + q_{n2} r_1 n\mu\}}{\lambda_1 \lambda_2 + (\lambda_1 r_1 + \lambda_2 r_2)n\mu},$$

$$p_{n2} := \frac{\lambda\{q_{n1} r_2 n\mu + q_{n2}(\lambda_1 + r_2 n\mu)\}}{\lambda_1 \lambda_2 + (\lambda_1 r_1 + \lambda_2 r_2)n\mu},$$

and, if $n \neq 0$,

$$q_{n-1,i} := \frac{n\mu}{\lambda} p_{ni} \quad \text{for} \quad i = 1, 2.$$

Step 4. $n := n - 1$. If $n \geq 0$, return to step 3; otherwise go to step 5.

Step 5. Compute the normalization constant

$$\gamma = \sum_{n=0}^{c-1} (p_{n1} + p_{n2}) + \frac{\rho}{1 - \rho} \frac{\lambda_2 q_{c-1,1} + \lambda_1 q_{c-1,2}}{\lambda_1 r_2 + \lambda_2 r_1}.$$

Next normalize p_{ni} and q_{ni} by

$$p_{ni} := \frac{1}{\gamma} p_{ni}, \qquad q_{ni} := \frac{1}{\gamma} q_{ni} \qquad \text{for} \quad n = 0, \ldots, c-1 \quad \text{and} \quad i = 1, 2.$$

Step 6. The delay probability Π_W, the probability P_B of having all servers busy and the average queue size $E(L_q)$ are approximated by

$$\Pi_W^{\text{app}} = 1 - \sum_{n=0}^{c-1} (q_{n1} + q_{n2}),$$

$$P_B^{\text{app}} = 1 - \sum_{n=0}^{c-1} (p_{n1} + p_{n2}),$$

$$E^{\text{app}}(L_q) = \frac{1}{1-\rho} \left\{ \tfrac{1}{2}(1 + c_A^2) + \tfrac{1}{2}\rho^2 \Pi_W^{\text{app}}(1 + c_S^2) - P_B^{\text{app}} \right.$$

$$\left. + \rho \left(\sum_{i=1}^{2} q_{c-1,i} \right) \frac{\lambda(1 + c_S^2)}{2c\mu} - \frac{\lambda_2 \sum_{n=0}^{c-1} p_{n1} + \lambda_1 \sum_{n=0}^{c-1} p_{n2}}{\lambda_1 r_2 + \lambda_2 r_1} \right\}.$$

The following remarks are in order. The algorithm is exact for the special case of exponential service in which case the algorithm closely resembles the method of Neuts (1982) for the Ph/M/c queue. Also, the algorithm is exact for the case of $c = 1$ server (cf. also subsection 4.2.3). For the case of H_2 input ($c_A^2 \geqslant 1$) the quantity p_{ni} approximates the time-average probability that n customers are in the system and the arrival process is in phase i, while the quantity q_{ni} approximates the probability that at a service completion epoch n customers are left behind and the arrival process is in phase i. Since the fraction of arrivals finding n customers present must be equal to the fraction of departures leaving n customers behind, we have that $q_{n1} + q_{n2}$ approximates the probability that a customer finds upon arrival n other customers present. For K_2 input with $\tfrac{1}{2} < c_A^2 < 1$ the calculated numbers p_{ni} and q_{ni} have no probabilistic meaning, but notwithstanding the quantities calculated in step 6 are valid approximations to Π_W, P_B and $E(L_q)$. Also, the above algorithm may be used to compute approximations for the $E_2/G/c$ queue by taking K_2 input with c_A^2 sufficiently close to 0.5 (say, $c_A^2 = 0.50001$). Finally, it should be pointed out that for the particular case of the M/G/c queue the algorithm modifies the earlier approximate algorithm (4.190) by replacing the constants β_{j-k} by the simpler constants α_{j-k}; this modification for the M/G/c queue yields again the Erlang delay probability approximation to Π_W, while the resulting approximation to $E(L_q)$ is the same as the appealing approximation $\tfrac{1}{2}(1 + c_S^2)E_{\text{exp}}(L_q)$ which was found in Nozaki and Ross (1978) among others and is known to be a practically useful approximation when c_S^2 is not too large.

In Table 4.26 we give the exact and approximate values of Π_W, P_B and $E(L_q)$ for several GI/G/c queueing systems. The interarrival-time and service-time distributions include the E_2 distribution and the H_2 distribution with the

Table 4.26 Exact and approximate values for the GI/G/c queue

		$(c_A^2, c_S^2) = (0.5, 0.5)$			$(c_A^2, c_S^2) = (2, 0.5)$			$(c_A^2, c_S^2) = (2, 2)$		
		Π_W	P_B	$E(L_q)$	Π_W	P_B	$E(L_q)$	Π_W	P_B	$E(L_q)$
$c = 2$ $\rho = 0.2$	exa	0.0166	0.040	0.003	0.1585	0.095	0.038	0.1378	0.088	0.054
	app	0.0183	0.045	0.003	0.1554	0.092	0.036	0.1418	0.091	0.060
$c = 2$ $\rho = 0.5$	exa	0.2096	0.290	0.118	0.4891	0.379	0.560	0.4537	0.366	0.809
	app	0.2166	0.301	0.118	0.4861	0.374	0.542	0.4562	0.371	0.844
$c = 10$ $\rho = 0.5$	exa	0.0100	0.014	0.006	0.1031	0.081	0.128	0.0842	0.067	0.133
	app	0.0126	0.018	0.007	0.0970	0.075	0.108	0.0901	0.073	0.167
$c = 10$ $\rho = 0.7$	exa	0.1328	0.155	0.178	0.3420	0.302	1.007	0.3195	0.284	1.313
	app	0.1467	0.172	0.179	0.3364	0.294	0.933	0.3243	0.292	1.453
$c = 25$ $\rho = 0.7$	exa	0.0247	0.029	0.035	0.1405	0.124	0.429	0.1215	0.108	0.472
	app	0.0305	0.360	0.037	0.1335	0.117	0.370	0.1285	0.116	0.575
$c = 25$ $\rho = 0.8$	exa	0.1272	0.140	0.291	0.3171	0.294	1.640	0.2978	0.277	2.117
	app	0.1432	0.158	0.294	0.3107	0.286	1.508	0.3036	0.284	2.367
$c = 50$ $\rho = 0.8$	exa	0.0394	0.043	0.094	0.1690	0.157	0.896	0.1500	0.139	1.026
	app	0.0480	0.053	0.099	0.1613	0.148	0.783	0.1576	0.148	1.229
$c = 50$ $\rho = 0.9$	exa	0.2698	0.282	1.332	0.4649	0.448	5.394	0.4508	0.436	7.519
	app	0.2931	0.307	1.336	0.4598	0.442	5.107	0.4550	0.441	8.091

gamma normalization and a squared coefficient of variation of 2. The number c of servers is varied as 2, 10, 25 and 50 and the server utilization ρ has two values depending on c.

Two-moment approximations

Practical useful two-moment approximations can be given for both the average delay in queue and the average conditional waiting time. The following approximation to $E(W_q)$ is suggested in Kimura (1985):

$$E_{app}(W_q) = (c_A^2 + c_S^2) \left\{ \frac{1 - c_A^2}{W_q(D/M/c)} + \frac{1 - c_S^2}{W_q(M/D/c)} + \frac{2(c_A^2 + c_S^2 - 1)}{W_q(M/M/c)} \right\}^{-1}$$

(4.225)

where $W_q(M/M/c)$, $W_q(M/D/c)$ and $W_q(D/M/c)$ denote the exact values of $E(W_q)$ for the respective queueing systems $M/M/c$, $M/D/c$ and $D/M/c$, having the same server utilization and the same average service rate as the original queueing system $GI/G/c$. This two-moment approximation is to be used only when both $0 \leqslant c_A^2 \leqslant 1$ and $0 \leqslant c_S^2 \leqslant 1$. The exact value of $W_q(M/M/c)$ is trivial to compute from (4.166). Rather than computing the exact values of $W_q(M/D/c)$ and $W_q(D/M/c)$, the following excellent approximations due to Cosmetatos (1975) may be used:

$$W_q^{app}(M/D/c) = \tfrac{1}{2}\{1 + C(c, \rho)\} W_q(M/M/c)$$

(4.226)

and

$$W_q^{app}(D/M/c) = \frac{W_q(D/M/1)}{W_q(M/M/1)} \{1 - 4C(c, \rho)\} W_q(M/M/c),$$

(4.227)

where

$$C(c, \rho) = (1 - \sigma)(c - 1)\frac{\sqrt{4 + 5c} + 2}{16c\rho}.$$

(4.228)

Here $W_q(M/M/1)$ and $W_q(D/M/1)$ denote the exact values of $E(W_q)$ in the queueing systems $M/M/1$ and $D/M/1$ with the same server utilization and the same average service rate as the original queueing system $GI/G/c$. The exact value of $W_q(M/M/1)$ is given by (4.166) with $c = 1$, while the general approximation formula (4.104) provides the following accurate approximation to $W_q(D/M/1)$:

$$W_q^{app}(D/M/1) = \tfrac{1}{2}e^{-2(1-\rho)/3\rho} W_q(M/M/1).$$

(4.229)

A two-moment approximation to the average conditional waiting time

Table 4.27   Two-moment approximations to $E(W_q)$ and TW

		$c_A^2 = 0.1, c_S^2 = 0.5$		$c_A^2 = 0.5, c_S^2 = \frac{1}{3}$		$c_A^2 = 0.5, c_S^2 = 0.75$		$c_A^2 = 0.75, c_S^2 = 0.5$	
		$E(W_q)$	TW	$E(W_q)$	TW	$E(W_q)$	TW	$E(W_q)$	TW
$c = 2$	exa	0.034	0.397	0.096	0.481	0.152	0.686	0.192	0.685
$\rho = 0.5$	app	0.043	0.439	0.081	0.514	0.120	0.701	0.155	0.682
$c = 2$	exa	0.412	0.827	0.673	1.094	1.025	1.611	1.085	1.613
$\rho = 0.8$	app	0.438	0.874	0.664	1.120	0.991	1.631	1.053	1.605
$c = 10$	exa	0.035	0.178	0.069	0.237	0.106	0.331	0.121	0.336
$\rho = 0.8$	app	0.039	0.178	0.066	0.228	0.096	0.328	0.112	0.324
$c = 10$	exa	0.160	0.327	0.256	0.444	0.387	0.642	0.408	0.648
$\rho = 0.9$	app	0.167	0.325	0.253	0.433	0.375	0.639	0.400	0.634
$c = 25$	exa	0.042	0.136	0.074	0.184	0.112	0.260	0.122	0.264
$\rho = 0.9$	app	0.045	0.130	0.072	0.173	0.106	0.256	0.118	0.254
$c = 25$	exa	0.149	0.255	0.229	0.351	0.345	0.510	0.358	0.514
$\rho = 0.95$	app	0.153	0.250	0.227	0.339	0.337	0.505	0.354	0.503

$TW = E(W_q | W_q > 0)$ is given by

$$TW^{app} = \begin{cases} (1 - \frac{1}{2}\rho - \frac{1}{2}\rho^2)\left(\dfrac{1 - c_S^2}{c + 1} + \dfrac{c_S^2}{c}\right)E(S) \\ \quad + \dfrac{(c_A^2 - 1)(1 + \rho) + (3\rho - \rho^3)(1 + c_S^2)}{4(1 - \rho)c}E(S), & \text{if } 0 \leqslant c_A^2, c_S^2 \leqslant 1, \\ (1 + \rho)\left(\dfrac{1 - c_S^2}{c + 1} + \dfrac{c_S^2}{c}\right)E(S) + \dfrac{\rho^2(c_A^2 + c_S^2)}{2(1 - \rho)c}E(S), & \text{otherwise.} \end{cases}$$

$$(4.230)$$

This approximation was obtained in Seelen and Tijms (1984) by using low-traffic and heavy-traffic results for multi-server queues and by making sure that the approximation is in agreement with the Pollaczek–Khintchine formula for the particular case of the M/G/1 queue. To illustrate the performance of the above two-moment approximations, we give in Table 4.27 the exact and approximate values of $E(W_q)$ and TW for several GI/G/c queueing systems. The interarrival-time and service-time distributions considered are the E_k distribution for $k = 2$, 3 and 10 and a mixture of E_1 and E_2 distributions with the same scale parameters. We have used (4.225) up to (4.229) for the calculation of the approximate value of $E(W_q)$.

To conclude, we remark that the approximations discussed in this section provide other examples of an appropriate combination of theoretical results and heuristic reasoning. In this book we have given many examples in which practical useful results were obtained by simple heuristic arguments backed up by sound probabilistic methods. It is our conviction that such a hybrid approach will prove to be successful in many other applied probability problems which are too complex for an exact solution.

EXERCISES

4.1 Consider the following variant of the M/G/1 queueing system. Suppose that after an idle period the server must first warm up before it can start the service of the customer who finds the system empty upon arrival. The warming-up time Z has a general probability distribution with finite first two moments. Modify the recursion scheme in subsection 4.2.1 for the computation of the probability distribution of the number of customers in the system. Show that the probability p_0 of having an empty system is given by $(1 - \rho)/\{1 + \lambda E(Z)\}$ and use generating functions to verify that the average number of customers in the system equals

$$E(L) = \rho + \frac{\lambda^2 E(S^2)}{2(1 - \rho)} + \frac{\lambda E(Z) + \frac{1}{2}\lambda^2 E(Z^2)}{1 + \lambda E(Z)}.$$

4.2 Consider the M/G/1 queueing system controlled by the N-policy discussed in example 1.12 in section 1.5 of chapter 1. Denoting by $p_{0k}(p_{1k})$ the time-average probability that k customers are present and the server is off (on), verify the recursion equations

$$p_{0k} = \frac{1-\rho}{N}, \qquad\qquad 0 \leqslant k \leqslant N-1,$$

$$p_{1k} = \frac{\lambda(1-\rho)}{N} A_{Nk} + \lambda \sum_{j=1}^{k} (p_{0j} + p_{1j}) A_{jk}, \qquad k \geqslant 1,$$

with the convention that $p_{0k} = 0$ for $k \geqslant N$, where A_{jk} is given by (4.30) with $A_{Nk} = 0$ for $k < N$.

4.3 Consider the $M/G/1$ queueing system with server vacations as discussed in example 1.13 in section 1.5 of chapter 1. Assume now that the length V of each server vacation has a general probability distribution function $V(t)$ with density $v(t)$. Denoting by $p_{0k}(p_{1k})$ the time-average probability that k customers are in the system and the server is on vacation (available for service), verify the recursion equations

$$p_{0k} = \frac{1-\rho}{E(V)} \int_0^\infty e^{-\lambda t} \frac{(\lambda t)^k}{k!} \{1 - V(t)\}\, dt, \qquad k \geqslant 0,$$

$$p_{1k} = \frac{1-\rho}{E(V)} \sum_{v=1}^{k} a_v A_{vk} + \lambda \sum_{j=1}^{k} (p_{0j} + p_{1j}) A_{jk}, \qquad k \geqslant 1,$$

where $a_v = \int_0^\infty e^{-\lambda t}(\lambda t)^v/v!\, v(t)\, dt$ and A_{jk} is given by (4.30). (*Hint*: take as the cycle the time elapsed between two consecutive epochs at which either the server becomes idle or finds upon return from vacation that the system is empty.)

4.4 Suppose an $M/G/1$ queueing system in which the service time of a customer depends on the queue size at the moment the customer enters service. The service time of a customer has the probability distribution function $B_1(x)$ when R or less customers are present when the customer enters service; otherwise faster service with the service-time distribution function $B_2(x)$ is provided. Denoting by $p_{1k}(p_{2k})$ the time-average probability that k customers are present and service according to $B_1(B_2)$ is provided, verify the recursion scheme

$$p_{1k} = \lambda p_0 A_{1k}^{(1)} + \lambda \sum_{j=1}^{\min(k,R)} p_{1j} A_{jk}^{(1)}, \qquad 1 \leqslant k \leqslant R,$$

$$p_{2k} = \lambda \sum_{j=R+1}^{k} (p_{1j} + p_{2j}) A_{jk}^{(2)}, \qquad k > R,$$

where

$$A_{jk}^{(i)} = \int_0^\infty e^{-\lambda t} \frac{(\lambda t)^{k-j}}{(k-j)!} \{1 - B_i(t)\}\, dt.$$

4.5 For each of the M/G/1 variants considered in the exercises 4.1–4.4, prove the relationships (4.8) and (4.9) between the queue size L_q and the waiting time W_q when service is in order of arrival. Use the relation (4.9) and the generating function of the queue size to obtain an expression for the Laplace transform $E(e^{-sW_q})$. In particular, for the server vacation queueing model of exercise 4.3, verify from a decomposition of the Laplace transform that the waiting time W_q can be represented as an independent sum of the waiting time in the standard M/G/1 queue without server vacation and the residual life of the vacation time, so that

$$P\{W_q \leqslant x\} = \int_0^x W_0(x-y)\frac{1}{E(V)}\{1-V(y)\}\,\mathrm{d}y, \quad x \geqslant 0,$$

where $W_0(x)$ is the probability distribution function of the waiting time in the standard M/G/1 queue (see also Fuhrmann and Cooper, 1985, for similar decomposition results for several other M/G/1 queueing models with server vacations.)

4.6 Consider the $G^X/G/1$ queue denoting the general single-server queue with batch arrivals, that is each arrival consists not of a single customer but of a random number of customers. The customers are served one at a time and the service times of the customers are independent of each other. Argue heuristically that the time-average probability p_0 of having an empty system is given by

$$1 - p_0 = \lambda E(B)E(S)$$

and that the average queue size $E(L_q)$ and the average delay $E(W_q)$ per customer are related to each other by

$$E(L_q) = \lambda E(B)E(S)E(W_q),$$

where λ, $E(B)$ and $E(S)$ denote the average arrival rate of batches, the average batch size and the average service time of a customer.

4.7 Let us consider a $G^X/M/1$ queue with batch arrivals and exponential services. A batch consists of l customers with probability ϕ_l, $l \geqslant 1$. Let p_j be the time-average probability of having j customers in the system and let π_j be the customer-average probability that an arriving batch finds upon arrival j other customers present. Denoting by λ the average arrival rate of batches and denoting by μ the average exponential service rate, argue heuristically that

$$\lambda \sum_{k=0}^{j-1} \pi_k \sum_{l \geqslant j-k} \phi_l = \mu p_j \quad \text{for } j = 1, 2, \ldots.$$

4.8 Let us consider the $M^X/G/1$ queue where the batches of customers arrive according to a Poisson process with rate λ and the batch size B has the discrete probability distribution $\{\phi_l, l \geqslant 1\}$. Denoting by p_k the time-average probability of having k customers in the system, verify the recursion scheme

$$p_k = \lambda p_0 \sum_{l=1}^{k} \phi_l A_{lk} + \lambda \sum_{j=1}^{k} \left(\sum_{i=0}^{j} p_i \sum_{l \geq j+1-i} \phi_l \right) A_{jk}, \qquad k \geq 1,$$

with $p_0 = 1 - \lambda E(B)E(S)$. Here A_{jk} is again defined as the expected amount of time that k customers are present during a service time which starts when j customers are present. Use generating functions and results for the compound Poisson process in section 1.6 of chapter 1 to verify that the average queue size $E(L_q)$ is given by

$$E(L_q) = \frac{\rho^2}{1-\rho} \frac{(1 + c_S^2)}{2} + \frac{\rho}{2(1-\rho)} \left\{ \frac{E(B^2)}{E(B)} - 1 \right\}$$

with $\rho = \lambda E(B)E(S)$. For the important case in which the service time has the generalized Erlangian distribution function (4.31), verify that the constants A_{jk} (depending on j and k only through $k - j$) can effectively be computed from the recursion scheme

$$A_{jk} = \sum_{i=1}^{r} q_i \alpha_{k-j}^{(i)}$$

with

$$\alpha_0^{(i)} = \frac{1}{\lambda + \mu} + \frac{\mu}{\lambda + \mu} \alpha_0^{(i-1)}, \qquad 1 \leq i \leq r,$$

$$\alpha_n^{(i)} = \frac{\lambda}{\lambda + \mu} \sum_{l=1}^{n} \phi_l \alpha_{n-l}^{(i)} + \frac{\mu}{\lambda + \mu} \alpha_n^{(i-1)}, \qquad n \geq 1, \quad 1 \leq i \leq r,$$

where $\alpha_n^{(0)} = 0$ by convention. Also, for the case of this service time distribution, modify the analysis in subsection 4.2.2 in order to calculate the waiting-time probabilities when service is in order of arrival. Here use the renewal-theoretic result that an arbitrary customer belongs to a batch of size j with probability $j\phi_j / \sum_k k\phi_k$. (*Hint*: use the equation $E(N_j) = \sum_{i=0}^{j} \{\lambda \sum_{l \geq j+1-i} \phi_l\} E(T_i)$ stating that the expected number of downward transitions from the set of states $\{j + 1, j + 2, \ldots\}$ to the set of states $\{0, 1, \ldots, j\}$ during a cycle equals the expected number of upward transitions from the set $\{0, 1, \ldots, j\}$ to the set $\{j + 1, j + 2, \ldots\}$ during a cycle, where state j corresponds to the situation that j customers are present.)

4.9 Users submit requests for service to a computing centre according to a Poisson process with rate λ. The handling time of each request takes a constant time D. Upon service completion of a request the user is satisfied with the answer with probability p; otherwise another request is submitted immediately. The computing centre can handle only one request at a time. Determine the average amount of time a user has to wait before obtaining a satisfying answer.

4.10 Let us consider a M/M/1 queue with p priority classes and the preemptive-resume queue discipline. That is, the service of a customer is interrupted whenever a customer of higher priority arrives. As soon as no customers of higher priority are present any more, the interrupted service is continued from the point at which it was left off. The customers of class j have priority over customers in the classes k with $k > j$, and within each of the classes the customers are served in order of arrival. The customers of class j arrive according to a Poisson process with rate λ_j and have exponentially distributed service times with mean β_j. Denote by $L^{(j)}$ and $W^{(j)}$ the average number of class j customers in the system and the average amount of time spent by a class j customer in the system. Use the memoryless property of the exponential distribution and the property 'Poisson arrivals see time averages' to verify that the quantities $L^{(j)}$ and $W^{(j)}$ can recursively be computed from the equations

$$L^{(j)} = \lambda_j W^{(j)}, \qquad\qquad j = 1, \ldots, p,$$

$$W^{(j)} = \sum_{i=1}^{j} L^{(i)} \beta_i + \beta_j + \sum_{i=1}^{j-1} \lambda_i W^{(j)} \beta_i, \qquad j = 1, \ldots, p.$$

(An extension of the above results to the multi-server preemptive priority queue with exponential services can be found in Buzen and Bondi, 1983.)

4.11 Suppose a single-server queueing system with Poisson arrivals at rate λ, where the service times of the customers become exactly known upon their arrivals and are sampled independently from a probability distribution function $F(x)$ with positive density $f(x)$ and finite first two moments $E(S)$ and $E(S^2)$. It is assumed that $\rho = \lambda E(S)$ is less than 1. The customers requiring a service time smaller than or equal to some positive number v are called class 1 customers, the other customers are called class 2 customers. The class 1 customers have non-preemptive priority over class 2 customers. Verify that the overall average delay in queue of a customer is given by

$$E(W_q) = \frac{\lambda E(S^2)}{2(1-\rho)} \left\{ 1 - \lambda \int_0^v s f(s) \, ds \right\}^{-1} \{1 - \rho F(v)\}.$$

Show that the optimal value of v satisfies the equation $v = E(S)\{1 + \lambda \int_0^v F(s) \, ds\}$. For the case of uniformly distributed service times, make some numerical comparisons with first-come–first-served queue discipline when the server utilization is varied.

4.12 Consider a single-server queueing system at which customers of the types 1 and 2 arrive according to independent Poisson processes with respective rates λ_1 and λ_2. Suppose that for each customer of type i a cost of $c_i > 0$ is incurred for each unit of time the customer has to wait in queue. Verify that the non-preemptive priority queue discipline giving customers of type 1 priority over customers of type 2 has lower average cost per unit time than the other non-preemptive queue discipline only if $c_1/\alpha_1 > c_2/\alpha_2$, where α_i denotes the average service time of a customer of type i. How can we generalize this result to the case of n types of customers? (*Note*: extensions of this problem are discussed in Meilijson and Weiss, 1977.)

4.13 Suppose a single-server queueing system at which customers arrive according to a Poisson process. A customer finding upon arrival j other customers present joins the system with probability $1/(j + 1)$; otherwise he balks from joining the queue. Discuss how for the case of generally distributed service times the state probabilities can be computed.

4.14 Consider a computer system in which a fixed number K of jobs circulate as shown in Figure 4.4. Here station 0 represents the central processing unit (CPU), while the other stations can be thought of as various input/output peripheral devices. In this multi-programming model the CPU and the input/output devices are assumed to be single-server stations which provide exponential services on the basis of the first-come–first-served discipline. The average service time of a job at station i equals $1/\mu_i (i = 0, \ldots, M)$. Upon completion of service at the CPU the job returns to the CPU queue with probability q_0 or requires service at the peripheral device i with probability $q_i (i = 1, \ldots, M)$. The return of a job from the CPU directly back to the CPU may represent the completion of an old job and the insertion of a new job to replace it.

For this model, denote by $\lambda_i(K)$, $EL_i(K)$ and $EW_i(K)$ the average arrival rate of jobs at station i, the average number of jobs present at station i and the average amount of time spent at station i by a job at each visit. Using mean value analysis, verify that the performance measures can recursively be computed from the relations

$$EW_i(K) = \mu_i^{-1}\{1 + EL_i(K - 1)\}, \qquad i = 0, \ldots, M,$$

$$K = \lambda_0(K)\left\{EW_0(K) + \sum_{i=1}^{M} q_i EW_i(K)\right\},$$

$$EL_i(K) = \begin{cases} \lambda_0(K)EW_0(K), & i = 0, \\ q_i \lambda_0(K)EW_i(K), & i = 1, \ldots, M, \end{cases}$$

where $EL_i(0) = 0$ for $i = 0, \ldots, M$ by convention.

4.15 Consider the single-server queueing system with state-dependent Markovian input dealt with in subsection 4.2.2 and give an algorithm for the computation of the waiting-time probabilities when the service time has an H_2 distribution and service is in order of arrival.

4.16 Extend to the $GI/E_r/1/N$ queueing system the algorithm given in subsection 4.2.3 for the calculation of the waiting-time probabilities. Also, relate for this queueing system the time-average probabilities to the customer-average probabilities by using an equation generalizing (4.23).

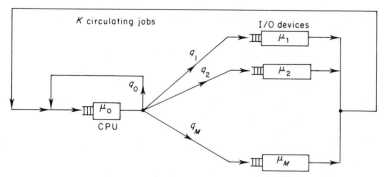

Figure 4.4 Central server model of multi-programming

4.17 Consider a single-server queueing system with batch arrivals occurring according to a renewal process with the interarrival-time density $g(t)$. A batch consists of j customers with probability $\alpha_j, j \geq 1$. The service time of a customer is a finite mixture of Erlangian distributions with the same scale parameters. The queue discipline is first-come–first-served for customers belonging to different batches; within a batch customers are served in random order. Extend the algorithm in subsection 4.2.3 for the computation of the distribution function of the delay in queue of an arbitrary customer. (*Hint*: use the renewal-theoretic result that an arbitrary customer comes from a batch of size j with probability $\beta_j = j\alpha_j/\sum_k k\alpha_k$, not with probability α_j!.)

4.18 For the buffer overflow model in subsection 4.3.2, derive the result that the average amount that overflows per unit time is given by

$$\lambda \left[E(S) - \frac{(\sigma/\lambda)\{\rho - \pi(K)\}}{1 - \pi(K)} \right].$$

4.19 Suppose a single-server queueing system at which customers arrive according to a Poisson process with rate λ. The queue discipline is first-come–first-served. The service time of a customer depends on its actual waiting time in the queue and has a Weibull distribution with mean $\alpha(w)$ and standard deviation $\sigma(w)$ when the customer delay in queue is w. It is assumed that $\alpha(w)$ and $\sigma(w)$ are both piecewise continuous in w. Apply the discretization method discussed in subsection 4.3.2 to derive an algorithm for calculating the distribution function of the waiting time of a customer. For the particular case of an exponentially distributed service time with mean $\alpha(w) = \alpha_0$ for $w = 0$ and $\alpha(w) = \alpha_1$ for $w > 0$, use an up-and downcrossings technique to set up an integro-differential equation for the waiting time density and find the analytical solution of this equation.

4.20 Customers arrive at a single-server station according to a Poisson process with rate λ and the service times of the customers are independent random variables having a common exponential distribution with mean $1/\mu$. Service is provided on the basis of the first-come–first-served discipline. However, none of the customers can stay longer than τ time units in the system with τ a given constant. Both a customer whose service is not begun a time τ after its arrival and a customer whose service is started but not completed a time τ after its arrival become a lost customer. Establish an analogy between this queueing model and the buffer overflow model studied in subsection 4.3.2. Verify that the long-run fraction of customers that is lost and the long-run fraction of time the system is empty are given by

$$\pi_{\text{lost}} = \frac{(1 - \lambda/\mu)e^{-(\mu - \lambda)\tau}}{1 - (\lambda/\mu)e^{-(\mu - \lambda)\tau}} \quad \text{and} \quad {}^{''}P_{\text{empty}} = \frac{1 - \lambda/\mu}{1 - (\lambda/\mu)e^{-(\mu - \lambda)\tau}}.$$

Are the assumptions of exponentially distributed interarrival times and exponentially distributed service times essential for establishing the analogy with the buffer overflow model of subsection 4.3.2? (*Note*: the multi-server extension of the above queueing model is dealt with in Gnedenko and Kovalenko, 1968.)

4.21 Consider a central blood bank at which units of fresh blood arrive according to a Poisson process with rate λ. The demand process for the units of blood is a Poisson process with rate μ. Any demand occurring while the system is out of stock is lost. Each unit of blood has a fixed lifetime of τ time units, by the end of which it becomes outdated and is discarded. The units of blood are issued according to first-in–first-out. A fixed cost of $p > 0$ is incurred for each demand that is lost and there is a fixed cost of $\theta > 0$ for each unit that is discarded for being over-age. Using the lack of memory of the demand process, establish an analogy between this model and the one dealt with in exercise 4.20. Next, supposing that the arrival rate λ can be controlled, verify that the value of λ for which the average costs per unit time are minimal is obtained by minimization of the function

$$g(\lambda) = \frac{\mu - \lambda}{1 - (\lambda/\mu)e^{-(\mu - \lambda)\tau}} \left(\frac{\lambda}{\mu}e^{-(\mu - \lambda)\tau} + \frac{p}{\theta} \right).$$

(This problem is taken from Nahmias, 1980; see also Cosmetatos and Prastacos, 1985, for related work when replenishments occur periodically.)

4.22 Show how to calculate the time-average probabilities p_j and the waiting-time probabilities $P\{W_q > x\}$ for the $M^X/M/c$ queue with batch arrivals (for the waiting-time distribution use the hint given in exercise 4.17).

4.23 Extend the results (4.176) and (4.181) to the $M^X/D/c$ queue with batch arrivals. Use the recursion relation (1.48) in section 1.6 of chapter 1.

4.24 Use analytical results from exercise 4.8 or exercise 4.25 below to investigate how good the corresponding approximations (4.198) and (4.201) are for the $M^X/G/c$ queue (to compute the waiting time percentiles for the particular case of the $M^X/D/c$ queue, use the generalization of the recursion scheme (4.182) in combination wih the conjectured expansion $P\{W_q > x\} \approx Ae^{-Bx}$ for x large enough, where the coefficients A and B could be calculated by applying the generalized recursion scheme for two sufficiently large values of x that are multiples of the service time D).

4.25 Consider the $M^X/E_2/c$ queue. Use the continuous-time Markov chain approach to develop an algorithm for computing the microstate probabilities and the waiting-time probabilities when service is in order of arrival. (*Hint*: obtain the waiting-time probabilities as a weighted sum of conditional waiting-time probabilities, where the conditional waiting-time distributions are calculated as first-passage time distributions in a modified continuous-time Markov chain in which new arrivals are disregarded).

4.26 Let us consider a multi-server queueing system with server breakdowns. A busy server is subject to breakdowns occurring according to a Poisson process with rate $\eta = \frac{1}{4}$; a server cannot break down whenever idle. Each service interruption caused by a breakdown takes a fixed time $\tau = \frac{1}{2}$. If a service is interrupted by a breakdown, then upon completion of the interruption the service is continued from the point at which it was interrupted. Customers arrive according to a Poisson process with rate $\lambda = 1.5$, the service time of a customer has mean $E(S) = 1$, a squared coefficient of variation $c_S^2 = \frac{1}{4}$ and the number of servers is $c = 2$. Calculate an approximation to the average time spent by a customer in the system. (*Answer*: 3.2.)

4.27 Customers with items to repair arrive at a repair facility according to a Poisson process with rate λ. The repair time of an item has a uniform distribution on $[a, b]$. There are ample repair facilities so that each defective item immediately enters repair. If the repair time of an item takes longer than τ time units with τ a given number between a and b, the customer gets after a time τ a loaner for his defective item until the item returns from repair. A sufficiently large supply of loaners is available. Identify an appropriate queueing model in order to show that the average number of loaners which are out equals $\frac{1}{2}\lambda(b - \tau)^2/(b - a)$. (This problem and the next one are based on Karmarkar and Kubat, 1983.)

4.28 Suppose that a repair facility has a supply of N serviceable units that can be temporarily loaned to customers bringing items for repair. The units for repair arrive according to a Poisson process with rate λ. The repair time of each unit is exponentially distributed with mean $1/\mu$. There are ample repair facilities so that each defective item immediately enters repair. A customer bringing an item for repair gets a loaner, when available; otherwise the customer waits until either a loaner becomes free or the repair of his item is completed, whichever occurs first. The loaners are provided to waiting customers on a first-come–first-served basis; a loaner is returned by a customer upon completion of repair of his item.

Determine an explicit expression for the average amount of time a customer has to wait until a loaner is provided or the repair of his item is completed. (*Hint*: denoting by $E(W_j)$ the conditional average waiting time of a customer finding upon arrival $N+j$ units in repair, verify the recursion equation

$$E(W_j) = \frac{1}{(N+j)\mu+\mu} + \frac{(N+j)\mu}{(N+j)\mu+\mu} E(W_{j-1}),$$

showing that $E(W_j) = (j+1)/\{\mu(N+j+1)\}$ for $j \geqslant 0$.)

4.29 Prove the result (4.204) for the $M^X/G/\infty$ queue with batch arrivals.

4.30 Suppose a large stackyard at which batches of containers arrive according to a Possion process with rate $\lambda = 1$ and the batch size has mean $\alpha = 5$ and standard deviation $\beta = 1$. The holding times of the various containers at the stackyard are independent of each other and have a mean of $m = 10$. Using the result (4.204) and fitting a gamma distribution to the moments in (4.204), calculate approximately the required capacity of the yard such that the overflow probability for an arriving batch is less than $v = 0.05$ when the time that a container is stored on the stackyard has

(a) a uniform distribution on $(0, 2m)$ (*answer*: 75),
(b) an exponential distribution (*answer*: 73),
(c) an H_2 distribution with the normalization of balanced means and having a squared coefficient of variation of 4 (*answer*: 70)

(This problem is taken from Van Hee, 1984. The 'counterintuitive' finding that the required yard capacity decreases when the coefficient of variation of the holding time increases is in agreement with theoretical results for the infinite-server queueing model with batch arrivals discussed in Wolff, 1977.)

4.31 At a telephone enquiry bureau with several telephone operators calls arrive at peak periods according to a Poisson process. The time to handle a call is approximately exponentially distributed with a mean of $1/\mu = 2$ minutes. A maximum of 10 calls can be held when all operators are busy answering other calls. Write a computer program to determine a planning graph showing for different call rates values of the required number of operators to give 1 call lost in 10 and a delay in queue not more than 30 seconds in 95 per cent of the accepted calls.

4.32 Let us consider the GI/M/c queueing system with the queue discipline of first-come–first-served. For this queueing system the waiting time W_q has an exponential distribution (see section 4.1). For the particular cases of the D/M/c queue and the M/M/c queue with the same average service times α, make some numerical comparisons with respect to the minimum number of servers required in order to achieve $P\{W_q > \alpha \,|\, W_q > 0\} \leqslant 0.05$ when the arrival rate λ is varied.

BIBLIOGRAPHIC NOTES

The queueing theory literature is voluminous. A good account of the basic theory is provided by the introductory texts of Cooper (1981), Allen (1978), Cox and Smith (1961) and Gross and Harris (1985). The books Hayes (1984), Kleinrock (1976), Kobayashi (1978) and Sauer and Chandy (1981) also discuss queueing networks and their applications to computer science. A book emphasizing the (approximate) analysis of the time-dependent behaviour of queues is Newell (1971). Tablebooks on queueing models are Hillier and Yu (1981), Kühn (1976) and Seelen, Tijms and Van Hoorn (1985).

A thorough treatment of most of the background material in section 4.1 can be found in the books by Franken *et al.* (1983) and Heyman and Sobel (1982). In the subsections 4.2.1 and 4.2.2 the regenerative approach of the recursive comput-ation of the state probabilities in single-server queueing models with (state-dependent) Markovian input comes from Hordijk and Tijms (1976) and Tijms and Van Hoorn (1981). Other applications of this fertile approach can be found in Federgruen and Green (1984), Federgruen and Tijms (1980), Schellhaas (1985) and Van Hoorn (1981). In subsection 4.2.3 the embedded Markov chain algorithm for the waiting-time probabilities in the $GI/G/1$ queue is based on Bux(1979); see also Ramaswami and Lucantoni (1985) for an alternative algorithm. In section 4.3 the approximate analysis of single-server queueing models with limited access of arrivals is based on De Kok and Tijms (1985a, 1985b). The treatment of approximate methods for multi-server queues in section 4.4 is largely taken from the papers of Tijms, Van Hoorn and Federgruen (1981), Tijms and Van Hoorn (1982) and Van Hoorn and Seelen (1986).

REFERENCES

Allen, A. O. (1978). *Probability, Statistics and Queueing Theory with Computer Science Applications,* Academic Press, New York.

Boxma, O. J. (1985). 'Response times in cyclic queues—the influence of the slowest server', Reprint No. 380, University of Utrecht, Utrecht.

Boxma, O. J., Cohen, J. W., and Huffels, N. (1979). 'Approximations of the mean waiting time in an $M/G/s$ queueing system', *Operat. Res., 27,* 1115–1127.

Brill, P. H., and Posner, M. J. M. (1977). 'Level crossings in point processes applied to queues: single-server case', *Operat. Res., 25,* 662–674.

Burman, D. Y., and Smith, D. R. (1983). 'A light-traffic theorem for multi-server queues', *Math. Operat. Res., 8,* 15–25.

Bux, W. (1979). 'Single server queues with general interarrival and phase type service time distributions', in *Proceedings 9th International Teletraffic Congress,* Torremolinos, paper 413.

Buzen, J. P., and Bondi, A. B. (1983). 'The response time of priority classes under preemptive resume in $M/M/m$ queues', *Operat. Res., 31,* 456–465.

Cohen, J. W. (1976). *On Regenerative Processes in Queueing Theory,* Lecture Notes in Mathematical Economics and Mathematical Systems, Vol. 121, Springer–Verlag, Berlin.

Cohen, J. W. (1977). 'On up- and downcrossings', *J. Appl. Prob.*, **14**, 405–410.
Cohen, J. W. (1979). 'The multiple phase service network with generalized processor sharing', *Acta Informatica*, **12**, 245–289.
Cohen, J. W. (1982). *The Single-Server Queue*, 2nd ed., North-Holland, Amsterdam.
Cooper, R. B. (1981). *Introduction to Queueing Theory*, Edward Arnold, London.
Cosmetatos, G. P. (1975). 'Approximate explicit formulae for the average queueing time in the processes (M/D/r) and (D/M/r)', *INFOR*, **13**, 328–331.
Cosmetatos, G. P. (1976). 'Some approximate equilibrium results for the multiserver queue M/G/r', *Operat. Res. Quart.*, **27**, 615–620.
Cosmetatos, G. P., and Prastacos, G. P. (1985). 'Approximate analysis of the D/M/1 queue with customer impatience', *RAIRO*, **19**, 133–142.
Cox, D. R. (1955). 'The statistical analysis of congestion', *J. Roy. Statist. Soc. A.*, **118**, 324–335.
Cox, D. R., and Smith, W. L. (1961). *Queues*, Chapman and Hall, London.
Crommelin, C. D. (1932). 'Delay probability formulae when the holding times are constant', *P.O. Elect. Engr. J.*, **25**, 41–50.
De Kok, A. G., and Tijms, H. C. (1985a). 'A queueing system with impatient customers', *J. Appl. Prob.*, **22**, 388–396.
De Kok, A. G., and Tijms, H. C. (1985b). 'A two-moment approximation for a buffer design problem requiring a small rejection probability', *Perform. Evaluation*, **7**, 77–84.
De Kok, A. G., and Van Ommeren, J. C. W. (1986). 'Asymptotic results for buffer systems under heavy load', Report 157, Vrije Universiteit, Amsterdam.
De Smit, J. H. A. (1983). 'A numerical solution for the multi-server queue with hyperexponential service times', *Operat. Res. Letters*, **2**, 217–224.
De Soua e Silva, E., Lavenberg, S. S., and Muntz, R. R. (1984). 'A perspective on iterative methods for the approximate analysis of closed queueing networks', in *Mathematical Computer Performance and Reliability* (Eds. G. Iazeolla, P. J. Courtois and A. Hordijk), pp. 225–244, North-Holland, Amsterdam.
Federgruen, A., and Green, L. (1984). 'An M/G/c queue in which the number of servers required is random', *J. Appl. Prob.*, **21**, 583–601.
Federgruen, A., and Tijms, H. C. (1980). 'Computation of the stationary distribution of the queue size in an M/G/1 queueing system with variable service rate', *J. Appl. Prob.*, **17**, 515–522.
Franken, P., König, D., Arndt, U. and Schmidt, V. (1983). *Queues and Point Processes*, Wiley, New York.
Fredericks, A. A. (1982). 'A class of approximations for the waiting time distribution in a GI/G/1 Queueing system', *Bell Syst. Techn. J.*, **61**, 295–325.
Fuhrmann, S. W., and Cooper, R. B. (1985). 'Stochastic decompositions in the M/G/1 queue with generalized vacations', *Operat. Res.*, **23**, 1117–1129.
Gavish, B., and Schweitzer, P. J. (1977). 'The Markovian queue with bounded waiting time', *Management Sci.*, **23**, 1349–1357.
Gnedenko, B. V., and Kovalenko, I. N. (1968). *Introduction to Queueing Theory*, Israel Program for Scientific Translations, Jerusalem.
Gross, D., and Harris, C. M. (1985). *Fundamentals of Queueing Theory*, 2nd. ed., Wiley, New York.
Haji, R. and Newell, G. F. (1971). 'A relation between stationary queue and waiting time distribution', *J. Appl. Prob.*, **8**, 617–620.
Hayes, J. F. (1984). *Modelling and Analysis of Computer Communications Networks*, Plenum Press, New York.
Heyman, D. P. (1980). 'Comments on a queueing inequality', *Management Sci.*, **26**, 956–959.

Heyman, D. P. and Sobel, M. J. (1982). *Stochastic Models in Operations Research*, Vol. I, McGraw-Hill, New York.

Hillier, F. S. and Yu, O. S. (1981). *Queueing Tables and Graphs*, North-Holland, Amsterdam.

Hokstad, P. (1979). 'A single-server queue with constant service time and restricted accessibility', *Management Sci.*, **25**, 205–208.

Hokstad, P. (1986). 'Bounds for the mean queue length of the $M/K_2/m$ queue', *European J. Operat. Res.*, **23**, 108–117.

Hordijk, A., and Tijms, H. C. (1976). 'A simple proof of the equivalence of the limiting distributions of the continuous time and the embedded process of the queue size in the M/G/1 queue', *Statistica Neerlandica*, **30**, 97–100.

Ishikawa, A. (1979). 'On the equilibrium distribution for the queuing system $GI/E_k/m$', *TRU Math.*, **15**, 47–66.

Ishikawa, A. (1984). 'Stationary waiting time distribution in a $GI/E_k/m$ queue', *J. Operat. Res. Soc. Japan*, **27**, 130–149.

Iversen, V. B. (1983). 'Decomposition of an $M/D/r.k$ queue with FIFO into k $E_k/D/r$ queues with FIFO', *Operat. Res. Letters*, **2**, 20–21.

Jagerman, D. L. (1985). 'Calculation of Laplace transforms', Research Report, Bell Laboratories, Holmdel, New Jersey.

Karmarkar, U. S., and Kubat, P. (1983). 'The value of loaners in product support', *IIE Trans.*, **15**, 5–11).

Kella, O., and Yechiali, U. (1985). 'Waiting times in the nonpreemptive-priority M/M/s queue', *Stochastic Models*, **1**, 257–262.

Kimura, T. (1985). 'Heuristic approximations for the mean waiting time in the GI/G/s queue', Report No. B.55, Tokyo Institute of Technology, Tokyo.

Kleinrock, L. (1976). *Queueing Systems*, Vol. 2: *Computer Applications*, Wiley, New York.

Klincewicz, J. C., and Whitt, W. (1984). 'On approximations for queues, II: shape constraints', *AT & T Bell Lab. Techn. J.*, **63**, 139–161.

Kobayashi, H. (1978). *Modelling and Analysis, An Introduction to System Performance Evaluation Methodology*, Addison-Wesley, Reading, Massachusetts.

Köllerström, J. (1974). 'Heavy traffic theory for queues with several servers, I', *J. Appl. Prob.*, **11**, 544–552.

Krämer, W., and Langenbach-Belz, M. (1976). 'Approximate formulae for the delay in the queueing system GI/G/1', in *Proceedings 8th International Teletraffic Congress*, Melbourne, pp. 235–1/8.

Kühn, P. J. (1972). 'On the calculation of waiting times in switching and computer systems', 15th Report on Studies in Congestion Theory, Institute of Switching and Data Technics, University of Stuttgart, Stuttgart.

Kühn, P. J. (1976). *Tables on Delay Systems*, Institute of Switching and Data Technics, University of Stuttgart, Stuttgart.

Martin, J. (1972). *System Analysis for Data Transmission*, Prentice-Hall, Englewood Cliffs.

Meilijson, I., and Weiss, G. (1977). 'Multiple feedback at a single-server station', *Stoch. Proc. Appl.*, **5**, 195–205.

Miyazawa, M. (1986). 'Approximations of the queue length distribution of an M/GI/s queue by the basic equations', *J. Appl. Prob.* (to appear).

Muckstadt, J. A. (1978). 'Some approximations in multi-level, multi-echelon inventory systems for recoverable items', *Naval Res. Logist. Quart.*, **25**, 377–394.

Nahmias, S. (1980). 'Queueing models for controlling perishable inventories', in *Proceedings First International Symposium on Inventories*, Publishing House of the Hungarian Academy of Sciences, Budapest, pp. 449–457.

Neuts, M. F. (1981). *Matrix-Geometric Solutions in Stochastic Models—an Algorithmic Approach*, The John Hopkins University Press, Baltimore.

Neuts, M. F. (1982). 'Explicit steady-state solutions to some elementary queueing models', *Operat. Res., 30*, 480–489.

Neuts, M. F. (1985). 'The caudal characteristic curve of queues', *Adv. Appl. Prob., 17* (to appear).

Newell, G. F. (1971). *Applications of Queueing Theory*, Chapman and Hall, London.

Nozaki, S. A. and Ross, S. M. (1978). 'Approximations in finite capacity multi-server queues with Poisson arrivals', *J. Appl. Prob., 15*, 826–834.

Ott, T. J. (1984). 'The sojourn time distribution in the M/G/1 queue with processor sharing', *J. Appl. Prob., 21*, 360–378.

Page, E. (1972). *Queueing Theory in O.R.*, Butterworths, London.

Parikh, S. C. (1977). 'On a fleet sizing and allocation problem', *Management Sci., 23*, 972–977.

Ramaswami, V., and Lucantoni, D. M. (1985). 'Stationary waiting time distributions in queues with phase-type service and in quasi-birth-and death processes, *Stochastic Models, 1*, 125–136.

Reiser, M., and Lavenberg, S. S. (1980). 'Mean value analysis of closed multichain queueing networks', *J. ACM, 27*, 313–322.

Ross, S. M. (1983). *Stochastic Processes*, Wiley, New York.

Sauer, C. H., and Chandy, K. M. (1981). *Computer Systems Performance Modelling*, Prentice-Hall, Englewood Cliffs, New Jersey.

Schassberger, R. (1973). *Warteschlangen*, Springer-Verlag, Berlin.

Schellhaas, H. (1985). 'Computation of the state probabilities in a class of semi-regenerative queueing models', in *Proceedings Volume on Semi-Markov Processes and their Applications* (Ed. J. Janssen), Plenum, New York.

Schweitzer, P. J., and Kindle, K. W. (1985). 'An iterative aggregation–disaggregation algorithm for solving linear equations', *Appl. Math. Comput.* (to appear).

Seelen, L. P. (1986). 'An algorithm for Ph/Ph/c queues', *European J. Operat. Res., 23*, 118–127.

Seelen, L. P. and Tijms, H. C. (1984). 'Approximations for the conditional waiting times in the GI/G/c queue', *Operat. Res. Letters, 3*, 183–190.

Seelen, L. P. and Tijms, H. C. (1985). 'Approximations to the waiting time percentiles in the M/G/c queue', *Proceedings 11th International Teletraffic Congress*, Kyoto, pp. 1.4.4.1–1.4.4.5.

Seelen, L. P., Tijms, H. C. and Van Hoorn, M. H. (1985). *Tables for Multi-Server Queues*, North-Holland, Amsterdam.

Sevcik, K. C. and Mitrani, I. (1981). 'The distribution of queueing network states at input and output instants', *J. ACM, 28*, 358–371.

Sherbrooke, C. C. (1968). 'METRIC: a multi-echelon technique for recoverable item control', *Operat. Res., 16*, 122–141.

Silver, E. A. and Smith, S. A. (1977). 'A graphical aid for determining optimal inventories in a unit inventory replenishment system', *Management Sci., 24*, 358–359.

Sobel, M. J. (1980). 'Simple inequalities for multiserver queues', *Management Sci., 26*, 951–956.

Stidham, S., Jr. (1974). 'A last word on $L = \lambda W$', *Operat. Res., 22*, 417–421.

Stoyan, D. (1983). *Comparison Methods for Queues and Other Stochastic Models*, Wiley, New York.

Takács, L. (1962). *Introduction to the Theory of Queues*, Oxford University Press, New York.

Takahashi, Y. (1981). 'Asymptotic exponentiality of the tail of the waiting time distribution in a Ph/Ph/c queue', *Adv. Appl. Prob., 13*, 619–630.

Takahashi, Y. and Takami, Y. (1976). 'A numerical method for the steady state probabilities of a GI/G/c queueing system in a general class, *J. Operat. Res. Soc. Japan, 19*, 147–157.

Tijms, H. C. and Van Hoorn, M. H. (1981). 'Algorithms for the state probabilities and waiting times in single-server queueing systems with random and quasi-random input and phase-type service times', *OR Spektrum*, **2**, 145–152.

Tijms, H. C. and Van Hoorn, M. H. (1982). 'Computational methods for single server and multi-server queues with Markovian input and general service times', in *Applied Probability-Computer Science, The Interface* (Eds. R. L. Disney and T. J. Ott), Vol. II, pp. 71–102, Birkhauser, Boston.

Tijms, H. C., Van Hoorn, M. H., and Federgruen, A. (1981). 'Approximations for the steady-state probabilities in the M/G/c queue', *Adv. Appl. Prob.*, **13**, 186–206.

Van Hee, K. M. (1984). 'Models underlying decision support systems for port terminal planning', *Wissenschaftliche Zeitschrift Technische Hochschule Leipzig*, **8**, 161–170.

Van Hoorn, M. H. (1981). 'Algorithms for the state probabilities in a general class of single server queues with group arrivals', *Management Sci.*, **27**, 1178–1187.

Van Hoorn, M. H. (1984). *Algorithms and Approximations for Queueing Systems*, CWI Tract No. 8., CWI, Amsterdam.

Van Hoorn, M. H. (1985). 'Numerical analysis of multi-server queues with deterministic service and special phase-type arrivals', *Zeitschrift für Operat. Res. A*, **29** (to appear).

Van Hoorn, M. H. and Seelen, L. P. (1983). 'The SPP/G/1 queue: a single server queue with a switched Poisson process as input process', *OR Spektrum*, **5**, 207–218.

Van Hoorn, M. H. and Seelen, L. P. (1986). 'Approximations for the GI/G/c queue', *J.Appl. Prob.*, **23** (to appear).

Whitt, W. (1984a). 'Minimizing delays in the GI/G/1 queue', *Operat. Res.*, **32**, 41–51.

Whitt, W. (1984b). 'Heavy-traffic approximations for service systems with blocking', *AT & T Bell Lab. Techn. J.*, **63**, 689–708.

Wolff, R. W. (1977). 'The effect of service time regularity on system performance', in *Computer Performance* (Eds. K. M. Chandy and M. Reiser), pp. 297–304, North-Holland, Amsterdam.

Appendixes

APPENDIX A. THE COMPUTATION OF PROBABILITIES AND EXPECTATIONS BY CONDITIONING

In many applied probability problems it is only possible to compute certain probabilities and expectations by using appropriate conditioning arguments. This appendix summarizes a number of useful results involving conditional probabilities and conditional expectations.

First we state the law of total probability for the computation of some event A. Suppose that $B_1, B_2, \ldots, B_n$ are mutually exclusive events such that the event A can only occur if one of the events B_k occurs. The *law of total probability* states that

$$P(A) = \sum_{k=1}^{n} P(A|B_k)P(B_k). \tag{A.1}$$

More generally, the probability $P(A)$ can be computed as the weighted average of conditional probabilities having as weights the probabilities of the possible outcomes of some random variable Y. Thus for a random variable Y having a discrete probability distribution

$$P(A) = \sum_{y} P(A|Y = y)P(Y = y), \tag{A.2}$$

and for a random variable Y having a continuous distribution with density $f(y)$

$$P(A) = \int_{-\infty}^{\infty} P(A|Y = y)f(y)\,dy. \tag{A.3}$$

The relation (A.2) is a special case of (A.1), while the relation (A.3) can be considered as a 'continuous' version of (A.2) by noting that the probability density $f(y)$ of a random variable Y has the interpretation $f(y)\Delta y \approx P(y < Y \leqslant y + \Delta y)$ for Δy small.

Similarly, the computation of an expectation can often be considerably facilitated by conditioning on the outcomes of some appropriate random variable. The *law of total expectation* states that, for any two random variables

391

X and Y,

$$E(X) = \sum_{y} E(X \mid Y = y)P(Y = y) \tag{A.4}$$

when Y has a discrete distribution and

$$E(X) = \int_{-\infty}^{\infty} E(X \mid Y = y)f(y)\mathrm{d}y \tag{A.5}$$

when Y has a continuous distribution with probability density $f(y)$. Here it is assumed that the relevant expectations exist.

As an application of the above results we give some useful relations for computing the expectation and variance of the sum of a random number of random variables. Suppose that $X_1, X_2, \ldots$ is a sequence of independent and identically distributed random variables whose first two moments are finite. Also, let N be a non-negative and integer-valued random variable having finite first-two moments. If the random variable N is independent of the random variables $X_1, X_2, \ldots$, then

$$E\left(\sum_{k=1}^{N} X_k \right) = E(N)E(X_1), \tag{A.6}$$

$$\mathrm{var}\left(\sum_{k=1}^{N} X_k \right) = E(N)\,\mathrm{var}(X_1) + \mathrm{var}(N)E^2(X_1), \tag{A.7}$$

where $E^2(X_1)$ denotes the squared expectation of X_1. It is noteworthy that the relation (A.6) is also true under the weakened assumption that for each n the event $\{N = n\}$ is independent of $X_{n+1}, X_{n+2}, \ldots$. This result is known as *Wald's equation*.

To illustrate the above concepts consider the following reliability problem. A computer system has a built-in redundancy in the form of a standby unit to support some operating unit. When the operating unit fails, its tasks are taken over immediately by the standby unit if available. The failed unit immediately enters repair. The system goes down when the operating unit fails while the other unit is still in repair. The lifetime of an operating unit has a probability density $g(x)$ with finite mean μ_L. The repair time of a failed unit has the probability distribution function $H(x)$. The lifetimes and repair times are mutually independent. Supposing that at time 0 an operating unit is installed and the standby unit is in good condition, we wish to find the expected time until the system goes down for the first time. To do so denote by L the lifetime of the operating unit installed at time 0 and denote by $L_1, L_2, \ldots$ the lifetimes of the subsequent operating units. Also, let $R_1, R_2, \ldots$ be the repair times associated with successive failures. Then the time until the first system breakdown is distributed as $L + L_1 + \cdots + L_N$ where N denotes the first n for which $R_n > L_n$. The random variables $L_1, \ldots, L_N$ and N are mutually dependent, but the event $\{N = n\}$ is independent of $L_{n+1}, L_{n+2}, \ldots$ for each n. Thus, by using Wald's equation,

$$E(\text{time until first system breakdown}) = E(L) + E(L_1)E(N)$$
$$= \{1 + E(N)\}\mu_L.$$

To calculate $E(N)$, let $q = P\{R_1 > L_1\}$ and note that N has the geometric distribution $P\{N = n\} = (1 - q)^{n-1}q$, $n = 1, 2, \ldots$. Hence

$$E(N) = \frac{1}{q}$$

The probability q follows by using the law of total probability,

$$q = P\{R_1 > L_1\} = \int_0^\infty P\{R_1 > L_1 | L_1 = x\}g(x)\,dx$$

$$= \int_0^\infty \{1 - H(x)\}g(x)\,dx.$$

APPENDIX B. USEFUL PROBABILITY DISTRIBUTION FUNCTIONS

This appendix discusses a number of important distributions which have been found useful for describing random variables in inventory, reliability and queueing applications. These distributions include the gamma, lognormal and Weibull distributions.

We first introduce some general concepts. Let X be a positive random variable with the probability distribution function $F(t)$ having finite mean $E(X)$ and finite standard deviation $\sigma(X)$. The random variable X may represent the lifetime of some item or the time to complete some task. The *coefficient of variation* of the positive random variable X is defined by

$$c_X = \frac{\sigma(X)}{E(X)}.$$

In applications one often works with the squared coefficient of variation c_X^2 rather than with c_X. The (squared) coefficient of variation is a measure of the variability of the random variable X. For example, the deterministic distribution has $c_X^2 = 0$, the exponential distribution has $c_X^2 = 1$ and the Erlang-k distribution has the intermediate value $c_X^2 = 1/k$.

Assuming that the probability distribution function $F(t)$ of the random variable X has a density $f(t)$, the *failure* (or *hazard*) *rate function* $r(t)$ of X is defined by

$$r(t) = \frac{f(t)}{1 - F(t)}$$

for those values of t for which $F(t) < 1$. The failure rate has a useful probabilistic interpretation. Thinking of the random variable X as the lifetime of some item, it follows from the relation

$$P\{X \in (t, t + \Delta t] \,|\, X > t\} \approx \frac{f(t)\Delta t}{1 - F(t)} \qquad \text{for small} \quad \Delta t$$

that $r(t)\Delta t$ gives approximately the probability that an item of age t will fail in the next time Δt. The failure rate function uniquely determines the corresponding probability distribution function by

$$1 - F(t) = \exp\left\{ - \int_0^t r(x)\,\mathrm{d}x \right\}, \qquad t \geqslant 0.$$

A random variable with increasing (decreasing) failure rate has the property that its coefficient of variation is smaller (larger) than or equal to 1. The failure rate is a concept that enables us to discriminate between distribution functions on physical considerations.

We now discuss the gamma, lognormal and Weibull distributions for a positive random variable X and list without proof relevant properties of these distributions.

The gamma distribution

The density $f(t)$ is given by

$$f(t) = \frac{\lambda^\alpha t^{\alpha - 1}}{\Gamma(\alpha)} \mathrm{e}^{-\lambda t}, \qquad t \geqslant 0, \tag{A.8}$$

where the shape parameter α and the scale parameter λ are both positive. Here $\Gamma(\alpha)$ is the complete gamma function defined by

$$\Gamma(\alpha) = \int_0^\infty \mathrm{e}^{-t} t^{\alpha - 1}\,\mathrm{d}t, \qquad \alpha > 0,$$

and having the property $\Gamma(\alpha + 1) = \alpha\Gamma(\alpha)$ for any $\alpha > 0$. The probability distribution function $F(t)$ may be written as

$$F(t) = \frac{1}{\Gamma(\alpha)} \int_0^{\lambda t} \mathrm{e}^{-u} u^{\alpha - 1}\,\mathrm{d}u, \qquad t \geqslant 0.$$

The latter integral is known as the incomplete gamma function. It is important to note that efficient numerical procedures are available for the computation of the gamma functions (cf. Abramowitz and Stegun, 1965, and Gautschi, 1979).

If the shape parameter α is a positive integer k, the gamma distribution is the well-known Erlang-k (E_k) distribution for which

$$f(t) = \lambda^k \frac{t^{k-1}}{(k-1)!} e^{-\lambda t} \quad \text{and} \quad F(t) = 1 - \sum_{j=0}^{k-1} e^{-\lambda t} \frac{(\lambda t)^j}{j!}, \quad t \geq 0.$$

The Erlang-k distribution has a very useful interpretation. A random variable with an Erlang-k distribution can be represented as the sum of k independent random variables having a common exponential distribution.

The mean and the squared coefficient of variation of the gamma distribution are given by

$$E(X) = \frac{\alpha}{\lambda} \quad \text{and} \quad c_X^2 = \frac{1}{\alpha}.$$

Since $E(X)$ and c_X^2 can assume arbitrarily positive values for the gamma density, a unique gamma distribution can be fitted to each positive random variable with given first two moments. To characterize the shape and the failure rate of the gamma density, we distinguish between the cases $c_X^2 < 1$ ($\alpha > 1$) and $c_X^2 \geq 1$ ($\alpha \leq 1$). The gamma density is always *unimodal*, that is the density has only one maximum. For the case of $c_X^2 < 1$ the density first increases to the maximum at $t = (\alpha - 1)/\lambda > 0$ and next decreases to zero as $t \to \infty$, whereas for the case of $c_X^2 \geq 1$ the density has its maximum at $t = 0$ and thus decreases from $t = 0$ on. The failure rate function is increasing from zero to λ if $c_X^2 < 1$ and is decreasing from infinity to zero if $c_X^2 > 1$. The exponential distribution ($c_X^2 = 1$) has a constant failure rate λ and is a natural boundary between the cases $c_X^2 < 1$ and $c_X^2 > 1$.

The lognormal distribution

The density $f(t)$ is given by

$$f(t) = \frac{1}{\alpha t \sqrt{2\pi}} e^{-\{\ln(t) - \lambda\}^2 / 2\alpha^2}, \quad t > 0, \quad \text{(A.9)}$$

where the shape parameter α is positive and the scale parameter λ may assume each real value. The probability distribution function $F(t)$ equals

$$F(t) = \Phi\left(\frac{\ln(t) - \lambda}{\alpha}\right), \quad t > 0,$$

where $\Phi(x) = (1/\sqrt{2\pi}) \int_{-\infty}^{x} e^{-u^2/2} du$ is the standard normal probability distribution function. The mean and the squared coefficient of variation of the lognormal distribution are given by

$$E(X) = e^{\lambda + \alpha^2/2} \quad \text{and} \quad c_X^2 = e^{\alpha^2} - 1.$$

Thus a unique lognormal distribution can be fitted to each positive random variable with given first two moments. The lognormal density is always unimodal with a maximum at $t = e^{\lambda - \alpha^2} > 0$. The failure rate function first increases and next decreases to zero as $t \to \infty$ and thus the failure rate is only decreasing in the long-life range.

The Weibull distribution

The density $f(t)$ is given by

$$f(t) = \alpha \lambda (\lambda t)^{\alpha - 1} e^{-(\lambda t)^{\alpha}}, \qquad t > 0, \qquad (A.10)$$

with the shape parameter $\alpha > 0$ and scale parameter $\lambda > 0$. The probability distribution function $F(t)$ equals

$$F(t) = 1 - e^{-(\lambda t)^{\alpha}}, \qquad t \geq 0.$$

The mean and the squared coefficient of variation of the Weibull density are given by

$$E(X) = \frac{1}{\lambda} \Gamma\left(1 + \frac{1}{\alpha}\right) \qquad \text{and} \qquad c_X^2 = \frac{\Gamma(1 + 2/\alpha)}{\{\Gamma(1 + 1/\alpha)\}^2} - 1.$$

A unique Weibull distribution can be fitted to each positive random variable with given first two moments. For that purpose a non-linear equation in α must be solved numerically. The Weibull density is always unimodal with a maximum at $t = \lambda^{-1}(1 - 1/\alpha)^{1/\alpha}$ if $c_X^2 < 1$ ($\alpha > 1$) and at $t = 0$ if $c_X^2 \geq 1$ ($\alpha \leq 1$). The failure rate function is increasing from 0 to infinity if $c_X^2 < 1$ and is decreasing from infinity to zero if $c_X^2 > 1$.

The gamma and Weibull densities are similar in shape, and for $c_X^2 < 1$ the lognormal density takes on shapes similar to the gamma and Weibull densities. For the case of $c_X^2 \geq 1$ the gamma and Weibull densities have their maximum value at $t = 0$ so that most outcomes will be small and very large outcomes occur only occasionally. The lognormal density tends to zero as $t \to 0$ faster than any power of t, and thus the lognormal distribution will typically produce fewer small outcomes than the other two distributions. This latter fact explains the popular use of the lognormal distribution in actuarial studies. The differences between the gamma, Weibull and lognormal densities become most significant in their tail behaviour. The densities for large t go down like $e^{-\lambda t}$, $e^{-(\lambda t)^{\alpha}}$ and $e^{-(\ln t)^2/2\alpha^2}$. Thus for a given mean and coefficient of variation the lognormal density has always the longest tail. The gamma density has the second longest tail only if $\alpha > 1$, that is only if its coefficient of variation is less than one. In Figure A.1 we illustrate these facts by drawing the gamma, Weibull and lognormal densities for $c_X^2 = 0.25$, where $E(X) = 1$ is taken.

To conclude this appendix, we discuss several useful extensions of Erlangian (exponential) distributions. In many queueing and inventory

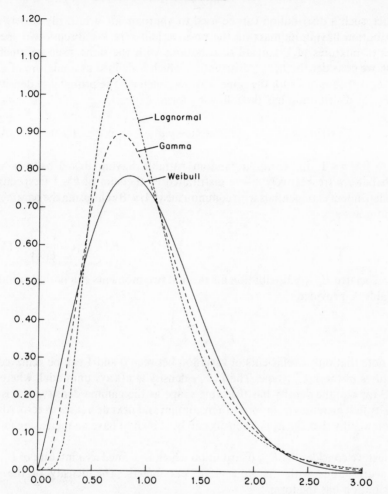

Figure A.1 The gamma, lognormal and Weibull densities with $E(X) = 1$ and $c_X^2 = 0.25$

applications there is a very substantial (numerical) advantage in using these distributions rather than other distributions.

Generalized Erlangian distributions

An Erlang-k (E_k) distributed random variable can be represented as the sum of k independent exponentially distributed random variables with the same means. A generalized Erlangian distribution is one built out of a sum or a mixture of exponentially distributed components, or a combination of both. A particularly convenient distribution arises when these components have the same means.

In fact, such a distribution can be used to approximate arbitrarily closely any distribution having its mass on the positive half-axis. We discuss two special cases of mixtures of Erlangian distributions with the same scale parameters. First, we consider the $E_{k-1,k}$ distribution which is defined as a mixture of E_{k-1} and E_k distributions with the same scale parameters. The probability density of an $E_{k-1,k}$ distribution has the following form:

$$f(t) = p\mu^{k-1}\frac{t^{k-2}}{(k-2)!}e^{-\mu t} + (1-p)\mu^k\frac{t^{k-1}}{(k-1)!}e^{-\mu t}, \quad t \geq 0, \qquad (A.11)$$

where $0 \leq p \leq 1$. In words, a random variable having this density is with probability p (respectively $1-p$) distributed as the sum of $k-1$ (respectively k) independent exponentials with common mean $1/\mu$. By choosing the parameters p and μ as

$$p = \frac{1}{1+c_X^2}[kc_X^2 - \{k(1+c_X^2) - k^2c_X^2\}^{1/2}] \quad \text{and} \quad \mu = \frac{k-p}{E(X)}, \quad (A.12)$$

the associated $E_{k-1,k}$ distribution fits the first two moments of a positive random variable X provided

$$\frac{1}{k} \leq c_X^2 \leq \frac{1}{k-1}.$$

We note that only coefficients of variation between 0 and 1 can be achieved by mixtures of the $E_{k-1,k}$ type. The $E_{k-1,k}$ density is always unimodal, where for $k \geq 3$ ($c_X^2 \leq \frac{1}{2}$) the density has the same shape as the gamma density, that is the density first increases from zero to a maximum and next decreases to zero. Also it is noteworthy that the $E_{k-1,k}$ density can be shown to have an increasing failure rate.

Next we consider the $E_{1,k}$ distribution which is defined as a mixture of E_1 and E_k distributions with the same scale parameters. The density of the $E_{1,k}$ distribution has the form

$$f(t) = p\mu e^{-\mu t} + (1-p)\mu^k\frac{t^{k-1}}{(k-1)!}e^{-\mu t}, \quad t \geq 0, \qquad (A.13)$$

where $0 \leq p \leq 1$. By choosing

$$p = \frac{2kc_X^2 + k - 2 - (k^2 + 4 - 4kc_X^2)^{1/2}}{2(k-1)(1+c_X^2)} \quad \text{and} \quad \mu = \frac{p+k(1-p)}{E(X)}, \quad (A.14)$$

the associated $E_{1,k}$ distribution fits the first two moments of a positive random variable X provided

$$\frac{1}{k} \leq c_X^2 \leq \frac{k^2+4}{4k}.$$

It should be noticed that the $E_{1,k}$ distribution with $k \geq 3$ can also achieve coefficients of variation greater than 1, while for $c_X^2 = 1$ the $E_{1,k}$ distribution with the above specification of p and μ is different from the exponential distribution when $k \geq 3$. Also, it can be shown that for $k \geq 3$ the $E_{1,k}$ density with the above choice for p and μ is bimodal (i.e. has two maxima) when

$$\frac{1}{k} < c_X^2 < \frac{2p_0 + (k+1)k(1-p_0)}{\{p_0 + k(1-p_0)\}^2} - 1 \quad \text{with} \quad p_0 = \frac{(k-2)^{k-2}}{(k-1)! + (k-2)^{k-2}}.$$

For use in applications, the $E_{k-1,k}$ density is in general more suited than the $E_{1,k}$ density since the $E_{k-1,k}$ density is always unimodal and has a similar shape to the frequently occurring gamma density. The $E_{1,k}$ density may be useful in sensitivity analysis. For both theoretical and practical purposes it is often easier to work with mixtures of Erlangian distributions than with gamma distributions, since mixtures of Erlangian distributions with the same scale parameters allow for the probabilistic interpretation that they represent a random sum of independent exponentials with the same means.

A commonly used representation of a positive random variable with the coefficient of variation greater than 1 is a mixture of two exponentials with different means. The distribution of such a mixture is called a *hyperexponential* (H_2) distribution of order two. The density of the H_2 distribution has the form

$$f(t) = p_1 \mu_1 e^{-\mu_1 t} + p_2 \mu_2 e^{-\mu_2 t}, \qquad t \geq 0, \tag{A.15}$$

where $0 \leq p_1, p_2 \leq 1$. Note that always $p_1 + p_2 = 1$, since the density $f(t)$ represents a probability mass of 1. In words, a random variable having the H_2 density is distributed with probability p_1 (respectively p_2) as an exponential variable with mean $1/\mu_1$ (respectively $1/\mu_2$). The hyperexponential density has always a coefficient of variation of at least 1 and is unimodal with a maximum at $t = 0$. The failure rate function of the hyperexponential distribution is decreasing. The hyperexponential distribution is not uniquely determined by its first two moments. In applications the H_2 distribution with *balanced means* is often used, that is the normalization $p_1/\mu_1 = p_2/\mu_2$ is used. The parameters of the unique H_2 density having balanced means and fitting the first two moments of a positive random variable X with $c_x^2 \geq 1$ are

$$p_1 = \tfrac{1}{2}\left(1 + \sqrt{\frac{c_X^2 - 1}{c_X^2 + 1}}\right), \qquad p_2 = 1 - p_1,$$

$$\mu_1 = \frac{2p_1}{E(X)} \quad \text{and} \quad \mu_2 = \frac{2p_2}{E(X)}. \tag{A.16}$$

In what follows we discuss another normalization we believe to be more natural; this alternative normalization leads to an H_2 density having in addition the same third moment as the gamma density with mean $E(X)$ and squared coefficient of variation c_X^2. The H_2 density (A.15) requires that the weights p_1

and p_2 satisfy $0 \leqslant p_1, p_2 \leqslant 1$. However, the right side of (A.15) may also represent a valid probability density without this requirement provided p_1, p_2, μ_1 and μ_2 are suitably chosen. Such a density belongs to the important class of the so-called K_2 densities. Here a K_2 density is defined as a probability density whose Laplace transform is the ratio of a polynomial of degree lower than 2 to a polynomial of degree 2 (cf. Cox, 1955). A K_2 density has always a squared coefficient of variation of at least $\frac{1}{2}$ and has a unimodal shape. The unique K_2 density with $c_X^2 = \frac{1}{2}$ is the E_2 density. A K_2 density with $c_X^2 \geqslant 1$ is always an H_2 density and thus has the representation (A.15) with $0 \leqslant p_1, p_2 \leqslant 1$, whereas a K_2 density of the form (A.15) and with $\frac{1}{2} < c_X^2 < 1$ has either p_1 or p_2 negative. In the latter case the K_2 density has in general no obvious probabilistic interpretation with the exception of the density with the parameter values

$$\mu_1^{-1} = \tfrac{1}{2} E(X)(1 + \sqrt{2c_X^2 - 1}), \qquad \mu_2^{-1} = E(X) - \mu_1^{-1}, \qquad p_1 = \frac{\mu_2}{\mu_2 - \mu_1},$$

$$p_2 = 1 - p_1,$$

corresponding to the sum of two independent exponentials with different means $1/\mu_1$ and $1/\mu_2$ (cf. also Whitt, 1982). Another useful K_2 density of the form (A.15) and having mean $E(X)$ and squared coefficient of variation c_X^2 is obtained by choosing

$$\mu_1 = \frac{2}{E(X)}\left(1 + \sqrt{\frac{c_X^2 - \frac{1}{2}}{c_X^2 + 1}}\right), \qquad \mu_2 = \frac{4}{E(X)} - \mu_1,$$

$$p_1 = \frac{\mu_1\{\mu_2 E(X) - 1\}}{\mu_2 - \mu_1}, \qquad\qquad p_2 = 1 - p_1. \tag{A.17}$$

This particular K_2 density covers any value of $c_X^2 > \frac{1}{2}$ and has the remarkable property that its third moment as well is the same as that of the gamma density with mean $E(X)$ and squared coefficient of variation c_X^2. The unique K_2 density having this property will therefore be called the K_2 density with the *gamma normalization*.

Finally, we mention the *shifted exponential* distribution. This useful distribution corresponds to the sum of a positive constant and an exponential random variable. The density of the shifted exponential distribution is given by

$$f(t) = \mu e^{-\mu(t - D)}, \qquad t \geqslant D. \tag{A.18}$$

The coefficient of variation of this density is always between 0 and 1. A shifted exponential distribution can be fitted to the first two moments of a positive random variable X with $c_X^2 \leqslant 1$ by taking

$$\mu = \frac{1}{c_X E(X)} \quad \text{and} \quad D = E(X) - \frac{1}{\mu}. \tag{A.19}$$

APPENDIX C.   SOME RESULTS FROM LAPLACE TRANSFORM THEORY

This appendix gives a brief outline of some results from Laplace transform theory that are useful in applied probability problems.

Suppose that $f(x)$ is a continuous real-valued function in $x \geqslant 0$ such that $|f(x)| \leqslant A e^{Bx}$, $x \geqslant 0$, for some constants A and B. The *Laplace transform* of $f(x)$ is defined by the integral

$$f^*(s) = \int_0^\infty e^{-sx} f(x) \, dx \qquad (A.20)$$

as a function of the complex variable s with $\text{Re}(s) > B$. The integral always exists when $\text{Re}(s) > B$. The following results can easily be verified from the definition of $f^*(s)$. Here it is assumed that the various integrals exist.

(a) If the function $f(x) = ag(x) + bh(x)$ is a linear combination of the functions $g(x)$ and $h(x)$ with Laplace transforms $g^*(s)$ and $h^*(s)$, then

$$f^*(s) = ag^*(s) + bh^*(s). \qquad (A.21)$$

(b) If $F(x) = \int_0^x f(y) \, dy$, then

$$\int_0^\infty e^{-sx} F(x) dx = \frac{f^*(s)}{s}. \qquad (A.22)$$

If $f(x)$ has a derivative $f'(x)$, then

$$\int_0^\infty e^{-sx} f'(x) dx = sf^*(s) - f(0). \qquad (A.23)$$

(c) If the function $f(x)$ is given by a convolution

$$f(x) = \int_0^x g(x - y) h(y) \, dy$$

of two functions $g(x)$ and $h(x)$ with Laplace transforms $g^*(s)$ and $h^*(s)$, then

$$f^*(s) = g^*(s) h^*(s). \qquad (A.24)$$

(d) If the function $f(x)$ is a probability density of a non-negative random variable X, then

$$f^*(s) = E(e^{-sX})$$

and

$$E(X^k) = (-1)^k \lim_{s \to 0} \frac{d^k f^*(s)}{ds^k}, \qquad k = 1, 2, \ldots. \qquad (A.25)$$

In specific applications requiring the determination of some unknown function

$f(x)$ it is often possible to obtain the Laplace transform $f^*(s)$ of $f(x)$. A very useful result is that the function $f(x)$ is uniquely determined by its Laplace transform $f^*(s)$. In principle the function $f(x)$ can be obtained by inversion of its Laplace transform. Extensive tables are available for the inverse of basic forms of $f^*(s)$; see, for example, Abramowitz and Stegun (1965). An inversion formula that is sometimes helpful in applications is the Heaviside formula. Suppose that

$$f^*(s) = \frac{P(s)}{Q(s)},$$

where $P(s)$ and $Q(s)$ are polynomials in s such that the degree of $P(s)$ is smaller than that of $Q(s)$. It is no restriction to assume that $P(s)$ and $Q(s)$ have no zeros in common. Let $s_1, \ldots, s_k$ be the distinct zeros of $Q(s)$ in the complex plane. For ease of presentation, assume that each root s_j is simple (i.e. has multiplicity one). Then it is known from algebra that $P(s)/Q(s)$ admits the partial fraction expansion

$$\frac{P(s)}{Q(s)} = \frac{r_1}{s - s_1} + \frac{r_2}{s - s_2} + \cdots + \frac{r_k}{s - s_k},$$

where $r_j = \lim_{s \to s_j}(s - s_j)P(s)/Q(s)$ and so $r_j = P(s_j)/Q'(s_j)$, $1 \leqslant j \leqslant k$. It is now easily seen that the inverse of the Laplace transform $f^*(s) = P(s)/Q(s)$ is given by

$$f(x) = \sum_{j=1}^{k} \frac{P(s_j)}{Q'(s_j)} e^{s_j x}. \tag{A.26}$$

This result is readily extended to the case in which some of the roots of $Q(s) = 0$ are not simple. For example, the inverse of the Laplace transform

$$f^*(s) = \frac{P(s)}{(s - a)^m},$$

where $P(s)$ is a polynomial of degree lower than m, is given by

$$f(x) = e^{ax} \sum_{j=1}^{m} \frac{P^{(m-j)}(a)}{(m-j)!} \frac{x^{j-1}}{(j-1)!}.$$

Here $P^{(n)}(a)$ denotes the nth derivative of $P(x)$ at $x = a$.

If the Laplace transform $f^*(s)$ of a function $f(x)$ cannot be inverted in simple terms, it is sometimes possible to find the asymptotic behaviour of $f(x)$ as $x \to \infty$ by studying the Laplace transform $f^*(s)$ as $s \to 0$. The following limiting result holds:

$$\lim_{x \to \infty} f(x) = \lim_{s \to 0} s f^*(s), \tag{A.27}$$

provided both limits exist.

To illustrate the above results, consider the following integro-differential equation in the unknown function $q(x)$:

$$q'(x) = -\frac{\lambda}{\sigma}\{1 - B(x)\} + \frac{\lambda}{\sigma}q(x) - \frac{\lambda}{\sigma}\int_0^x q(x-y)b(y)\mathrm{d}y, \qquad x > 0. \quad (A.28)$$

Here $B(x)$ is the probability distribution function of a positive random variable with probability density $b(x)$ and finite mean m, while λ and σ are positive constants such that $\lambda m/\sigma < 1$. The integro-differential equation (A.28) appears in section 1.9 and the function $q(x)$ represents among others the complementary waiting-time distribution in the standard single-server queueing system with Poisson input. From physical considerations we know that $q(x)$ is decreasing in x and has limit 0 as $x \to \infty$. Using Laplace transform theory, $q(x)$ can be explicitly determined when $b(x)$ is a K_2 density, that is the Laplace transform of $b(x)$ is the ratio of a polynomial of degree lower than 2 to a polynomial of degree 2.

Denoting by $q^*(s)$ and $b^*(s)$ the Laplace transforms of $q(x)$ and $b(x)$, it follows by using (A.20) to (A.24) that for all s with $\mathrm{Re}\,(s) > 0$,

$$sq^*(s) - q(0) = -\frac{\lambda}{\sigma}\left\{\frac{1}{s} - \frac{b^*(s)}{s}\right\} + \frac{\lambda}{\sigma}q^*(s) - \frac{\lambda}{\sigma}q^*(s)b^*(s). \quad (A.29)$$

The unknown $q(0)$ follows by letting $s \to 0$ in both sides of (A.29) and applying (A.27) with f replaced by q. Noting that $\lim_{s\to 0}\{1 - b^*(s)\}/s = m$ (use L'Hôpital's rule), we obtain $q(\infty) - q(0) = -\lambda m/\sigma$ and so, by $q(\infty) = 0$,

$$q(0) = \frac{\lambda m}{\sigma}. \quad (A.30)$$

Suppose now that the probability density $b(x)$ is given by

$$b(x) = p\mu e^{-\mu x} + (1-p)\mu^2 x e^{-\mu x}, \qquad x \geqslant 0,$$

where $0 \leqslant p \leqslant 1$, that is $b(x)$ is a mixture of E_1 and E_2 densities with the same scale parameters. Then

$$b^*(s) = \frac{p\mu}{s+\mu} + \frac{(1-p)\mu^2}{(s+\mu)^2}. \quad (A.31)$$

It is a matter of straightforward algebra to derive from (A.29) to (A.31) and the relation $m = (2-p)/\mu$ that

$$q^*(s) = \frac{\lambda(2-p)(2\mu+s)/(\sigma\mu) - \lambda/\sigma}{(s+\mu)^2 - (\lambda/\sigma)(s+2\mu-p\mu)}.$$

Next, by using the inversion formula (A.26), we obtain

$$q(x) = A_1 e^{-b_1 x} + A_2 e^{-b_2 x}, \qquad x \geqslant 0, \quad (A.32)$$

where

$$b_1 = \mu - \frac{\lambda}{2\sigma} + \frac{1}{2\sigma}\{\lambda^2 + 4\lambda\mu\sigma(1-p)\}^{1/2},$$

$$b_2 = 2\left(\mu - \frac{\lambda}{2\sigma}\right) - b_1,$$

$$A_1 = \frac{-\lambda\{(2-p)(2-b_1/\mu)-1\}}{\sigma(b_1-b_2)},$$

$$A_2 = \frac{\lambda\{(2-p)(2-b_2/\mu)-1\}}{\sigma(b_1-b_2)}.$$

In the same way we obtain an explicit expression for $q(x)$ for the special case of

$$b(x) = p_1\mu_1 e^{-\mu_1 x} + p_2\mu_2 e^{-\mu_2 x}, \qquad x \geqslant 0.$$

For this K_2 density (cf. appendix B), we find

$$q(x) = C_1 e^{-d_1 x} + C_2 e^{-d_2 x}, \qquad x \geqslant 0, \tag{A.33}$$

where

$$d_1 = \frac{1}{2}\left(\mu_1 + \mu_2 - \frac{\lambda}{\sigma}\right) + \frac{1}{2}\left\{\left(\mu_1 + \mu_2 - \frac{\lambda}{\sigma}\right)^2 - 4\mu_1\mu_2 + \frac{4\lambda}{\sigma}(p_1\mu_2 + p_2\mu_1)\right\}^{1/2},$$

$$d_2 = \mu_1 + \mu_2 - \frac{\lambda}{\sigma} - d_1,$$

$$C_1 = \frac{\lambda}{\sigma}(d_1 - d_2)^{-1}\left\{d_1\left(\frac{p_1}{\mu_1} + \frac{p_2}{\mu_2}\right) - \left(\frac{p_1\mu_2}{\mu_1} + \frac{p_2\mu_1}{\mu_2}\right)\right\},$$

$$C_2 = \frac{\lambda}{\sigma}(d_1 - d_2)^{-1}\left\{\frac{p_1\mu_2}{\mu_1} + \frac{p_2\mu_1}{\mu_2} - d_2\left(\frac{p_1}{\mu_1} + \frac{p_2}{\mu_2}\right)\right\}.$$

To conclude this appendix, we give some results for the generating function, being a special case of the Laplace transform. It will be demonstrated that the generating function may enable us to determine asymptotic estimates for the tail of a discrete probability distribution. Suppose that $\{f_n, n = 0, 1, \ldots\}$ is a sequence of non-negative numbers such that $\sum_{n=0}^{\infty} f_n \leqslant 1$. The generating function (or z-transform) of this sequence is defined by

$$F(z) = \sum_{n=0}^{\infty} f_n z^n \qquad \text{for} \qquad |z| \leqslant 1,$$

with z being a complex variable (note that $F(z)$ can be interpreted as a Laplace transform by taking $z = e^{-s}$). In many applications the domain in which $F(z)$ is defined may be extended to the whole complex plane. To be specific, assume that $F(z)$ can be written as

$$F(z) = \frac{G(z)}{H(z)}, \tag{A.34}$$

where $G(z)$ and $H(z)$ are analytic functions in the whole complex plane. Suppose now that the denominator $H(z)$ has a finite number of distinct simple zeros $z_1, \ldots, z_k$. It is no restriction to assume that none of these zeros is a zero of the numerator $G(z)$; otherwise cancel out common zeros. Since $F(z)$ is bounded in $|z| \leqslant 1$, we have $|z_j| > 1$ for all $1 \leqslant j \leqslant k$. Assuming that $G(z)/H(z)$ is bounded in z for all $|z|$ sufficiently large, it is known from complex function theory (see, for example, section 7.4 in Whittaker and Watson, 1952) that $F(z)$ admits the partial fraction expansion

$$F(z) = F(0) + \sum_{j=1}^{k} r_j \left(\frac{1}{z - z_j} + \frac{1}{z_j} \right),$$

for all values of z except the poles z_j themselves, where

$$r_j = \lim_{z \to z_j} (z - z_j) F(z) = \frac{G(z_j)}{H'(z_j)}, \qquad 1 \leqslant j \leqslant k.$$

Noting the geometric series expansion $1/(1 - z/z_j) = 1 + \sum_{n=1}^{\infty} (z/z_j)^n$ for $|z| < |z_j|$, it follows that

$$\sum_{n=0}^{\infty} f_n z^n = F(0) - \sum_{j=1}^{k} \frac{r_j}{z_j} \sum_{n=1}^{\infty} \left(\frac{z}{z_j} \right)^n, \qquad |z| \leqslant 1.$$

Thus, by equating the coefficients of z^n in the two power series, we find

$$f_n = \sum_{j=1}^{k} \frac{-r_j}{z_j^{n+1}}, \qquad n = 1, 2, \ldots .$$

This exact expression enables us to give an extremely useful asymptotic estimate for f_n. Suppose that the roots $z_1, \ldots, z_k$ of $H(z) = 0$ are (re)numbered such that z_1 is smaller in absolute value than the other roots. Then, as $n \to \infty$, each term $z_j^{-(n+1)}$ with $j \neq 1$ tends faster to zero than the term $z_1^{-(n+1)}$. Thus,

$$f_n \approx \frac{-G(z_1)}{H'(z_1)} z_1^{-n-1} \qquad \text{for large} \qquad n. \tag{A.35}$$

Note that the root z_1 is real and larger than 1. In many practical applications the asymptotic expansion (A.35) yields useful estimates already for relatively small values of n. It is noteworthy to point out that the asymptotic expansion (A.35) also applies when some of the roots of $H(z) = 0$ are not simple, as long as the root z_1 having the smallest absolute value of the roots is simple.

APPENDIX D. SUCCESSIVE OVERRELAXATION METHODS FOR SOLVING LINEAR EQUATIONS

In Markov chain applications one often has to solve a system of linear equations of the form

$$x_i = \sum_{j=1}^{N} p_{ji} x_j, \qquad i = 1, \ldots, N, \tag{A.36}$$

$$\sum_{i=1}^{N} x_i = 1, \tag{A.37}$$

where the non-negative constants p_{ji} satisfy $p_{ii} < 1$ for $i = 1, \ldots, N$. Assuming the regularity condition posed on the Markov chains in chapter 2, this system of linear equations has a unique non-negative solution and, moreover, any solution to the balance equations (A.36) is uniquely determined up to a multiplicative constant, a constant which is found from the normalizing equation (A.37).

A suitable method for solving the linear equations arising in Markov chain analysis is the iterative method of successive overrelaxation which generates a sequence of vectors $x^{(0)} \to x^{(1)} \to x^{(2)} \to \cdots$ converging towards a solution of the balance equations. This method is easy to program and enables us to solve efficiently quite large systems of linear equations for Markov chains. In what follows we first describe the standard method of successive overrelaxation with a fixed relaxation factor and next describe an overrelaxation method with a dynamically adjusted relaxation factor. To apply successive overrelaxation, we first rewrite the balance equations (A.36) into the more convenient form

$$x_i = \sum_{\substack{j=1 \\ j \neq i}}^{N} a_{ij} x_j, \qquad i = 1, \ldots, N, \tag{A.38}$$

where

$$a_{ij} = \frac{p_{ji}}{1 - p_{ii}}, \qquad i, j = 1, \ldots, N, \quad j \neq i.$$

The standard successive overrelaxation method uses a fixed relaxation factor ω for speeding convergence. The method starts with an initial approximation vector $x^{(0)} \neq 0$. In the kth iteration of the algorithm an approximation vector $x^{(k)}$ is found by a recursive computation of the components $x_i^{(k)}$ for $i = 1, \ldots, N$ such that the calculation of the new estimate $x_i^{(k)}$ uses both the new estimates $x_j^{(k)}$ for $j < i$ and the old estimates $x_j^{(k-1)}$ for $j \geq i$. The components $x_i^{(k)}$ are recursively computed from

$$x_i^{(k)} = (1 - \omega) x_i^{(k-1)} + \omega \left(\sum_{j=1}^{i-1} a_{ij} x_j^{(k)} + \sum_{j=i+1}^{N} a_{ij} x_j^{(k-1)} \right), \qquad i = 1, \ldots, N, \tag{A.39}$$

with $\sum_{j=1}^{0} = 0$ by convention. The iterative scheme is stopped when the approximation vector $x^{(k)}$ has converged sufficiently. The stopping criterion can be taken to be

$$\sum_{i=1}^{N} |x_i^{(k)} - x_i^{(k-1)}| \leqslant \varepsilon \sum_{i=1}^{N} |x_i^{(k)}|, \qquad (A.40)$$

where ε is some small tolerance number. The specification of ε depends typically on the particular problem considered and the accuracy required in the final answers. In addition to the stopping criterion, it may be helpful to use extra accuracy checks in the form of relations other than (A.36) for the equilibrium probabilities of the underlying Markov chain application. An extra accuracy check may prevent a decision upon premature termination of the algorithm when the tolerance number ε in (A.40) is not chosen sufficiently small. Suppose the algorithm has converged to the vector x^*, then the desired probabilities p_i, being the unique solution to (A.36) and (A.37), are next computed from

$$p_i = \frac{x_i^*}{\sum_{j=1}^{N} x_j^*}, \qquad 1 \leqslant i \leqslant N.$$

Notice that the normalizing equation (A.37) is used only at the very end of the algorithm. In applying successive overrelaxation it is highly recommended that all of the balance equations (A.36) are used rather than omitting one redundant equation in (A.36) and substituting the normalizing equation (A.37) for it.

The convergence speed of the successive overrelaxation method may dramatically depend on the choice of the relaxation factor ω, and even worse the method may diverge for some choices of ω. A suitable value of ω has to be determined experimentally. Usually $1 \leqslant \omega \leqslant 2$. The 'optimal' value of the relaxation factor ω usually depends heavily on the parameters of the particular problem considered. It is pointed out that the iteration method with $\omega = 1$ is the well-known Gauss–Seidel method, being convergent in nearly all practical cases, and that the optimal value of ω may be close to 1. Also, the ordering of the states may have a considerable effect on the convergence speed of the successive overrelaxation algorithm. In general one should order the states such that the upper diagonal part of the matrix of coefficients is as sparse as possible. In specific applications the transition structure of the Markov chain often suggests an appropriate ordering of the states.

The standard successive overrelaxation method has the drawback that a good choice of ω is usually not clear beforehand and has to be determined by trial and error. This drawback can be circumvented by using an overrelaxation method in which the relaxation factor ω is dynamically adjusted. Following Seelen (1986), we now describe such a method. The following shorthand notation is useful. For a given value of ω, the associated operator B_ω adds to each vector $x = (x_1, \ldots, x_N)$ a vector $B_\omega x$ whose components $(B_\omega x)_i$ are recursively defined by

$$(B_\omega x)_i = (1-\omega)x_i + \omega\left\{\sum_{j=1}^{i-1} a_{ij}(B_\omega x)_j + \sum_{j=i+1}^{N} a_{ij}x_j\right\}, \qquad i=1,\ldots,N,$$

(cf. equation A.39). Thus any solution to (A.38) is an eigenvector of B_ω with an associated eigenvalue 1. Letting $\lambda_1(\omega)$ be the eigenvalue having the largest absolute value among the eigenvalues of B_ω that are not equal to 1, it can be shown that the standard successive overrelaxation method with a fixed relaxation factor ω converges only if $|\lambda_1(\omega)| < 1$. Moreover, the standard overrelaxation method has the best convergence rate for that value of ω for which $|\lambda_1(\omega)|$ is smallest. Assuming that $\lambda_1(\omega)$ is real and satisfies $|\lambda_1(\omega)| < 1$, it is in general possible to obtain an estimate of $|\lambda_1(\omega)|$ after some iterations of the overrelaxation method with the relaxation factor ω by using the result

$$\lim_{k\to\infty} \frac{\|x^{(k)} - x^{(k-1)}\|/\|x^{(k)}\|}{\|x^{(k-1)} - x^{(k-2)}\|/\|x^{(k-1)}\|} = |\lambda_1(\omega)|,$$

where

$$\|x\| = \sum_{i=1}^{N} |x_i|.$$

The estimate for $|\lambda_1(\omega)|$ provides a method for formulating a successive overrelaxation method in which the relaxation factor is dynamically adjusted in order to search for that value of ω for which $|\lambda_1(\omega)|$ is smallest. We now state such an algorithm. This algorithm is nearly as effective as the standard successive overrelaxation method using the (unknown) optimal value of the relaxation factor.

Successive overrelaxation algorithm with a dynamic relaxation factor

Step 0. Initialize $\omega = 1.20$ and choose an initial approximation vector $x^{(0)} \neq 0$.
Step 1. $\omega_{\text{old}} := 0$, $\lambda(\omega_{\text{old}}) := 1$, $k := 1$, $r^{(1)} := 0$, $x^{(1)} := B_\omega x^{(0)}$.
Step 2. $k := k+1$, $x^{(k)} := B_\omega x^{(k-1)}$. If $\|x^{(k)} - x^{(k-1)}\| < \varepsilon \|x^{(k)}\|$ where ε is a prespecified accuracy number, then go to step 4. Otherwise, compute

$$r^{(k)} := \frac{\|x^{(k)} - x^{(k-1)}\|/\|x^{(k)}\|}{\|x^{(k-1)} - x^{(k-2)}\|/\|x^{(k-1)}\|}$$

and perform the following test. If $r^{(k)} \geq 1$ or $k \geq 10$, then ω is likely to be too large; decrease ω according to $\omega := 1 + \frac{1}{2}(\omega - 1)$, let $x^{(0)} := x^{(k)}$ and go to step 1. If $r^{(k)} < 1$ and $r^{(k)}$ has sufficiently converged according to $|(r^{(k)} - r^{(k-1)})/r^{(k)}| < 0.025$, then go to step 3; otherwise return to step 2.

Step 3. $\lambda(\omega):= r^{(k)}$. Test for one of the following four possibilities:

(a) $\omega > \omega_{\text{old}}$ and $\lambda(\omega) > \lambda(\omega_{\text{old}})$;   (b) $\omega > \omega_{\text{old}}$ and $\lambda(\omega) \leqslant \lambda(\omega_{\text{old}})$;

(c) $\omega < \omega_{\text{old}}$ and $\lambda(\omega) > \lambda(\omega_{\text{old}})$;   (d) $\omega < \omega_{\text{old}}$ and $\lambda(\omega) \leqslant \lambda(\omega_{\text{old}})$.

For cases (a) and (d),

$$\omega_{\text{old}}:= \omega, \lambda(\omega_{\text{old}}):= \lambda(\omega), \qquad \omega:= 1 + 0.85(\omega - 1),$$

whereas for cases (b) and (c),

$$\omega_{\text{old}}:= \omega, \qquad \lambda(\omega_{\text{old}}):= \lambda(\omega), \qquad \omega:= 1 + 1.25(\omega - 1).$$

Next,

$$x^{(0)}:= \frac{x^{(k)} - \lambda(\omega)x^{(k-1)}}{1 - \lambda(\omega)}, \qquad k:= 1, \qquad r^{(1)}:= 0,$$

$$x^{(1)}:= B_\omega x^{(0)},$$

and go to step 1.

Step 4. Stop the algorithm and compute the desired probabilities p_i from

$$p_i:= \frac{x_i^{(k)}}{\sum_{j=1}^N x_j^{(k)}}, \qquad 1 \leqslant i \leqslant N.$$

To conclude this section, we briefly mention an alternative iterative method that has the advantage of providing at each iteration lower and upper bounds on the quantities to be calculated. In many applications the state probabilities of the underlying Markov chain are calculated with the only purpose of obtaining a single performance measure (e.g. the blocking probability or the average queue size in a queueing application). In such situations the problem can be reduced to calculating the average cost per unit time in a single Markov chain on which an appropriate cost structure is superimposed. A convenient method to calculate the average cost is the (modified) value-iteration algorithm from section 3.4, where the data transformation from section 3.5 should be applied first when the underlying Markov chain has a continuous-time parameter. This value-iteration method can be extended to calculate simultaneously all of the state probabilities of the Markov chain such that lower and upper bounds on the state probabilities are provided at each iteration (see Van Der Wal and Schweitzer, 1987).

REFERENCES

Abramowitz, M., and Stegun, I. (1965). *Handbook of Mathematical Functions,* Dover, New York.

Cox, D. R. (1955). 'A use of complex probabilities in the theory of stochastic processes', *Proc. Camb. Phil. Soc.,* **51**, 313–319.

Gautschi, W. (1979). 'Algorithm 542, incomplete gamma functions', *ACM Trans. Math. Software*, **5**, 482–489.

Seelen, L. P. (1986). 'An algorithm for Ph/Ph/c queues', *European J. Operat. Res.*, **23**, 118–127.

Van Der Wal, J., and Schweitzer, P. J. (1987). 'Iterative bounds on the equilibrium distribution of a finite Markov chain', *Prob. Engineering and Inform. Sci.*, **1** (to appear).

Whitt, W. (1982). 'Approximating a point process by a renewal process, I: two basic methods', *Operat. Res.*, **30**, 125–147.

Whittaker, E. T., and Watson, G. N. (1952). *A Course of Modern Analysis*, 5th ed., Cambridge University Press, Cambridge.

Author Index

411

Subject Index

(*continued from front*)